The Civilian Legacy of the Roman Army

History of Warfare

VOLUME 144

The titles published in this series are listed at *brill.com/hw*

The Civilian Legacy of the Roman Army

Military Models in the Post-Roman World

Edited by

Luca Loschiavo

BRILL

LEIDEN | BOSTON

Cover illustrations: a. Ostrogothic belt buckle, inv. Or. 20. With kind permission of the Musei Civici del Castello Visconteo in Pavia (https://museicivici.comune.pv.it/site/home.html).
b. Map: 'The Germanic Kingdoms and the East Roman Empire in 486' from William Shepherd's *Historical Atlas*. New York: Henry Holt and Company, 1911. https://maps.lib.utexas.edu/maps/historical/history_shepherd_1911.html. Last accessed 12 January 2024.

The Library of Congress Cataloging-in-Publication Data is available online at https://catalog.loc.gov
LC record available at https://lccn.loc.gov/2024012169

Typeface for the Latin, Greek, and Cyrillic scripts: "Brill". See and download: brill.com/brill-typeface.

ISSN 1385-7827
ISBN 978-90-04-69347-0 (hardback)
ISBN 978-90-04-69801-7 (e-book)
DOI 10.1163/9789004698017

This book is printed on acid-free paper and produced in a sustainable manner.

Contents

Preface

The importance of the army in Roman political and social history is a well-known and widely studied topic. The role of the army as a vehicle of 'Romanisation' for the people who lived in the provinces of the empire or in the wide belt of the *limes* is equally well known. By contrast, not so well known is the impact that the organizational models employed by the army (hierarchies, administration, payroll, food acquisition, storage and distribution, logistics, justice, tax collection and so on) had on the rising Barbarian and Romano-Barbarian societies, which progressively replaced the great construction of the Western Empire during the centuries (4th–7th) that saw the passage from Late Antiquity to Early Middle Ages.

Increasingly open to the recruitment of non-Roman 'manpower', for centuries, the Imperial army was also the meeting place of different cultures (today we would use the expression 'melting pot'). However, the coexistence of people of different ethnic origin was only one problem for Roman generals. Continuously forced to adapt to different situations and environments, the Roman army had to develop the capacity of finding multiple and innovative solutions that allowed it to maintain the control of the *Urbs* over very large territories with diminished troops, so that real force would be used as little as possible.

Sometimes Roman officers imposed on the occupied territories models taken from Latin tradition (the use of a common language and writing, ownership structures, the separation between administration and religion, etc.). In other cases, they realized that it would be more useful and effective to adapt solutions already spread among local populations (simplified negotiating practices, symbolic formalism to strengthen juridical obligations, reliance on divine judgment, gift economy, military and paramilitary clientele, etc.).

Quite often the groups of barbarians that penetrated the borders dealt with these very models, and they eventually adopted them at the moment of settling within the Empire, replacing their ancestral models of leadership with the more elastic Roman techniques of governance. One can believe that the army constituted an important social and organizational reference model for the barbarian populations, especially as it was distinct, at least in part, from the complex and aristocratic sphere of *civilitas*. In any case, the great flexibility and adaptability of the solutions used by the Roman army allowed the 'new *domini*' not only to maintain power for a long time and spread throughout medieval Europe, but also to leave a lasting cultural imprint. It is precisely the resilience of such solutions at a time when the empire fades away and the military dimension of power assumes less aristocratic forms than those of the late-antique *civilitas*, and becomes predominant in the new realms, that represents the principal reason for this type of investigation today.

Starting from this point of view – analysing the defining moment when Antiquity came to end and a new age begun, whose influence in some ways is still visible today – some scholars, from a variety of different disciplines and countries, gathered for the first time in Cagliari in October 2012. Their perspective ranged from political, social and economic history to law and anthropology. The purpose of that first meeting was to reconstruct the *status quaestionis* in different areas: the results are now collected in the volume *Civitas, Iura, Arma. Organizzazioni militari, istituzioni giuridiche e strutture sociali all'origine dell'Europa* (*secc. III–VIII*), Fabio Botta and Luca Loschiavo eds. (Lecce 2015).

This research group – enriched with new specialists (esp. scholars of Roman and medieval history, Germanic philologists and linguists, and historians of religion and iconography) – continued to work on the subject and to share their knowledge. They all pay close attention to cultural and specifically lexical aspects. One of the main purposes of the research group is to avoid the generation of confusion and misunderstanding through use of the same terms – which are recurrent through various centuries and yet assume different meanings in different contexts – among scholars from different disciplines, hindering (as in fact happened in the past) a truly fruitful dialogue.

This volume collects and presents the results of that research. It certainly does not claim to be the definitive word on the subject. However, it is intended to give an account of the richness of the problem and the usefulness of a multidisciplinary study conducted in this way. The very core of many contributions remains the legal issues with the greatest social impact: marriage (Castagnino), personal status and integration processes (Poly, Marotta), hereditary successions (Ruggiero), dispute resolution (Kerneis), taxes and military obligation (Wood, Osaba), land distribution and creation of new settlements (Trisciuoglio and Porena), legal pluralism (Loschiavo), maintenance of order and punishments (Esders), collective criminal responsibility (Botta). Much importance – as just mentioned – is also given to the linguistic developments starting from military terminology and the linguistic solutions adopted within the army (Haubrichs, Falluomini). Also taken into account is the complex historiographical topic of 'ethnogenesis' and the formation-transformation of the core-traditions of the barbarian people before and after contact with the Roman world (Borri), the meaning of military ideology for the newcomers (especially the Lombards: Gasparri) and the resilience and re-adaptation of Roman military power-symbols, such as the *cingulum* to which the precious belt buckle of Ostrogothic origin reproduced on the cover of this book certainly belonged (Verardi). The concluding chapter (Pohl) offers an account of the current historiographical debate and provides new insights for the prosecution of research.

The principal aim of this book is to indicate to the scholarly community how interesting and fruitful this path could be in order to learn more about the origins of the civilisation that began after the fall of the Roman world. It is therefore hoped that this research can continue and that other scholars would like to join our research group and provide their specific contribution.

•••

Many people and institutions deserve to be mentioned for their help in the realisation of this volume. Jean-Michel Carrié, Gisella Bassanelli Sommariva, Fausto Goria and Arnaldo Marcone participated at different times in the research group, giving punctual advice and substantial support of various kinds. Elvira Migliario and Saverio Lomartire with their papers (unfortunately not published here) enriched the 2019 Roman conference. Guido Rossi was invaluable in the last editorial phase. The Departments of Law and Humanities of the Roma Tre University provided financial and logistical support. Laura Aldovini, Director of the Civic Museums of Pavia, has kindly and generously granted the opportunity to use the image of the above-mentioned Ostrogothic buckle kept at the museum for the cover.

Personally, I would especially like to thank all the authors of the chapters collected here for their commitment and patience during the various stages of the project. To their knowledge and dedication, we owe all the good and interesting things this book has to offer to reader.

Rome, December 2023

Maps

Contributors

Francesco Borri (University of Venice)

Fabio Botta (University of Cagliari)

Francesco Castagnino (University of Pavia)

Stefan Esders (Freie Universität in Berlin)

Carla Falluomini (University of Perugia)

Stefano Gasparri (University of Venice)

Wolfgang Haubrichs (University of Saarland)

Soazick Kerneis (University of Paris Ouest-Nanterre)

Luca Loschiavo (University of Teramo)

Valerio Marotta (University of Pavia)

Esperanza Osaba (Universidad del Pais Vasco – Bilbao)

Walter Pohl (University of Vienna)

Jean-Pierre Poly (University of Paris Ouest-Nanterre)

Pierfrancesco Porena (Roma Tre University)

Iolanda Ruggiero (University of Siena)

Andrea Trisciuoglio (University of Turin)

Andrea A. Verardi (Pontificia Università Gregoriana in Rome)

Ian Wood (University of Leeds)

CHAPTER 1

Transformation of the Military in the Late Antique West

Ian Wood

1 From Standing Army to Army of Obligation

The Roman Empire boasted of the peace it had established: the *Pax Romana*.[1] But it has also been observed that '[war] was a reality of the Roman world, and a major preoccupation of its rulers, and to some extent the primary purpose of the Roman state was, quite simply, the financing of combat.'[2] The Empire depended on a massive standing army. This constitutes our point of departure. With regard to the situation in the sixth and seventh centuries '[it] has been argued that, subsequent to the end of the 'Roman Peace', warfare had become a significant aspect of contemporaries' everyday experience.'[3] Whether or not this claim is justified (and that depends on how one rates the conflicts of the late sixth and seventh centuries, which were rarely large-scale),[4] there is no doubt that the majority of males in the early medieval West were potential warriors, who could be called upon to fight[5] – even if few of them were involved in any sustained campaigns. To quote Laury Sarti, '[as] every man was now a potential warrior, it had also become more difficult to distinguish a civil man from the military.'[6] Or to quote a phrase of Stefan Esders, 'soldiers were no longer referred to as *milites*, but simply as *homines*.'[7] In other words

1 Greg Woolf, 'Roman Peace', in *War and Society in the Roman World*, eds. J. Rich and G. Shipley (London, 1993), pp. 171–194.

2 Meaghan McEvoy, *Child Emperor Rule in the Late Roman West,* AD *367–455* (Oxford, 2013) p. 29; A. Doug Lee, *War in Late Antiquity: a social history* (Oxford, 2007), pp. 21–49.

3 Laury Sarti, *Perceiving War and the Military in Early Christian Gaul (ca. 400–700* AD (Leiden, 2013), p. 82.

4 I would argue that the seventh century was extraordinarily peaceful, and that it was dominated not by the military, but by the Church: Ian Wood, *The Transformation of the Roman West* (Leeds, 2018), p. 82.

5 Stefan Esders, 'Late Roman Military Law in the Bavarian Code', in *La forge du droit. Naissance des identités juridiques en Europe* (IVe-XIIIe *siècles*), Clio@Themis, 10 (March 2016), esp. fn. 41–2.

6 Sarti, *Perceiving War*, p. 40.

7 Esders, 'Late Roman military law', p. 20.

 | DOI:10.1163/9789004698017_002

we begin in a world that depended on a standing army, even if it was not a major presence outside the frontier regions, and we end in one where the military depended upon obligations that fell largely on what Bernard Bachrach has termed farmer-soldiers, who were to be found everywhere.[8] The obligations themselves are spelt out for those living on fiscal land already in the *Pactus pro tenore pacis* of the Merovingian kings Childebert and Chlothar (that is probably of Childebert I (511–58) and Chlothar I (511–61)). The *Lex Ribvaria*, which seems to be of the early seventh century, deals more generally with those owing military obligation.[9]

2 Late-Roman Recruitment

The shift from a standing army to armies of obligation has been explained in a number of ways. Historians have noted the difficulties experienced by the Empire in recruiting troops to replace those retiring or killed. There are numerous laws on deserters in the *Theodosian Code*.[10] Three laws, of 367, 373 and 381, deal with those mutilating themselves in order to avoid military service.[11] There are a further three laws from 397 that point to a crisis in recruitment.[12] In 406 such was the danger posed by Radagaisus that slaves were encouraged to enlist, with a promise of subsequent manumission,[13] while freeborn men taking up arms were to be paid 10 *solidi*.[14] In 410 recruits were demanded from officials in Africa, Sardinia, Sicily and Corsica, but not from regions that had been devastated by the barbarians.[15] And in 416 (or possibly 403) there was legislation against procurators harbouring those avoiding recruitment.[16]

8 Bernard S. Bachrach, 'Grand Strategy in the Germanic Kingdoms: Recruitment of the rank and file', in *L'Armée romaine et les barbares du IIIe au VIIe siècle*, eds. F. Vallet – M. Kazanski (Paris, 1993), pp. 55–63.

9 *Pactus pro tenore pacis*, ch. 3 (ed. Boretius, p. 7); *Lex Ribvaria*, 68 (ed. Beyerle – Buchner, p. 199). See Elisabeth Magnou-Nortier, *Aux origines de la fiscalité moderne. Le système fiscal et sa gestion dans le royaume des Francs* (Geneva, 2012), pp. 457–459.

10 *Codex Theodosianus*, VII 18, 1 (a. 365), 2 (a. 379), 3 (a. 380), 4 (a. 380), 5 (a. 381), 6 (a. 382), 7 (a. 383), 8 (a. 383/391), 9 (a. 396), 10 (a. 400), 11 (a. 403), 12 (a. 403), 13 (a. 403), 14 (a. 403), 15 (a. 406), 16 (a. 413), 17 (a. 413); *Novella Valentiniani*, III, VI,1.

11 *Codex Theodosianus*, VII, 13, 4, 5, 10.

12 *Codex Theodosianus*, VII, 13, 11–14.

13 *Codex Theodosianus*, VII, 13, 16.

14 *Codex Theodosianus*, VII, 13, 17.

15 *Codex Theodosianus*, VII, 13, 20.

16 *Codex Theodosianus*, VII, 13, 21.

The Novels of Valentinian III show the same problems in the 440s: in 440 the emperor instructed the Roman population to prepare to defend itself against a possible Vandal attack, and to this end he restored the right to use weapons.[17] There was a similar, but now lost, Novel of Majorian.[18] This, one might note, is comparable to instructions given by Honorius after 406, recorded by Zosimus,[19] and also by Gildas.[20] In 443 senators and landowners were required to supply recruits.[21] In the following year *illustres* had to provide the funds for three recruits each, and lower ranking officials funds for one or one third of a recruit.[22] Here the problem was not just recruits, but also the ability to pay for them, which had already been a problem in 412 (or perhaps 409).[23] The importance of payment is best illustrated by an anecdote in the *Vita Severini*, by Eugippius, where the soldiers sent to Italy 'at the time when the Roman Empire was still in existence' to collect the *stipendia* for the garrison at Batava were killed on their return journey: as a result the garrison disbanded.[24] A little while later Onulf, the brother of Odoacer, ordered the withdrawal of the remaining Romans to Italy[25] – although, as Stefan Esders has pointed out, there are reasons for thinking that some elements of Roman military organisation survived to provide the basis of the *ducatus Baiuvariorum*.[26] The transfer of frontier troops to deal with crises within the Empire is something already attested by Claudian, in his account of Stilicho's Gothic Wars.[27] Benjamin Isaac has analysed the meaning of the terms *limes* and *limitanei*,[28] while A.H.M. Jones provided a convenient table of the numerous groups of *limitanei* transferred to the Western *comitatus*.[29]

The result of the recruitment crisis can be seen most clearly in the description of the imperial forces at the Battle of the Catalaunian Plains given by Jordanes, where we hear that alongside the Visigoths Aetius led an army of Franks, Sarmatians, Armoricans, *Liticiani*, Burgundians, Saxons, Riparians, and

---

17 *Novella Valentiniani*, III, IX.
18 *Novella Majoriani*, VIII.
19 Zosimus, *Historia Nova*, VI, 10.
20 Gildas, *De Excidio Britonum*, 18.
21 *Novella Valentiniani*, III, VI, 2.
22 *Novella Valentiniani*, III, VI, 3.
23 *Codex Theodosianus*, XI, 18, 1.
24 Eugippius, *Vita Severini*, 20, 1.
25 Eugippius, *Vita Severini*, 44, 5.
26 Esders, *Late Roman Military Law*.
27 Claudian, *De Bello Getico*, ll. 412–422.
28 Benjamin Isaac, 'The meaning of the terms *limes* and *limitanei*', in *Journal of Roman Studies*, 78 (1988), pp. 125–147.
29 A. Hugo M. Jones, *The Later Roman Empire, 284–602* (Oxford, 1964), pp. 1435–6.

Olybriones, 'quondam milites Romani', as well as Celts and Germanic peoples.[30] In other words, by the year 451 there was little in the way of a standing army in the Roman West.

3 Federates

In addition to the problem of recruits, historians have pointed to an increasing reliance on the use of federate troops. Indeed Wolfgang Liebeschuetz has argued that in place of recruits the Empire was happy to take cash payments from Romans in order to pay federates.[31] The employment of barbarians was, of course, a longstanding feature of Roman policy.[32] It was criticised, however, by both Synesius[33] and Zosimus.[34] But such criticisms had no effect.

The federates themselves Liebeschuetz divided into two types: small groups enlisted for specific campaigns,[35] and complete peoples, such as the Visigoths who in 376, at least, were seen as potential members of the Roman forces.[36] Constantius employed Athaulf against Jovinus,[37] and against the Vandals.[38] At the Catalaunian Plains the Visigoths and the Burgundians were clearly fulfilling the role that the Empire hoped they would take on. The smaller groups of federates are more difficult to trace, although we do hear of 30,000 men recruited by Stilicho, who transferred their allegiance to Alaric after the execution of

30 Jordanes, *Getica*, 191.

31 John Hugo Wolfgang Liebeschuetz, 'The End of the Roman Army in the Western Empire', in *War and Society in the Roman World*, eds. J. Rich and G. Shipley (London, 1993), at p. 274. [See also *Codex Theodosianus*, VII, 13, 13–14]. Sarti, *Perceiving War*, p. 24, citing Ammianus Marcellinus, *Rerum Gestarum*, XIX, 11, 7; XXXI, 4, 4 and Ausonius, *Precatio consulis*, ll. 29–31; *Pacatus panegyric*, 12, 3; 32, 3–5. See also Avshalom Laniado, *Ethnos et droit dans le monde protobyzantin, Ve-VIe siècle* (Geneva, 2015).

32 John Hugo Wolfgang Liebeschuetz, *Barbarians and Bishops: Army, Church and State in the Age of Arcadius and Chrysostom* (Oxford, 1990), pp. 14–23; Luca Loschiavo, *L'età del passaggio: all'alba del diritto comune europeo (secoli III–VII)*, 2nd ed. (Rome, 2019), pp. 138–141.

33 Synesius, *De Regno*; Id., *De Providentia*. Alan Cameron – Jacqueline Long, *Barbarians and Politics at the Court of Arcadius* (Berkeley, 1993).

34 Zosimus, *Historia Nova*, IV, 30.

35 Liebeschuetz, 'The End of the Roman Army', pp. 266–267.

36 Ammianus Marcellinus, *Rerum gestarum*, XXXI, 4, 4.

37 Olympiodorus, fr. 18; Liebeschuetz, 'The End of the Roman Army', p. 268.

38 Hydatius, *Chronicon*, 70; Jordanes, *Getica*, 163–166. Liebeschuetz, 'The End of the Roman Army', p. 268.

the *magister militum*.[39] Another group that changed sides were the Visigothic federates who deserted Castinus in 422.[40]

4 *Buccellarii*

Such federates clearly overlap with a third group that has been presented as indicative of the shift in the late Roman military: the *buccellarii*, and more generally the *satellites* of individual commanders and subsequently of major landowners. Tellingly a fragment of Olympiodorus states that 'in the time of Honorius the name *buccellarius* was given not only to Roman soldiers but also to certain Goths. Similarly the name *foederati* was given to a diverse and mixed body of men.'[41] These groups are not always easy to differentiate. According to Olympiodorus, Galla Placidia was supported by a troop of barbarians as a result of her marriages to Athaulf and Constantius.[42] Whether these men would have been better described as *buccellarii* or federates is unclear: Olympiodorus uses the unhelpful term πλῆθος.[43]

It is important to stress the Roman origins of the *buccellarii*, despite the fact that Jill Harries has argued that 'the provisions relating to *bucellarii* (sic)' in the *Codex Euricianus* are 'entirely alien to the ordered world of the Roman lawyer'.[44] In fact the *buccellarii* are best seen in the context of the rise of the Roman military *patronus*.[45] Quite apart from the comment of Olympiodorus, we have the evidence of Gregory of Tours, apparently dependent on the lost earlier history of Renatus Profuturus Frigeridus, who says that Valentinian III was killed by Occila, the *buccellarius* of Aetius.[46] In addition, *buccellarii* appear in a law of Anthemius from 468 preserved in the *Corpus Iuris Civilis*.[47]

---

39 Zosimus, *Historia Nova*, V, 35, 6.

40 Hydatius, *Chronicon*, 77.

41 Olympiodorus, fr. 7, 4. 8.

42 Olympiodorus, fr. 38.

43 On the meaning of this term, see Botta in this volume p. 19 fn. 94.

44 Jill Harries, 'Not the Theodosian Code: Euric's Law and late fifth-century Gaul', in *Society and Culture in Late Antique Gaul: revisiting the sources*, eds. R. Mathisen – Danuta Shanzer (Aldershot, 2001), at p. 42, citing *Codex Euricianus*, 310 to which one can add *Leges Visigothorum*, V, 3, 1.

45 Loschiavo, *L'età del passaggio*, pp. 26, 141–143. See also Noel Lenski, 'Schiavi armati e formazione di eserciti privati nel mondo tardoantico', in *Ordine e sovversione nel mondo greco e romano*, ed. G. Urso (Pisa, 2009), at pp. 158–66.

46 Gregory of Tours, *Decem Libri Historiarum*, II, 8.

47 *Codex Justinianus*, IX, 12, 10 (468).

It can be difficult to determine whether the forces led by a Roman military commander were *buccellarii* or regular troops. Wolfgang Liebeschuetz implies that most of the forces led by rebellious generals in the early part of the fifth century were *buccellarii*,[48] but they could just as well have been detachments of the regular army. Our sources were often less concerned with the nature of the troops than they were with the leadership. When he comes to the period after the Sack of Rome Orosius announces that he is going to talk of 'the string of usurpers as briefly as possible' ('de catalogo tyrannorum quam brevissime loquar'), and he concludes his discussion with a comment on Constantius' destruction of 'this entire list of usurpers and insubordinate *duces*' ('omnem catalogum ... vel manifestorum tyrannorum vel inoboedientium ducum'), which was made possible by the piety of Honorius.[49] In his account he makes no distinction between regular troops, federates and *buccellarii*. In dealing with the same period Olympiodorus, followed by Sozomen, talks of the 'many usurpers'.[50] And the *Ravenna Annals* are concerned almost entirely with usurpers and natural disasters.[51] The civil wars of the fifth century may well have been a major factor in the implosion of the standing army.

5 *Patrocinium*

The use that military men made of *patrocinium* to build up a following of *satellites*, *plebs* and *servi*, as did Firmus,[52] was clearly well advanced by the end of the fourth century.[53] Increasingly, however, it was not just military leaders, but also aristocrats who raised warbands. Shortly after 410 Verinianus and Didymus created an army of slaves and *coloni* to combat the usurping emperor Constantine III: Constans, who was sent to deal with them recruited a group of barbarian federates whom he named *Honoriaci*.[54] At this point in time the private army of the Roman aristocrats was no match for the federates, who would have been much more used to fighting. But later in the century, when Ecdicius

48 Liebeschuetz, 'The End of the Roman Army', pp. 269–70.

49 Orosius, *Historia adversus Paganos*, VII 42, 4–15.

50 Olympiodorus, fr. 5.

51 Bernhard Bischoff – Wilhelm Koehler, 'Eine illustrierte Ausgabe der spätantiken Ravennater Annalen', in *Studies in Memory of A. Kingsley Porter*, ed. W. Koehler (Cambridge, Mass., 1939), pp. 125–138.

52 Ammianus Marcellinus, *Reur. gestarum*, XXIX, 5, 34–6.

53 Dick Whittaker, 'Landlords and warlords in the later Roman Empire', in *War and Society in the Roman World*, eds. J. Rich – G. Shipley (London, 1993), at pp. 288–289.

54 Orosius, *Historia adversus Paganos*, VII, 40, 5–6; Sozomen, *Ecclesiastical History*, IX, 11.

put together a band of 18 *satellites*, who doubtless had their own retainers, to relieve the Visigothic siege of Clermont, the Romans were the victors.[55] By this time even bishops might surround themselves with armed followers. Hilary of Arles was denounced by Valentinian III for raising an *abominabilis tumultus*.[56]

6 The Armies of the Successor States

The changes in the Roman military are relatively well known, even if the exact balance between the different types of troops and the chronology of the disappearance of the standing army can be debated. The development of the armies of the early Successor States is very much less clear. Lellia Cracco Ruggini has demonstrated from the *Variae* of Cassiodorus that the Ostrogothic army in Italy continued to be supported by the *annona militare*.[57] Not surprisingly similar arrangements seem to have been in place under the Exarchate, to judge from the letters of Gregory the Great.[58] Jean Durliat has argued that the same is true for Merovingian Gaul and Visigothic Spain: 'L'armée est donc toujours une armée de métier, rétribuée selon les méthodes antiques, en Espagne comme par ailleurs.'[59] But while it is unquestionable that certain aspects of Roman organisation survived into the seventh century and beyond, this is an extreme statement. Walter Goffart, however, has seen the military obligations of the Franks as being essentially a tax-substitute.[60]

Although Frankish kings certainly had followers, *antrustiones*,[61] who bear some comparison with members of the old *comitatus*, or equally with the *buccellarii* of the fifth century, by the end of the sixth century the Merovingian

55 Sidonius Apollinaris, *Ep*. III, 3, 3–4. Whittaker, *Landlords and warlords*, pp. 295–297. See also Lenski, *Schiavi armati*.

56 *Novella Valentiniani*, XVII, 1 (445). Whittaker, *Landlords and warlords*, p. 291.

57 Cassiodorus, *Variae*, I, 31: II, 5, 12, 26: III, 40: V, 35: X, 27: XI, 16: XII, 18, 26. Lellia Cracco Ruggini, *Economia e società nell'"Italia annonaria": rapporti fra agricoltura e commercio dal IV al VI secolo d.C.* (Milan, 1961), pp. 207, 233, 268, 281, 284, 291, 320, 324, 336. See also Jean Durliat, 'Armée et société vers 600. Le problème des soldes', in *L'armée romaine et les barbares du IIIe au VIIe siècle*, eds. F. Vallet – M. Kazanski (Condé-sur-Noireau, 1993), at pp. 33–34.

58 Durliat, 'Armée et société vers 600'.

59 Durliat, 'Armée et société vers 600', pp. 34, 36. For a clear analysis of the evidence for recruitment, Guy Halsall, *Warfare and Society in the Post-Roman West, 450–900* (London, 2003), pp. 40–70.

60 Walter Goffart, 'Frankish military duty and the fate of Roman taxation', in *Early Medieval Europe* 16 (2008), pp. 166–190.

61 Sarti, *Perceiving War*, p. 28.

army was for the most part not a permanent body, but was raised on a city basis by a *dux* or a *comes civitatis*. Thus in Gregory of Tours we hear on one occasion of the men of Tours, Poitiers, Angers and Nantes,[62] and on another of the men Bourges, Saintes, Périgueux and Angoulême.[63] Sometimes Gregory refers not to cities, but to their territories: thus he talks of men of Touraine, Poitou, the Bessin, Maine and Anjou,[64] and also of the Champagne.[65] But here we must still be dealing with arrangements made using the administrative system of the *civitas*. Exactly how the armies were raised is unclear, although we can gain some idea from the evidence of attempts (usually successful) for Churches to gain exemptions.[66] As for those who did serve, it would seem that they did not think that they should serve for long periods: when the men of Orléans and Blois were sent to keep watch over Eberulf at Tours they did so for two weeks, and then decided to return home.[67]

How these city forces were supported when they were in the field is unclear. Gregory of Tours refers on several occasions to plundering committed by armies.[68] And there is the famous story of Clovis and the division of booty taken after the defeat of Syagrius.[69] More interesting is the account of the king's march to Vouillé, when he issued an order to his troops telling them to requisition only fodder and water.[70] Troops needed more than that to survive, but how they were provisioned is left unstated.

In late seventh-century Visigothic Spain both Wamba and Ervig issued recruitment laws, in the latter's case the insistence was that each aristocrat should bring one in ten of their slaves to the army.[71] Otherwise, we can perhaps come a little closer to understanding how the armies of the Successor States were raised if we turn to the eighth-century evidence. There we find the Carolingians and also the Anglo-Saxon rulers demanding the fulfilment of obligations. In 749, at the Synod of Gumley king Æthelbald of Mercia confirmed to all

62 Gregory of Tours, *Decem Libri Historiarum*, VI, 31.
63 Gregory of Tours, *Decem Libri Historiarum*, VIII, 30. For the men of Bourges see also VI, 31.
64 Gregory of Tours, *Decem Libri Historiarum*, V, 26.
65 Gregory of Tours, *Decem Libri Historiarum*, X, 3.
66 Gregory of Tours, *Decem Libri Historiarum*, IV, 2; Bachrach, 'Grand Strategy', p. 57.
67 Gregory of Tours, *Decem Libri Historiarum*, VII, 21.
68 Gregory of Tours, *Decem Libri Historiarum*, VI, 31; VIII, 30: X, 3.
69 Gregory of Tours, *Decem Libri Historiarum*, II, 27.
70 Gregory of Tours, *Decem Libri Historiarum*, II, 37.
71 *Leges Visigothorum*, IX, II, 8 (Wamba), 9 (Ervig); Halsall, *Warfare and Society*, pp. 60–61; Esperanza Osaba García, 'Lex Visigothorum 9, 2: *De his qui ad bellum non vadunt aut de bello refugiunt*', in *Civitas, Iura, Arma. Organizzazioni militari, istituzioni giuridiche e strutture sociali alle origini dell'Europa* (*sec. III–VIII*), eds. F. Botta – L. Loschiavo (Lecce, 2015), pp. 159–191 and Ead. in this volume.

churches and monasteries immunity from obligations ('a publicis vectigalibus et ab omnibus operibus oneribus'), with the exception of bridge-building and fortress-maintenance ('instructionibus pontium, vel necessariis defensionibus arcium contra hostes').[72] Subsequently, in a charter dated either to 793 or 796 we find military service listed alongside the obligations relating to bridges and fortresses: 'preter expeditionalibus causis et pontium structionum et arcium munimentum quod omni populo necesse est ab eo opere nullum excussatum esse'.[73]

7 *Munera Sordida*

As has often been noted, these military demands seem to have been inspired by earlier Merovingian custom.[74] More important for our present concerns, it is likely that the obligations demanded by Merovingian kings ultimately derived from the *munera sordida* of the Roman World.[75] The key here is the reference to bridge-building, which is one of the *munera* of the Later Roman Empire. A law of 423 states '[therefore], no class of men, by merit or any high rank or veneration, shall be exempt from the construction and repair of roads and bridges. We also gladly ascribe the property of the divine imperial household and the venerable churches to such a laudable list. The judges of all the provinces shall be notified of this law, so that they may know that what antiquity decreed should be assigned to the public roads must be furnished without the exception of any person on the grounds of either reverence or high rank.'[76] In addition there is a law of 407 or 412 which states that 'all persons, regardless of any privilege, shall be compelled to provide for the construction of the walls as well as for the purchase and transport of supplies in kind for the needs of Illyricum.'[77] One might add that a Novel of Valentinian III from 440 explicitly denied any exemption from public services.[78]

72 Arthur West Haddan – William Stubbs, *Councils and Ecclesiastical Documents relating to Great Britain and Ireland*, vol. 3, *The English Church, 595–1066* (Oxford, 1871), pp. 386–7.

73 "Electronic Sawyer" 139, esawyer.lib.cam.ac.uk.

74 Richard P. Abels, *Lordship and Military Obligation in Anglo-Saxon England* (London, 1988), pp. 52–53, with p. 228, n. 60.

75 On the survival of *munera sordida*, Magnou-Nortier, *Aux origines de la fiscalité*, pp. 72–78, 421–423.

76 *Codex Theodosianus*, XV, 3, 6 (423).

77 *Codex Theodosianus*, XI, 17, 4 (407/412); see also XV, 1, 34 (396), 36 (397), 49 (412/407/408), 51 (413).

78 *Novella Valentiniani*, X (440).

It would seem that some aspects of the *munera sordida* of the Later Empire were still enforced in the Merovingian kingdom. This, however, does not provide a full explanation for the rise of the obligation of military service in the post-Roman World. The *Theodosian Code* talks of obligations to work on roads, bridges and walls, but not to serve in the army, even if in legislation from the year 406 there was an encouragement to enlist,[79] as we have noted. And, as we have also seen, demands were made on senators, landowners and officials to supply recruits in 443–4.[80] It may be that such demands, in the closing years of imperial rule underlie the military obligations of later centuries, and that they continued alongside the obligations that Goffart has seen as being central to the accommodation of the barbarians.[81] If we assume that the military obligations of the post-Roman period derived not from Germanic tradition, for which there is no evidence, but from Roman law, we need to consider the role of the military leaders and the kings who succeeded them.

8 The Military Oath

In becoming federates the barbarians submitted themselves to late-Roman military jurisdiction, swearing an oath of loyalty to the Emperor, as is apparent from Zosimus' account of the conflict between Fravitta and Eriulf.[82] Such oaths were clearly Roman in origin, and, despite traditional views, as expressed for instance by Maurizio Lupoi, we do not need to invoke Germanic tradition to understand the relations between sovereign and people.[83] Vegetius even provides a description of the military oath, sworn before God, Christ, the Holy Spirit and the emperor.[84] The federates' experience of Roman law was thus, first and foremost, of Roman military law. This is an area of law that is not easily reconstructed, despite the dedication of Book VII of the *Theodosian Code* to the topic *De re militari*: indeed, it was never a free-standing area of jurisprudence, but was simply made up of those aspects of law that were concerned with the military. Certain elements of military law have, however, been

79 *Codex Theodosianus*, VII, 13, 16–17.

80 *Novella Valentiniani*, III, VI, 2–3.

81 Goffart, 'Frankish military law'.

82 Zosimus, *Historia Nova*, IV, 56; Stefan Esders, 'Rechtliche Grundlagen frühmittelalterlicher Staatlichkeit', in *Der Mittelalterliche Staat – Europäische Perspektiven*, eds. W. Pohl – V. Wieser (Vienna, 2009), at p. 424.

83 Maurizio Lupoi, *Alle radici del mondo guiridico europeo. Saggio storico comparativo* (Rome, 1994), pp. 335–49.

84 Vegetius, *De re militari*, II, 5; Magnou-Nortier, *Aux origines de la fiscalité*, pp. 181–183.

elucidated in recent years.[85] Even though the laws of the *Theodosian Code* tell us little about legal practices within the military, with regard to the practicalities of law Stefan Esders has drawn particular attention to the issues of the oath of loyalty[86] and to the crime of treason (*laesa maiestatis*).[87]

The oath of loyalty would become a central feature of the political system of the Successor States.[88] Athalaric affirmed the importance of the *sacramentum* of the people in the consensus of the Ostrogothic kingdom.[89] Sixth-century Merovingian kings also emphasised the oaths of loyalty that were made to them, as when Chilperic attacked Praetextatus of Rouen for inciting the people so that 'none kept his faith with me': 'nullus mecum fidem habitam custodiret'.[90] And Gregory of Tours relates that the people of his diocese once swore an oath to king Charibert.[91] Despite the relative insignificance of the *sacramentum* in the Burgundian sphere,[92] the origins of the oath of loyalty in the Successor States would seem to take us back to the military arrangements of the later Roman Empire, and to the oaths sworn by soldiers and federates. Luca Loschiavo has noted the influence of Roman military law in concepts of treason, desertion, espionage, disobedience in Lombard law.[93]

---

85 Esders, 'Late Roman Military Law', p. 4, with the bibliography in n. 23; Id., 'Rechtliche Grundlagen', p. 424, with the bibliography in n. 3.

86 Esders, 'Rechtliche Grundlagen'.

87 Esders, 'Late Roman Military Law'.

88 Matthias Becher, *Eid und Herrschaft: Untersuchungen zum Herrschethos Karls des Grossen* (Sigmaringen, 1993). Stefan Esders, *Sacramentum Fidelitatis: Treueid, Militärwesen und Formierung mittelalterliche Staatlichkeit* (Bochum, 2003); Id., 'Rechtliche Grundlagen'; Id., ''Faithful believers': Oaths of allegiance in post-Roman societies as evidence for Eastern and Western Visions of Community', in *Visions of Community in the post-Roman World: The West, Byzantium and the Islamic World, 300–1100*, eds. W. Pohl – C. Gantner – R. Payne (Farnham, 2012), esp. at p. 361. See also, in the same volume, Helmut Reimitz, 'The Providential Past: Visions of Frankish Identity in the early medieval History of Gregory of Tours' *Historiae* (sixth-ninth century)', at p. 117.

89 Cassiodorus, *Variae*, VIII, 3.

90 Gregory of Tours, *Decem Libri Historiarum*, V, 18; Sarti, *Perceiving War*, p. 28. Magnou-Nortier, *Aux origines de la fiscalité*, pp. 325–72, provides a full examination of the evidence for early Merovingian oaths, which largely repeats arguments from her *Foi et Fidelité. Recherches sur l'évolution des liens de fidelité chez les Francs du VIIe au IXe siècle* (Toulouse, 1976).

91 Gregory of Tours, *Decem Libri Historiarum*, IX, 30.

92 Perhaps reflecting the position of the Gibichungs as agents of the emperor, rather than as independent rulers.

93 Luca Loschiavo, 'Looking at the Edict of Rothari: between German ancestral customs and Roman legal traditions', *Roma Tre Law Review* 2 (2019), at pp. 79–80.

9 Barbarians and Roman Military Office

But barbarian leaders did not just swear oaths: as Roman officials they also had to exact the oath from others. Already in the fourth century barbarian soldiers were achieving high military office within the Roman army, as Karl Friedrich Stroheker pointed out in a number of articles that were reprinted in his collection *Germanentum und Spätantike* in 1965.[94] The point has been reaffirmed more recently by Lellia Cracco-Ruggini.[95] In the fifth century several barbarian leaders, including Alaric I, wanted to hold Roman military office,[96] and several, including the Gibichungs Gundioc, Chilperic, and Gundobad, as well as Theodoric the Great and Clovis, did so.[97] The Roman Law experienced and enforced by these men as federates and as military officials was first and foremost military law.[98]

By the third quarter of the fifth century, however, barbarian *magistri militum* were not just controlling military forces: they were also acting as agents for the civilian state.[99] Thus, the Gibichung leader and *magister militum per Gallias* Gundioc found himself involved in an ecclesiastical dispute, when bishop Mamertus of Vienne intervened in the appointment of Marcellus of Die, a diocese which was subordinate to the bishop of Arles.[100] A little while later we find Chilperic, Gundioc's brother and, it would seem, his successor in office, working together with bishop Patiens of Lyon.[101] Chilperic also appears in the *Vita Patrum Iurensium*, in a passage that provides us with an extraordinary statement of the take-over of civilian authority by a barbarian general. When abbot Lupicinus intervened to secure the freedom of some poor

---

94 Karl Friedrich Stroheker, 'Zur Rolle der Heermeister fränkischer Abstammung im späten vierten Jahrhundert' and 'Alemannen im römischen Reichsdienst', both in Id., *Germanentum und Spätantike* (Stuttgart, 1965), pp. 9–29 and 30–53 respectively.

95 Lellia Cracco Rugini, 'Les généraux franques au IVe–Ve siècles et leurs groupes aristocratiques', in *Clovis, histoire et mémoire, vol. 1, Clovis et son temps, l'événement*, ed. M. Rouche (Paris, 1997), pp. 673–688.
Les généraux franques, pp. 673–688.

96 Christine Delaplace, *La fin de l'empire romain d'Occident. Rome et les Wisigoths de 382 à 534* (Rennes, 2015), pp. 119–122, 138–146.

97 Paul Barnwell, *Emperor, Prefects and Kings. The Roman West, 395–565* (London, 1992), pp. 82–83, 97–99, 135.

98 Loschiavo, 'Looking at the Edict', p. 80.

99 Loschiavo, 'Looking at the Edict', pp. 81–83.

100 *Epistolae Arelatenses Genuinae*, 19 (ed. Gundlach, pp. 28–29). See Ralph Mathisen, *Ecclesiastical Factionalism and Religious Controversy in fifth-century Gaul* (Washington D.C., 1989), p. 216.

101 Sidonius, *Ep.* VI, 12, 3.

people who were being enslaved, the oppressor accused him of prophesying the impending collapse of the Empire. This prompted Lupicinus' reply to his critic: 'Do you not see, you ignoble and miserable person, that law and justice have been thrown into confusion on account of your sins and those of your followers in your frequent exactions against the innocent, and that the authority of the purple-banded fasces has been changed under a judge clothed in skins?' ('Nonne cernis, degener et infelix, ius fasque confusum, ob tuis tuorumque crebra in innocentium pervasione peccatis, mutari muriceos pellito sub iudice fasces'.)[102] The result of this outburst was that Chilperic freed the men who were being enslaved. Despite the description of the ruler as being clothed in skins (which may be a reference to his military attire rather than simply an indication that he was a barbarian) both Lupicinus and the hagiographer clearly approved of Chilperic.

But it is not only barbarians holding Roman military office who were acting in civil suits: barbarian kings were doing the same. There are the famous accounts of the courts of Theodoric II and Euric by Sidonius Apollinaris.[103] And we also hear of a legal appeal against the unjust sentence of a Spaniard in the *Vita Orientii*,[104] and another in the *Vita Bibiani* against the imprisonment of *mediocres* and *nobiles* who had failed to pay the *iniunctio*,[105] both relating to the court of the Visigothic king Theodoric I. Like Athaulf in the anecdote related by Orosius, these men thought it wrong 'to deprive a state of laws (without which a state is not a state at all).'[106]

10 Early Barbarian Law-*Givers*

It is precisely this group of barbarian kings and *magistri militum* who were responsible for the earliest sub-Roman laws and law books. There may have been legislation by Theodoric I or II: Sidonius relates that at mealtimes the renegade senator Seronatus praised the laws of Theodoric over those of Theodosius – by which he presumably meant the *Codex Theodosianus*.[107] Famously Sidonius also praised his friend Syagrius as a Solon of the Burgundians. He must

102 *Vita Patrum Iurensium*, II, 94: translation adapted (following a suggestion by Noel Lenski) from Tim Vivian – Kim Vivian – Jeffrey Burton Russell, *The Lives of the Jura Fathers* (Kalamazoo, 1999), p. 145.

103 Sidonius Apollinaris, *Epp.* I, 2; VIII, 9, 5.

104 *Vita Prima Orientii*, 3, p. 63.

105 *Vita Bibiani*, 4.

106 Orosius, *Historia adversus Paganos*, VII, 43, 6.

107 Sidonius, *Ep.* II, 1, 3.

have been offering legal advice to either Gundioc or Chilperic.[108] Sidonius' own epitaph states that *Leges barbaros dedit furori*, or *Leges barbarico dedit furori*.[109] In either reading he was acting as he says Syagrius had done. Although it is possible that Sidonius was involved in Visigothic legislation, it is more likely, given the structure of the epitaph, and the chronology of his career, that his legal activity was at the Gibichung court.[110] Another of Sidonius' correspondents, Leo, may have been involved in Euric's legislation,[111] which is attested in Isidore's *History*,[112] and which may be preserved in the *Codex Euricianus*.[113] The involvement of Romans in the lawgiving of the Successor States takes us, of course, on to the advisers of Alaric II, especially Anianus *spectabilis*,[114] to Cassiodorus and Theodoric the Great (another *magister militum*), and to the Gallo-Romans Parthenius and Asclepiodatus, both of whom worked for sixth-century Merovingians.[115]

11 Military Law in the Barbarian Codes

Against this background the presence of military law in the codes of the Successor States is no surprise. The procedures to which the *magistri militum* and federate leaders were accustomed were military, but the men themselves were extending their authority over aspects of society that had hitherto been civilian, and with the guidance of civilian officials. Some issues necessarily related to the relations between soldiers and the rest of the population – most obviously the clauses to be found in most of the codes relating to the

108 Sidonius, *Ep.* V, 5, 3.

109 Christian Luetjohann, *Praefatio in Sidonius*, in Sidonius Apollinaris, *Epistulae et carmina* (MGH, *Auct. Antiquiss.*, VIII) (Berlin, 1887), p. VI.

110 Ian Wood, 'Sidonius and the Burgundians', in *Academica Libertas: Essais en honneur de professeur Javier Arce*, eds. D. Moreau – R. González Salinero (Turnhout, 2019), at p. 367.

111 Sidonius, *Ep.* VIII, 3, 3; Harries, *Not the Theodosian Code*, pp. 48–49.

112 Isidore, *Historia Gothorum*, 51.

113 *Legum Codicis Euriciani*, ed. Zeumer, *Leges Visigothorum*.

114 *Breviarium Alarici, Prologus*; Ralph Mathisen, 'D'Aire-sur-l'Adour à Agde: les relations entre la loi séculière et la loi canonique à la fin du Royaume de Toulouse', in *Le Bréviare d'Alaric. Aux origines du code civil*, eds. M. Rouche – B. Dumézil (Paris, 2008), at p. 43.

115 Patrick Wormald, 'The Decline of the Western Empire and the Survival of its Aristocracy', in *Journal of Roman Studies* 66 (1976), pp. 217–226.

settlement of the barbarians.[116] But Alexander Callander Murray has emphasised the military basis of much Frankish administrative organisation.[117]

12 *Pactus Legis Salicae*

It is worth pausing briefly on whether the *Pactus Legis Salicae* might be interpreted from a military point of view. Elisabeth Magnou-Nortier has argued that the first forty-two clauses of the *Pactus Legis Salicae* should be read as 'un code pénal à l'usage des Francs Saliens, inspiré par la discipline militaire de l'armée romaine en matière de réquisitions et de répression d'actes de violences perpétrés par des soldats …'[118] Karl Ubl, currently the leading interpreter of the *Pactus*, does not address this precise issue in his overview of the origins of *Lex Salica*, published in 2009, which provided a statement of the *status quaestionis* up to that date.[119]

It is worth noting, however, that Ubl, like Hermann Nehlsen, recognises that the *Pactus* as we have it is a text that has undergone modifications.[120] In addition, it is highly likely that it includes laws that were issued at precise moments in time, as well as laws which may have reflected traditional practice.[121] Most obviously, as an example of a specific piece of legislation, there is the law on slaves taken overseas.[122] In other words, it is almost certain that the *Pactus* as a whole contains clauses that had previously been issued as edicts or had been included in earlier codes. The manuscripts may also include clauses and phrases that were added by later redactors. It is, therefore, clear that no single clause dates the complete *Pactus*, despite scholarly attempts to the contrary.

116 P. Porena and Y. Rivière, eds., *Expropriations et confiscations dans les royaumes barbares* (École française de Rome) (Rome, 2013).

117 Alexander Callander Murray, 'From Roman to Frankish Gaul: *Centenarii* and *centenae* in the administration of the Merovingian kingdom', in *Traditio* 44 (1980), pp. 59–100; Id., 'The position of the *grafio* in the constitutional history of Merovingian Gaul', in *Speculum* 64 (1986), pp. 787–805. See also the contribution of Stefan Esders in this volume.

118 Elisabeth Magnou-Nortier, 'Remarques sur la genèse du *Pactus Legis Salicae*', in *Clovis, histoire et mémoire*, vol. 1, *Clovis et son temps: l'événement*, ed. M. Rouche (Paris, 1997), at p. 502.

119 Karl Ubl, 'L'origine contestée de la loi salique. Une mise au point', *Revue de l'Institut français d'histoire en Allemagne*, 1 (2009), pp. 208–234 [pp. 1–18] –https://doi.org/10.4000/ifha.365. I cite the text by paragraph number, as in the online edition. See also Magali Coumert, *La Loi Salique: retour aux manuscrits* (Turnhout, 2023).

120 Ubl, *L'origine contestée*, pp. 208–234, § 26.

121 Ian Wood, *The Merovingian Kingdoms, 450–751* (London, 1994), p. 110.

122 *Pactus Legis Salicae*, 39, 2.

The clause that has been used most frequently to date the *Pactus* is number 47, which refers to those subject to Salic law who live between the Loire on the one side, and the *Silva Carbonnaria* on the other. They are given 40 days to come to court to settle disputed possession, whereas those living beyond those boundaries are given 80. Ubl (like most scholars) originally argued that this only makes sense between 507 (the battle of Vouillé) and Clovis' death four years later.[123] More recently, however, he has argued that the boundaries make no sense for Clovis' time, and that the clause must be an addition of the later sixth or seventh century.[124] Certainly there are strong reasons for thinking that the Loire and *Silva Carbonnaria* make little sense for the years 507–511. They may, however, make some sense in the years immediately after Clovis' victory over Syagrius in 486. The *Silva Carbonnaria* might easily have been the north-eastern frontier of the kingdom of Tournai, which was the power centre of Clovis' father, Childeric, to judge by his burial.[125] In the high Middle Ages Tournai itself would seem to have lain on the western edge of the Charbonnière.[126] To the east there were the Rhineland Thuringians, who Clovis defeated in 491 according to Gregory of Tours (though the precise date may well be suspect).[127] Although we do not know the extent of Syagrius' area of rule,[128] it is not unlikely that it stretched at some point to the Loire. A geographical unit running from the *Carbonnaria* to the Loire may, then, make sense between c.486 and c.491. If one objects that the presence of Franks outside those limits only makes sense after 507, it is worth noting that, according to Gregory, Clovis demanded the return of fugitives who had crossed the Loire in the aftermath of his victory over Syagrius.[129] And, in any case, the interpolation of one phrase is just as likely as the interpolation of an entire clause: the phrase dealing with those living outside the defined bounds could be an addition.

Certainly the *Pactus* as a whole in its surviving form is unlikely to date from the period after Clovis' conversion, given its reference to a votive boar.[130] This

123 Ubl, 'L'origine contestée'.

124 Karl Ubl, *Sinnstiftungen eines Rechtsbuchs. Die Lex Salica im Frankenreich* (Ostfildern, 2017), p. 65.

125 Dieter Quast, ed., *Das Grab des fränkischen Königs Childerich in Tournai und die Anastasis Childerici von Jean-Jacques Chifflet aus dem Jahre 1655* (Mainz, 2015).

126 Paolo Charruadas, 'L''ombre' de la forêt charbonnière', in *La forêt en Lotharingie medieval*, eds. M. Pauly – H. Pettiau (Luxembourg, 2016), at p. 89.

127 Gregory of Tours, *Decem Libri Historiarum*, II, 27.

128 Edward James, 'Childéric, Syagrius et la disparition du royaume de Soissons', in *Revue archéologique de Picardie* 3–4 (1988), pp. 9–12.

129 Gregory of Tours, *Decem Libri Historiarum*, II, 27.

130 *Pactus Legis Salicae*, 2, 16.

scarcely makes sense after Clovis' conversion and baptism. We do know that the king issued orders defending Church property to his army en route to Vouillé.[131] Moreover, as Ubl has rightly noted, we have legislation from Clovis in the period between 507 and 511, in the form of his letter to the Aquitanian bishops and it is totally dissimilar in style from the *Pactus*.[132]

In other words, the territory defined by the Charbonnière and the Loire may have been an identifiable unit between c. 486 and c.491, and clause 47 of the *Pactus* might reflect that. The one clause does not date the Code as a whole, but it does suggest that some clauses contained within it could be linked to moments of conquest. This could support Magnou-Nortier's argument for the existence of a military code within the *Pactus*. And Ubl has insisted that the Code needs to be understood in its military as well as its juridical context.[133] The aftermath of Syagrius' defeat would also help explain the low status of the Romans within the Code – and it is worth remembering that it is Romans and not Goths (or indeed the Alamans or Burgundians, who were also defeated by Clovis) who are given low wergelds in the *Pactus*.[134]

13 Gibichung Legislation

It is, however, in the Burgundian legislation that one can perhaps best see the coalescence of military and civilian authority, which went hand in hand with the erosion of a distinction between *miles* and *homo*. Fortunately we have some indication of the contexts in which the two Gibichung compilations were made. According to Gregory of Tours, following his victory over his brother Godigisel and the execution of the senatorial and Burgundian rebels, Gundobad issued *leges mitiores* to the Burgundians to prevent the oppression of Romans.[135] Some of these laws would seem to be included in the *Liber Constitutionum*, which survives in a recension issued by Sigismund in 517 – in all probability the first 41 clauses were drawn from an earlier and shorter code.[136] Of the clauses in this opening section of the *Liber Constitutionum*, several do indeed deal with Romano-barbarian relations, and might be categorised as

131 Gregory of Tours, *Decem Libri Historiarum*, II, 37.

132 Ubl, *Sinnstiftungen eines Rechtsbuchs*, p. 55.

133 Ubl, *Sinnstiftungen eines Rechtsbuchs*, p. 66.

134 *Pactus Legis Salicae*, 11, 1, 9; 42, 4; 104, 9; 117, 2.

135 Gregory, *Decem Libri Historiarum*, II, 33.

136 Ian Wood, 'Burgundian law-making, 451–534', in *Italian Review of Legal History* 3 (2017), pp. 1–27 – https://doi.org/10.13130/2464-8914/12878, at pp. 9–10.

leges mitiores[137] – that is to say that they may be the remnants of a law-book compiled by Gundobad. In addition to the laws mentioning *hospitalitas* or *hospites*,[138] clause 22 forbids barbarian *patrocinium* in Roman lawsuits: clause 28 deals with the rights of Romans and Burgundians in forest land: 31 with the right to plant vineyards in common land: and 38 with the obligation of Burgundians as well as Romans to provide for legates

14 The Oath and Trial by Battle

Perhaps yet more interesting is a clause which falls outside this initial group, and which is firmly dated to 502.[139] Here, because oath-taking was not taken seriously by members of *populi nostri* – a phrase that in the law book usually means Burgundian, or perhaps more generally the non-Roman followers of the ruler[140] – they were instead to be subject to trial by battle. This law is important for numerous reasons. First, it shows clearly that oaths were not a form of attestation that Burgundians necessarily took seriously. It, therefore, implies that oath-taking was a Roman rather than a barbarian practice.[141] Second, although it is tempting to assume that trial by battle was a barbarian means of dispute resolution, in this instance it may be better to see it as a military procedure, just as other types of ordeal seem to have been.[142]

137 Ian Wood, 'The legislation of *magistri militum*: the laws of Gundobad and Sigismund', in *La forge du droit. Naissance des identités juridiques en Europe* (*IVe–XIIIe siècles*), Clio@Themis, 10 (march 2016) – http://www.cliothemis.com. *Liber Constitutionum*, XXII, XXVIII, XXXI, XXXVIII, LIV, LV, LXXXIV.

138 *Liber Constitutionum*, I, XIII; see also LIV, *Constitutiones Extravagantes* XXI, 12. Clarifying the meaning of the term *hospitalitas* are the essays by Trisciuoglio and Porena in this volume.

139 *Liber Constitutionum*, XLV.

140 Wood, *The legislation of magistri militum*.

141 Ian Wood, 'Disputes in late fifth- and sixth-century Gaul: some problems', in *The Settlement of Disputes in Early Medieval Europe*, eds. W. Davies – P. Fouracre (Cambridge, 1986), pp. 7–22.

142 Soazick Kerneis, 'Marcher au chaudron dans l'Empire romain: Genèse de l'ordalie (IIe–IVe siècle après J.-C.', in *Les Justices de l'Invisible*, eds. R. Verdier – N. Kálnoky – S. Kerneis (Paris, 2013), pp. 255–268; Ead., 'Les jugements des soldats et la fabrique des premières lois barbares (Ve siècle)', in *Ravenna Capitale, Giudizi, giudici e norme processuali in Occidente nei secoli IV–VIII*, vol. 1 (Santarcangelo di Romagna, 2015), pp. 211–224.

Certainly trial by battle appears in other codes of the Successor States, including those of Lombard Italy.[143] It also appears in the *Lex Alamannorum*.[144] The most interesting parallel to the reference to trial by battle in the *Liber Constitutionum*, however, comes in the second clause of the *Lex Baiuwariorum*, where an accusation that a man had attempted to kill the *dux* had to be supported by three witnesses: but if two of the witnesses disagreed, they should 'resort to the judgment of God' and 'go to the field': 'exeant in campo et cui Deus dedit victoriam, illi credatur'.[145]

15 *Crimen Laesae Maiestatis*

Stefan Esders has recently argued strongly that the Bavarian Code contains a good deal of Roman military law, which had been enforced in the frontier region of the Empire and was subsequently drawn on in the Frankish *ducatus* of Bavaria. Key to his argument is the Roman *crimen laesae maiestatis*, which seems to underpin the Bavarian legislation protecting the *dux*.[146] Interestingly the so-called *Lex Romana Burgundionum*, which is better called by the name in the manuscripts – the *Forma et Expositio Legum* – also has a clause: 'Crimina vero maiestatis haec sunt, quibus legibus designantur, id est: salus principis, traditio regionis aut adeptio tyrannidis.'[147] Ludwig Rudolf von Salis could find no source for this clause, despite the words *quibus legibus designantur*. Since the legislation on the *crimen laesae maiestatis* in the Bavarian Code seems to reflect Roman military law, the same origin is surely likely for the clause in the *Forma et Expositio*.

143 *Edictum Rothari*, 164–5, 198, 202, 228, 365; *Leges Grimualdi*, 1, 2, 4, 7; *Leges Liutprandi*, 11, 56, 68, 71, 118, 121; *Capitula Adelchis*, 6, 7; *Codex Gothanus*, 5.

144 *Lex Alamannorum*, 81.

145 *Lex Baiuwariorum*, II; Esders, 'Late Roman Military Law', p. 13. For the *Lex Baiuvariorum* see also the new edition of Roman Deutinger (*Lex Baioariorum: Das Recht der Bayern*). See also Harald Siems, 'Herrschaft und Konsens in der 'Lex Baiuvariorum' und den 'Decreta Tassilonis'', in *Recht und Konsens im frühen Mittelalter*, eds. v. Epp – Ch.H.F. Meyer (Ostfildern, 2017), pp. 299–360.

146 Esders, 'Late Roman Military Law'.

147 *Forma et Expositio Legum*, VII, 6, p. 130. For a comparison with the Bavarian law on the *crimen maiestatis*, see Siems, 'Herrschaft und Konsens', pp. 310–313. For a recent comment on the *Forma et Expositio Legum* see Detlef Liebs, 'Geltung kraft Konsenses oder kraft königlichem Befehl? Die *lex Romana* unter Westgoten, Burgundern und Franken', in *Recht und Konsens im frühen Mittelalter*, eds. v. Epp – Ch.H.F. Meyer (Ostfildern, 2017), esp. 72–76.

It is worth speculating a little further on the Roman code, the *Forma et Expositio Legum*, of the Gibichungs. Some of the clauses may have been set down by Syagrius or even Sidonius. But the compilation as a whole may belong to the same context as that of the hypothetical recension of the *Liber Constitutionum*, which seems to have been issued in c.500. Not only would the clause on the *crimen maiestatis* fit very neatly into legislation following the suppression of the uprising of Godegisel, but the subsequent clause on violence, with its emphasis on *collectis turbis* and *tumultum* could also fit the same period of disorder. If this is correct, Gundobad had two law-books compiled in c.500, which constitute the juridical actions of a military man in response to an attempted coup.

16 Conclusion

To return to the question of the transformation of the Roman military, it would seem that not only are we dealing with the evaporation of the Roman army and its replacement with a force of farmer-soldiers, but we are also dealing with a transformation of authority, in which civilian rule came to be exercised by military men, with the result that there was no longer any distinction between civil and military law, which, in any case, had never been a free-standing area of jurisprudence, but rather those laws which deal with soldiers.

This, however, does not mean that we enter a period dominated by war – except in the regions affected by Justinian's expeditions to the West. There was certainly plenty of violence in late sixth- and seventh-century Gaul and Spain, but it was small scale.[148] Even a blood feud might be called a *bellum … inter cives* by Gregory of Tours.[149] Although it is clear that armies on the move could cause a great deal of destruction, we do not know how large those armies were. Procopius' account of the seizure of the Visigothic throne in c.531 by Theudis provides an interesting, and plausible figure: the Goth, who was in fact an Ostrogothic agent of Theodoric, married a Hispano-Roman heiress, and raised a force of 2,000 from her estates, which was enough to allow him to seize power – and also to discourage the Ostrogothic king from intervening against him. He did, however, continue to pay tribute to his Italian overlord.[150] Writing about the year 589 John of Biclar stated that the Visigothic *dux* Claudius put to flight 60,000 Franks with an army of scarcely 300.[151] The number of

148 Wood, *The Transformation of the Roman West*, pp. 81–83.

149 Gregory of Tours, *Decem Libri Historiarum*, IX, 19. See Sarti, *Perceiving War*, pp. 41–43.

150 Procopius, *Wars*, V, XII (ed. Dewing, pp. 50–54).

151 John of Biclar (Iohannes Biclarensis), s.a. 589.

Franks is clearly a huge exaggeration: that of the Visigoths may be plausible, although it is surely derived from the size of the army of Gideon that defeated the Midianites.[152] Admittedly Gregory of Tours tells us that the army raised in Bourges numbered 15,000[153] – though this sounds an unlikely contingent from one city, which surely housed a population of tens, rather than hundreds, of thousands. The men of Carcassonne were able to put to flight what was supposedly a large-scale army made up of the men of Bourges, Saintes, Périgueux and Angoulême.[154] So too the *Cordubenses cives* could defeat the army of the Visigothic king Agila.[155] It would be surprising if any of these cities produced forces numbering more than a few hundred, or that the combined armies of a group of cities reached far beyond the very low thousands. It is worth remembering that a military invasion (*heriraita*) in the *Lex Baiuwariorum* need only number 42 warriors.[156] For the Anglo-Saxons an army was anything more than 35 men.[157]

Although this was a martial society violence was probably mostly low-level, even if it was endemic. Because the Successor States were initially ruled over by men who were either federates or barbarians holding Roman military office, their government was deeply influenced by the traditions of Roman military law. This was not, of course, the only influence on the government of the Successor States: Roman imperial law, vulgar law and provincial law were also drawn on, as were the practices of the old imperial administration. Increasingly important, to the extent that it was the dominant factor by the seventh century, was the Church, whose authority, as exercised by hundreds of individual bishops, and whose wealth, in terms of property acquired above all from the late fifth century onwards, meant that it was more than an equal to kings and secular authority. Moreover, despite the fact that this was a world where all men were potential warriors, after the settlement of the Lombards, in most of western Europe this was not a period of large-scale military combat. Paradoxically, the Roman Empire, where the military was largely to be found on the frontiers, was more concerned with its army than were the Successor States, most of whose male members were only potential warriors, who were rarely called upon to fight.

---

152 Judges, VII, 7. Roger Collins, *Visigothic Spain, 409–711* (Oxford, 2004), p. 68.

153 Gregory of Tours, *Decem Libri Historiarum*, VI, 31.

154 Gregory of Tours, *Decem Libri Historiarum*, VIII, 30.

155 Isidore, *Historia Gothorum*, 45.

156 *Lex Baiuwariorum*, IV, 23; Esders, 'Late Roman Military Law', p. 18.

157 Laws of Ine, 13, 1; Peter Sawyer, *The Age of the Vikings*, 2nd ed., (London, 1971), p. 123.

Bibliography

Primary Sources

Ammianus Marcellinus, *Rerum gestarum libri*, ed., J.C. Rolfe, 3 vols. (Cambridge, Mass., 1935–9)

Ausonius, *Precatio consulis*, ed., H.G. Evelyn-White, 2 vols (Cambridge, Mass., 1921)

Capitula Adelchis, see *Leges Langobardorum*

Cassiodorus Senatoris, *Variae*, ed. Th. Mommsen [MGH, *Scriptores – Auct. ant.* XII] (Berlin, 1894) – Engl. transl. by M.S. Bjornlie, *The Variae. The Complete Translation.* (Oakland, CA, 2019)

Claudianus, *De Bello Getico*, ll. 412–22, ed. M. Platnauer (Cambridge, Mass., 1922)

Codex Euricianus, see *Leges Visigothorum*

Codex Gothanus, see *Historia Langobardorum Codicis Gothani*

Codex Justinianus, ed. P. Krüger in T. Mommsen – P. Krüger, *Corpus Iuris Civilis*, II (Berlin, 1954[11]) – Engl. transl. B.W. Frier, *The Codex of Justinian. A New Annotated Translation*, I–III, (Cambridge, 2016)

Codex Theodosianus: *Theodosiani libri XVI cum constitutionibus Sirmondianis et leges Novellae ad Theodosianum pertinentes*, ed. Th. Mommsen – P.M. Meyer, 2 vols. (Berlin, 1905) – Engl. transl. by C. Pharr *The Theodosian Code and Novels and the Sirmondian Constitutions. A Translation with commentary, glossary, and bibliography*, [Princeton, 1952], 10th Print (Union, N.J., 2001)

Edictum Rothari, see *Leges Langobardorum*

Eugippius, *Vita Severini*, ed. P. Régerat, *Eugippe, Vie de saint Séverin*, [Sources Chrétiennes, 374] (Paris, 1991)

Forma et Expositio Legum, see *Leges Burgundionum*

Gildas, *De Excidio Britonum*, ed. M. Winterbottom, *Gildas, The Ruin of Britain and other documents* (Chichester, 1978)

Gregorius Turonensis, *Historiarum libri decem*, ed. Rudolf Buchner, *Ausgewählte Quellen zur Geschichte des Mittelalters*, 2 (Darmstadt, 1974)] – Engl. transl. by L. Thorpe, *Gregory of Tour, The history of Franks* (London, 1974)

Historia Langobardorum Codicis Gothani [MGH, *Scr. rer. Lang.* Hannoverae 1878, pp. 7–11]

Hydatius, *Chronicon*, ed. R. Burgess, *The Chronicle of Hydatius and the Consularia Constantinopolitana: two contemporary accounts of the final years of the Roman Empire* (Oxford, 1993)

Isidorus Hyspalensis, *Historia Gothorum*, ed. Th. Mommsen, in *Chronica Minora*, II [MGH, *Auct. Ant.*, XI] (Berlin, 1894)

John of Biclar (Iohannes Biclarensis), *Chronacon*, ed. C. Cardelle de Hartmann, *Corpus Christianorum, Series Latinorum* 173A (Turnhout, 2001)

Jordanes, *Getica*, ed. Th. Mommsen, [MGH *Auct. Antquiss.* 5,1] (Berolini, 1882)

Laws of Ine, ed. F.L. Attenborough, *The Laws of the earliest English Kings* (Cambridge, 1922)
Leges Burgundionum, ed. L.R. de Salis [MGH, *Leges nat. Germ.* II.1] (Hannoverae, 1892)
Leges Grimualdi, see *Leges Langobardorum*
Leges Langobardorum, ed., F. Bluhme, in [MGH *Leges* IV] (Hannoverae, 1868)
Leges Liutprandi, see *Leges Langobardorum*
Leges Visigothorum, ed. K. Zeumer [MGH, *Leges nat. Germ.* I] (Hannoverae et Lipsiae, 1902)
Lex Alamannorum, ed., K.A. Eckhardt [MGH, *Leges nat. Germ.*, V,1] (Hannoverae, 1966) – Engl. transl. by Th.J. Rivers, *Laws of the Alamans and Bavarians* (Philadelphia, 1977)
Lex Baioariorum: Das Recht der Bayern, ed. R. Deutinger (Regensburg, 2017)
Lex Baiuwariorum, ed. E. von Schwind [MGH, *Leges nat. Germ.*, V,2] (Hannoverae, 1926) – Engl. transl. by Th.J. Rivers, *Laws of the Alamans and Bavarians* (Philadelphia, 1977)
Lex Ribvaria, eds., F. Beyerle – R. Buchner [MGH, *Leges Nat. Germ.* III.2] (Hannoverae, 1954)
Liber Constitutionum, see *Leges Burgundionum* – Engl. transl. K. Fisher Drew, *The Burgundian code: book of constitutions or law of Gundobad* (Philadelphia, 1972)
Novella Majoriani, see *Codex Theodosianus*
Novella Valentiniani, see *Codex Theodosianus*
Olympiodorus, ed., R.C. Blockley, *The Fragmentary Classicising Historians of the Later Roman Empire, Eunapius, Olympiodorus, Priscus and Malchus*, vol. 2 (Liverpool, 1983)
Orosius, *Historia adversus Paganos*, ed., M.-P. Arnaud-Lindet, *Orose, Histoires contre les Païens*, 3 vols. (Paris, 2003)
Pacati panegyricus, eds., C.E.V. Nixon – B. Saylor Rodgers – R.A.B. Mynors, *In praise of later Roman Emperors: the panegyrici latini* (Berkeley, 1994)
Pactus Legis Salicae, ed., K.A. Eckhardt [MGH, *Leg. Nat. Germ.*, 4.1] (Hannoverae, 1962)
Pactus pro tenore pacis, ed., A. Boretius, *Captilaria Merowingica* [MGH, *Capitularia regum Francorum* I] (Hannover, 1883), pp. 4–7
Procopius Caesariensis, *De bellis – Tôn Polemôn Logoi*, ed. (and Engl. transl.) H.B. Dewing, *History of the Wars* (Cambridge, Mass., 1919)
Sidonius Apollinaris, *Epistulae et carmina*, ed., C. Luetjohann [MGH, *Auct. Ant.*, 8] (Berlin, 1887)
Sozomenus, *Historia ecclesiastica*, ed. A.-J. Festugière – B. Grillet – G. Sabbah, *Sozomène, Histoire Ecclésiatique*, 4 vols., [Sources Chrétiennes, 516] (Paris, 2008)
Synesius, *De Regno*; Id., *De Providentia*, ed. J. Lamoureux – N. Aujoulat, *Synésios de Cyrène, Opuscules*. vols. 2–3 (Paris, 2008)
Vegetius, *Epitoma rei militaris*, ed. A. Önnerfors (Stuttgart, 1995)

Vita Bibiani, ed. B. Krusch, [MGH, *Script. rerum Meroving.* III] (Hannoverae, 1896)
Vita Patrum Iurensium, ed. F. Martine, *Vie des Pères du Jura*, [Sources Chrétiennes, 142] (Paris, 1968)
Vita Prima Orientii, 3, AASS May 1st (Paris, 1866), pp. 62–63
Zôsimos, *Istoriai Neai – Historia Nova*, ed. F. Paschoud, *Zosime, Histoire Nouvelle*, I–VI (Paris, 1971–1989)

Secondary Literature

Abels, Richard P., *Lordship and Military Obligation in Anglo-Saxon England* (London, 1988)
Bachrach, Bernard S., 'Grand Strategy in the Germanic Kingdoms: Recruitment of the rank and file', in *L'Armée romaine et les barbares du IIIe au VIIe siècle*, eds. F. Vallet – M. Kazanski (Paris, 1993): pp. 55–63
Barnwell, Paul. *Emperor, Prefects and Kings. The Roman West, 395–565* (London, 1992)
Becher, Matthias, *Eid und Herrschaft: Untersuchungen zum Herrschethos Karls des Grossen* (Sigmaringen, 1993)
Bischoff, Bernhard and Wilhelm Koehler, 'Eine illustrierte Ausgabe der spätantiken Ravennater Annalen", in *Studies in Memory of A. Kingsley Porter*, ed. W.R.W. Koehler (Cambridge, Mass., 1939), pp. 125–138
Cameron, Alan and Jacqueline Long, *Barbarians and Politics at the Court of Arcadius* (Berkeley, 1993)
Charruadas, Paolo, 'L''ombre' de la forêt charbonnière', in *La forêt en Lotharingie medieval*, eds. M. Pauly – H. Pettiau (Luxembourg, 2016), pp. 87–136
Collins, Roger, *Visigothic Spain, 409–711* (Oxford, 2004)
Coumert, Magali. *La Loi Salique: retour aux manuscrits* (Turnhout, 2023)
Cracco Ruggini, Lellia, *Economia e società nell'"Italia annonaria": rapporti fra agricoltura e commercio dal IV al VI secolo d.C.* (Milan, 1961)
Cracco Rugini, Lellia, 'Les généraux franques au IVe–Ve siècles et leurs groupes aristocratiques', in *Clovis, histoire et mémoire, vol. 1, Clovis et son temps, l'événement*, ed. M. Rouche (Paris, 1997), pp. 673–688
Delaplace, Christine, *La fin de l'empire romain d'Occident. Rome et les Wisigoths de 382 à 534* (Rennes, 2015)
Durliat, Jean, 'Armée et société vers 600. Le problème des soldes', in *L'armée romaine et les barbares du IIIe au VIIe siècle*, eds. F. Vallet – M. Kazanski (Condé-sur-Noireau, 1993), pp. 31–38
Esders, Stefan, *Sacramentum Fidelitatis: Treueid, Militärwesen und Formierung mittelalterliche Staatlichkeit* (Bochum, 2003)
Esders, Stefan, 'Rechtliche Grundlagen frühmittelalterlicher Staatlichkeit', in *Der Mittelalterliche Staat – Europäische Perspektiven*, eds. W. Pohl – V. Wieser (Vienna, 2009), pp. 423–432

Esders, Stefan, "Faithful believers': Oaths of allegiance in post-Roman societies as evidence for Eastern and Western Visions of Community", in *Visions of Community in the post-Roman World: The West, Byzantium and the Islamic World, 300–1100*, eds. W. Pohl – C. Gantner – R. Payne (Farnham, 2012), pp. 357–374

Esders, Stefan, 'Late Roman Military Law in the Bavarian Code', in *La forge du droit. Naissance des identités juridiques en Europe (IVe–XIIIe siècles)*, Clio@Themis, 10 (March 2016), http://www.cliothemis.com 2016

Goffart, Walter, 'Frankish military duty and the fate of Roman taxation', in *Early Medieval Europe* 16 (2008), pp. 166–190

Haddan, Arthur West and William Stubbs, eds., *Councils and Ecclesiastical Documents relating to Great Britain and Ireland*, vol. 3, *The English Church, 595–1066* (Oxford, 1871)

Halsall, Guy, *Warfare and Society in the Post-Roman West, 450–900* (London, 2003)

Harries, Jill, 'Not the Theodosian Code: Euric's Law and late fifth-century Gaul', in *Society and Culture in Late Antique Gaul: revisiting the sources*, eds. R. Mathisen – Danuta Shanzer (Aldershot, 2001), pp. 39–51

Isaac, Benjamin, 'The meaning of the terms *limes* and *limitanei*', in *Journal of Roman Studies*, 78 (1988), pp. 125–47

James, Edward, 'Childéric, Syagrius et la disparition du royaume de Soissons', in *Revue archéologique de Picardie* 3–4 (1988), pp. 9–12

Jones, A. Hugo M., *The Later Roman Empire, 284–602* (Oxford, 1964)

Kerneis, Soazick, 'Marcher au chaudron dans l'Empire romain: Genèse de l'ordalie (IIe–IVe siècle après J.-C.', in *Les Justices de l'Invisible*, eds. R. Verdier – N. Kálnoky – S. Kerneis (Paris, 2013), pp. 255–268

Kerneis, Soazick, 'Les jugements des soldats et la fabrique des premières lois barbares (Ve siècle)', in *Ravenna Capitale, Giudizi, giudici e norme processuali in Occidente nei secoli IV–VIII*, vol. 1 (Santarcangelo di Romagna, 2015), pp. 211–224

Laniado, Avshalom, *Ethnos et droit dans le monde protobyzantin, Ve–VIe siècle* (Geneva, 2015)

Lee, A. Doug, *War in Late Antiquity: a social history* (Oxford, 2007)

Lenski, Noel, 'Schiavi armati e formazione di eserciti privati nel mondo tardoantico', in *Ordine e sovversione nel mondo greco e romano*, ed. G. Urso (Pisa, 2009), pp. 145–175

Liebeschuetz, John Hugo Wolfgang, *Barbarians and Bishops: Army, Church and State in the Age of Arcadius and Chrysostom* (Oxford, 1990)

Liebeschuetz, John Hugo Wolfgang. 'The End of the Roman Army in the Western Empire', in *War and Society in the Roman World*, eds. J. Rich and G. Shipley (London, 1993), pp. 265–276

Liebs, Detlef, 'Geltung kraft Konsenses oder kraft königlichem Befehl? Die *lex Romana* unter Westgoten, Burgundern und Franken', in *Recht und Konsens im frühen Mittelalter*, eds. V. Epp – Ch.H.F. Meyer (Ostfildern, 2017), pp. 62–85

Loschiavo, Luca, *L'età del passaggio: all'alba del diritto comune europeo* (*secoli* III–VII), 2nd ed. (Rome, 2019)

Loschiavo, Luca, 'Looking at the Edict of Rothari: between German ancestral customs and Roman legal traditions', *Roma Tre Law Review* 2 (2019), pp. 65–90

Luetjohann, Christian, *Praefatio in Sidonium*, in Sidonius Apollinaris, *Epistulae et carmina* (MGH, *Auct. Antiquiss.*, VIII) (Berlin, 1887)

Lupoi, Maurizio, *Alle radici del mondo guiridico europeo. Saggio storico comparativo* (Rome, 1994)

Magnou-Nortier, Elisabeth, *Foi et Fidelité. Recherches sur l'évolution des liens de fidelité chez les Francs du* VII*e au* IX*e siècle* (Toulouse, 1976)

Magnou-Nortier, Elisabeth. 'Remarques sur la genèse du *Pactus Legis Salicae*', in *Clovis, histoire et mémoire*, vol. 1, *Clovis et son temps: l'événement*, ed. M. Rouche (Paris, 1997), pp. 495–538

Magnou-Nortier, Elisabeth, *Aux origines de la fiscalité moderne. Le système fiscal et sa gestion dans le royaume des Francs* (Geneva, 2012)

Mathisen, Ralph, *Ecclesiastical Factionalism and Religious Controversy in fifth-century Gaul* (Washington D.C., 1989)

Mathisen, Ralph, 'D'Aire-sur-l'Adour à Agde: les relations entre la loi séculière et la loi canonique à la fin du Royaume de Toulouse', in *Le Bréviare d'Alaric. Aux origines du code civil*, eds. M. Rouche – B. Dumézil (Paris, 2008), pp. 41–52

McEvoy, Meaghan, *Child Emperor Rule in the Late Roman West,* AD *367–455* (Oxford, 2013)

Murray, Alexander Callander, 'From Roman to Frankish Gaul: *Centenarii* and *centenae* in the administration of the Merovingian kingdom', in *Traditio* 44 (1980), pp. 59–100

Murray, Alexander Callander, 'The position of the *grafio* in the constitutional history of Merovingian Gaul', in *Speculum* 64 (1986), pp. 787–805

Osaba García, Esperanza, 'Lex Visigothorum 9, 2: *De his qui ad bellum non vadunt aut de bello refugiunt*', in *Civitas, Iura, Arma. Organizzazioni militari, istituzioni giuridiche e strutture sociali alle origini dell'Europa* (*sec.* III–VIII), eds. F. Botta – L. Loschiavo, pp. 159–191 (Lecce, 2015)

Porena, Pierfrancesco – Yann Rivière, eds., *Expropriations et confiscations dans les royaumes barbares* (École française de Rome) (Rome, 2013)

Quast, Dieter, ed., *Das Grab des fränkischen Königs Childerich in Tournai und die Anastasis Childerici von Jean-Jacques Chifflet aus dem Jahre 1655* (Mainz, 2015)

Reimitz, Helmut, 'The Providential Past: Visions of Frankish Identity in the early medieval History of Gregory of Tours' *Historiae* (sixth-ninth century)', in *Visions of Community in the post-Roman World: The West, Byzantium and the Islamic World, 300–1100*, eds. W. Pohl – C. Gantner – R. Payne (Farnham, 2012), pp. 109–125

Sarti, Laury, *Perceiving War and the Military in Early Christian Gaul* (*ca. 400–700* AD (Leiden, 2013)

Sawyer, Peter. *The Age of the Vikings*, 2nd ed., (London, 1971)

Siems, Harald, 'Herrschaft und Konsens in der 'Lex Baiuvariorum' und den 'Decreta Tassilonis'', in *Recht und Konsens im frühen Mittelalter*, eds. V. Epp – Ch.H.F. Meyer (Ostfildern, 2017), pp. 299–360

Stroheker, Karl Friedrich, 'Zur Rolle der Heermeister fränkischer Abstammung im späten vierten Jahrhundert', in Id., *Germanentum und Spätantike*, pp. 9–29

Stroheker, Karl Friedrich, 'Alemannen im römischen Reichsdienst', in Id., *Germanentum und Spätantike*, pp. 30–53

Stroheker, Karl Friedrich, *Germanentum und Spätantike* (Stuttgart, 1965)

Ubl, Karl, 'L'origine contestée de la loi salique. Une mise au point', in *Revue de l'Institut français d'histoire en Allemagne*, 1 (2009), pp. 208–234 pp. 208–234 [pp. 1–18] – https://doi.org/10.4000/ifha.365

Ubl, Karl, *Sinnstiftungen eines Rechtsbuchs. Die Lex Salica im Frankenreich* (Ostfildern, 2017)

Vivian, Tim, Kim Vivian and Jeffrey Burton Russell, eds., *The Lives of the Jura Fathers* (Kalamazoo, 1999)

Wormald, Patrick, 'The Decline of the Western Empire and the Survival of its Aristocracy', in *Journal of Roman Studies* 66 (1976), pp. 217–26

Whittaker, Dick, 'Landlords and warlords in the later Roman Empire', in *War and Society in the Roman World*, eds. J. Rich – G. Shipley (London, 1993), pp. 277–302

Wood, Ian, 'Disputes in late fifth- and sixth-century Gaul: some problems', in *The Settlement of Disputes in Early Medieval Europe*, eds. W. Davies – P. Fouracre (Cambridge, 1986), pp. 7–22

Wood, Ian, *The Merovingian Kingdoms, 450–751* (London, 1994)

Wood, Ian, 'The legislation of *magistri militum*: the laws of Gundobad and Sigismund', in *La forge du droit. Naissance des identités juridiques en Europe* (IVe–XIIIe *siècles*), Clio@Themis, 10 (march 2016) – http://www.cliothemis.com 2016

Wood, Ian, 'Burgundian law-making, 451–534', in *Italian Review of Legal History* 3 (2017), pp. 1–27 – https://doi.org/10.13130/2464-8914/12878

Wood, Ian. *The Transformation of the Roman West* (Leeds, 2018)

Wood, Ian, 'Sidonius and the Burgundians', in *Academica Libertas: Essais en honneur de professeur Javier Arce*, eds. D. Moreau – R. González Salinero (Turnhout, 2019), pp. 365–371

Woolf, Greg. 'Roman Peace', in *War and Society in the Roman World*, eds. J. Rich and G. Shipley (London, 1993), pp. 171–194

PART 1

The Words of the Soldiers

CHAPTER 2

Hospitalitas I: The *Munus Hospitalitatis* and Its Limits

Andrea Trisciuoglio

1 Introduction

The temporary billeting in private *domus* and *praedia* of troops and (civil or military) officials in transit, evoked by the term *hospitalitas* (or *hospitium*, or *metatum*),[1] is traditionally associated with the broader issue of taxation in Late Antiquity, given that the sources regard the billeting service as a special *munus*.[2]

The official *Codices* stress the importance of this institute by devoting three titles to it. In the Theodosian Code, these are: 7.8 (*De metatis*), 7.9 (*De salgamo hospitibus non praebendo*), and 7.11 (*Ne comitibus et tribunis lavacra praestentur*). They respectively correspond to the following titles in the Code of

1 On *hospitalitas* in the aforementioned sense, see A. Hugo M. Jones, *The Later Roman Empire 284–602: A Social, Economic and Administrative Survey*, II (Oxford 1964), pp. 631*f.*; Sergio Roda, '*Militaris impressio* e proprietà senatoria nel tardo impero', in *Studi tardoantichi* [Hestíasis. Studi di tarda antichità offerti a S. Calderone], vol. IV (Messina 1987), pp. 229 *ff.*; Émilienne Demougeot, 'Une lettre de l'empereur *Honorius* sur l'*hospitium* des soldats', in Id., *L'empire romain et les barbares d'Occident (IVᵉ–VIIᵉ siècle). Scripta varia* (Paris 1988[2]); Vincenzo Giuffrè, *Letture e ricerche sulla 'res militaris'*, vol. II (Napoli 1996), pp. 409 *ff.*; Benjamin Isaac, *The Limits of Empire: The Roman Army in the East*, rev. ed. (Oxford 1999), pp. 299 *ff.*; A. Doug Lee, 'The Army' [ch. 7], in *CAH. The Late Empire, A.D. 337–425*, eds. A. Cameron – P. Garnsey, vol. XIII (Cambridge 2000), p. 230; Claudia Tavolieri, 'La fiscalità nel mondo antico. Il caso dell'*hospitium militum*', in *Zeitschrift für antikes Christentum*, 11.3 (2008), pp. 452 *ff.*; Michael Kulikowski, '*Hospitalitas*', in eds. Y. Le Bohec *et alii*, *The Encyclopedia of the Roman Army*, vol. II (Oxford 2015), pp. 489 *f.*

2 See esp. CTh. 7.8.3; CTh.7.8.16; see also D. 50.4.3.13. By its very nature, *hospitalitas* translated into temporary services that, as we shall see, were not very physically (or economically) taxing – clearly, when they were lawfully applied. Therefore, at least from a descriptive standpoint it seems more reasonable to include them among *munera extraordinaria*, rather than among *sordida*: see Pasquale Voci, *Nuovi studi sulla legislazione romana del tardo impero* (Padova 1989), pp. 133, 144. By contrast, in the light of D.50.4.3.14, there can be no doubt as to the fact that the *munus hospitalitatis* was of a patrimonial rather than personal nature: see Thomas Drew-Bear, Werner Eck and Peter Herrmann, '*Sacrae Litterae*', in *Chiron* 7 (1977), p. 379.

 | DOI:10.1163/9789004698017_003

Justinian: 12.40 (*De metatis et epidemeticis*),[3] 12.41 (*De salgamo hospitibus non praebendo*), and 1.47 (*Ne comitibus rei militaris vel tribunis lavacra praestentur*). The last of these actually includes only the first of two constitutions featured under title 7.11 in the Theodosian Code.[4]

After the Theodosian Code, *hospitalitas* was extensively discussed in Nov. Theod. 25 (= C. 12.40.9) from the year 444. Moreover, the exemption from this *munus* of provincial *domus* belonging to *ex militantes* 'in sacris nostris scriniis' is discussed in a constitution issued by Leo (C. 12.19.9), and dealing with similar matters as a Greek constitution by an anonymous emperor that extends more generally this exemption (for a single *domus*) to former members of the (civilian or military) *militia* and *ex* lawyers (C. 10.56.1.2 = B. 54.7.33 [Hb. 5, 134]).[5] References to *hospitalitas* are further to be found in Nov. Iust. 130, devoted to the *transitus militum*,[6] and, apparently, in Nov. Iust. 134.1.[7] Possible archaeological evidence pertaining to the logistic solution under discussion has also been identified.[8]

This institute is a very ancient one and – not least on the basis of the endurance of certain terms – can probably be traced back to *hospitium*, the hospitality on Roman soil implemented and regulated in detail in international relations already in the Republican Age.[9] We also know that certain

3 For this title, the Justinianic compilers extensively drew upon the Theodosian Code: 7 of the 12 constitutions featured were already included in CTh.7.8, and one was found in CTh.13.3; see the table in Anna Maria Giomaro, *Il Codex repetitae praelectionis. Contributi allo studio dello schema delle raccolte normative da Teodosio a Giustiniano* (Roma 2001), pp. 492 *f.* We shall be returning to the issue of the Justinianic addition of *epidemeticis* to the rubric *infra*, § 2.

4 However, the fine for double the amount set at the end of C.1.47.1 is based on CTh.7.11.2.

5 See Avshalom Laniado, 'La Vie d'Hypatius de Gangres (BHG 759a), Jean Malalas et l'impôt du xylélaion', in *Analecta Bollandiana* 115 (1997), p. 145. Concerning the exemption of veterans, cf. D. 50.4.18.29; Esther Domínguez López, '*Veterani*: ciertos privilegios fiscales', in eds. A. Fernández de Buján *et alii*, *Hacia un derecho administrativo, fiscal y medioambiental romano* III (Madrid 2016), pp. 528 *f.*

6 Cf. the mention of *prometatores* and *metata* in *capita* 6 and 9 of the Novella, discussed in Salvatore Cosentino, 'Sul transito dei soldati (in merito a Iust. Nov. CXXX)', in *Bulgaria Mediaevalis* 2 (2011) [Studies in honour of Professor Vassil Gjuzelev], pp. 25 *ff.* and by Philip Rance, 'The farmer and the soldier should be friends: Justinian's legislation on the provisioning of soldiers (Novel 130)', in *Journal of Late Antiquity* 12.2 (2019), pp. 380 *ff.* Concerning the *prometatores* (προμητάτορες) mentioned here, cf. Theodore of Hermopolis, *Brev. Nov.* 130.8 (ed. Zachariae, *Anekdota* III, p. 140).

7 Owing to the mention of *epidemeticum*; on the meaning of this term, see *infra*, § 2.

8 See – in relation to a *domus* in Trento – Claudia Tavolieri, 'Origine militare e fiscale dell'*hospitalitas*. L'esempio dell'*Epistula* 62 (19) inviata da Ambrogio a Vigilio di Trento', in *Zeitschrift für antikes Christentum* 12.3 (2008), pp. 533 *ff.*

9 On the expression *ius hospitale* used in CTh.7.8.1 and before in Liv.45.20.6, see Massimo Palazzi, '*Foedus cum natione*'. *I trattati alle radici dell'Europa. Aspetti delle relazioni fra Impero romano e popolazioni barbariche fra II e V secolo* (Gallarate 2014), p. 203.

professional categories (philosophers, orators, grammarians, and physicians) had already been exempted from the *munus hospitalitatis* by a provision of Antoninus Pius[10] which, in relation to physicians and liberal arts teachers, was later confirmed by a constitution issued by Constantine *ad populum* in 333 (CTh. 13.3.3 = C. 10.53.6.pr.).[11] An exemption in favour of the *inviti* senators of Rome established by a *senatus consultum* of uncertain date mentioned in a rescript by Septimius Severus and Caracalla of 204 can further be traced back to the age of the Principate.[12] Aside from these privileged categories also *clerici*[13] and *picturae professores*[14] were added later on.

Several literary sources for the fourth-sixth centuries refer to this subject, but such references are often far from clear.[15] I would like to draw attention to at least one passage from emperor Aurelian's biography in the *Historia Augusta*, which suggests that this emperor was especially active in repressing the abuses of army members provisionally hosted by provincials. This repression takes a particularly severe form (quartering) in the case of adultery with the host's

---

10 Cf. D. 27.1.6.8 (Mod. 2 *excusat.*); however, the exemption of these subjects had already been established by Vespasian and Hadrian: see *Edictum Vesp. de privil. med. et magistr.* (ed. in FIRA. I².73, p. 421); D.50.4.18.30 (Arc. Char. *l. sing. de mun. civ.*); see also Isaac, *The Limits of Empire*, p. 299.

11 The professional exemptions from the *munus hospitalitatis* established by Constantine were later confirmed by Theodosius II in 427: cf. CTh.13.3.18 = C.12.40.8.

12 This rescript is known from several Latin and Greek inscriptions: see Drew-Bear – Eck – Herrmann, '*Sacrae Litterae*', pp. 355 *ff.*; Dieter Knibbe – Reinhold Merkelbach, 'Allerhöchste Schelte (Zwei Exemplare der *Sacrae Litterae* aus Ephesos)', in *Zeitschrift für Papyrologie und Epigraphik* 31 (1978), pp. 229 *ff.*; Christopher P. Jones, 'The *Sacrae Litterae* of 204: Two Colonial Copies', in *Chiron* 14 (1984), pp. 93 *ff.*; Denis Feissel, 'Le rescrit de Valérien à Apellas (I. Smyrna, 604) d'après une copie inédite de John Covel', in Id., *Documents, droit, diplomatique de l'Empire romain tardif* (Paris 2010), pp. 110 *f.*

13 Cf. CTh. 16.2.8 = C. 1.3.1 (a. 343); Tavolieri, 'La fiscalità', pp. 460 *f.*

14 Cf. CTh. 13.4.4 (a. 374); C. 12.40.8 (a. 427).

15 Cf. Symmachus, *epp.* 2.52; 6.72; 9.48 (see – also with regard to dating – Roda, '*Militaris impressio*', pp. 215 *ff.*; see also Giovanni Alberto Cecconi, *Commento storico al libro II dell'epistolario di Q. Aurelio Simmaco* (Pisa 2002), pp. 323 *f.*); S. Ambr., *epist.* 62.6 (v. Tavolieri, *Origine militare*, pp. 504 *ff.*); Pseudo Joshua Stylites, *Chron.* 86, 93–96 (on which see *infra*, § 3); Procopius, *Bell. Vand.* 1.21; *Anecd. sive Hist. arc.* 23.22; probably also relevant here is Malalas, *Chron.* XVIII (ed. Dindorf, 437, p. 17 *f.*), on account of its mention of the Gothic ξυλέλαιον (see Laniado, 'La Vie d'Hypatius', pp. 136 *ff.*; Fritz Mitthof, *Annona militaris. Die Heeresversorgung im spätantiken Ägypten. Ein Beitrag zur Verwaltungs- und Heeresgeschichte des Römischen Reiches im 3. Bis 6. Jh. N. Chr.* [Firenze 2001], p. 213). Demougeot, 'Une lettre', p. 75, notes that the *munus* in question seems to be practically unknown to Ammianus Marcellinus in his *Historiae*.

wife.[16] We then have a letter from emperor Honorius, which can be dated to 412 and is addressed to soldiers (*seniores, iuniores, speculatores Britannici*). It was transcribed into a (rather corrupt) tenth-century manuscript for the purpose of celebrating the Spanish city of Pamplona.[17] According to Charles Perrat's reconstruction of the text, followed by Demougeot, reference is made to an imperial concession of *hospitium* in the following terms:

> agnoscimus, ut ubi ubi (*sic!*) vivendi degendique tempus extiterit omni alacritate atque virtute habeatis hospitii obsequium.
>
> We grant that where there is the chance to live and dwell, you may enjoy the privilege of hospitium in a prompt and perfect manner. [18]

However, we are evidently dealing with a different kind of *hospitium* from the temporary billeting of troops: this concession would appear to have a stable character; as such, it seems closer to the kind of hospitality granted to barbarian *foederati* for defensive purposes in occupied territories.[19] The latter institute – according to the most recent and convincing interpretations[20] – would not appear to be rooted in the constitution that Arcadius issued in 398 (then incorporated into CTh. 7.8.5) despite the fact that this constitution reserves one-third of the space for the hosting of Ostrogoths. It suffices to read the eulogy of Liberius (a patrician and former praetorian prefect of Italy),

16 Cf. Scriptores Historiae Augustae, *Aurel.* 7.3*ff.*; Tavolieri, 'La fiscalità', p. 453. Adultery committed by a member of the army is also punished from a procedural standpoint by the later *lex* of 393 – included in CTh.9.7.9 and addressed to Gildo, *comes et magister utriusque militiae per Africam* at the time – which denies the soldier in question the *privilegium fori*.

17 See – also with regard to dating and to the categories of *milites* mentioned – Christophe Archan, 'L'*epistula Honorii* de 412: établissements barbares dans l'armée impériale', in *El ejército, la paz y la guerra.* Jornadas de la Sociedad de historia del derecho, eds. J.J. De los Mozos Touya - I.S. León-Borja (Valladolid-Segovia 2006), pp. 153*ff.*

18 See Demougeot, 'Une lettre', p. 81. A very different reconstruction of the text is provided by Hagith Sivan, 'An Unedited Letter of the Emperor Honorius to the Spanish Soldiers', in *Zeitschrift für Papyrologie und Epigraphik* 61 (1985), p. 274: 'ut ubi alibi vivendi degendique tempus extiterit omni alacritate atque virtute abeatis, hospitiis obsequamini'; in my view, however, this is influenced by a questionable prejudicial assessment of the *hospitium* evoked in Honorius' letter, as the temporary billeting of an army member; see the author's commentary, *op. cit.*, p. 282 and fn. 38.

19 See Archan, 'L'*epistula Honorii*', pp. 153 and 160, according to whom the imperial *beneficium* that Honorius granted to army members consisted in 'des terres à la fin de leur service'.

20 See Pierfrancesco Porena, *L'insediamento degli Ostrogoti in Italia* (Rome 2012), pp. 9*ff.* and esp. 152; *contra* Lorenzo Fascione, 'Barbari e lavoro della terra in Occidente da Teodosio I (382) a Odoacre (476)', in *Historia et ius* 11 (2017), p 3.

written by Cassiodorus on Theodoric's behalf in a letter to the Roman senate,[21] to realise significant differences with regard to the permanent settlement of the Ostrogoths in Italy, particularly concerning the aims of these two kinds of *hospitalitas.*

Particular attention to the limits of *hospitalitas* clearly emerges from the juridical material chiefly preserved in the official *Codices*, which I have just briefly outlined. Indeed, the legislative profiles considered by the emperors concern, first of all, exceptions to the general obligation for subjects (especially provincials) to provide *hospitalitas* for army soldiers; therefore, the measure adopted is intended to define *immunitates*, whether total or partial – some of them, as we shall see, are justified by the particular destination of certain buildings or *praedia* (and not merely in economic terms). Furthermore, the imperial lawgiver's focus is on the kind of (supply of) services which the host, bearing the burden of the *munus*, owes to the soldier he is hosting.[22]

By contrast, in the light of the aforementioned *leges*, it is difficult to grasp some of the organisational aspects of the temporary billeting of troops (evidently, mobile units of *palatini* and *comitatenses*[23]), with regard to which I will suggest some interpretations. First of all, we need to focus on the role of the urban authorities. Were these in any way involved in the identification of billets? A second problem concerns the relationship between *metatores* and *mensores*: are they separate officials with different duties as regards the identification of billeting areas, or – as some have argued[24] – do they essentially play the same role? One last question, on which the legal sources prove rather

21 Cf. Cass., *Var.* 2.16.5. On the *laus Liberii* see Pierfrancesco Porena, 'Voci e silenzi sull'insediamento degli Ostrogoti in Italia', in *Expropriations et confiscations dans les royaumes barbares. Une approche régionale*, eds. P. Porena and Y. Rivière (Rome 2012), pp. 254*ff.*, and the clarification on p. 259 fn. 41.

22 This essentially agrees with the threefold thematic division drawn, in relation to the Theodosian Code, by Hagith Sivan, 'On *foederati, hospitalitas,* and the settlement of the Goths in AD 418', in *American Journal of Philology* 108.4 (1987), p. 767: 'The legislation falls into three categories: laws dealing with abuses of *hospitium* by civilians and soldiers; laws granting exemption from *hospitalitas*; and laws detailing the actual process of dividing a house between a 'host' and a 'guest' '. In my view, the distinction that Sivan draws immediately afterwards between urban and rural *hospitalitas* can serve as a foundation for correct interpretation of these laws.

23 Not troops of *limitanei*: see Roda, '*Militaris impressio*', p. 229.

24 See (in the wake of Jacques Cujas) Lauretta Maganzani, *Gli agrimensori nel processo privato romano* (Rome 1997), p. 38 and fn. 95; along much the same lines, see Tavolieri, 'La fiscalità', pp. 455, 457 and fn. 55 (where the author regards the terms *mensores* and *metatores* as interchangeable in relation to CTh.7.8.4 = C.12.40.1).

reticent, concerns the host's ability to avoid the burden of the *munus* on the basis of an *adaeratio* agreement signed with the official in charge.

2 Immunitates Pertaining to the *Munus Hospitalitatis*

Leaving aside data more clearly pertaining to the *hospitalitas* owed to civil servants,[25] it may be noted first of all, in general terms, that exemptions could concern either properties or persons. In relation to the first kind of immunity, mention should be done first of all of Valentinian I's provision addressed to *magister officiorum* Remigius (a. 368–373) to order the expulsion of those staying in a Jewish synagogue as a *hospes*. This provision states explicit that the *munus hospitalitatis* ought to apply to private *domus* and not to places of worship in general.[26] A second instance of the exemption of property, recalled in the aforementioned constitution issued by Arcadius in 398 (CTh. 7.8.5), which we shall discuss in this section, is provided by *ergasteria*, that is places of trade.[27] These were to remain available to their owners or managers, unless no space could be found in their *domus* that could be converted into a stable. It seems to me that this provision clearly reflects a concern that the billeting of soldiers might disrupt the rhythm of urban trade. Finally, mention ought to be made of the rather unique case of the *praedia Gildonis* located in Africa, which are the focus of two constitutions under title 7.8 of the Theodosian Code (*leges* 7 and 9).[28] These *praedia* coincided with the landed patrimony of Gildo[29] and his allies ('*satellites*'), property so vast that when it was confiscated it required the

25 Reference to *hospitalitas iudicum* is made, for instance, in CTh.7.8.6 and CTh.7.8.11. On the forms of *hospitalitas* shown to provincial governors, Giovanni Alberto Cecconi, 'Crisi e trasformazioni del governo municipale in Occidente fra IV e VI secolo', in *Die Stadt in der Spätantike - Niedergang oder Wandel?*, eds. J-U. Krause and Chr. Witschel (Stuttgart 2006), p. 290.

26 Cf. CTh.7.8.2 = C.1.9.4. Godefroy, *ad h.l.*, does not hesitate to state, since CTh.7.13.5, that this measure concerned soldiers deployed in Gaul; on this text, see esp. Roda, '*Militaris impressio*', p. 230; Giorgio Barone-Adesi, *L'età della «lex Dei»* (Napoli 1992), pp. 155 *f.* On the *magister officiorum Remigius* see Manfred Clauss, *Der magister officiorum in der Spätantike (4–6. Jahrhundert). Das Amt und sein Einfluß auf die kaiserliche Politik* (München 1980), pp. 186 *f.*

27 C.12.40.10.6 (impp. Valent. III and Marc., a. 450[?]–452) clarifies with regard to *ergasteria*: '*quae in plateis vel angiportis esse noscuntur*'.

28 Other mentions of the *praedia* (*possessiones*) *Gildonis* may be found in CTh.9.42.16 (a. 399) and CTh.9.42.19 (a. 405); see Voci, *Nuovi studi*, pp. 28 *f.*

29 On Gildo, see Yves Modéran, 'Gildon, les Maures et l'Afrique', in *MEFRA. Antiquité* 101.2 (1989), esp. pp. 865 *ff.*

establishment within the *res privata* of a separate department headed by the *comes Gildoniaci patrimonii*.[30] If I am not mistaken, this is the only case in the legal sources under consideration where the imperial chancellery expressly spells out the reasons for the immunity:

> quo possint conductores facilius inveniri
>
> so that lessees may be found more easily.[31]

The *munus*, therefore, would probably have made it difficult to find tenants for the confiscated estates in that part of the world; for this reason, addressing Sapidianus (probably the *vicarius* of Africa), Honorius decided to make the *domus* formerly part of Gildo's patrimony exempt from the *munus*. In 413 (CTh. 7.8.10) the same emperor extended this exemption to all African *praedia*, be they private or part of the imperial *domus* (hence, not just those confiscated from Gildo).[32] He authorised the *domini* and those working for them (procurators and tenants) to immediately expel the *metatores* by the lawful use of force.[33] It is worth emphasising that this constitution also occurs in the Code of Justinian (C. 12.40.5.1), albeit without the reference to Africa. Assuming that this omission reflects a conscious choice on the compilers' part, it is conceivable therefore that in the Justinianic era a general exception was made for landed estates,[34] and particularly for those of the *res privata* leased out

30 Cf. *Not. Dign., Occ.*12.5; Modéran, 'Gildon', p. 859; Roland Delmaire, *Largesses sacrées et res privata. L'aerarium impérial et son administration du IVe au VIe siècle* (Rome 1989), pp. 218*f.*

31 CTh.7.8.9, a. 409 (transl. Pharr). See Giuffré, *Letture*, p. 411; Modéran, 'Gildon', p. 862; Jean Peyras, 'La *potestas occupandi* dans l'Afrique romaine', in *Dialogues d'histoire ancienne* 25.1 (1999), p. 151. A different *ratio legis* would appear to lie behind the previous CTh.7.8.7 (a. 400), which exempted Gildo's country estates (but not his *domus*) from *metatum*, setting a fine of five gold pounds for any offending guests. The sentence 'quisque praedium nostrum manendi causa inportunus intraverit' indeed suggests a desire on the emperor's part to avoid the extended billeting of troops, which might lead to abuses against tenants; for an interpretation along these lines, see Delmaire, *Largesses*, p. 686.

32 Concerning the extension of the exemption to all private *praedia* in Africa, see Godefroy, *ad h.l.*, n. f.; see also Demougeot, 'Une lettre', p. 77 and fn. 17; *contra*, however, see Voci, *Nuovi studi*, p. 145 fn. 23 (the text here erroneously has 'cost. 12' instead of 'cost. 10'). On CTh.7.8.10, see Roda, '*Militaris impressio*', p. 235, and especially – for the important lexical clarifications it provides – Jean Peyras, 'Écrits d'arpentage et hauts fonctionnaires géomètres de l'antiquité tardive', in *Dialogues d'histoire ancienne* 21.2 (1995), pp. 176*ff.*

33 On *metatores* cf. *infra*, in the present section.

34 Along these lines, Peyras, 'Écrits', p. 183 fn. 142; Roland Delmaire, 'Du Code Théodosien au Code Justinien. L'adaptation de lois anciennes à des situations nouvelles', in *Société,*

to private individuals.[35] Here too, it seems reasonable to conclude that the ultimate *ratio* of this provision lay in the defence of public utility, which the forced coexistence of soldiers and farmers would have somehow jeopardised. Prolonged billeting of soldiers would inevitably have made farming more difficult, and this in turn would have negatively affected the payment of rent to the imperial coffers, in addition to creating possible supply problems.

It must further be noted that an individual subject to the *munus hospitalitatis* usually had the right to reserve part of his property for his own use; therefore, we can certainly speak of a general 'partial' property exemption. The fundamental law by Arcadius (CTh. 7.8.5.pr.) previously recalled, found in C. 12.40.2.pr.,[36] reserves two thirds of the dwelling to the owner, prescribing its ideal subdivision into three parts and allowing the owner to choose one third for himself. This required a measuring of dwelling spaces, presumably by the *mensores* under the orders and supervision of the *magister officiorum*,[37] the high-ranking official to whom constitutions concerning *hospitalitas* are addressed. As previously mentioned, the *mensores* have been regarded as soldiers in charge of logistics, with a function equivalent, in terms of the organising of billets, to those of *metatores*,[38] even though Vegetius clearly attributes different duties to the two.[39] However, if we consider passages of the constitutions mentioning

---

Économie, Administration dans le Code Théodosien, eds. S. Crogiez-Pétrequin, P. Jaillette (Lille 2012), p. 168.

35 In the case of estates belonging to the *res privata*, tenants and farmers' peace of mind was ensured by their immunity from *munera extraordinaria* (including the burden of *hospitalitas*: cf. *supra*, § 1 fn. 2), already laid out by a constitution of 343: CTh. 11.16.5 (= C. 11.75.1).

36 Cf. also C. 12.40.10.6 (impp. Valent. III and Marc., a. 450[?]-452).

37 See – also with regard to the organisation of the *schola* of *mensores* (cf. CTh. 6.34.1, a. 405) – Clauss, *Der magister officiorum*, pp. 19 *f.*

38 Cf. *supra*, fn. 24.

39 Cf. Veg., *epit. rei mil.* 2.7: 'Metatores qui praecedentes locum eligunt castris ... Mensores qui in castris ad podismum demetiuntur loca, in quibus tentoria milites figant, vel hospitia in ciuitatibus praestant'; Vegetius' work has been dated to the years between 386 and 388 AD: see Jocelyne Nelis-Clément, *Les beneficiarii: militaires et administrateurs au service de l'empire (Ier s. a.C. - VIe s. p.C.)* (Bordeaux 2000), p. 64 fn. 34. The *metator* is thus entrusted with the delicate task of selecting the place where to set up camp – cf. Isid., *Etym.*10.179: 'Metatores appellantur qui castra designant'. Once this place has been selected, the *mensor* must then measure out and assign the spaces for soldiers' tents; in an urban context, he instead assigns soldiers their *hospitia* (inside *domus*). On this distinction, already highlighted by Godefroy in his commentary on the Theodosian Code (*ad* CTh.7.8.4), see Pascal Arnaud, 'Les *mensores* des légions: *mensores agrarii* ou *mensores frumentarii*?', in *La hiérarchie (Rangordnung) de l'armée romaine sous le Haut-Empire*. Actes du Congrés de Lyon (15–18 septembre 1994), ed. Y. Le Bohec (Paris 1995), pp. 252 *f.* Concerning the distinction between *metator* and *mensor* (in this case, however, as the person entrusted

mensores, the latter would indeed appear to be in charge of executive tasks in urban contexts: in CTh. 7.8.4 *mensores* are to physically mark out houses destined to serve as billets with the image of a hand (the traditional symbol of *hospitium*[40]), by writing on the front door the name of the soldier to be housed there;[41] in CTh. 7.8.5 the *iniquitas* of the *mensores* would appear to be related to the division of the *domus hospitalis* into separate spaces for the host and his guests. By contrast, the *metatores* – as suggested by the largely metaphorical use of the term in late-antique Christian literature[42] – would appear to have been explorers in charge of identifying places where the *castrum* could be set up for the stationing of troops – yet, apparently, without the use of pre-existing buildings such as private *domus*. Indeed, the only mention of *metatores* that in my view can be detected in tit. 7.8 of the Theodosian Code seems to associate them with preliminary preparation work, presumably for the establishment of a *castrum*, in relation to a *praedium* or *possessio* (country estate) that could be public or private or belong to the *res privata* (cf. CTh. 7.8.10 = C. 12.40.5).[43] Nov. Iust. 130.6 (a. 545) further mentions *pro-metatores* (προμητάτορες) travelling ahead of the troops and selecting suitable billeting places, according to their superiors' orders. It must be said, however, that the connection between the *metator* and the setting up of a *castrum* in an open field is not confirmed by Nov. Theod. 25.2 (a. 444) = C. 12.40.9.pr.; this text suggests that *metatores* were in charge of identifying lodgings to be provided for soldiers by the owners of *domus* in Constantinople, which is to say in an urban centre, notwithstanding the possibility for these owners to establish an agreement with the *metatores* – which according to legal historians involved *adaeratio*.[44] However, there is no

---

with measuring out soldiers' grain rations), see Yann Le Bohec, *L'armée romaine sous le Bas-Empire* (Paris 2006), p. 92.

40 See Mattia Milani, 'La mano destra in Roma antica', in *Il corpo in Roma antica. Ricerche giuridiche*, ed. L. Garofalo, vol. II (Pisa 2017), pp. 106 *ff.*

41 According to the *lex*, erasure of the hand symbol made one liable to the charge of forgery.

42 See the sources quoted in Maganzani, *Gli agrimensori*, p. 34 fn. 73. See also Jean-Marie Auwers and Jean-Bernard Demoulin, '*Metator*: un cas d'appropriation chrétienne de terme institutionnel romain', in *Revue d'histoire ecclésiastique* 98.3–4 (2003), pp. 398 *ff.*

43 The idea that, according to law, the *metator*'s activity concerned country estates (not urban buildings), to the detriment of landowners (see Roda, '*Militaris impressio*', p. 235) seems to me to be confirmed also by the following passage of the *lex*, which refers to a *destinatio 'ad agrum'* of the *metator* ordered by the high-ranking official: 'Administrantem vero eiusque officii proceres, quorum praecepto inhibitam personam [scil. metator] ad agrum aliquem destinarit ...'.

44 On this Novella, see Auwers – Demoulin, 'Metator', p. 397. For an interpretation of the agreement in question as one concerning adaeration, see Johannes Steineken, *Disputatio inauguralis juridica de jure metatorio* (Basileae 1680), s.p. (Thesis XV).

trace of this agreement anywhere in title 7.8 of the Theodosian Code, whereas rather detailed textual confirmation is to be found for the *adaeratio metatorum* (ἀπαργυρισμὸς μητάτων)[45] in the Justinianic era (C. 1.4.26.16, a. 530). Furthermore, from the aforementioned Novella by Theodosius II no connection emerges with the *epidemeticum* (ἐπιδημητικόν), and it is far from certain that this term describes the *adaeratio* rate to free oneself from the burden of the *metatum*.[46] This interpretation can be traced back at least to Jacques Cujas and Jacques Godefroy,[47] but the occurrences of the term are actually quite few and its meaning is rather disputed. In Latin, we find it in rubric of C. 12.40 (*de metatis et epidemeticis*), as well as in the version of Nov. Iust. 134.1 provided by the *Authenticum*;[48] in Greek, the term occurs not just in Nov. 134.1 (a. 556), but also in Bas. 57.5.12, which transmits C. 12.40.12 (a. 530),[49] the last constitution of this title and the only one that justifies the rubric through its reference to *epidemetica*. On the basis of C. 12.40.12, some scholars have further maintained that the term refers to a fraud against *metatum*, based on the sale of the exemption rights that the laws assigned to certain categories of dignitaries.[50] Finally, according to one subjective interpretation, the 'epidemetics' were those seeking

45 In Bas. 6.1.98 (= C.1.47.1), by contrast, the term 'ἀργυρισμός' is used for *adaeratio* pertaining to the heating of private baths.

46 This contribution can be traced back to the *vectigal praetorium*, which already in the Republican Age provincials would pay to praetors, to avoid having to provide winter lodgings for soldiers (see Cicero, *Ad Atticum* 5.21).

47 Cf. Cujas, ad l. XII Cod.: 'epidemetica, id est adaerationes metatorum'; Godefroy, ad CTh. 7.8: 'Epidemetica sunt, quae milites vel iudices hac specie exigebant a provincialibus, ne illos hospitio gravarent: atque adeo quibus sese provinciales redimebant ab onere hospitium suscipiendorum'. Latin and Greek dictionaries agree: see Heinrich Eduard Dirksen, v. '*epidemeticum*', in *Manuale latinitatis fontium iuris civilis Romanorum* (Berolini 1837), p. 324; Charles du Fresne Du Cange, *Glossarium ad scriptores mediae et infimae graecitatis* (Graz 1958), p. 421 (v. ἐπιδημιτικά); Henricus Stephanus, *Thesaurus Graecae Linguae*, IV (repr. Graz 1954) col. 1573 (v. ἐπιδημητικός).

48 'Nulli vero liceat civili aut militari iudici circuire provinciam sine causa necessaria. Si vero aliqua necessitas vocaverit tale aliquid fieri, expensis propriis hoc facere iubemus tam iudices quam competentes eis et oboediens eis officium, neque vero angariis aut his quae vocantur epidemeticis aut alio quolibet dispendio gravare nostros subiectos …'.

49 C.12.40.12: 'Μηδεὶς ἄρχων ἐν τῷ περινοστεῖν ζητείτω προφάσει ἐπιδημητικῶν … χρυσίον' [transl. Krüger, p. 477 fn. 4: 'Nullus praeses in circumeundo provinciarum aurum requirat hospitii obtentu …'; engl. transl. (eds. B.W. Frier *et alii*, Cambridge 2016, III, pp. 2973*f.*): *'No governor shall, while traveling about in his province, ask for any money on the pretext that it is commutation money for quartering* (*prophasei epidêmêticôn*)*…'*]. *Iugenda* with C.1.4.26.16.

50 See Giovanni Rotondi, *Gli atti in frode alla legge nella dottrina romana e nella sua evoluzione posteriore* (Turin 1911), p. 114 and fn. 123.

to avoid the obligation of *metatum*.[51] What seems rather certain to me is that the term spread especially in the Justinianic era – I have already recalled how the rubric of CTh. 7.8 only mentions *metata*. However, with regard to the two alternative interpretations of the term suggested by scholars – a *munus* in kind or an *adaeratio* rate – it is important to bear in mind Caimi's remark concerning the list of levies 'in cash' provided by John Lydus (*de mag*. 3.70). There is no trace of *epidemetica* here, but mention is made of μητατορικά, which on the lexical level are more directly connected to *metata*.[52] *Pace* Caimi, this silence weighs more in favour of the essential synonymity between *metata* (idiomatic Latin) and *epidemetica* (idiomatic Greek), in the metonymic sense of *munus hospitalitatis*. This would explain the complete silence concerning *epidemetica* in the Latin constitutions in C. 12.40, while it is conceivable that (lawfully paid rather than extorted) adaeration rates were referred to as μητατορικά. But aside from these initial remarks, which would be worth investigating further, the fact remains that, given the freedom to establish the kind of exemption agreements mentioned in Nov. Theod. 25.2, *metatores* enjoyed a degree of liberty in the identification of *metata* that *mensores* probably did not.[53]

As far as provincial *civitates* are concerned, in the light of the juridical sources considered, not much can be said as to whether town councils were granted the authority to select and assign billets – an authority which might have set them in competition with the *metatores*. Even in the letter by Symmachus (*ep*. 9.48)[54] probably addressed to the (consular) governor of *Flaminia et Picenum* – which according to Sergio Roda bears witness to the deep involvement of the town council of *Ariminum* (Rimini) in the selection of *metata*, to the detriment of the prerogatives of the civilian-military bureaucratic authorities[55] – no reference is to be found to *metatores* and their duties.[56] Mention is instead made of the intermediary role played by town councils in CTh. 7.11.1 = C. 1.47.1 (a. 406), but with special reference to the citizens' duty to provide private baths for military personnel, a privilege which in this provision is only

51 See Lorenzo Fascione, *Fraus legi. Indagini sulla concezione della frode alla legge nella lotta politica e nella esperienza giuridica romana* (Milan 1983), p. 232.

52 See James Caimi, '*Ioannis Lydi de magistratibus* III 70. Note esegetiche e spunti in tema di fiscalità e legislazione protobizantina', in *Rivista di studi bizantini e slavi* 1 (1981), pp. 322 *ff*., esp. 346 *f*.

53 A similar interpretation is adopted by Roda, '*Militaris impressio*', p. 238, yet without taking a stance regarding *metatores*.

54 On this letter, see Sergio Roda, *Commento storico al libro IX dell'epistolario di Q. Aurelio Simmaco* (Pisa 1981), pp. 175 *ff*.

55 Cf. Roda, '*Militaris impressio*', pp. 216, 227 *f*., esp. 238 *ff*.

56 Indeed, the *epistula* does not clarify who selected the *domus* inherited by Petronius' sons.

granted by emperors to *comites illustres* and *magistri militum*. However, with the supply of auxiliary services for members of the army, we are dealing with a sphere that has little to do with the choice of suitable billeting places, a choice which must have implied strategic and logistic assessments that presumably fell outside the purview of local councils.

What cannot be explained by invoking public interest concerns, strictly connected to the intended use of the buildings, is the exemption granted to the *fabricenses* (armourers) of Antioch in CTh. 7.8.8 (a. 405).[57] This should rather be understood as a privilege generally granted[58] to a professional class that played a vital role for the Empire's fate; therefore, this kind of immunity concerned individuals more than properties. Indeed, in this case the exempted buildings were not workshops, but the *fabricenses*' private dwellings.

This is one of the numerous cases in which immunity – which takes various forms (depending, for instance, on the number of houses owned or their location) – simply represents a privilege that the emperor granted to certain socio-professional categories. Among these we may recall, first of all, the *inviti senatores* of CTh. 7.8.1, who – as we have seen[59] – had already been exempted during the Principate; former high-ranking imperial officials (*praefecti praetorio, magistri equitum*, and *peditum, comites consistoriani, praepositi sacri cubiculi*), whose exemption from the *munus* was actually limited to a single *domus* of their own choosing (CTh. 7.8.3); and, finally, minor *ex officiales* within the imperial administration (the heads of the *sacra scrinia*), whose provincial *domus* – as has already been recalled – were made exempt by emperor Leo (C. 12.19.9).[60] Also fully exempted were the homes of physicians, teachers, and clergymen.[61]

An altogether different case is represented by those buildings in Constantinople suited to housing, if only temporarily, the numerous soldiers who resided in the city in the Justinianic era.[62] Theodosius II repeatedly turned his

57 On this *lex*, see Maria Grazia Bianchini, *Caso concreto e «lex generalis». Per lo studio della tecnica e della politica normativa da Costantino a Teodosio II* (Milan 1979), pp. 35 *ff.*, 115 *f.*; on the *fabricenses* of Antioch, see John H.W.G. Liebeschuetz, *Antioch. City and Imperial Administration in the Later Roman Empire* (Oxford 1972), p. 58.

58 Unless, that is, it was necessary to provide hospitality for the imperial *comitatus* travelling through Antioch: see, again, CTh. 7.8.8.

59 Cf. *supra*, § 1 and fn. 12.

60 See Agatina Stefania Scarcella, *La legislazione di Leone I* (*Milan 1997*), pp. 366 *f.*

61 Cf. *supra*, § 1, fnn. 11 and 13.

62 Procopius (*Hist. arc.* 23.24) reports that under Justinian around 70,000 barbarian soldiers were billeted in houses in Constantinople. As late as the fourth and fifth centuries, the capital still had no stable barracks or military garrisons; the city's defence was entrusted to travelling units. On Constantinople's status as an 'unarmed city', see Gilbert Dagron,

attention to the capital: with CTh. 7.8.13 (a. 422) = C. 12.40.7,[63] he authorised the stationing of *milites* 'ex procinctu redeuntes[64] vel proficiscentes ad bella' (*returning from combat service or setting out for war* [Pharr]) in the towers along the recently reconstructed city walls. In this case, the hosts are not owners but, presumably, tenants (referred to as *possessores* in the constitution), who were also charged with the upkeep of the capital's defensive structures.[65] Private citizens thus had the opportunity to make use of public places without having to pay any rent, in return for personal obligations serving the public interest – an arrangement possibly already alluded to in CTh. 7.8.5 of 398.[66] On the other hand, *hospitalitas* only concerned one third of the available area on the ground floor (*pedeplanus*), which evidently was easier to access and depart from for the soldiers, and not the upper floors of the towers, which presumably continued to be used by the tenants. Between 427 and 435, Theodosius II issued three imperial constitutions, addressed to *magistri officiorum* in office, to confirm the immunity granted to dignitaries and their heirs in relation to a limited number of *domus* located *in alma Urbe* (generally private dwellings; see CTh. 7.8.14; CTh. 7.8.15; CTh. 7.8.16). This privilege, extended to *consulares patricii*, is confirmed in Nov. Theod. 25.1 (*de metatis*) of 444.[67]

The latest relevant provision is Nov. 130, issued by Justinian in 545 – particularly ch. IX.[68] Here the safeguarding of subjects in relation to the *munus hospitalitatis* would appear to be strengthened with respect to the picture that

---

Costantinopoli. Nascita di una capitale (330–451), ital. transl. A. Serafini, (Turin 1991), pp. 107*ff.*

63 See René Rebuffat, 'La loi et la ville. Greniers et enceintes', in *L'Africa romana. Trasformazione dei paesaggi del potere nell'Africa settentrionale fino alla fine del mondo antico*, Atti del XIX convegno di studio, Sassari, 16–19 dicembre 2010, eds. M.B. Cocco, A. Gavini, A. Ibba, vol. I (Rome 2012), pp. 134*f.*; Rene Pfeilschifter, *Der Kaiser und Konstantinopel: Kommunikation und Konfliktaustrag in einer spätantiken Metropole* (Berlin-Boston 2013), pp. 225*f.*

64 Godefroy, *ad h.l.*, believes this may refer to the soldiers returning from the Persian campaign, which ended in 422.

65 The same can be inferred from CTh.15.1.51 (a. 413), which is already viewed in relation to the constitution examined in the commentary by Godefroy, *ad h.l.*; see also Dagron, *Costantinopoli*, pp. 109*f.* and fnn. 172, 173. On CTh.15.1.51, see Marialuisa Navarra, '*Utilitas publica-utilitas singulorum* tra IV e V sec. d.C. Alcune osservazioni', in *Studia et documenta historiae et iuris*, 63 (1997), p. 280.

66 See Francesco Sitzia, *Studi sulla superficie in epoca giustinianea* (Milan 1979), pp. 52*ff.*; Ana Zaera Garcia, *La superficies en derecho romano* (Madrid 2017), pp. 166*f.*

67 For *consulares patricii*, therefore, an exemption was now granted for three houses in Constantinople, rather than one (as for *inlustres*, cf. CTh.7.8.14 and 15) or two (as for *viri consulares*, cf. CTh. 7.8.16); C. 12.40.10.pr. (impp. Valent. III and Marc., 450[?]- 452). On the constitutions recalled here, see the succinct remarks in Delmaire, *Largesses*, p. 51.

68 See Cosentino, 'Sul transito', pp. 31, 38; Rance, 'The farmer', pp. 399*f.*

emerges from the *Codex*, insofar as a complete exemption would appear to be granted to the main dwelling of the encumbered person (cf. the expression … εἶς τὰς αὐθεντικὰς οἰκήσεις, ἐν αἷς οἱ δεσπόται τῶν οἰκημάτων εὑρεθεῖεν μένοντες / *in the main dwellings in which the owners of the houses are known to reside*), his everyday living quarters, without any mention being made of a division of the *domus* between host and guest, as in C. 12.40.2.pr. (= CTh. 7.8.5.pr.).[69] Apparently, then, a functional criterion governed the selection of billeting places, one which took domestic requirements more into account, rather than the merely quantitative criterion – based on the three-fold division of the *domus* – which had been devised by Arcadius' chancellery,[70] and which remained in force even fifty years later, at time of Valentinian III and Marcian.[71]

3 Services Owed by the *Hospes*

In a constitution issued in Ravenna in 413 (CTh. 7.8.10.2 = C. 12.40.5.2–4), Honorius specifies that *hospitalitas*, which by nature is of limited duration, cannot include the supply of food for the soldiers and their animals, and that violations (the strong expression '*mos sceleratus*' is used) would be punished with heavy fines. To do away with such shameful customs, the emperor himself establishes a punishment even for any host who agrees to supply food *contra legem*. Certainly, this legal principle is far more ancient,[72] and is consistent with late-imperial military logistics. According to the latter, living expenses, including those of *comitatenses*, had to be covered by military rationing (and not occasionally by hosts) through provincials' levies in kind, which over time generally – and to a large extent, under the reign of Theodosius I – become open to *adaeratio* (with the exception of bread and wine).[73]

69 It may also be significant that in *Breviarium Novellarum* 130.11 (ed. Zachariae, *Anekdota* III, 140) Theodore of Hermopolis does not recall the crucial *lex* 2 of C. 12.40, but only – as far as *immunitates* are concerned – *leges* 4, 5, 8, 10, and 11. Might this be because Nov. 130.9 had abrogated C. 12.40.2.pr.?

70 Cf. *supra*, in the present section.

71 Cf. C. 12.40.10.6; the addressee of the constitution, *magister officiorum Iohannes Vincomalus*, was in office in the years 450(?)-452: Fergus Millar, *A Greek Roman Empire: Power and Belief under Theodosius II (408–450)* (Berkley - Los Angeles - London 2006), p. 199.

72 Concerning the *caenaticum*, see CTh. 7.4.12 (= C. 12.37.3), issued some fifty years earlier and addressed to the *magister militum* (a. 364, impp. Valentinian and Valens).

73 With regard to the years 313–438 and tit. 7.4 of the Theodosian Code (*De erogatione militaris annonae*), see Chantal Vogler, 'La rémunération annonaire dans le Code Théodosien', in *Ktema* 4 (1979), esp. pp. 301*f.* (on *comitatenses*) and 306–310.

The same kind of spontaneous supplying of goods on the host's part was further forbidden with regard to *salgamum* by a *lex* by Costantius II, incorporated into CTh. 7.9.2 (*tit.*: *de salgamo hospitibus non praebendo*). Sometime before, in the *pars Occidentis*, by evoking and encouraging '*spontanea humanitas*', Constans (CTh. 7.9.1) had probably dealt with this subject in different terms, by allowing provincial hosts to provide oil, firewood (for cooking and heating), and other items required for a short stay, including a mattress (*culcita*)[74] – everything, that is, which could succinctly be described as *salgamum*.[75] This contradiction is no longer to be found in the Code of Justinian, where title C. 12.41, still bearing the rubric '*de salgamo hospitibus non praebendo*', only incorporates CTh. 7.9.2, thereby forbidding any supplementary services, even if spontaneously offered by the host.[76] If we then look at the CIC. as a whole, we find that the Digest quite coherently includes a text from the first of the *libri Opinionum* of Ulpian (D. 1.18.6.5) which – within the framework of a vivid portrait of the actual situation – assigns provincial governors the task of punishing any episodes of abuse whereby – '*sub praetextu adventus ... militum*' / *under the pretext of the arrival of soldiers* – provincials have been deprived of their *lumen unicum* and the *brevis suppellex* (I should think that this includes not just hosts, but also people living near billeting areas).[77]

Imperial legislation even concerned itself with the private baths of *domus*. Any attempt to make it available to guests is forbidden in Africa by a

---

74 Evidently, both constitutions also forbid supplies extorted by soldiers through the use of violence. On these *leges*, see Demougeot, 'Une lettre', p. 76 fn. 10; Andrea Trisciuoglio, '*Humanitas* e cura delle esigenze umane nel Codice Teodosiano', in *Ius Romanum* 2 (2018), pp. 352*f.* The date of CTh. 7.9.2 is uncertain: Mommsen argues for 340 or 361. Also uncertain is who issued CTh. 7.9.1, which is generically addressed *ad provinciales*: John Dillon, *The Justice of Constantine. Law, Communication, and Control* (Michigan 2012), p. 53 fn. 74, attributes it to Constantius II.

75 On the meaning of this term (as exclusively related to food), see Charles du Fresne Du Cange, *Glossarium mediae et infimae latinitatis*, t. VI (Parisiis 1846), p. 38 (v. '*salgama*'): 'Quod nomen traductum postea a militibus ad rerum ad victum necessariarum praebitiones quas ab hospitibus suis exigebant'; along the same lines, see Steineken, *Disputatio*, s.p. (Thesis XVI). Papyrological evidence suggest that oil and salt too was included in the concept of *salgamum*: see Mitthof, '*Annona militaris*', p. 212; this scholar (p. 214) also rightly recalls the twofold function of oil in Late Antiquity of foodstuff and fuel. On the basis of CTh. 7.9.3 (which mentions a mattress), the term is instead assigned a broader meaning by Laniado, 'La Vie d'Hypatius', p. 142.

76 CTh. 7.9.3 (a. 393, imp. Theodosius I) and CTh. 7.9.4 (a. 416, imp. Theodosius II) essentially reproduce the same provision as CTh. 7.9.2. In all likelihood, it is for this reason that the Justinian compilers omitted them in title 12.41 of the *Codex*.

77 On D. 1.18.6.5, see Bernardo Santalucia, *I «libri opinionum» di Ulpiano, II (Milan 1971)*, pp. 19*f.*

constitution that Honorius issued in Ravenna in 414 (CTh. 7.8.12 = C. 12.40.6).[78] Any attempt to reap a profit from the use of private baths by requesting an *adaeratio* rate from provincials is punished with a fee amounting to double the sum charged; however, private baths must be made available to *comites illustres* and *magistri militum*, according to a provision issued by Arcadius (CTh. 7.11.1 = C. 1.47.1) in 406. CTh. 7.11.2 (a. 417), which completes the title (*Ne comitibus et tribunis lavacra praestentur*), merely records the Theodosian application of this provision, insofar as it sets a fine for double the amount (three tremisses a day) unlawfully exacted for three years by the *duces* of the Euphrates border as a contribution for the baths (and the firewood required to heat the water). Given the specific historical circumstances of this *lex*, it must have been of little interest to the compilers of the *Codex Justinianus*, if not in relation to the fine it provided;[79] from a more general perspective, it hardly introduced any changes compared to CTh. 7.11.1. It is unsurprising, therefore, that it is no longer included in the corresponding Justinianic title C. 1.47 (*Ne comitibus rei militaris et tribunis lavacra praestentur*).[80]

Up until this point we have been discussing what supplies were not owed by the host. Now it is time to clarify what the host did owe. I would argue that this amounted only to lodging in a protected place, the *metatum*, and nothing else. However, historical events do not always confirm this picture. For example, in Edessa, in the years 503–505, which is to say during the war against the Persians, something occurred that can hardly be explained in the light of the laws presumably in force under the reign of emperor Anastasius. Among other anomalies, Joshua the Stylite (*Chron.* 86, 93–96) – or whoever might be the author of the text transmitted under his name[81] – recounts that one *dux Romanus*[82] authorised the city's landowners to furnish the soldiers (Gothic mercenaries) billeted there with a bed (one for every two men), bedsheets, 200 pounds (91 kg) of wood a month, and a certain amount of oil – thereby clearly contravening the explicit provisions in the *lex de salgamo non praebendo* (CTh. 7.9.2 = C.

78 Peyras, 'Écrits', pp. 177 fn. 118 and 179, infers from the aforementioned constitution of 414 that *hospitalitas* in all its forms was abolished in Africa.

79 Cf. *supra*, § 1 fn. 4.

80 On CTh. 7.11.1–2, see esp. Ramsay MacMullen, *Soldier and Civilian in the Later Roman Empire* (Cambridge Massachusetts 1963), pp. 84*f.* and fn. 30.

81 On the disputed authorship of the *Chronicon*, see esp. André Guillou, 'La cosiddetta cronaca siriaca di Giosuè lo Stilita', in Id., *La civiltà bizantina dal IV al IX secolo. Aspetti e problemi* (Bari 1977), p. 369.

82 *Romanus* was probably *comes rei militaris* at that time (a. 505): see A. Hugo M. Jones, John R. Martindale and John Morris. *The Prosopography of the Later Roman Empire*. vol. 2, *AD 395–527* (Cambridge 1980), p. 948.

12.41.1), discussed above.[83] It is difficult to account for the *dux*'s decision simply in the light of legislative parameters allegedly still in force under Anastasius. Clearly, we cannot rule out that an exception was made to the restrictions on *salgamum* on account of the anomalous situation at Edessa in those years, and of the strong pressure exerted by the Gothic mercenaries, who even attempted to murder the *dux Romanus*.

4 Conclusions

The legislation concerning the *munus hospitalitatis* in the timeframe considered (fourth-sixth cent. AD) seems to reflect two underlying trajectories, which are constantly visible and which converge towards the goal of separation in the invariably delicate relations between the army and the local population. On the one hand, what emerges indirectly yet rather clearly is the *principle of the logistic self-sufficiency of the Roman army*, which relied as little as possible on the occasional collaboration of (provincial) private citizens, and then exclusively in relation to billeting: a logistic-strategic need that could not be met in any other way, on account of the lack of any stable military structures spread across the imperial territory. As far as food and other items necessary to everyday life (*salgamum*) were concerned, soldiers could not ask their hosts for anything, but were expected to rely on military rationing. On the other hand, we find resolute and repeated legislative responses to any possible forms of harassment and abuse that might arise from the encounter between soldiers and the civilian population,[84] especially in case of cohabitation in *metata*. This legislative orientation is possibly reflected in the ideal model of military command embodied by Pescennius Niger, as portrayed by Aelius Spartianus in the *Historia Augusta* (a collection of biographies written between the late fourth

83 Concerning the aforementioned passages from Joshua the Stylite's (in Syriac) *Chronicon*, and other references to violations of the late-antique laws on *munus hospitalitatis* known to us (billeting in workshops and clergymen's houses), see Jones, *The Later Roman Empire*, pp. 631*f.*; Guillou, 'La cosiddetta cronaca', pp. 374, 381; Roda, '*Militaris impressio*', p. 236 fn. 49; Laniado, 'La Vie d'Hypatius', pp. 141*ff.*; Frank R. Trombley and John W. Watt, *The Chronicle of Pseudo-Joshua the Stylite* (Liverpool 2000), pp. 103*ff.*, 111*ff.* (annotated engl. transl.); Nigel Pollard, *Soldiers, Cities, and Civilians in Roman Syria* (Michigan 2000), p. 108.

84 See Jean-Michel Carrié, 'L'esercito: trasformazioni funzionali ed economie locali', in *Società romana e impero tardoantico. Istituzioni, ceti, economie*, ed. A. Giardina (Bari 1986), p. 486.

and early fifth centuries).[85] The measures here range from an acknowledgement of self-defence (the expulsion of *metatores* in CTh. 7.8.10 = C. 12.40.5) to punishments that vary depending on the culprit and his military rank: fines (cf. CTh. 7.8.7, CTh. 7.8.10), expulsion from the army (cf. CTh. 7.8.5), and a generic *gravis vexatio* (cf. CTh. 7.9.2). As far as *metatores* are concerned, *supplicium* and *exilium* are foreseen in the event of monetary extortion (cf. Nov. Iust. 130.6).

With regard to the definition of unlawful behaviour, any spontaneous offer to provide non-mandatory services on the host's part is also punished, in an effort to uproot this *mos sceleratus* (CTh. 7.8.10), following an initial stage in which the display of *spontanea humanitas* is allowed, to supplement the providing of *salgamum* to hosted soldiers (CTh. 7.9.1).

Apparently, *metatores* could ask for money from subjects burdened with the *munus hospitalitatis*, by lawfully establishing *adaeratio* pacts, which may have been allowed only in the absence of higher logistic and strategic requirements. Indeed, it must be supposed that billeting places were not always interchangeable.[86] Such agreements are only mentioned in Nov. Theod. 25, in relation to the *domus* in Constantinople, which presumably could easily be replaced as billets with other *domus*. The general provision to separate the army from the civilian population as far as possible is also visible in the setting of spatio-temporal limits to soldiers' dwelling in private buildings: domestic spaces had to be shared with the owners according to very detailed instructions (cf. CTh. 7.8.5; Nov. Iust. 130.9); stays ought to be as brief as possible (CTh. 7.8.10).

Another aspect of the legislation on the *munus hospitalitatis* concerns *immunitates*. Certainly, what comes into play here are the privileges granted to socio-professional categories that were perceived as important on account of their rank (*illustres*) or function (*fabricenses*, individuals practising the liberal arts, and clergymen). However, we also find certain exemptions, which I have described as 'property exemptions', that may be explained by considerations of a different nature: respect for places of worship (synagogues); respect for

85 Cf. *Script. Hist. Aug.*, Ael. Spart., *Pesc. Nig.* 3.6: 'Numquam sub eo (scil. Pescenn. Niger) miles provinciali lignum, oleum, operam extorsit', where the specific mention of wood and oil among the goods never extorted from provincials under Pescennius Niger actually matches the pair of terms mentioned in all four of the constitutions of title 7.9 of the Theodosian Code, issued between AD 340 and 416 (cf. *supra*, § 3). Considering this textual comparison, Soverini's invitation to read the *Historia Augusta* in the light of late-antique Latin and its use of certain terms seems particularly apt: see Paolo Soverini, 'Per lo studio del testo della *Historia Augusta*: tra analisi linguistica e critica storica', in *Atti dell'Accademia ligure di scienze e lettere* 46 (1989), pp. 434 *ff*.

86 Consider the need for billets to be located close to rationing centres, as mentioned by Carrié, 'L'esercito', p. 467.

urban trade (*ergasteria*). In this case too what emerges is the attempt to establish a delicate balance, or harmony, between the *utilitas publica* connected to the functioning of the army in the Late Empire[87] and a *utilitas privatorum* which takes a variety of forms.

Bibliography

Primary Sources

Ambrosius, *Epistulae*

Authenticum, see Iustinianus, *Novellae*

Basilicorum libri LX, ed. K.W.E. Heimbach, 6 voll. (Lipsiae 1833–1870); *etiam* ed. Series A, Textus librorum I–LX, voll. I–VIII, eds. H.J. Scheltema – D. Holwerda – N. van der Wal (Groningen – Djakarta – Gravenhage 1955–1988)

Cassiodorus Senator, *Variae*, ed. Th. Mommsen [MGH, *Scriptores – Auct. ant.* XII] (Berlin 1894); *etiam* Flavio Magno Aurelio Cassiodoro Senatore, *Varie*, (dir. A. Giardina), eds. A. Giardina – G.A. Cecconi – I. Tantillo – F. Oppedisano, 6 vols. – Engl. transl. by M.S. Bjornlie, *The Variae. The Complete Translation* (Oakland, CA, 2019)

Cicero (Marcus Tullius), *Epistulae ad Atticum*

Codex Iustinianus see Iustinianus, *Codex*

Codex Theodosianus: *Theodosiani libri XVI cum constitutionibus Sirmondianis et leges Novellae ad Theodosianum pertinentes*, ed. Th. Mommsen – P.M. Meyer, 2 vols. (Berlin 1905) – Engl. transl. by C. Pharr *The Theodosian Code and Novels and the Sirmondian Constitutions. A Translation with commentary, glossary, and bibliography*, [Princeton, 1952], 10th Print (Union, N.J. 2001)

Cujas, Jacques (Cuiacius), *Opera* XIII voll. (Prati 1836–44)

Digesta see Iustinianus, *Digesta*

Epistula Honorii, ed. J. M. Lacarra, *Textos navarros del códice de Roda, Estudios de Edad Media de la Corona de Aragon 1* (1945) [*De laude Pampilone epistola*], pp. 268–9

Epitome Iuliani, Iuliani Epitome latina novellarum Iustiniani, ed. G. Hänel (Lipsiae 1873)

Godefroy, Jacques (Gothofredus), *Codex Theodosianus cum Perpetuis Commentariis*, 6 voll. (Leipzig 1738 = Hildesheim - Zürich - New York 1975)

Historia Augusta, ed. D. Magie, *The Scriptores Historiae Augustae*, with an english translation by D. Magie (London 1921)

---

87 Cf. Jean Gaudemet, '*Utilitas publica*', in *Revue historique de droit français et étranger* 29 (1951), pp. 482 *f.*; Navarra, *Utilitas publica-utilitas singulorum*, p. 273 fn. 24 (with detailed references to late-antique *leges*).

Ioannes Lydus, *De magistratibus Reipublicae Romanae libri tres*, ed. *Jean le Lydien, Des magistratures de l'État romain*, I-II, texte établi, trad. M. Dubuisson – J. Schamp, (Paris 2006)

Iustinianus, *Codex*, ed. P. Krüger in T. Mommsen – P. Krüger, *Corpus Iuris Civilis*, II (Berlin 1954[11]) – Engl. transl. eds. B.W. Frier *et alii, The Codex of Justinian. A New Annotated Translation, with Parallel Latin and Greek Text*, voll. 3 (Cambridge 2016)

Iustinianus, *Digesta*, ed. T. Mommsen – P. Krüger, *Corpus Iuris Civilis*, I (Berlin 1963[17]) - English translation edited by Alan Watson, 4 vols. (Philadelphia, Pennsylvania 1985)

Iustinianus, *Novellae*, ed. R. Schöll – G. Kroll, in *Corpus Iuris Civilis*, III (Berlin 19638) –Engl. transl. D.J.D. Miller – P. Sarris, *The novels of Justinian. A complete annotated English translation* (Cambridge-New York 2018)

Livius, *Ab Urbe condita*

Malalas (Johannes), *Chronicon*, ed. L. Dindorf, 2 vols. (Bonnae 1831)

Notitia Dignitatum: accedunt Notitia urbis Constantinopolitanae et laterculi prouinciarum, ed. O. Seeck (Berlin 1876 = Frankfurt 1983); *etiam* ed., C. Neira Faleiro, *La Notitia Dignitatum. Nueva edición critica y comentario histórico* (Madrid 2005); ed. in progress, I. Maier, *The compilation 'notitia dignitatum', Cnd* (//www.notitiadignitatum.org) (Units: Hoffmann, 1969–1970)

Novellae Iustiniani, see Iustinianus, *Novellae*

Novellae Theodosiani, see *Codex Theodosianus*

Procopius Caesariensis, *Anecdota sive Historia arcana*, ed. (and engl. transl.) H.B. Dewing, *Anecdota or Secret history* (London – New York 1914)

Procopius Caesariensis, *De bellis – Tôn Polemôn Logoi*, ed. (and engl. transl.) H.B. Dewing, *History of the Wars* (Cambridge, Mass. 1919)

Pseudo-Joshua the Stylite, *Chronicon*, ed. W. Wright, *The chronicle of Joshua the Stylite*: composed in Syriac A.D. 507 (Cambridge 1882) – Engl. transl. F.R. Trombley – J.W. Watt, *The Chronicle of Pseudo-Joshua the Stylite* (Liverpool 2000)

Scriptores Historiae Augustae, ed. D. Magie, *The Scriptores Historiae Augustae*, with an english translation by D. Magie (London 1921)

Symmachus, Quintus Aurelius, *Epistulae* ed. O. Seeck [MGH, *Auct. ant.* VI.1 (Berlin 1883), *etiam* ed. J.-P. Callu, with french transl., 4 voll. (Paris 1972 – 2002)

Theodore of Hermopolis, *Breviarium Novellarum*, ed. C.E. Zachariae von Lingenthal, *Anekdota*, vol. III (Lipsiae 1843)

Vegetius, *Epitoma rei militaris*, ed. A. Önnerfors (Stuttgart 1995)

Secondary Literature

Archan, Christophe. 'L'*epistula Honorii* de 412: établissements barbares dans l'armée impériale', in *El ejército, la paz y la guerra*. Jornadas de la Sociedad de historia del derecho, eds. J.J. De los Mozos Touya - I.S. León-Borja (Valladolid-Segovia 2006), pp. 153–164

Arnaud, Pascal. 'Les *mensores* des légions: *mensores agrarii* ou *mensores frumentarii*?', in *La hiérarchie (Rangordnung) de l'armée romaine sous le Haut-Empire*. Actes du Congrés de Lyon (15–18 septembre 1994), ed. Y. Le Bohec (Paris 1995), pp. 251–256

Auwers, Jean-Marie and Jean-Bernard Demoulin. '*Metator*: un cas d'appropriation chrétienne de terme institutionnel romain', in *Revue d'histoire ecclésiastique* 98.3–4 (2003), pp. 389–417

Barone-Adesi, Giorgio. *L'età della «lex Dei»* (Napoli 1992)

Bianchini, Maria Grazia. *Caso concreto e «lex generalis». Per lo studio della tecnica e della politica normativa da Costantino a Teodosio II* (Milan 1979)

Caimi, James. '*Ioannis Lydi de magistratibus* III 70. Note esegetiche e spunti in tema di fiscalità e legislazione protobizantina', in *Rivista di studi bizantini e slavi* 1 (1981), pp. 317–361

Carrié, Jean-Michel. 'L'esercito: trasformazioni funzionali ed economie locali', in *Società romana e impero tardoantico. Istituzioni, ceti, economie*, ed A. Giardina (Bari 1986), pp. 449–488

Cecconi, Giovanni Alberto. *Commento storico al libro II dell'epistolario di Q. Aurelio Simmaco* (Pisa 2002)

Cecconi, Giovanni Alberto. 'Crisi e trasformazioni del governo municipale in Occidente fra IV e VI secolo', in *Die Stadt in der Spätantike - Niedergang oder Wandel?*, eds. J-U. Krause and Chr. Witschel (Stuttgart 2006), pp. 285–318

Clauss, Manfred. *Der magister officiorum in der Spätantike (4.-6. Jahrhundert). Das Amt und sein Einfluß auf die kaiserliche Politik* (München 1980)

Cosentino, Salvatore. 'Sul transito dei soldati (in merito a Iust. Nov. CXXX)', in *Bulgaria Mediaevalis* 2 (2011), pp. 25–38

Cujas, Jacques (Cuiacius), *Opera* XIII voll. (Prati 1836–44)

Dagron, Gilbert. *Costantinopoli. Nascita di una capitale (330–451)*, transl. A. Serafini, (Turin 1991)

Delmaire, Roland. *Largesses sacrées et res privata. L'aerarium impérial et son administration du IVe au VIe siècle* (Rome 1989)

Delmaire, Roland. 'Du Code Théodosien au Code Justinien. L'adaptation de lois anciennes à des situations nouvelles', in *Société, Économie, Administration dans le Code Théodosien*, eds. S. Crogiez-Pétrequin, P. Jaillette (Lille 2012), pp. 165–182

Demougeot, Émilienne. 'Une lettre de l'empereur *Honorius* sur l'*hospitium* des soldats', in Id., *L'empire romain et les barbares d'Occident (IVe - VIIe siècle). Scripta varia* (Paris 1988[2]), pp. 75–100

Dillon, John. *The Justice of Constantine. Law, Communication, and Control* (Michigan 2012)

Dirksen, Heinrich Eduard. *Manuale latinitatis fontium iuris civilis Romanorum* (Berolini 1837)

Domínguez López, Esther. '*Veterani*: ciertos privilegios fiscales', in eds. A. Fernández de Buján *et alii*, *Hacia un derecho administrativo, fiscal y medioambiental romano* III, (Madrid 2016), pp. 521–535

Drew-Bear, Thomas, Werner Eck and Peter Herrmann. '*Sacrae Litterae*', in *Chiron* 7 (1977), pp. 355–383

Du Cange, Charles du Fresne. *Glossarium mediae et infimae latinitatis*, t. VI (Parisiis 1846)

Du Cange, Charles du Fresne. *Glossarium ad scriptores mediae et infimae graecitatis* (Graz 1958)

Fascione, Lorenzo. *Fraus legi. Indagini sulla concezione della frode alla legge nella lotta politica e nella esperienza giuridica romana* (Milan 1983)

Fascione, Lorenzo. 'Barbari e lavoro della terra in Occidente da Teodosio I (382) a Odoacre (476)', in *Historia et ius* 11 (2017), pp. 1–24

Feissel, Denis. 'Le rescrit de Valérien à Apellas (I. Smyrna, 604) d'après une copie inédite de John Covel', in Id., *Documents, droit, diplomatique de l'Empire romain tardif*, (Paris 2010), pp. 101–116

Gaudemet, Jean. '*Utilitas publica*', in *Revue historique de droit français et étranger*, 29 (1951), pp. 465–499

Giomaro, Anna Maria. *Il Codex repetitae praelectionis. Contributi allo studio dello schema delle raccolte normative da Teodosio a Giustiniano* (Roma 2001)

Giuffrè, Vincenzo. *Letture e ricerche sulla 'res militaris'*, vol. II (Napoli 1996)

Godefroy, Jacques (Gothofredus), *Codex Theodosianus cum Perpetuis Commentariis*, 6 voll. (Hildesheim - Zürich - New York 1975, repr. ed. Leipzig 1738)

Guillou André, 'La cosiddetta cronaca siriaca di Giosuè lo Stilita', in Id., *La civiltà bizantina dal IV al IX secolo. Aspetti e problemi* (Bari 1977), pp. 369–383

Isaac, Benjamin. *The Limits of Empire: The Roman Army in the East*, rev. ed. (Oxford 1999)

Jones, A. Hugo M. *The Later Roman Empire 284–602: A Social, Economic and Administrative Survey*, vol. II (Oxford 1964)

Jones, A. Hugo M., John R. Martindale and John Morris. *The Prosopography of the Later Roman Empire*. vol. 2, *AD 395–527* (Cambridge 1980)

Jones, Christopher P. 'The *Sacrae Litterae* of 204: Two Colonial Copies', in *Chiron* 14 (1984), pp. 93–99

Knibbe, Dieter and Reinhold Merkelbach. 'Allerhöchste Schelte (Zwei Exemplare der *Sacrae Litterae* aus Ephesos)', in *Zeitschrift für Papyrologie und Epigraphik* 31 (1978), pp. 229–232

Kulikowski, Michael. v. '*hospitalitas*', in eds. Y. Le Bohec *et alii*, *The Encyclopedia of the Roman Army*, vol. II (Oxford 2015), pp. 489–490

Laniado, Avshalom. 'La Vie d'Hypatius de Gangres (BHG 759a), Jean Malalas et l'impôt du xylélaion', in *Analecta Bollandiana* 115 (1997), pp. 133–146

Le Bohec, Yann. *L'armée romaine sous le Bas-Empire* (Paris 2006)
Lee, A. Doug. 'The Army' [ch. 7], in CAH. *The Late Empire, A.D. 337–425*, eds. A. Cameron – P. Garnsey, vol. XIII (Cambridge 2000)
Liebeschuetz, John H.W.G. *Antioch. City and Imperial Administration in the Later Roman Empire* (Oxford 1972)
MacMullen, Ramsay. *Soldier and Civilian in the Later Roman Empire* (Cambridge Massachusetts 1963)
Maganzani, Lauretta. *Gli agrimensori nel processo privato romano* (Rome 1997)
Milani, Mattia. 'La mano destra in Roma antica', in ed. L. Garofalo, *Il corpo in Roma antica. Ricerche giuridiche*, vol. II (Pisa 2017), pp. 25–121.
Millar, Fergus. *A Greek Roman Empire: Power and Belief under Theodosius* II *(408–450)* (Berkley - Los Angeles - London 2006)
Mitthof, Fritz. *Annona militaris. Die Heeresversorgung im spätantiken Ägypten. Ein Beitrag zur Verwaltungs- und Heeresgeschichte des Römischen Reiches im 3. bis 6. Jh. N. Chr.*, I (Firenze 2001)
Modéran, Yves. 'Gildon, les Maures et l'Afrique', in MEFRA. *Antiquité* 101.2 (1989), pp. 821–872
Navarra, Marialuisa. '*Utilitas publica-utilitas singulorum* tra IV e V sec. d.C. Alcune osservazioni', in *Studia et documenta historiae et iuris* 63 (1997), pp. 269–291
Nelis-Clément, Jocelyne. *Les beneficiarii: militaires et administrateurs au service de l'empire (Ier s. a.C.–VIe s. p.C.)* (Bordeaux 2000)
Palazzi, Massimo. *«Foedus cum natione». I trattati alle radici dell'Europa. Aspetti delle relazioni fra Impero romano e popolazioni barbariche fra II e V secolo* (Gallarate 2014)
Peyras, Jean. 'Écrits d'arpentage et hauts fonctionnaires géomètres de l'antiquité tardive', in *Dialogues d'histoire ancienne* 21.2 (1995), pp. 149–204
Peyras, Jean. 'La *potestas occupandi* dans l'Afrique romaine', in *Dialogues d'histoire ancienne* 25.1 (1999), pp. 129–157
Pfeilschifter, Rene. *Der Kaiser und Konstantinopel: Kommunikation und Konfliktaustrag in einer spätantiken Metropole* (Berlin-Boston 2013)
Pollard, Nigel. *Soldiers, Cities, and Civilians in Roman Syria* (Michigan 2000)
Porena, Pierfrancesco. *L'insediamento degli Ostrogoti in Italia* (Rome 2012)
Porena, Pierfrancesco. 'Voci e silenzi sull'insediamento degli Ostrogoti in Italia', in *Expropriations et confiscations dans les royaumes barbares. Une approche régionale*, eds. P. Porena and Y. Rivière (Rome 2012), pp. 227–278
Rance, Philip. 'The farmer and the soldier should be friends: Justinian's legislation on the provisioning of soldiers (Novel 130)', in *Journal of Late Antiquity* 12.2 (2019), pp. 380–421
Rebuffat, René. 'La loi et la ville. Greniers et enceintes', in *L'Africa romana. Trasformazione dei paesaggi del potere nell'Africa settentrionale fino alla fine del mondo antico*, Atti del XIX convegno di studio, Sassari, 16–19 dicembre 2010, eds. M.B. Cocco, A. Gavini, A. Ibba, vol. I (Rome 2012), pp. 121–140

Roda, Sergio. *Commento storico al libro IX dell'epistolario di Q. Aurelio Simmaco. Introduzione, commento storico, testo, traduzione e indici* (Pisa 1981)

Roda, Sergio. '*Militaris impressio* e proprietà senatoria nel tardo impero', in *Studi tardoantichi* [Hestíasis. Studi di tarda antichità offerti a S. Calderone], vol. IV (Messina 1987), pp. 214–241

Rotondi, Giovanni. *Gli atti in frode alla legge nella dottrina romana e nella sua evoluzione posteriore* (Turin 1911)

Santalucia, Bernardo. *I «libri opinionum» di Ulpiano*, II (Milan 1971)

Scarcella, Agatina Stefania. *La legislazione di Leone* I (Milan 1997)

Sitzia, Francesco. *Studi sulla superficie in epoca giustinianea* (Milan 1979)

Sivan, Hagith. 'An Unedited Letter of the Emperor Honorius to the Spanish Soldiers', in *Zeitschrift für Papyrologie und Epigraphik* 61 (1985), pp. 273–287

Sivan, Hagith. 'On *foederati, hospitalitas,* and the settlement of the Goths in AD 418', in *American Journal of Philology* 108.4 (1987), pp. 759–772

Soverini, Paolo. 'Per lo studio del testo della *Historia Augusta*: tra analisi linguistica e critica storica', in *Atti dell'Accademia ligure di scienze e lettere* 46 (1989), pp. 434–444

Steineken, Johannes. *Disputatio inauguralis juridica de jure metatorio* (Basileae 1680)

Stephanus, Henricus. *Thesaurus Graecae Linguae*, vol. IV (repr. Graz 1954)

Tavolieri, Claudia. 'La fiscalità nel mondo antico. Il caso dell'*hospitium militum*', in *Zeitschrift für antikes Christentum*, 11.3 (2008), pp. 448–463

Tavolieri, Claudia. 'Origine militare e fiscale dell'*hospitalitas*. L'esempio dell'*Epistula* 62 (19) inviata da Ambrogio a Vigilio di Trento', in *Zeitschrift für antikes Christentum* 12.3 (2008), pp. 504–543

Trisciuoglio, Andrea. '*Humanitas* e cura delle esigenze umane nel Codice Teodosiano', in *Ius Romanum* 2 (2018), pp. 349–357

Trombley, Frank R. and John W. Watt. *The Chronicle of Pseudo-Joshua the Stylite* (Liverpool 2000)

Voci, Pasquale. *Nuovi studi sulla legislazione romana del tardo impero* (Padova 1989)

Vogler, Chantal. 'La rémunération annonaire dans le Code Théodosien', in *Ktema* 4 (1979), pp. 293–319

Zaera García, Ana. *La superficies en derecho romano* (Madrid 2017)

CHAPTER 3

Hospitalitas II: The Changing Meaning of *Hospitalitas*

Pierfrancesco Porena

1 Hospitalitas between East and West

The practice of *hospitalitas* offered by private citizens to Roman functionaries and soldiers travelling around the immense territory of the Empire emerged as an inevitability. It was simply impossible to build a robust and stable 'state' logistical structure along all the arteries and in all the cities of the Roman provinces; therefore, civil functionaries and military officials and contingents – be they Roman or in the late Empire, barbarian – often along with their entourage, were forced to use temporary accommodation provided in various ways by private Roman citizens, apparently free of charge. Obviously, however, every form of short, medium or long-term lodging in these types of structures came at a cost, while also creating problems of cohabitation and exposing residents to abuse. Nonetheless, this necessity remained in effect through the fifth and sixth centuries, in the period when barbarian groups were moving and settling in the Empire.

In Part I, we have seen that the protection of the *praedia rustica* by the *metatum* endured in the Eastern Empire.[1] The first characteristic of the *metatum* was the division of the living spaces by percentage shares. The houses chosen were divided into three parts by the *metator*: the owner chose the first part for himself; the assigned guest chose and occupied the second part, and the third and remaining part was left for the owner. In the case of an *illustris* guest, however, the space was divided in half: the owner divided it but the

1 Cf. Mauricius, *Strategikon* 1.9 (ed. Dennis, p. 104): the armies must hurry through the countryside without disturbing the farmers who produce goods and pay taxes. Procopius Caesariensis, *Bellum Vandalicum* 1.1–11; 17.6–10; 20.1–2 and 18–23; 21.8–10 (533), who appears to be in line with the Justinian rescripts known in epigraphic form to protect the *praedia rustica* in the *Asiana* diocese, on which cf. Philip Rance, 'The farmer and the soldier should be friends: Justinian's legislation on the provisioning of soldiers (Novel 130)', in *Journal of Late Antiquity*, 12.2 (2019), pp. 405–414.

 | DOI:10.1163/9789004698017_004

guest had first choice as to which half he would occupy.[2] A second characteristic of the *metatum* concerned the duration of the *hospitalitas*, which had to be as brief as possible, a stop in a journey from a specific departure point to a specific arrival point, and not become a permanent stay.[3]

The Roman institution of *hospitalitas* / *metatum* – enshrined throughout the Late Roman legislative system – was obligatory and tended to be carried out as a flexible instrument to respond to the unavoidable and burdensome necessity of the internal movement of men in the service of the Empire. However – and this is a fundamental aspect – *hospitalitas* was intended to guarantee the property rights of Roman citizens as owners of the properties involved, with neither expropriation nor confiscation. It aimed to regulate, and usually to limit, the forms of temporary use of portions of the property by guests so that every other associated need of the guest during his stay in the owner's property was borne by the Roman administrative structure.[4] In short, *hospitalitas* rigorously safeguarded private property and strictly limited the time allowed for a portion of the property to be used by guests.

Another fundamental aspect to be considered is that virtually the only object or asset involved in this dynamic is the *domus*, or anyway, the urban residence, and more rarely the rural one. There is no reference to income-producing lands in the sources regarding *hospitalitas* / *metatum*. Simply, it involves temporary accommodation. Where possible, this temporary lodging was completely supported by taxation as far as the essential needs of the civilian or military guest were concerned. Guests could not advance any request for provisions, make any payment or, especially, receive anything produced by the property where they are temporarily quartered. For example, they could not advance any claim to its production or income.

In analyzing the relationship between *hospitalitas* and barbarian settlements, it is important to keep in mind the existence of a one-third share of the private property as the upper limit for temporary occupation by the civilian or military official. In the awareness of this burdensome *munus*, the percentage calculated by the *metatores* must have been an actual, clearly understood reality fully accepted by both guests and hosts.

This regime of *hospitalitas* / *metatum* remained in effect and was strictly regulated only in the eastern part of the empire, however, at least until the reign of Justinian. In this regard, along with the Novel 25 of Theodosius II of 444 (*De metatis*) and Code of Justinian C. 12.40 (*De metatis et epidemeticis*), the

2 On the important constitution of Arcadius *Codex Theodosianus* 7.8.5 = *Codex Iustinianus* 12.40.2 (February sixth 398 AD) see above (Part I).

3 Cf. *Codex Theodosianus* 7.8.10 (Honorius, June 12th 412 AD), § 2 (Engl. transl. p. 167).

4 See above (Part I).

most relevant document is the Novel 130 of Justinian, addressed to the praetorian prefect of the East (for the first time) Petrus Barsymes on March 1st 545 AD concerning financial support for the armies marching through the provinces.[5] The *Novella* shows the application of the *metatum* to the armies, where the eastern regulations tend to focus on the relative immunity from *hospitalitas* for civil dignitaries and on the privileges of the residents of Constantinople.

There is no doubt that the eastern laws continued to impose rigid rules with widespread application concerning the accommodation of dignitaries, officials and Late Roman troops in cities and along the major transit routes of the *pars Orientis*, with no interruption or distinction between the *comitatensis* units and the permanently recruited or occasionally hired *foederati* barbarian troops. The system of imperial accommodation remained unchanged, and its disadvantages were constantly being obviated by the Augusti.[6] These sovereigns controlled the movement of the armed units in the provinces and the capital.

The itineraries of troops through provincial cities were calculated, supplied and overseen by special staff accompanying the troops (*delegatores*, who were probably *praefectiani*, and *metatores*) and responsibility for any eventual infractions was divided through the long chain of military command and civilian supervision. Indeed, there were hundreds of people responsible for the application of this clearly-defined regulation, and justice for eventual infractions and abuses was administered by the praetorian prefect (who could impose severe punishments).[7] Specifically, in § 9 of *Novella* 130, soldiers were forbidden to quarter in the principal residence of the citizen, which had to remain the exclusive possession of the owner who alone had rights to enjoy it.[8] The meticulous *Novella* suggests that the infractions may have been common, but Justinian's text was perentory: the Roman army and barbarian contingents in the service of Constantinople had no margin of freedom.[9] They had to move within an 'sterile corridor', devised to avoid interaction with civilian subjects.

---

5 Cf. Salvatore Cosentino, 'Sul transito dei soldati (in merito a Iust. Nov. CXXX)', in *Bulgaria Mediaevalis*, 2 (2011), pp. 25*ff.*, text completed by an italian translation and a large historical commentary; see also Rance, 'The farmer and the soldier'. On the recipient of the law cf. *PLRE* III/B, 999–1002.

6 See below at note 21.

7 Cf. now Giorgia Maragno, *'Punire e sorvegliare'. Sanzioni in oro, imperatori, burocrazia* (Napoli 2020).

8 *Novella Iustiniani* 130.9 (Engl. transl. p. 866).

9 On the famous passage of (pseudo-)Joshua the Stylite (*Chronicon* 86.93–96) see Frank R. Trombley and John W. Watt, *The Chronicle of Pseudo-Joshua the Stylite* (Liverpool 2000), pp. 103*ff.*, 111*ff.* That infraction of the rules of the *metatum* is, however, set in the context of a war emergency on the Persian front in 505.

The situation was quite different in the West. The last western constitution with references to the institution of *hospitalitas* dates to 414 AD, twenty-four years before the promulgation of the Theodosian Code.[10] In the succeeding years, in the West, neither the Theodosian Code nor the western *Novellae*, nor the substantial collection of official documents of the Ostrogothic kings in the *Variae* by Cassiodorus (507–537 AD), nor the Acts of the councils, nor other sources such as historians, written correspondence and homilies make any reference to the institution of *hospitalitas* / *metatum* described in the Late Roman regulations as temporary accommodation during travel in provincial territories. A comparison between the Novel 130 of Justinian and some of Cassiodorus' *Variae* reveals a completely different situation in the reign of the Ostrogoths several years before the promulgation of the *Novella*.

In 508 AD, King Theoderic reimbursed the Italian subjects who had been devastated by the passage of the Ostrogothic troops, his own army, on their way to Gaul.[11] The king himself urges (the *comes*?) Wandil to control the troops when they reach Gaul to keep them from pillaging the inhabitants of Avignon in Provence.[12] Shortly thereafter, Theoderic was forced to send money from the royal estate via his *Comes Privatarum* to the (Gothic) army in Pannonia to prevent them from pillaging in order to support themselves.[13] In this same period, the king ordered the praetorian prefect of Italy to waive the land tax of the inhabitants of the province of *Alpes Cottiae* for the year 509/510 AD as the Ostrogothic army had devastated their fields during its march to Gaul.[14]

In 523/526 AD, in the context of the crisis in the alliances established by Theoderic and during the war against the Burgundians, a letter from the king to Verane, *saio*, testifies to the risk for the populations of moving a contingent of the Gepid allies from Pannonia through *Venetia* and *Liguria* to Provence.[15] In the subsequent letter, the king is aware that without the *solidi* assigned to each soldier for his needs on the march, the Gepids would plunder the goods

10 *Codex Theodosianus* 7.8.12. The dating (between 412 and 423 AD) and the textual restitution of Honorius' epistle to some of his units preserved in the margin of an epistle in praise of Pamplona in Codex 78, Bibl. Acad. Hist. Madrid, 'Rotense', is problematic. See Trisciuoglio, in this volume at p. 34.

11 Cassiodorus, *Variae* 2.8 (King Theoderic to Severus, venerable Bishop - 508 AD), (Engl. transl., p. 86).

12 Cassiodorus, *Variae* 3.38 (King Theoderic to Wandil - 508 AD ?) (Engl. transl., p. 149).

13 Cassiodorus, *Variae* 4.13.2 (King Theoderic to Senarius, *vir illustris Comes Privatarum* - 509/510 AD) (Engl. transl., p. 174).

14 Cassiodorus, *Variae* 4.36.2 (King Theoderic to Faustus, *praefectus praetorio* – before September 1st 509) (Engl. transl., p. 189).

15 Cassiodorus, *Variae* 5.10 (King Theoderic to Veranus, *saio* - 523/526 AD) (Engl. transl., p. 212).

of his subjects.[16] In the same period of the war, the king was also aware that the military ration was opportune to keep the army from looting the goods of people in the provinces.[17]

It might be assumed that the difficulty of controlling the Ostrogothic soldiers and their *foederati* troops coincided with periods of war, but that was not the case. As part of a normal convocation of Ostrogothic soldiers residing in *Picenum* and *Samnium* to Ravenna for the distribution of awards and gifts, Theoderic specified that they must not damage or oppress and rob the civil population during their journey.[18]

As a rule, the soldiers of the Ostrogothic army (and other barbarian allies) tyrannized the population during their movements through the Italian provinces, and outside of Italy as well, while the king seemed unable to control them. In the Ostrogothic reign, where, for example, the legislation concerning the *cursus publicus* was solid and the administrative structures quite similar to those of the late Empire, laws concerning the *metatum* seem to have disappeared long before.[19] The Ostrogothic army established itself as a proud independent community with its own sense of identity, escaping the control of their king and the administrative officials distributed throughout the territory.[20] In contrast, Justinian was apparently able to control the behaviour of his soldiers and civil

16 Cassiodorus, *Variae* 5.11 (King Theoderic to the Gepids sent to Gaul- 523/526 AD) (Engl. transl., pp. 212–213).

17 Cassiodorus, *Variae* 5.13 (King Theoderic to Eutropius and Agroecius - 523/526 AD) (Engl. transl., p. 214).

18 Cassiodorus, *Variae* 5.26.2 (King Theoderic to All Goths Settled in Picenum and Samnium - 523/526 AD) (Engl. transl., p. 226).

19 On *cursus publicus* in Ostrogothic Italy cf. Cassiodorus, *Variae* 1.29; 4.47; 5.5; 11.14, and see Lucietta Di Paola, *Viaggi, trasporti, e istituzioni. Studi sul «cursus publicus»* (Messina 1999), pp. 104–114; on the structures of the *cursus publicus* cf. now Basso – Zanini, *Statio amoena*. On the administration of the Ostrogothic kingdom cf. the commentaries to Cassiodorus' *Formulae* by Francesco M. Petrini (*Variae* Book VI) and Giovanni A. Cecconi (*Variae* Book VII) in *Flavio Magno Aurelio Cassiodoro Senatore, Varie*, vol. 3; Friederike Gatzka, *Cassiodor, Variae 6. Einführung, Übersetzung und Kommentar* (Berlin 2019) and the articles by Michael S. Bjornlie, 'Governmental Administration' and Jonathan J. Arnold, 'Ostrogothic Provinces: Administration and Ideology', both in *A Companion to Ostrogothic Italy*, eds. J.J. Arnold – M.S. Bjornlie – K. Sessa (Leiden-Boston 2016), pp. 73*ff.* and pp. 47*ff.* respectively.

20 Cf. Peter Heather, 'Merely an Ideology? – Gothic Identity in Ostrogothic Italy' and Samuel J.B. Barnish, 'Cuncta Italiae membra componere: Political Relations in Ostrogothic Italy', both in *The Ostrogoths. From the Migration Period to the Sixth Century*, eds. S.J.B. Barnish – F. Marazzi (Woodbridge 2007), pp. 31 *ff.* and 317 *ff.* respectively; Hans-Ulrich Wiemer, 'Theoderich und seine Goten. Aufstieg und Niedergang einer Gewaltgemeinschaft', in *Gewaltgemeinschaften. Von der Spätantike bis ins 20. Jahrhundert*, ed. W. Speitkamp (Göttingen 2013), pp. 15 *ff.*; Id., 'Die Goten in Italien. Wandlungen und Zerfall einer Gewaltgemeinschaft', in *Historische Zeitschrift*, 296 (2013), pp. 593*ff.*; Brian Swain, 'Goths

functionaries. Despite certain deficiencies, the Byzantine armies were victorious in all of Justinian's wars. Behind this difference between East and West lies a diverging evolution in the relationship between armies and private property in the two areas.[21]

2 *Hospitalitas* in Modern Theories of Barbarian Settlement in the West

The transformation of the institution of *hospitalitas* in the West in the fifth and sixth centuries introduces into the historiographic conversation the question of whether this legal institution lies at the base of the forms of settlement of barbarian groups and the constitution of the so-called Roman-barbarian kingdoms. In 1844, Ernst Theodor Gaupp (1796–1859) published his *Die germanischen Ansiedlungen und Landtheilungen in den Provinzen des Römischen Westreiches*,[22] where he hypothesized that the settlement of barbarian groups in the West between the fifth and sixth centuries was carried out by expropriating a third of the lands held by Romans, and that this percentage was determined on the basis of the Late Roman regulation of *hospitalitas*. According to Gaupp, the land expropriated from the Romans would be assigned to barbarian soldiers (Visigoth, Burgundian and Ostrogoth) in lots, called *sortes*, while the unassigned land was taxed at the traditional rate of one third of its income (*tertiae*).

---

and Gothic Identity in the Ostrogothic Kingdom', in *A Companion to Ostrogothic Italy*, eds. J.J. Arnold – M.S. Bjornlie – K. Sessa (Leiden-Boston 2016), pp. 203*ff.*

21 Cf. the important comparative and diachronic analysis by Chris Wickham, *Framing the early Middle Ages. Europe and the Mediterranean, 400–800* (Oxford 2005); Gilles Bransbourg, 'Le grand écart. Les destins divergents de l'Orient et de l'Occident à travers le prisme fiscal de l'Empire tardif', in *Libera curiositas. Mélanges d'histoire romaine et d'Antiquité tardive offerts à J.-M. Carrié*, eds. C. Freu – S. Janniard – A. Ripoll (Turnhout 2016), pp. 67*ff.*; cf. also Heather, 'East & West in the Fifth Century', in *Governare e riformare l'impero al momento della sua divisione. Oriente, Occidente, Illirico*, eds. U. Roberto – L. Mecella (Roma 2015); John Bintliff, 'Aspects of settlement change in late Antiquity from regional survey evidence', in *New cities in Late Antiquity: documents and archaeology*, ed. E. Rizos (Turnhout 2017), pp. 13*ff.*

22 Ernst Th. Gaupp, *Die germanischen Ansiedlungen und Landtheilungen in den Provinzen des Römischen Westreiches in ihrer völkerrechtlichen Eigenthümlichkeit und mit rücksicht auf verwandte Erscheinungen der alten Welt und des späteren Mittelalters* (Breslau 1844). Successful restatement of Gaupp's theory by Ferdinand Lot, 'Du régime de l'hospitalité', in *Revue Belge de Philologie et d'Histoire*, 7 (1928), pp. 975*ff.*

This reconstruction was well-accepted in the 20th century until 1980, when it was rebutted by Walter Goffart in his acclaimed *Barbarians and Romans A.D. 418–584: The Techniques of Accommodation*. Goffart rejected both the legal validity of *hospitalitas* as the basis for the barbarian settlement and the fact of the expropriation of one third of the private property belonging to Romans, then redistributed among barbarians. On the contrary, he proposed that there was a simple division of land taxes of the Romans, so that the Roman property was not expropriated; instead, roman owners 'hosted' barbarians on portions of Roman lands, legally property of Romans, but the taxes concerning these portions of Roman lands were paid directly to the barbarians at the rate established by the *sors* owed by each barbarian.[23] This dynamic would have been the same for all settlements of barbarian groups in the Empire.

While Goffart's reconstruction met with some agreement, there are numerous critiques from various historiographic schools because of the difficulty of reconciling it with the data found in the sources.[24] Certain elements of the hypothesis concerning access to the tax payments run counter to the historic processes of the period.[25] In my opinion, Goffart's theory about the institution

---

23 Walter Goffart, *A.D. 418–584: the Techniques of Accommodation* (Princeton 1980), pp. 58–102. The scholar confirmed his hypothesis in his recent works: *Barbarian tides. The migration age and the later Roman Empire* (Philadelphia 2006); 'The Technique of Barbarian Settlement in the Fifth Century: A Personal, Streamlined Account with Ten Additional Comments', in *Journal of Late Antiquity*, 3/1 (2010); 'Le début (et la fin) des «sortes Vandalorum»', in *Expropriations et confiscations dans les royaumes barbares. Une approche régionale*, eds. P. Porena – Y. Rivière (Roma, 2012); 'Administrative Methods of Barbarian Settlement in the Fifth Century. The Definitive Account', in *Gallien in Spätantike und Frühmittelalter. Kulturgeschichte einer Region*, eds. S. Diefenbach – G.M. Müller (Berlin 2013).

24 The issue has been systematically discussed in Porena – Rivière, *Expropriations et confiscations*. For an analysis of the different opinions, for, against and neutral to the Goffart hypothesis, see the overviews by Guy Halsall, *Barbarian migrations and the Roman West, 376–568* (Cambridge-New-York 2007), pp. 422–447 (partially revised in Id., 'The Ostrogothic Military', in *A Companion to Ostrogothic Italy*, eds. J.J. Arnold – M.S. Bjornlie – K. Sessa (Leiden-Boston 2016), pp. 177*ff.*); Christine Delaplace, *La fin de l'Empire romain d'Occident. Rome et les Wisigoths de 382 à 531* (Rennes 2015), pp. 29–31, and the commentary by Pierfrancesco Porena to Cassiodorus, *Variae* 2.16, in *Flavio Magno Aurelio Cassiodoro Senatore, Varie*, I.

25 The tax system controlled by the praetorian prefectures appears to be functioning e.g. in Ostrogothic Italy, cf. Salvatore Cosentino, 'Fine della fiscalità, fine dello stato romano ?', in *Le trasformazioni del V secolo. L'Italia, i barbari e l'Occidente romano*, eds. P. Delogu – S. Gasparri (Turnhout 2010); Domenico Vera, 'Stato, fisco e mercato nell'Italia gotica secondo le Variae di Cassiodoro: fra ideologia politica e realtà', in *Agoranomes et édiles. Institutions des marchés antiques*, eds. L. Capdetrey – C. Hasenohr (Bordeaux 2012); Luca Loschiavo, 'Oltre la milizia. Fisco e civilitas per i Goti di Teoderico', in *Cittadinanze*

of *hospitalitas* is correct, but not his overall reconstruction of the mechanisms of barbarian settlement in the West.

There is no doubt that the Late Roman practice of *hospitalitas*, regulated through the legislative 'package' known by the Romans as *metatum*, could not have formed the legal foundation for the stable settlement of barbarian groups in the provinces of the West. As we have seen, *hospitalitas* referred to the simple temporary accommodation for soldiers and paid officials, usually in or near cities, and even more importantly, it was intended to protect in all respects the property rights of Roman citizens over their own properties. On the other hand, the concrete settling of barbarian groups in the West – of all the groups, that is, including Visigoths, Alans, Suebi, Burgundians, Franks, Vandals, Ostrogoths, Lombards and many other smaller groups – led everywhere to the establishment of a new and different minority of barbarian warriors who were also landowners.[26] Their assets included not only urban *praedia* and residences but also composite properties, especially productive lands generating income, at times quite substantial, for the barbarian elites. These properties came from the expropriation of Roman owners in ways that varied in both time and space, with no homogeneous process.

To sum up, there was indeed expropriation – whether partial or total – despite what Goffart claims, and this dynamic is the valid aspect of Gaupp's reconstruction. However, the actual expropriations were not a natural and direct outcome of the legislation regarding *hospitalitas*, as Gaupp believed; on the contrary, a closer analysis of the *metatum* shows that its legal aim was the very opposite – to protect in all respects the property of the citizens against the claims of transiting officials and soldiers. From this point of view, Goffart is correct.

---

medievali. Dinamiche di appartenenza a un corpo comunitario, ed. Sara Menzinger (Roma, 2017). On the importance of taxation and its evolution between the fifth and 7th centuries cf. Wickham, *Framing*, chap. 3. Andrey Young, 'Hospitalitas: Barbarian settlements and constitutional foundations of medieval Europe', in *Journal of Institutional Economics*, 14/4, follows Goffart's hypothesis (allocation of tax quotas to barbarians).

26 Cf. the conclusions by Wickham in his *Framing*; Peter Heather, 'Elite militarisation and the post-roman West', in *Istituzioni, carismi ed esercizio del potere* (*IV–VI secolo d.C.*), eds. G. Bonamente – R. Lizzi Testa (Bari 2010); Bruno Dumézil, *Servir l'état barbare dans la Gaule franque: du fonctionnariat antique à la noblesse médiévale, IVe–IXe siècle* (Paris 2013); Pierfrancesco Porena, 'La posizione dell'elemento militare nell'Impero romano e i 'regni romano-barbarici'', in *Civitas, iura, arma: organizzazioni militari, istituzioni giuridiche e strutture sociali alle origini dell'Europa* (secc. *III–VIII*), eds. F. Botta – L. Loschiavo (Lecce, 2015).

3 Metamorphosis of *Hospitalitas*

That said, *hospitalitas*, understood as the practice of accommodation of soldiers, was not totally irrelevant to barbarian settlement in the West. In this contribution we hypothesize that *hospitalitas / metatum* was transformed by a 'short circuit' that occurred in the fifth century in certain areas with a higher density of Roman properties, which distorted its legal substance and upended its purpose. It seems highly probable that in these areas, the shares of the division of temporary accommodation inspired the 'maximum shares' of Roman properties to be assigned to barbarians. In order to evaluate this historical dynamic, however, it is necessary to put into context, as much as possible, the process of the stabilization of barbarian groups in the West in the fifth century. For lack of space, it is not possible here to examine the stages of the settlement of the various barbarian groups in the West in that period; the issue is too broad and analysis is based on very fragmentary sources.[27] We can only give a brief diachronic overview of the aspects of settlements related to these shares, and in particular to the one-third share (*tertiae*) in the documentation.

Traditionally, the ideal settlement of barbarian individuals and groups, according to the Romans, involved the total submission of the barbarians and their association with the land established in a subordinate regime. The Roman government had always avoided that the barbarians, inside the Empire, should

27 On the question of the migration of peoples in Late Antiquity and the current historiographical debate cf. the synthesis by Rosalba Arcuri, 'Etnogenesi, «entelechia barbarica» e attuali orientamenti storiografici sulla Völkerwanderungszeit', in *Koinonia*, 37 (2013), pp. 107 *ff.*; reconstruction of the historical process by Heather, *Empires and Barbarians: The Fall of Rome and the Birth of Europe* (Oxford 2009), pp. 577–617; focus on historiographical aspects by Iann Wood, *The Modern Origins of the Early Middle Ages* (Oxford 2013); Id., 'The Uses and Abuses of the Barbarian Invasions in the Nineteenth and Twentieth Centuries', in *Manufacturing Middle Ages: entangled history of medievalism in nineteenth-century Europe*, eds. P. Geary – G. Klaniczay (Leiden 2013), pp. 51 *ff.*; Id., *The Transformation of the Roman West* (Leeds 2018); Id., 'Responses to Migration and Migrants in the Fifth- and Sixth-Century West', in *Le migrazioni nell'Alto Medioevo* (Atti Settimane del CISAM, LXVI) (Spoleto 2019), pp. 177 *ff.*; Walter Pohl, 'Völkerwanderung', in *Migrationen im Mittelalter: ein Handbuch*, ed. M. Borgolte (Berlin, 2014), pp. 231 *ff.*; 'Von der Ethnogenese zur Identitätsforschung', in *Neue Wege der Frühmittelalterforschung. Bilanz und Perspektiven*, eds. W. Pohl – M. Diesenberger – B. Zeller (Wien 2018), pp. 9 *ff.*; 'Dinamiche etniche nel corso delle migrazioni', in *Le migrazioni nell'Alto Medioevo* (Atti Settimane del CISAM, LXVI) (Spoleto 2019), pp. 1 *ff.*; Patrick Geary, 'Political Identity, Ethnic Identity, Genetic Identity: the Dangers of Conceptual Confusion', in *Neue Wege der Frühmittelalterforschung. Bilanz und Perspektiven*, eds. W. Pohl – M. Diesenberger – B. Zeller (Wien 2018), pp. 35 *ff.*; in detail and extensively Misha Meier, *Geschichte der Völkerwanderung. Europa, Asien und Afrika vom 3. bis zum 8. Jahrhundert n.Chr.* (München 2019), chap. I.

enjoy freedom of movement or acquire lands at the expense of the assets of tax-paying Romans.[28] Indeed, barbarian settlement from outside the empire had always been tightly controlled.[29]

From the second to the fourth century, Romans preferred that barbarians enter the militarized border provinces, or at times further inland, but not settle as large compact ethnic groups in the productive provinces of more central areas of the empire. The barbarians were admitted and distributed in the Empire according to demographic, economic and military needs planned by the Romans, within which there was no room to manoeuvre or bargain. These barbarians were mostly prisoners of war who had surrendered to the Romans or been captured by force, or, as barbarian groups allocated usually after an unfavourable treaty sealed out of necessity for their own survival or after being defeated militarily, or *dediticii* who had turned themselves in voluntarily, or, finally, mercenaries and volunteers destined to 'denationalize' themselves, often from ethnic groups who were 'clients' of the Romans.[30]

---

28 Trade across the frontier was tightly controlled; the lack of Roman and other citizenship must have limited the barbarians' negotiating activity aimed at acquiring property in the provinces. In general, on the control of relations along the frontiers see the recent Slootjes – Peachin, Slootjes, Daniëlle and Michael Peachin, eds., *Rome and the Worlds beyond its frontiers* (Leiden 2016).

29 Cf. Yves Modéran, 'L'établissement de barbares sur le territoire romain à l'époque impériale', in *La mobilité des personnes en Méditerranée da l'Antiquité à l'époque moderne. Procédures de contrôle et documentation d'identification*, ed. by Claude Moatti (Roma, 2004), 337–397; Alessandro Barbero, *Barbari. Immigrati, profughi, deportati nell'impero romano* (Roma 2006); Sebastian Schmidt-Hofner, 'Barbarian migrations and socio-economic challenges to the Roman landholding elite in the fourth century CE', in *Journal of Late Antiquity*, 10.2 (2017).

30 *Panegyrici Latini* 8 (5).1.4; 8.4; 9.1–4; 21 (ed. Mynors, pp. 215, 220–221, 229). *Codex Theodosianus* 5.6.3 (April 12th 409 AD); *Codex Theodosianus* 13.11.10 (April fifth 399 AD); *Codex Theodosianus* 7.20.12 (January 30th 400 AD). Cf. Arnaldo Marcone, 'Dal contenimento all'insediamento: i Germani in Italia da Giuliano a Teodosio Magno', in *Germani in Italia*, eds. B. Scardigli – P. Scardigli (Roma 1994), pp. 239*ff.*; on forms of wartime imprisonment in Late Antiquity, cf. Hervé Huntzinger, *La captivité de guerre en Occident dans l'Antiquité tardive (378–507) / War captivity in Late Antique Occident (378–507)* (Thèse Univ. Strasbourg, École Doctorale des Sciences de l'Homme et des Sociétés, oct. 2009). On the ideology underlying the control of the barbarian in the late Empire see e.g. Caterina C. Berardi, '«Barbari exsecrati gladios suos ad aratra conversi sunt» (Oros., adv. pag. 7, 41, 7): il tema della pacificazione dei barbari tra IV e V sec. d.C.', in *Auctores nostri: studi e testi di letteratura cristiana antica*, 5 (Bari, 2007). On *dediticii* still in fifth Century cf. Soazick Kerneis, 'Rome et les barbares: aux origines de la personnalité des lois', in *Civitas, iura, arma: organizzazioni militari, istituzioni giuridiche e strutture sociali alle origini dell'Europa (secc. III–VIII)*, eds. F. Botta – L. Loschiavo (Lecce 2015), pp. 103*ff.* On the different dynamics of integration through military recruitment in the fifth Century cf. Michael Vannesse, 'La militarizzazione dell'impero: il ruolo dei barbari (395–455 d.C.)', in *Potere e politica nell'età della famiglia teodosiana (395–455): i linguaggi dell'impero, le identità dei barbari*,

The barbarians and their families seem to have been bound to an inferior social status, equivalent to that of *colonus* (settler) or more correctly, of *inquilinus*; like the Roman *coloni*, they were free but they did not have *origo* in the Empire nor on the lands where they were employed as tenants. Except for particular types of lands, generally they did not own the land they cultivated and where they lived, which continued to belong to either Roman citizens, cities or the Emperor. Strictly speaking, they were not Roman citizens, though they appear to have been granted citizenship individually if enrolled in a military unit.[31] Like Roman *coloni*, they were obliged to pay rent to the owner, pay the *capitatio* to the Roman treasury and provide recruits for the Roman troops.[32]

It seems likely that, in some cases, the state of the municipal or imperial land on which they settled transformed them into *inquilini* subject to military service and inheritance tax; such was the case with the *laeti* and *terre laeticae*.[33] Interests converged on the settling of the barbarian *inquilini* and the

---

eds. I. Baldini – S. Cosentino (Bari 2013); Ralph Mathisen, 'The End of the Western Roman Empire in the Fifth Century CE: Barbarian Auxiliaries, Independent Military Contractors, and Civil Wars', in *The Fifth Century: Age of Transformation. Proceedings of the 12th Biennial Shifting Frontiers in Late Antiquity Conference*, eds. J.W. Drijvers – N. Lenski (Bari 2019).

31 See the contribution of Valerio Marotta in this volume.

32 Cf. Cam Grey, 'The «ius colonatus» as a model for the settlement of barbarian prisoners-of-war in the late Roman Empire?', in *Romans, barbarians, and the transformation of the Roman world. Cultural interaction and the creation of identity in Late Antiquity*, eds. R.W. Mathisen – D. Shanzer (Aldershot, 2011) with remarks by Jean-Michel Carrié, 'Problèmes de la terre dans l'Antiquité Tardive: réalités archéologiques, structures agraires et normes juridiques", in *Atti dell'Accademia romanistica costantiniana* (AARC), XXII. *Questioni della terra. Società, economia, normazioni, prassi. In onore di M. Bianchini* (Napoli, 2017), pp. 32–33 and Pasquale Rosafio, '*Constitutio de Scyris* e l'insediamento dei barbari nell'impero', in *Figures d'Empire, fragments de mémoire: pouvoirs et identités dans le monde romain impérial, IIe s. av. n. è.-VIe s. de n. è.*, eds. S. Benoist – A. Daguet-Gagey – C. Hoët-Van Cauwenberghe (Villeneuve-d'Ascq, 2011); Id., 'Nuove ricerche sul colonato', in *Uomini, istituzioni, mercati. Studi di storia per Elio Lo Cascio*, eds. M. Maiuro – G.D. Merola – M. De Nardis – G. Soricelli (Bari, 2019); Lorenzo Fascione, 'Barbari e lavoro della terra in Occidente da Teodosio I (382) a Odoacre (476)', *Historia et ius*, 11 (2017) – paper 9. On the meaning of *colonus* cf. Domenico Vera, 'Questioni di storia agraria tardoromana: schiavi, coloni, villae', in *Antiquité Tardive*, 20 (2012), p. 116, with the remarks by Carrié, 'Problèmes de la terre', p. 23; on *inquilinus* cf. Pasquale Rosafio, 'Inquilinus', in *Opus*, 3 (1984), pp. 121 *ff.* We do not enter into the debate on the existence of the colonate as a legal profile. The expression «living 'according to the laws of the Romans'», as the *recepti in leges* of the Latin Panegyrics (*Panegyrici Latini* 8 (5).21, ed. Mynors, p. 229), perhaps indicates the framing of the barbarians within the boundaries imposed by fiscal and military law. The link between juridical status and submission to agricultural services became much weaker in the barbarian kingdoms, where the term *mancipia* collects all forms of agrarian dependence (of freemen and slaves).

33 On the *laeti* cf. Valerio Marotta, 'Il problema dei 'laeti'. Fonti e storiografia', in *Civitas, iura, arma: organizzazioni militari, istituzioni giuridiche e strutture sociali alle origini*

laeti. Legislative Codes insist on the exploitation of the *agri deserti* in relation to this type of integration of barbarians in the imperial system: 'ownerless' (*vacantes, caducae*) lands, fiscally unproductive, required workers to upgrade and improve them; the hard work produced taxes and recruits thanks to the populations tied up to those lands.[34] The key point is that the barbarians, relocated as families of *coloni* – in any case, this refers to the *ius colonatus* – lacking self-determination as *deditici*, or settled on *terrae laeticae*, 'served' the lands of Roman owners as *coloni/inquilini* or were bound to their function as farmers, taxpayers and recruits as *laeti*. Settlement prevented them from acquiring any autonomy, freedom of movement or access to profitable forms of autonomous use of the land with full property rights to it.

At least until the end of the fourth century in the West – and even in the East – Roman civil and military authorities seem to have had full control of the situation, including the terms of access and allocation to barbarians. The primary feature of this dynamic of integrating barbarians into the productive and military structure of the later Empire is their immobility, in terms of both

---

dell'Europa (*secc.* *III–VIII*), eds. F. Botta – L. Loschiavo (Lecce 2015); Alain Chauvot, 'Julien et les lètes. À propos d'Ammien, XX,8,13', in *Libera curiositas. Mélanges d'histoire romaine et d'Antiquité tardive offerts à J.-M. Carrié*, eds. C. Freu – S. Janniard – A. Ripoll (Turnhout 2016); Poly, *Laeti and gentiles*, in this volume. The units of *laeti* and *gentiles* were directly controlled according to their service in the army: for them, the granting of land and the loss of autonomy was a function of their obligation to provide recruits to the Romans and to submit to the military requirements of the *magistri militum*. On the *terre laeticae* cf. also Giovanni Papa, 'Multae gentes ... quibus terrae laeticae administrandae sunt. A proposito di CTh.13.11.10', in *Studia et documenta historiae et iuris*, 82 (2016). The hereditary transmission of land ownership was linked to the performance of military functions (cf. Marotta, 'Il problema', pp. 150–157, by analogy with the *terrae obnoxiae* linked to the performance of menbers of late roman corporations).

34 On land tenure regimes cf. the recent important researches by Pierre Jaillette, 'La structure agraire dans l'Empire tardif au miroir du Code Théodosien: quelques observations', in *Antiquité Tardive* 20 (2012) and by Carrié, 'Problèmes de la terre', esp. pp. 42–45. On uncultivated lands cf. Pierre Jaillette, 'Les dispositions du Code Théodosien sur les terres abandonnées', in J.-L. Fiches, ed., *Le IIIe siècle en Gaule Narbonnaise. Données régionales sur la crise de l'Empire* (Sophia Antipolis 1996); Cam Grey, 'Revisiting the problem of agri deserti in the late Roman Empire', in *Journal of Roman Archaeology*, 20 (2007); Simona Tarozzi, 'Riforma dello Stato e gestione della terra. La questione degli agri deserti nella prospettiva dei Codici tardo antichi. Alcuni spunti di riflessione', *Mélanges de l'École Française de Rome. Antiquité* (*MEFRA*), 125 (2013). The lack of information on property acquired by veterans of barbarian origin during and at the end of their service seems significant. As it seems, the socio-economic impact of these investments was weak.

physical movement and property accumulation,[35] which is precisely the type of 'integration' the barbarian fighters of the fifth century sought to avoid.

However, from the early fifth century, after the disaster of Hadrianopolis in August 378 AD, with the internal reorganization of the Gothic *foederati* in Illyricum and with the intrusions of barbarian groups that began in 407 AD – and this is a key point – it became increasingly difficult for the Roman authorities to control their presence and settlement in the traditional terms. This was primarily due to the weakness of the Roman military in the face of the protracted military crisis, which became chronic in the West. Nonetheless, the Augusti refused to deviate from the traditional system of controlling the barbarians.

The barbarian groups that had penetrated within the borders of the western part of the Empire after Hadrianopolis were 'landless men', but they were organized militarily, resolute and supported by innovative processes of ethnogenesis providing them with a new sense of identity.[36] This complex barbarian world offered more than agricultural or military manpower; it was promoting new historical processes within the Empire. In the West, Honorius and Valentinian III, finding themselves in increasing difficulty, set about exploiting the military capacity of these groups and the conflicts within them so that, in their role as *foederati*, their mobility, which could not be prevented forcibly, would be stymied by their precarious socio-economic situation.

The Roman authority tried to keep barbarian groups with strong ethnic or identity connotation and a monarchical political structure, at least at the top, from becoming an armed minority of landowners. It was impossible for these well-organized fighting groups – expert warriors led by Roman and barbarian warlords – to be immobilized in the subordinate role of *coloni / inquilini* or *tirones*. Nonetheless, it was necessary to keep them from being able to harm Roman landowners by plundering their property or becoming owners of productive lands vielding agricultural income and thus competing with Roman citizens.

---

35 At the higher level of treaties (*foedera*) with the barbarian tribes close to the Empire's borders, the ideal of *pax* is resolved in a supplication to Augustus imposing warlike and often physical 'immobility' on the barbarian group, as can be seen from literary contexts and monetary iconography; cf. Peter Heather, 'Foedera and Foederati of the Fourth century', in *Kingdoms of the Empire. The Integration of Barbarians in Late Antiquity* (Leiden-New York-Köln, 1997); Massimo Palazzi, *«Foedus cum natione». I trattati alle radici dell'Europa. Aspetti delle relazioni fra impero romano e popolazioni barbariche fra II e V secolo* (Gallarate, 2014).

36 The historiographical and archaeological debate on the concept of ethnogenesis/identity of the barbarians is enormous; cf. recently the bibliography cited at note 27 and the contribution of Francesco Borri in this volume.

This dynamic may have had a slight impact in marginal, less Romanized areas, with little concentration of properties owned by Romans (e.g., north of the Loire, northern Spain, Mauritania), but the impact was strong in economically and physically vital provinces, generally closer to the Mediterranean. The transformation of warring barbarians into stable landowners of provincial lands could be, and would have been, a 'revolution', which followed distinct paths in the different provincial areas. The practice of *hospitalitas* was reinterpreted in this discontinuous process.

4 The Visigoths

The case of the Visigoths is emblematic of this process.[37] After the catastrophe of Hadrianopolis in August 378 AD, over the course of the next four difficult years of fighting and negotiating, Emperor Theodosius I slowly regained control of the situation. Unfortunately, we do not know the details of the treaty stipulated in October 382 AD between Theodosius I and the Goths who had settled in the diocese of Thrace. Vital space was assigned to the Goths, who served as allies and suffered grave losses fighting under Theodosius against Maximus in 388 AD and against Eugenius in 394 AD.

However, the convulsive history of the migrations of those Goths within the *pars Occidentis* beginning in 395 and especially from 401 to 418 AD, led by King Alaric and his successors, reveals very clearly that the Visigoths were intruders with no stable bases or productive lands. They did not own the lands on which they were settled in Thrace from 382 AD, so their settlement there had not transformed them into landowners with stable income. The Visigoths would never have roamed around fighting in the Eastern part – and pillaging, too, as a means to pressure Honorius – for over twenty years after the Battle of Frigidus (395–418 AD) through the Pannonias, Italy, Gallia Narbonensis, the Spains and as far as Aquitaine – if they had come into possession of stable productive lands and properties. Constantinople avoided ceding productive Roman lands to provide stable settlement to Gothic landowners and condemned the large barbarian group to seek its fortune in the West, where the military

37 On the long evolution of the increasingly autonomous Visigoth reality up to the settlement in Aquitaine we follow Maria Cesa, *Impero tardoantico e barbari: la crisi militare da Adrianopoli al 418* (Como, 1994); Peter Heather, *Goths and Romans, 332–489* (Oxford 1991), pp. 142–218; Id., 'La diplomazia romana e il problema goto: 376–418', in *Romani e barbari. Incontro e scontro di culture* (*Atti del Conv. di Bra, 11–13 aprile 2003*), ed. Silvia Giorcelli Bersani (Torino, 2004), 160–172; Delaplace, *La fin*, pp. 88–162; Meier, *Geschichte*, pp. 171–224.

crisis offered greater opportunity to negotiate. With the Goths and then the Visigoths, Theodosius I and his successors applied the traditional principle of identifying them as providers of military contingents (*foederati*), but not landowners competing with Roman taxpayers. Only in 418 AD, and especially from 425 AD on, were the Visigoths involved in an irreversible process of hegemony over large territories in Gaul, which must have increased their property.

The failed attempts of the Visigoths – or at least a part of them – and later the success of the Vandals in reaching Africa between 410 and 429 AD can be explained by the difficulty for those large groups of barbarians to have and keep stable bases from which they could obtain agricultural income. Without that, they aimed to settle in Africa, home of the most productive dioceses of the West, those with substantial agrarian income. This goal was particularly necessary for the Visigoths insofar as their numbers had grown significantly during the years 408–410 AD after the defeats and death of Stilicho, as the result of the incorporation of barbarian deserters devoted to the *patricius*, the remains of Radagaisus' troops and fugitive barbarian slaves in Italy. The ethnogenesis of the Visigoths took place through a long process of demographic growth that complicated settlement and intensified its urgency.

With regard to their stabilization, in 411/413, the Visigoths were once again in the service of Honorius against the usurpers Constantine III, Jovinus and Sebastian, but from 414 to 416, they were again in a precarious situation between Bordeaux and Narbonne, and then Barcelona, where King Athaulf paid with his life for his inability to stabilize his own army. In 416, the *patricius* Constantius stipulated an agreement with King Wallia that called for the liberation of Galla Placidia and the use of the Visigothic army as an expeditionary force in Spain, supported by 600,000 *modii* of grain[38] – the traditional system of supplying an army on a mission.

But in 418 AD, the agreement with King Wallia – a new agreement or a revision of the treaty of 416 AD – called for the settlement of Visigoths in the valley of the Garonne between Toulouse and Bordeaux (in the provinces of Novempopulana, Narbonensis Prima and Aquitania Secunda).[39] This may have occurred by the end of the year 419 AD, perhaps after other agreements negotiated between Constantius and the notables gathered in the Council of the

38 Olympiodorus, fr. 30 (ed. Blockley). Cf. Orosius, 7.43.11–12; Prosperus Tiro, *Epitoma Chronicon* 1259 (a. 416); Hydatius, *Chronicon* 60.

39 About the Visigoths after 410 AD and until the settlement of 418 AD, see Delaplace, *La fin*, pp. 155–162 and 167–184, where the previous extensive literature is also indicated.

Seven Provinces.[40] The sources refer specifically to the concession of productive lands, and not simply of accommodation or rations and payment in gold.[41] These lands were divided with the Romans in shares of one third (*tertiae*), a division that seems to have remained the standard for the Visigoths in the fifth century, as Sidonius Apollinaris wrote in 476 AD.[42]

---

40 The distribution of the Visigoths in the rural territories of cities in three neighbouring provinces suggests a desire to distribute the burdens of expropriation among owners-contributors spread over larger areas, and to recalculate among several administrative districts also the fiscal distributions linked to the taking possession of land by new occupants.

41 Philostorgius, *Historia Ecclesiastica* 12.4: '[…] μοῖράν τινα τῆς τῶν Γαλατῶν χώρας εἰς γεωργίαν ἀποκληρωσάμενοι (κλῆρος = *sors*)'; transl: (*The Visigoths*) *were given a part of the rural region of Gaul to divide into 'sortes'* (i.e. allocation of plots of land) *for cultivation* (i.r. to receive incomes from farming); Prosperus Tiro, *Epitoma Chronicon* 1271 (a. 419): 'Constantius patricius pacem firmat cum Wallia data ei ad inhabitandum secunda Aquitanica et quibusdam civitatibus confinum provinciarum'; *Chronica Gallica anni CCCCLII*, 73: *Aquitania Gothis tradita*; transl.: *The patrician Constantius signs a peace agreement with King Wallia and Aquitanica Secunda, together with some cities along the borders of the provinces, is assigned to the king to live*; Hydatius, *Chronicon* 69 (418 AD): 'Gothi, intermisso certamine quod agebant per Constantium ad Gallias revocati, sedes in Aquitanica a Tolosa usque ad Oceanum acceperunt'; transl.: *The Goths, having interrupted the military campaign they were conducting* (*in Spain*) *for Constantius, were recalled by the latter to Gaul, and places to settle in Aquitanica, from Toulouse to the Ocean, were granted to them.* The agreement seems to take the form of a *foedus*, according to the reference in the verses of Sidonius Panegyric for the Emperor Avitus (January 1st 456 AD), *Carmen* 7, vv. 306–315: the renewal in 439 of the foedus with the Visigoths by the future Augustus, then praetorian prefect of Gauls is mentioned in these verses.

42 *Codex Euricianus*, cc. 277 and 276 (ed. D'Ors, p. 21). This is a fundamental testimony on the taking possession of the share of the fund, the *traditio corporalis*, on which cf. Aude Laquerrière-Lacroix, 'La vicinitas à l'épreuve du droit: la mutation des iura vicinitatis dans l'Antiquité Tardive', in *Droit, religion et société dans le Code Théodosien*, eds. J.-J. Aubert – P. Blanchard (Neuchâtel, 2009); Francesco Fasolino, 'Forme di pubblicità degli atti di alienazione nel tardo antico: dall'intervento dei vicini all'*insinuatio apud acta*', in *Disciplina degli atti negoziali inter vivos nelle fonti di IV–VII secolo in Occidente* (Sant'Arcangelo di Romagna, 2019). *Lex Visigothorum* X,1,8 e X,1,16 (ed. Zeumer, pp. 385–386 and 389). In the attached letter to the poem in honour of King Euricus, composed in Bordeaux in 476 AD, Sidonius begs his high-ranking friend Lampridius to intercede for the restitution of at least the one-third share of the property of his mother-in-law, wife of the Emperor Avitus, a property evidently confiscated by the king and which Sidonius would be content to recover even if only according to the canonical division of one-third shares (*Epistulae* 8.9.2): '[…] necdum enim quicquam de hereditate socruali vel in usum tertiae sub pretio medietatis obtinui'; transl.: *I got nothing from my mother-in-law's inheritance* (properties likely assigned to a Gothic warrior) *not even the enjoyment of the third part, and by renouncing half of my inheritance*; cf. Ralph Mathisen, 'A Prosopography of Sidonius', in *The Edinburgh Companion to Sidonius Apollinaris*, eds. G. Kelly – J. van Waarden (Edinburgh, 2020), p. 128.

Paulinus Pellaeus' account of his misfortunes, apparently illustrating events of 414 AD, or at least prior to 418 AD, is relevant to the transition from temporary *hospitalitas* to long-term settlement. Paulinus describes the Visigoths as *hospites* in the city of Bordeaux in 414 AD, and thus still quartered in the Roman city according to the dynamics of the *metatum*. However, those same *foederati* troops also responded to the pressure of the Roman army by pillaging and burning Bordeaux and laying siege to the city of Bazas before withdrawing *en masse* beyond the Pyrenees. Bazas, in turn, was defended and liberated by the *foederati* Alans, who were briefly *hospites* there before moving to other areas where they hoped to be settled definitively through the assignment of *sortes*.[43]

At least until a stable solution was reached in 418 AD, the Visigoths, like the Burgundians and Alans, also lived in Gaul for a long time, quartered precariously according to the institution of the *hospitalitas* / *metatum*. However, the sources, though they are few, describe a perhaps precocious but certainly progressive and irreversible shift from precarious military hospitality to possession of shares of productive Roman properties and spaces for settlement of the families of barbarian soldiers.[44] In the case of the Visigoths, this solution

43 Paulinus Pellaeus, *Eucharistichon* 281–290: 'allicerent et contra animum suetudo quietis, / otia nota, domus specialia commoda plura, / omnibus heu nimium blandis magnisque refertae / deliciis cunctisque bonis in tempore duro, / hospite tunc etiam Gothico quae sola careret'; transl. *on the other hand, I was attracted, against my will, by the habit of tranquillity, the well-known comforts, the many particular comforts of my house, full, alas, of seductive delights for that hard age, all the more so since it was then the only house* (in Bordeaux) *that did not have to accommodate Goths*; and 395–398: 'Cuius non sero secuti / exemplum et nostri quos diximus auxiliares / discessere, fidem pacis servati parati / Romanis quoquo ipsos sors oblata tulisset'; transl.: *Soon this example was followed by ours, those whom we called auxiliary troops (Alani), who left the camp determined to respect the peace stipulated with the Romans wherever the lands (dwellings?) they had been given in fate had led them*; on the latter expression cf. *Novella Theodosii* 25 (444 AD): 'Cum inter multa privilegia, quae militaribus detulit viris antiqua provisio, habitaculis etiam eorum certae portionis domorum sorte prospexerit', [...]; transl. (Pharr, p. 512): *Since, among the many privileges which ancient provisions conferred upon military men, they also provided for their lodgings by the allotment of a definite portion of houses* [...]. On this historical context cf. Sergej Perevalov, 'Bazas 414: la rupture de l'alliance alano-gothique' (transl. from Russian), in *Dialogues d'Histoire Ancienne*, 26 (2000); Helmut Castritius, 'Barbaren ante portas: die gentes zwischen Beutemachen und Ansiedlung am Beispiel von Bazas', in *Millennium*, 6 (2009).

44 In his later years (ca. 425/455 AD) Paulinus of Pella himself mentions the situation of his properties as being divided with Goths who received agricultural income, cf. *Eucharisticon* 498–502: 'quae mihi iam derant natis abeuntibus a me, / non equidem paribus studiis nec tempore eodem, / succensis pariter sed libertatis amore, / quam sibi maiorem contingere posse putabant / Burdigalae, Gothico quamquam consorte colono'; transl. *These*

was negotiated with the Roman authority, as proven by an epistle by Honorius issued in Ravenna on April 17th 418 AD and directed to Agricola, the praetorian prefect of the Gauls. In it, Honorius called for the regular, annual meeting (between 13th August and 13th September) of the *concilium* of the *Septem provinciarum* diocese, expanded and directly affected by Visigoth settlement.[45] The stabilization of the Visigoths required the coordination of the aristocracies of the diocese under the leadership of the prefect of the Gauls. It is probably in anticipation of this novelty and of the debate within the Diocesan Council in Arles that Rutilius Namatianus returned from Rome to the region of Toulouse, where had owned large tracts of land.[46] The settlement of the Visigoths probably took place through expropriation by shares, which must have constituted limits on eventual forms of out-of-control settlements.

The Roman authority hoped to avoid the type of uncontrolled appropriations that had occurred a few years earlier, between 409 and 411 AD, during the first phase of the settlement of Vandals, Alans and Suebi and other troops in Spain.[47] The basis for the allocation to the Visigoths seems to have been

---

(hopes) were dashed when my sons left, not at the same time and with the same intent, but both out of a strong desire for independence which they believed they could achieve better in Bordeaux, even though they had to share their property with the Goths who had settled there. This is similar to the situation described by Sidonius years later for his properties probably in the same region (see above fn. 42). Cf. furthermore the *Passio S. Vincenti Agenatensis*, 6 (ed. de Gaiffier, p. 180): 'Nam veteres illius loci incolas gentis Goticae et generationis unus precipue dominatus sub hospitalitatis (*MS:* hostilitatis) sorte omnis proprietatis iure privaverat'; transl.: *It was precisely the exclusive domination of the Gothic population within a single generation that had deprived the inhabitants of those places of their entire property rights as an allotment of hospitality.*

45 *Epistula Arelatensis genuina* 8 (ed. Gundlach, p. 13). On Agricola cf. *PLRE* II, pp. 36–37.

46 On Rutilius Namatianus cf. *PLRE* II, pp. 770–771; on his journey cf. Cristina Corsi, 'Travelling first class ? Emperors, dignitaries and intellectuals on the roads of the Roman Empire', *L'Antiquité Classique*, 88 (2019).

47 Cf. the long description by Orosius, 7.40, 3–10 (409 AD), in part. 9–10: 'Igitur Honoriaci inbuti praeda et inlecti abundantia, quo magis scelus inpunitum foret atque ipsi sceleri plus liceret, prodita Pyrenaei custodia claustrisque patefactis cunctas gentes, quae per Gallias vagabantur, Hispaniarum provinciis inmittunt isdemque ipsi adiunguntur: ubi actis aliquamdiu magnis cruentisque discursibus, post graves rerum atque hominum vastationes, de quibus ipsos quoque modo paenitet, habita sorte et distributa usque ad nunc possessione consistunt'; transl.: *Therefore the Honoriaci, being full of prey and seduced by abundance, in order that their crime might remain more surely unpunished, and to enjoy greater licence, having abandoned the guard of the Pyrenees, and leaving the passes free, they brought into the Hispanic provinces all the peoples who wandered in Gaul, whom they themselves joined. Having for some time carried out great and cruel raids, and after great devastation of goods and men, for which they themselves are sorry, having made a division by lot and distributed the lands which they still possessed, they settled in Spain*; Hydatius,

inspired by the shares governing *hospitalitas*. The most likely hypothesis is that it was the Roman negotiators of Emperor Honorius – and so, the Roman authorities – who took the first step towards transforming the shares of *hospitalitas*, traditionally temporary urban accommodation, into shares for the possession of lands. The process may have been planned and realized by Emperor Honorius and by the *patricius* Constantius, in agreement with the praetorian prefect of the Gauls in Arles, in the period of the Hispanic war of the Visigoths under Wallia (416/418 AD). Regardless, the institution of *hospitalitas/metatum* underwent a metamorphosis, and the land allocated along the Garonne River represented an initial core of lands capable of producing stable income for the Goths, fresh from their victories in Spain.

It is difficult to say whether this Visigothic allocation in the 'hinterland' of Gaul – far from the Mediterranean and far from Arles, the capital of Gauls – was understood at the time as a non-definitive location. In any event, after the death of Constantius and Honorius (421/423 AD), that core Gothic territory represented an important innovation. Then, by the 440s and 450s, the division of productive lands in Gaul (*partitio*), in forms mediated to a degree with the local Roman aristocracy, had become the standard practice.[48] The land base distributed in 418/419 AD formed a core of resources that supported the

---

Chronicon 49 (411 AD): 'Subversis memorata plagarum grassatione Hispaniae provinciis, barbari ad pacem ineundam, Domino miserante, conversi, sorte ad inhabitandum sibi provinciarum dividunt regiones. Gallaeciam Vandali (Asdingi) occupant et Suevi sitam in extremitate Oceani maris occidua. Alani Lusitaniam et Carthaginiensem provincias, et Vandali cognomine Silingi Baeticam sortiuntur. Hispani per civitates et castella residui a plagis, barbarorum per provincias dominantium se subiciunt servituti'; transl. *While the provinces of Spain were being ruined by the blows of those scourges, the barbarians, by the mercy of God determined to negotiate peace, divided up the areas of the provinces to inhabit them. The Vandals (Asdingi) occupied Galicia, the Swabians its westernmost part, on the edge of the Ocean Sea. The Alans the provinces of Lusitania and Carthage, and the Vandals called Silingi divide the Betica into lots. The Hispanics (= Romans) scattered in the cities and castles, survived the scourges and submitted to the servitude of the barbarians who ruled in the provinces.*

48 *Chronica Gallica CCCCLII*, 124 (440/444 AD): 'Deserta Valentinae urbis rura Alanis, quibus Sambida praeerat, partienda traduntur'; transl.: *Abandoned lands of the city of Valence are assigned to the Alans, led by Sambida, to share them (with the Romans?)*; 127 (442 AD): 'Alani, quibus terrae Galliae ulterioris cum incolis dividendae a patricio Aetio traditae fuerant, resistentes armis subigunt et expulsis dominis terrae possessionem vi adipiscuntur'; transl.: *The Alans, who had been given by the patrician Aetius lands in Gaul further to share with the residents (Romans), making those opposing, acquired them with arms and expelled the owners (Romans) obtained by violence the ownership of the land*; 128 (443 AD): 'Sapaudia Burgundionum reliquiis datur cum indigenis dividenda'; transl.: *Savoy is given to the remains of the Burgundians, to be shared with the inhabitants born there (= Romans)*; Marius Aventicensis, *Chronicon* a. 456: 'Eo anno Burgundiones partem

Visigoths through the two phases of their creation of an autonomous kingdom in Gaul: consolidation from 418 to 439 AD, a phase rendered unstable by Aetius' policies in Gaul, and expansion, between 455 and 477 AD during the irreversible political crisis suffered by the Empire of the West after the death of Valentinian III.[49]

5 The Burgundians

The experience of fifth-century Gaul reveals more details of the evolution of *hospitalitas* into a new mechanism of settlement in areas that were increasingly isolated from the ancient centre of gravity of the Mediterranean.[50] The stratification of the forms of settlement of the Burgundians has been acutely analyzed by Ian Wood, who identified at least six phases in the settlement of Burgundian soldiers in the fifth century.[51] Compared to the large Visigoth army, the Burgundians around the mid-fifth century were a minor and weak fighting group. In 413 AD, the usurper Jovinus had granted them a settlement in Worms (Germania Prima between Mainz and Speyer) as *foederati intralimitanei*, but in 443 AD, they were decimated by Aetius and moved to Sapaudia, a more marginal area. After their enemy Aetius died in 454 AD and the Theodosian dynasty came to an end in the West, during the reign of the Gallic Emperor Avitus, the Burgundians and the Gallic senators, oppressed by the tributes they

---

Galliae occupaverunt terrasque cum Gallis senatoribus diviserunt'; transl.: *That year the Burgundians occupied part of Gaul and divided the land with the Gallic Senators.*

49 Ralph Mathisen and Hagith S. Sivan, 'Forging a New Identity: the Kingdom of Toulouse and the Frontiers of Visigothic Aquitania (418–507)', in *The Visigoths. Studies in Culture and Society*, ed. A. Ferreiro (Leiden-Boston-Köln 1999).

50 This is the agreeable part of the reasoning of Michael Kulikowski, 'Sundered Aristocracies, New Kingdoms, and the End of the Western Empire', in *Gallien in Spätantike und Frühmittelalter. Kulturgeschichte einer Region*, eds. S. Diefenbach – G. Müller (Berlin 2013).

51 Cf. Ian Wood, 'Ethnicity and the Ethnogenesis of the Burgundians', in *Typen der Ethnogenese unter besonderer Berücksichtigung der Bayern*, eds. H. Wolfram – W. Pohl (Wien, 1990) and above all Id., 'L'installation des Burgondes dans l'Empire romain: histoire événementielle', in *Expropriations et confiscations dans les royaumes barbares. Une approche régionale*, eds. P. Porena – Y. Rivière (Roma 2012) (with previous bibliography). In his opinion, «on peut distinguer dans la loi LIV du *Livre de Constitutions* au moins six phases: le régime de l'hospitalité; les donations des *agri cum mancipiis* de Gundioc; les donations de Chilpéric; celles de Gondebaud (qui sont trois phases distinctes); l'allocation des *terrae cum mancipiis*; les réformes de la loi LIV»; then he concludes (p. 89) that the settlement was «une série de donations de modalités variées». Recently Merle Eisenberg, 'A New Name for a New State: the Construction of the Burgundian Regio', in *The Fifth Century, age of transformation*, eds. J.W. Drijvers – N. Lenski (Bari 2019).

were forced to pay, agreed on a new form of settlement in the Rhône valley.[52] In 456/457, the Burgundians under King Gundioc, allied with the Visigoths, expanded further with new settlements in the province of Lugdunensis Prima and central-west Gaul.[53]

After the parenthesis of the reign of Emperor Majorian (458–460 AD), the Burgundian presence in central Gaul was consolidated through agreements and military successes for almost forty years, not least thanks to the protection of King Gundioc (*magister per Gallias* 461/463 AD) and his son, the future King Gundobad, provided by the *magister militum* Ricimer in the years 461–472 AD.[54] In July 472 AD the *magister* Gundobad defeated the emperor Anthemius in Rome and later deposed his magisterium: from that moment (474 AD) he ruled over Burgundians, though with powerful competition from his family members especially his brother Godegisel, who was only defeated in 500 AD after a bitter civil war. Thus, for approximately thirty years, Gundobad was forced to adopt a policy based on the assignation of properties to his Burgundian and foreign supporters, carry out confiscations at the expense of his Roman and barbarian enemies, establish alliances with Ostrogoths and Franks and exploit the reduction of the Visigoth army after the Battle of Vouillé in 507 AD – all operations that involved the redistribution of property and adjustments in the distribution of lands.[55] All this fluctuation in the system employed for settlements reflects the fluctuation in the size of the territory and number of the

52 Marius Aventicensis, *Chronicon* a. 456 (nt. 48); cf. Ralph W. Mathisen, *Roman Aristocrats in barbarian Gaul. Strategies for survival in an age of transition* (Austin Tex., 1993).

53 *Auctarium Haunaunense Prosperi* a. 457 (ed. Mommsen, p. 305): 'Gundiocus rex Burgundionum cum gente et omni praesidio annuente sibi Theudorico ac Gothis intra Galliam ad inhabitandum ingressus societate et amicitia Gothorum functus'; transl.: *Guondioc the king of the Burgundians with all his people and his army with the permission of Theoderic and the Goths entered Gaul to inhabit it, having made a treaty of alliance with the Goths*. Cf. Fredegarius, *Historia Francorum* 2.46.

54 Cf. Fabrizio Oppedisano, 'Il generale contro l'imperatore: la politica di Maioriano e il dissidio con Ricimero', in *Athenaeum*, 97 (2009); Friedrich Anders, *Flavius Ricimer. Macht und Ohnmacht des weströmischen Heermeisters in der zweiten Hälfte des 5. Jahrhunderts* (Bern – Frankfurt am Main 2010), pp. 333–370. Wood, 'L'installation', tends to date the beginning of the settlement practices mentioned in *Liber Constitutionum* LIV to 461 AD, at the same time as the *magisterium per Gallias* of Gundioc.

55 Wood, 'L'installation', pp. 87–89, assumes that law LIV dates from 500 AD and depicts the assignment of *agri cum mancipiis* after 474 AD, of *terrae cum mancipiis* later; the law would not recall the settlement patterns in Sapaudia (443 AD), on the Rhone valley (456/457 AD, above note 49), the assignments of Gundioc in Lyonnais in 461 AD, these of Chilpericus, but would focus on those of Gundobad after 474 AD. When the law was composed, it was necessary to make attention first to the rewards for the fighters in the campaign against Anthemius in 472 AD, then for the king's partisans in the civil war against Godegisel in 500 AD.

Burgundians and their allies who received lands starting around the mid-fifth century up to the beginning of the sixth century.

Also in the Burgundian case, at a certain point, most likely in the middle of the fifth century – the descent to Rhône valley in 456–457 AD seems to be the crucial turning point – the settlement on provincial lands was negotiated on the basis of shares based on the *hospitalitas / metatum*, that is, with property divided in one-third shares. In our opinion, the echo of the more ancient modes of settlement to which Gundobad refers in *Liber Constitutionum* LIV provides for the original division of property between Romans and Burgundians (*populus noster*) in shares of one third. [56]

It seems likely that the application of *tertiae* shares to lands to be divided between barbarians and Romans eliminated any definitive expropriation of the property of Romans. The Roman authority, which probably sought to defend the property of Roman taxpayers and maintain the barbarian warriors in a precarious position (as in the Visigothic settlement) or the Gallo-Roman aristocrats, who had to protect themselves against the barbarian soldiers even though they had chosen to give preference to a weak fighting group (as in the Burgundian settlement of 456/457), may have used the shares of the traditional *metatum*. They would have been applied in the allocation of productive lands to barbarians as they imposed a limit on the area that could be assigned to the barbarian *hospes* on the basis of a share well-known by both Romans and the barbarian warriors. Furthermore, such an allocation attested that the share which now was assigned to the barbarians as *sors* was originally property of Roman citizens. The *metatum* was one of the few legislative devices that provided instruments for dividing private property according to a well known and traditional criterion. During the fifth century, there was an illusory expectation that no Roman property, whether of homes or of rural residences and lands, would pass definitively to the barbarians, especially in the case of barbarian soldiers and heads of family who died without heirs.

---

56 On the inalienability of the *sortes* assigned to Burgundian warriors cf. *Liber Constitutionum* I, 1 (ed. De Salis, p. 41) '[…] decrevimus, ut patri etiam antequam dividat de communi facultate et de labore suo cuilibet donare liceat, absque terra sortis titulo adquisita […]'; transl.: *We decree that the father, before he divides, may donate from the common property and its revenues, except for the land he has received by way of sors*. The primary source on Burgundian shares related to shares stated in the *hospitalitas* system is *Liber Constitutionum* LIV, 1 (ed. De Salis, p. 88 f.): it regulates their relationship to subsequent regal donations based on different apportionments. The reference to the *hospitalitas* shares remains a guiding criterion in the retrospective of Gundobad's legislation issued after 500 AD, cf. *Liber Constitutionum* LIV, 2 (ed. De Salis, p. 90). In 508 or 524 AD, either King Gundobad or King Sigismund allocated land according to a 50% division with the Roman owner (*Liber Const., extravagantes* XXI,12; ed. De Salis, p. 121).

It should be pointed out that the terminology used in the sources regarding *sortes* and *tertiae* in the settlement of Visigoth and Burgundian groups in the Gauls – even when it mentions shares of a third of lands and slaves, parallel to the one-third shares of the Roman regime of *hospitalitas*, and even when it uses the very term "*hospitalitas*" – was not exactly proposing the older Roman institution of *hospitalitas*, designed to fully defend taxpayers' private property. Instead, this terminology describes the practice of actually transferring property from Romans to barbarians. In this context, during the fifth century, the *sortes* are the new full properties assigned by the barbarian king to his warriors; the *tertiae* are the portions of property of many Romans expropriated and assembled to constitute the *sortes*, which the barbarians could leave to their descendants.

This applied not only to shares of residential units, as provided by the *hospitalitas/metatum*, but also to shares of income-producing rural properties. This type of acquisition was a crucial new phenomenon since it made possible a definitive process of stabilization as well as the formation of a landowning military barbarian elite committed to the defence of provincial areas. This historical process accelerated after the end of the Theodosian dynasty in the West, when Ravenna's control over the prefecture of the Gauls had weakened. During the fifth Century the best-organized ethnic groups overcame the risk of being imprisoned in the regime of the *colonate* or of being bound to treaties that called for military service in temporary sites in exchange for *annonae*, a tax paid in kind. Their clear goal was to become landowners with a stable and regular agricultural income.

6 The Ostrogoths

Settlement on provincial lands according to shares inspired by *hospitalitas / metatum* was not a short-lived experiment; nor was it limited to the Gauls. It was the situation in Italy as well, a diocese where, during the fifth century, the concentration of rural properties was greater, and the control of the imperial authority was probably stronger than in south-central Gaul.

The settlement of the Ostrogoths led by king Theoderic in Italy, which began in 493 AD, the last great settlement of the fifth century, is relatively well-documented.[57] After Theoderic's victory and murder of Odoacer in March 493

57 For the settlement of the Ostrogoths in Italy with the assignment of concrete land properties, see Pierfrancesco Porena, *L'insediamento degli Ostrogoti in Italia* (Roma 2012); similar interpretations in Loschiavo, *Oltre la milizia*; Wiemer, *Die Goten in Italien*; Id., *Theoderich*, pp. 198–205; Meier, *Geschichte*, pp. 523–527. Michael Bjornlie, 'Law, ethnicity and taxes

AD, the Ostrogothic king commissioned the praetorian prefect of Italy Liberius to proceed with the settlement of family heads and the families of the fighters in the victorious Ostrogothic army in urban and rural *praedia* confiscated from Odoacer's defeated troops and expropriated from Romans (the latter in addition to lands already taken by confiscation). The settling process was carried out with admirable order and discipline, as the king explained in a letter to the Senate praising Liberius.[58] The most important characteristic of the settlement of the Ostrogoths in the Italian diocese, which started in 493 AD was the strict control exercised by the praetorian prefecture over confiscations and expropriations, distribution of land to the Goths and the assessment, levying and distribution of taxes related to the new property allocations.

In Italy, too, the Goths' new properties were denominated *sortes*, and they were assembled with shares of landed properties expropriated from Romans,

in Ostrogothic Italy. A case for continuity, Adaption and Departure', in *Early Medieval Europe* 22 (2014), pp. 160–162 and 166–167, proposes a settlement of the Ostrogoths on lands belonging to the *res privata* (reorganised in *patrimonium*), lands already assigned by Odoacer to his soldiers; Guy Halsall, 'The Ostrogothic Military', in *A Companion to Ostrogothic Italy*, eds. J.J. Arnold – M.S. Bjornlie – K. Sessa (Leiden-Boston 2016), thinks to a balance between real land assignments and tax rents according to Goffart's theory.

58 Cassiodorus *Variae* 2.16.4–5 (Senatui urbis Romae Theodericus rex - 507/511 AD): '[...] Iuvat nos referre quemadmodum in tertiarum deputatione Gothorum Romanorumque et possessiones iunxit et animos. Nam cum se homines soleant de vicinitate collidere, istis praediorum communio causam videtur praestitisse concordiae: sic enim contigit, ut utraque natio, dum communiter vivit, ad unum velle convenerit. En factum novum et omnino laudabile: gratia dominorum de cespitis divisione coniuncta est; amicitiae populis per damna creverunt et parte agri defensor adquisitus est, ut substantiae securitas integra servaretur. Una lex illos et aequabilis disciplina complectitur. Necesse est enim, ut inter eos suavis crescat affectus, qui servant iugiter terminos constitutos. Debet ergo Romana res publica et memorato Liberio tranquillitatem suam, qui nationibus tam praeclaris tradidit studia caritatis. [...]; transl.: *We are pleased to report how in the allocation of the thirds he merged the possessions and the feelings of the Goths and the Romans. For whereas men are wont to come into conflict because of problems of neighbourhood, for them the commonality of agrarian properties turns out to have been a reason for concord: thus it happens that one and the other nation, living in communion of goods, came to a single will. Here is a new and altogether praiseworthy fact: the favour of the owners met on the division of the land, the friendship between the peoples grew through losses, and with a portion of the field a defender was gained, so that the security of the whole estate was preserved intact. One law and one impartial rule binds them. For it is necessary that affection should grow sweet among those who mutually observe the established limits. The Roman community owes its tranquillity also to the remembered Liberius, who has transmitted the love of mutual charity to such illustrious nations* (for a different translation see Bjornlie, *The Variae*, pp. 93–94). On Liberius the patrician cf. *PLRE* II, pp. 676–681; Bruno Dumézil, 'Le patrice Liberius: développment et redéploiement d'un réseau dans la première moitié du VIe siècle', in *Échanges, communications et réseaux dans le haut Moyen Âge*, eds. A. Gautier – C. Martin (Turnhout 2011).

known as *tertiae* (perhaps within the maximum share of one third of the property of an individual Roman citizen in a given civic territory). In the opening of his letter praising the praetorian prefect in charge of settlement, Theoderic refers to that process as a *deputatio tertiarum*, an 'allocation of thirds'. As this part of the letter is clearly describing a distribution of assets and landed properties to the Goths, the *tertiarum deputatio*, whose method (*quemadmodum*) was praised by Cassiodorus-Theoderic, must have been a distribution of lands, and thus there must have been shares of land identified in the *tertiae*.

In the context of the king's letter to the Senate, the expression suggests that the *tertiae* is the limit to expropriations imposed by the prefect Liberius in defence of Roman property. The king intends to exalt the assignment (*deputatio*) to the Goths, not brutally, of the Romans' properties – a *deputatio praediorum* or *possessionum* or *agrorum* – nor does he want to celebrate the assignment of the Goths' new property – a *deputatio sortium*. Instead, he wants to present the successful transferral as an assignment to the barbarians of the maximum share of property that could be taken away from the lands of the Romans; this share (*tertiae*, always plural) constituted an insuperable limit to guarantee the integrity of Romans' assets and was probably based on a traditional measure for settlement processes in the West.

This term is important because it sheds light on the earlier and first settlement of barbarian warriors in Italy seventeen years before the settlement of the Ostrogoths encouraged by Theoderic and overseen by Liberius. In the case of the Ostrogoths, a passage from *Bellum Gothicum* by Procopius of Caesarea provides further clarification of the adjective-turned-noun *tertiae*. In a well-known section, the Byzantine historian describes the events that led to the first distribution of productive lands to barbarian fighters on Italian soil, a process carried out by Odoacer in 476 AD.[59] At that time, Italy was the only

---

59 Procopius, *Bellum Gothicum* 1(5).1.4–8 (on the year 476 AD): 'ὥστε αὐτοὺς ἀνέδην ἄλλα τε πολλὰ οὔ τι ἑκουσίους ἠνάγκαζον καὶ τελευτῶντες ξύμπαντας πρὸς αὐτοὺς νείμασθαι τοὺς ἐπὶ τῆς Ἰταλίας ἀγροὺς ἠξίουν. ὧν δὴ τὸ τριτημόριον σφίσι διδόναι τὸν Ὀρέστην ἐκέλευον, ταῦτά τε ποιήσειν αὐτὸν ὡς ἥκιστα ὁμολογοῦντα, εὐθὺς ἔκτειναν. ἦν δέ τις ἐν αὐτοῖς Ὀδόακρος ὄνομα, ἐς τοὺς βασιλέως δορυφόρους τελῶν· ὃς αὐτοῖς τότε ποιήσειν τὰ ἐπαγγελλόμενα ὡμολόγησεν, ἤνπερ αὐτὸν ἐπὶ τῆς ἀρχῆς καταστήσωνται. οὕτω τε τὴν τυραννίδα παραλαβὼν ἄλλο μὲν οὐδὲν τὸν βασιλέα κακὸν ἔδρασεν, ἐν ἰδιώτου δὲ λόγῳ βιοτεύειν τὸ λοιπὸν εἴασε. καὶ τοῖς βαρβάροις τὸ τριτημόριον τῶν ἀγρῶν παρασχόμενος τούτῳ τε τῷ τρόπῳ αὐτοὺς βεβαιότατα ἑταιρισάμενος τὴν τυραννίδα ἐς ἔτη ἐκρατύνετο δέκα'; transl. (Dewing, pp. 5–7): *so that the barbarians ruthlessly forced many other measures upon the Romans much against their will and finally demanded that they should divide with them the entire land of Italy. And indeed they commanded Orestes to give them the third part of this, and when he would by no means agree to do so, they killed him immediately. Now there was a certain man among the Romans named Odoacer, one of the bodyguards of the emperor, and he at that time agreed to carry*

diocese in the West where this type of allocation had not already taken place. Discussing the first violent demand by barbarians in Italy for settlement with allocation of landed properties, Procopius does not state that Odoacer's barbarians obtained a third of all the lands of Italy, which they had claimed in vain from Orestes, but simply that they received from Odoacer a one-third share of lands (τοῖς βαρβάροις τὸ τριτημόριον τῶν ἀγρῶν παρασχόμενος), enough to satisfy their needs (τούτῳ τε τῷ τρόπῳ αὐτοὺς βεβαιότατα ἑταιρισάμενος).

The Scirian king seems to have planned and carried out a settlement in Italy of barbarian soldiers with a one-third share, but limited expropriations to the immediate satisfaction of the soldiers in revolt. It is likely that this 'third', τὸ τριτημόριον, does not represent the portion of all lands expropriated in Italy, but rather the maximum fraction of property that could be expropriated from a Roman citizen to create all or part of the lot destined for the barbarian soldier (his *sors*).[60] If we consider that the adjective-turned-noun *tertiae* used by Theoderic/Cassiodorus in the letter to the Senate of Rome (*deputatio tertiarum*) implies a feminine plural noun, there could be a close linguistic connection between the noun composed by Procopius, τὸ τριτημόριον, that is, 'the third of three parts', and the *tertiae* of Cassiodorus.

Procopius's τριτημόριον presupposes the existence of τρίται μοῖραι, that is, 'third parts', of which only one, 'the third', τὸ τριτημόριον, can be expropriated and allocated; indeed, τρίται μοῖραι can be translated into Latin as *tertiae partes*. These *tertiae partes* would correspond well to the "third parts of property that can be expropriated from Roman citizens" of the *deputatio tertiarum* cited in the letter praising the ex-praetorian prefect Liberius. Due to a lexical simplification, which was not a new practice – indeed it was common – for the Late Roman administration, the Greek periphrasis to indicate "the third part of three parts of the property of Roman citizens that can be expropriated" would be reduced to the singular noun τριτημόριον, while the corresponding periphrasis in Latin would be reduced to the plural adjective-made-noun *tertiae*.

out their commands, on condition that they should set him upon the throne. And when he had received the supreme power in this way, he did the emperor no further harm, but allowed him to live thenceforth as a private citizen. And by giving the third part of the land to the barbarians, and in this way gaining their allegiance most firmly he held the supreme power securely for ten years. On Odoacer's reign cf. now Elena Caliri, *Praecellentissimus rex. Odoacre tra storia e storiografia* (Messina 2017); on the year 476 AD cf. Arnaldo Marcone, *L'ultimo anno dell'impero. Roma: 476* (Roma 2021).

60 There is no evidence to establish with certainty whether in 476 AD the one-third share expropriated in Italy corresponded to one-third of the total property of the individual expropriated Roman citizen included in a given administrative district, as seems likely, or to one-third of a production unit, or to one-third of the entire property of the individual expropriated Roman citizen.

This analysis suggests that there is a direct link, on an administrative and linguistic level, between the τριτημόριον assigned in Italy to Odoacer's warriors in 476 AD – the first settlement of barbarian fighters in Italy – and the *tertiae* assigned seventeen years later to Theoderic's warriors in 493 AD in the same diocese – at the time of the second settlement of barbarian warriors in the diocese. The maximum share of one third of the property that could be expropriated from Roman citizens referred to by Procopius (τριτημόριον) suggests that the Roman authority was already involved in the procedures of expropriation and distribution of properties in 476 AD. That share appears to be a reassuring aspect, although it is impossible to evaluate the degree and impact of Odoacer's expropriations. Similarly, it can be supposed that, given the strategic limits imposed by the defence of the diocese, the warrior-grantees of 476 AD were settled more or less in the same areas that would later host the Ostrogoths, who had the same military function, in 493 AD (i.e. in central and northern Italy, from Tuscia, Valeria, Flaminia to the Alps). This means that, in 476 AD in Italy, not all Roman owners had property expropriated and that expropriations were only carried out in certain civic territories, as in 493 AD and thereafter.

The crucial fact that emerges from Procopius' explanation of the events of 476 AD is Odoacer's establishment of a specific fraction of land that could be expropriated from Romans to form the new properties of the barbarians. That one-third share (τριτημόριον) seems to have been maintained by the prefect Liberius as the maximum limit for expropriation and as a generic term to indicate the actual allocations of landed properties to the Ostrogoths in 493 AD as well.

It is likely that the actual dynamic of settlement would often have reduced the calculation of the limit of one third of the property of a Roman citizen held in the city where the expropriations were taking place. The praetorian prefect could be flexible with the shares, depending on the assets of the Romans being expropriated and their location. He could compare the numbers of the expropriation of a given citizen with his holdings in other cities in order to avoid expropriating from the same citizen several times or beyond the established quota.

The existence of a limit on expropriations was a guarantee in itself. Secondly, it is possible that this quota was not so rigid and that the Roman citizen could have less than a third of his property expropriated. In short, the *tertiae* could have represented an imprecise upper limit to expropriations. Thirdly, given that the *sors* for fighting barbarians was drawn entirely within the territory of a single *civitas* under the supervision of the praetorian prefect, it seems likely that a given citizen would only suffer expropriation once, for the formation of a *sors*. In the case of those with properties in more than one city, which was quite common, it is likely that a given Roman citizen would have property expropriated in one civic territory but not in another, with the prefecture keeping a record of expropriations.

Alternatively, the Roman citizen could have property expropriated in different civic territories, but the entire expropriation would not exceed one third of his total assets. In any case, in the settlement of Odoacer's soldiers in Italy in 476 AD and then in that of Theoderic's soldiers in 493 AD, with the allocation to warriors of rural properties (*sortes*) that produced income, we see again the one-third share of division (*tertiae*) that had been the law and practice under *hospitalitas / metatum*. Examining the dynamics of the stabilization of the Visigoths and Burgundians in Gaul, it will be seen that the one-third share is the standard of reference each time a stable settlement of barbarians was negotiated with the Roman authority. The need for a maximum share of expropriation/occupation was greater in areas where the density of cities, of Roman property owners and of rural properties was higher, i.e., south-central Gaul and Italy.

7 The Vandals

The fact that the metamorphosis of the institution of *hospitalitas* – its legal content distorted and its purpose inverted – resulted from actions of the Roman authority is confirmed by the lack of references to one-third shares when the barbarian settlements occurred on Roman provincial lands outside the strict control of the Roman authority. The land of Roman citizens was appropriated by both the Alans sent north of the Loire by Aetius in 442 AD to oppose the Bagaudae and by the (Hasdingi and Silingi) Vandals, Suevi and Alans who penetrated the Iberian Peninsula in 409 AD in the period of the multiple usurpations by Constantine III and Constans, Maximus and Jovinus. These appropriations occurred primarily in areas with large open spaces and settlements where the Roman population had thinned out, also perhaps taking advantage of the small number of barbarians in proportion to the size of the Iberian Peninsula.

The settlements were divided between five different provinces to limit their impact and to separate the armed groups. While they may have been authorized by Maximus, it is possible that the actual taking of possession proceeded with a certain degree of autonomy, as the sources suggest. Indeed, it seems likely that the political crisis made it difficult to uphold limits to their occupation.[61] Luckily, there is much better documentation concerning the settlement of the Vandals in Africa.

61 *Chronica Gallica CCCCLII*, 127 (442 AD); Orosius, 7.40.3–10; Hydatius, *Chronicon* 49. On the context of this barbarian settlement in Spain, see Delaplace, *La fin*, and now Umberto Roberto, *Il secolo dei Vandali. Storia di un'integrazione fallita* (Palermo 2020)] pp. 36–41.

The confiscations and resulting allocations to the royal patrimony and Vandal soldiers carried out by the Vandals under Geiseric in the diocese of Africa between 439 and 442 AD is an (extreme) case of the total expropriation of entire properties completely outside the control of Roman authorities.[62] Geiseric had spent ten years in Mauretania Sitifensis and Numidia, where he became familiar with the structure of Roman society in the diocese of Africa. With this knowledge, he meticulously planned a 'Blitzkrieg' and conquered Carthage in October 439 AD in a period of the year that precluded any military reaction (an attack by his enemies from the Mediterranean was then impossible).

In a well-prepared and swift manoeuvre, he attacked the resident Roman landowning ruling class, that is, the civic aristocracy and the great *possessores* of the capital and the richest cities in Africa (Proconsularis Zeugitana and Byzacena). He expropriated their entire property, rather than taking shares, banished them and destroyed the land registers to make it impossible to return the confiscated properties to their original owners.[63] The consequences of these actions offer

62 The bibliography on the Vandals in Africa is very large; for the settlement we limit ourselves to referring to the detailed studies by Modéran, 'L'établissement'; Id., *Confiscations*; Id., *Les Vandales*, pp. 155–179; Roberto, *Il secolo*, pp. 77–79; 92–97.

63 Victor Vitensis, *Historia persecutionis africanae provinciae* 1.12–15 (439 AD): 'Quid multa? Post has truces impietatis insanias ipsam urbem maximam Carthaginem Geisericus tenuit et intravit, et antiquam illam ingenuam ac nobilem libertatem in servitutem redegit: nam et senatorum urbis non parvam multitudinem captivavit. Deinde proponit decretum, ut unusquisque auri, argenti, gemmarum vestimentorumque pretiosorum quodcumque haberet offeret, et ita in brevi avitas atque paternas opes tali industria abstulit rapax. Disponens quoque singulas quasque provincias, sibi Byzacenam, Abaritanam atque Getuliam et partem Numidiae reservavit, exercitui vero Zeugitanam vel Proconsularem funiculo hereditatis divisit, Valentiniano adhuc imperatore reliquas licet iam exterminatas provincias defendente. [...] Praeterea praecipere nequaquam cunctatus est Wandalis, ut episcopos atque laicos nobiles de suis ecclesiis vel sedibus nudos penitus aufugarent: quodsi optione proposita exire tardarent, servi perpetui remanerent. Quod etiam in plurimis factum est: multos enim episcopos et laicos, claros atque honoratos viros, servos esse novimus Wandalorum. [...] Tunc vero memoratae urbis episcopum, id est Carthaginis, Deo et hominibus manifestum, nomine Quodvultdeus, et maximam turbam clericorum navibus fractis inpositam nudos atque expoliatos expelli praecepit. Quos Dominus miseratione bonitatis suae prospera navigatione Neapolim Campaniae perducere dignatus est civitatem. Senatorum atque honoratorum multitudinem primo exilio crudeli contrivit, postea transmarina parte proiecit'; transl.: *But what is the point of speaking at length? After this cruel and impious madness, Genseric entered and occupied the main city itself, Carthage, and enslaved that ancient, generous and noble freedom (=decurions): in fact he captured a not inconsiderable number of the city's senators. Then he issued a decree, that each one should hand over all the gold, silver, gems and precious garments he had; and so in short by this expedient he rapaciously took possession of the property of their fathers and forefathers. In disposing also of the individual provinces, he set aside for himself Byzacena and Abaritana and Getulia and part of Numidia; for the army he then divided Zeugitana and Proconsular into hereditary lots, while to Valentinian, who was still*

evidence that the expropriations were not controlled by Romans, and what's more, that they were intended to undermine the power of Emperor Valentinianus

---

emperor, he left the other provinces, which were nevertheless devastated. [...] *Moreover, he did not hesitate at all to order the Vandals to expel the bishops and noble laymen naked from their churches and seats without exception: that if, despite the power granted to them, they were late in leaving, they should remain as servants for ever. This happened with regard to many: many bishops, in fact, and laymen, illustrious and honourable, we know to be servants of the Vandals.* [...] *(Genseric) even ordered that the bishop of the aforementioned city, Carthage, named Quodvultdeus, and a very large crowd of ecclesiastics, embarked on broken ships, naked and stripped, be chased away. But the Lord in the mercy of his goodness deigned to bring them to Naples, a city of Campania. A great multitude of senators and the honoured wore them down first with cruel exile, then relegated them to Transmarine territory.* Procopius, *Bellum Vandalicum* 1(3).5.11–17 (439 AD ff.): 'τῶν δὲ Λιβύων εἴ τι μὲν δόκιμον ἐτύγχανεν ὂν καὶ πλούτῳ ἀκμάζον, αὐτοῖς ἀγροῖς τε καὶ πᾶσι χρήμασιν ἐν ἀνδραπόδων μοίρᾳ παρέδωκε τοῖς παισὶν Ὀνωρίχῳ τε καὶ Γένζωνι. Θεόδωρος γὰρ ὁ νεώτατος ἐτελεύτα ἤδη, ἄπαις τὸ παράπαν ἄρρενός τε καὶ θήλεος γόνου. Λίβυας δὲ τοὺς ἄλλους ἀφείλετο μὲν τοὺς ἀγρούς, οἳ πλεῖστοί τε ἦσαν καὶ ἄριστοι, ἐς δὲ τὸ τῶν Βανδίλων διένειμεν ἔθνος, καὶ ἀπ' αὐτοῦ κλῆροι Βανδίλων οἱ ἀγροὶ οὗτοι ἐς τόδε καλοῦνται τοῦ χρόνου. τοῖς δὲ δὴ πάλαι κεκτημένοις τὰ χωρία ταῦτα πένεσθαί τε ὡς μάλιστα καὶ ἐλευθέροις εἶναι ξυνέβαινεν· ἦν δὲ αὐτοῖς ἐν ἐξουσίᾳ καὶ ὅποι βούλοιντο ἀπαλλάσσεσθαι. καὶ τὰ μὲν χωρία ξύμπαντα, ὅσα τοῖς τε παισὶ καὶ τοῖς ἄλλοις Βανδίλοις Γιζέριχος παραδεδώκει, οὐδεμιᾶς φόρου ἀπαγωγῆς ὑποτελῆ ἐκέλευσεν εἶναι. τῆς δὲ γῆς ὅση οἱ οὐκ ἀγαθὴ ἔδοξεν εἶναι, ἀφῆκε τοῖς πρότερον ἔχουσι, τοσαῦτα ἐνθένδε τῷ δημοσίῳ φέρεσθαι τάξας, ὥστε οὐδ' ὁτιοῦν περιῆν τοῖς τὰ χωρία τὰ σφέτερα αὐτῶν ἔχουσιν. ἔφευγον δὲ πολλοὶ καὶ ἐκτείνοντο. αἰτίαι γὰρ αὐτοῖς πολλαί τε καὶ χαλεπαὶ προσεφέροντο. πασῶν δὲ μία μεγίστη δὴ ἐδόκει εἶναι ὅτι χρήματά τις οἰκεῖα ἔχων ἀπέκρυπτεν. οὕτω τοὺς Λίβυας πᾶσα ἰδέα ξυμφορᾶς περιέστη'; transl. (Dewing, p. 51): *And among the Libyans all who happened to be men of note and conspicuous for their wealth he handed over as slaves, together with their estates and all their money, to his sons Honoric and Genzon. For Theodorus, the youngest son, had died already, being altogether without offspring, either male or female. And he robbed the rest of the Libyans of their estates, which were both very numerous and excellent, and day by day them among the nation of the Vandals, and as a result of this these lands have been called "Vandals' estates" up to the present time. And it fell to the lot of those who had formerly possessed these lands to be in extreme poverty and to be at the same time free men; and they had the privilege of going away wheresoever they wished. And Gizeric commanded that all the lands which he had given over to his sons and to the other Vandals should not be subject to any kind of taxation. But as much of the land as did not seem to him good he allowed to remain in the hands of the former owners, but assessed so large a sum to be paid on this land for taxes to the government that nothing whatever remained to those who retained their farms. And many of them were constantly being sent into exile or killed. For charges were brought against them of many sorts, and heavy ones too; but one charge seemed to be the greatest of all, that a man, having money of his own, was hiding it. Thus the Libyans were visited with every form of misfortune.* Procopius, *Bellum Vandalicum* 2(4).8.25 (534 AD): ἐπειδὴ δὲ τῶν ἐπὶ Λιβύης χωρίων τοὺς φόρους οὐκέτι ἦν ἐν γραμματείοις τεταγμένους εὑρεῖν, ἧπερ αὐτοὺς ἀπεγράψαντο ἐν τοῖς ἄνω χρόνοις Ῥωμαῖοι, ἅτε Γιζερίχου ἀναχαιτίσαντός τε καὶ διαφθείραντος κατ' ἀρχὰς ἅπαντα, Τρύφων τε καὶ Εὐστράτιος πρὸς βασιλέως ἐστάλησαν, ἐφ' ᾧ τοὺς φόρους αὐτοῖς τάξουσι κατὰ λόγον ἑκάστῳ. οἳ δὴ οὐ μέτριοι Λίβυσιν οὐδὲ φορητοὶ ἔδοξαν εἶναι; transl. (Dewing, pp. 277–279): *And since it was no longer possible to find the revenues*

III. Not only did the king and his Vandals obtain entire highly profitable agrarian properties, mostly in the Proconsularis Zeugitana – that is, in the heart of the defence of the kingdom – but those properties were exempt from taxation too. The flow of taxes for Ravenna was cut off, and this provided the Vandals in Africa with stability and wealth, while simultaneously weakening Valentinianus III and his successors and impoverishing also the Roman landowners left in the African provinces, who, on the contrary, were taxed by the crown.

The elimination of the only Roman élite – rich, powerful, tied to Ravenna and residing in the African provinces (Proconsularis Zeugitana and Byzacena) – and the persecution of the Catholic Church in Africa impeded the expression of opposition from the large Roman urban and rural population against the barbarian minority of new, recently-settled landowners. The Late Roman society on the far South side of the Mediterranean no longer had its own social and religious leadership. The religious distinction between the Vandals and the Romans of Africa did the rest, ensuring the separation of the few new rich Vandal occupiers from the socially subordinate mass of provincial Romans in Africa.

The most important aspect of this traumatic settlement process was that the Vandal properties (*sortes* = κλῆροι) were 100% intact productive units, not new farms built with the one-third shares (*tertiae*) expropriated from different Roman landowners, as occurred, for example, with the Visigoths and Burgundians in southern Gaul or with the Ostrogoths in Italy.

The nature of the Vandal confiscation set off a positive stabilization process in the kingdom. The productivity of these *sortes* was more than sufficient to support the Vandal army, stationed in only a few African provinces. Along with the tax immunity, these *sortes* enriched the new barbarian nobility, giving it a decisive economic dominance free of Roman competitors. After more than thirty years of precarious survival in the provinces of the West, the Vandals managed to escape any form of Roman control, which they knew all too well, overcoming the limits that Roman power had traditionally sought to impose

of the districts of Libya set down in order in documents, as the Romans had recorded them in former times, inasmuch as Gizeric had upset and destroyed everything in the beginning, Tryphon and Eustratius were sent by the emperor, in order to assess the taxes for the Libyans each according to his proportion. But these men seemed to the Libyans neither moderate nor endurable. Cf. Modéran, 'L'établissement', p. 117: «Genséric a bien détruit des registres fiscaux en 439, mais qui étaient ceux relatifs aux sortes Wandalorum, aux biens donnés à l'Église arienne, et à ceux remis à ses fils, tous, selon Procope (I, 5, 11–12), exemptés d'impôts et situés en majorité (selon Victor de Vita) en Proconsulaire». The expression *sortes Vandalorum* is found in the edicts of king Huneric copied by Victor Vitensis, *Historia persecutionis africanae provinciae* 2.39 and 3.4; cf. Ferrandus, *Vita Fulgentii* 1.1; for the fate of the Roman exiles who sailed from Africa and landed in Italy, cf. *Novellae Valentiniani* 12 (443 AD) and 34 (451 AD).

on barbarian groups – physical and material 'immobility'. Without local negotiations with the Roman civil administration, it was impossible to set quotas for the expropriation and redistribution of property based on the one-third percentage inherited from the regime of *hospitalitas*.

8 Conclusions

In the late Empire the institution of *hospitalitas / metatum* was the obligation to offer accommodation for brief periods to bureaucrats, soldiers and officials moving through the provinces, a practice conceived to defend the integrity of the private property of Roman citizens subject to this *munus*. In opposition to expropriations and confiscations, it was defined as a simple temporary lodging, limited to a one-third share of the private residential property of the citizen. This system of short-term lodging did not include the use of income-producing lands. The documentation dealing with stable, long-term settlements of barbarian groups on Roman property on the basis of one-third shares (*tertiae*) suggests that, in those places where there were negotiations with the Roman authority or with powerful Roman landowners, the barbarians obtained fractions of properties, including agrarian properties, that evoked the fractions of temporary occupation assigned by the Roman administration to the institution of *hospitalitas* (although the legal substance and purpose of the division for the barbarian settlement were different from that system).

The practice of one-third shares of lands and slaves to be divided between the old Roman owners whose property was expropriated and the new barbarian owners is only documented among the Visigoths, Burgundians – at least in one phase of their evolving stabilization – and the Ostrogoths. In effect, the one-third share is documented in areas with a high density of agrarian property owned by Romans and a large resident Roman population – in Gallia Narbonensis, Lyonnaise, Aquitania and Italy. The similarities in the terminology used in independent sources – concerning the Burgundians, Visigoths and Ostrogoths – suggest that a standard base model of division, with variants depending on time and space, prevailed in areas where there was strong competition with Romans for control and distribution of agricultural resources. In such places, the quota was a clearly understood concrete reality, basically accepted by both hosts and guests.

A 'short circuit' transformed the shares of the *metatum* into the *tertiae* of productive lands acquired by barbarians. This change may have been triggered by the Roman authority, which, after the most acute phase of the crisis – the decade 406–416 AD – probably oversaw the most substantial, critical and challenging settlement of the Visigoths in Aquitania using the parameters of the institution of *hospitalitas* practiced until that moment. This time, however, those parameters were used to assign productive rural properties, whose provenance could be traced, on a temporary but long-term basis at a one-third share to barbarian warriors and heads of family.

Over most of the century, from 418 to 493 AD, the process of settlement and appropriation of productive lands for the diverse groups in the various provincial areas of the West was neither unified nor uniform. The legal content of the *hospitalitas / metatum* was distorted and its purpose partially inverted. The episodes of uncontrolled and often violent expropriations by barbarian troops in Spain, by some of the Alans and especially by the Vandals confirm the risks of barbarian settlement without an attempt to impose legal and material boundaries. This metamorphosis of *hospitalitas* opened the way to the definitive passage of property to the barbarians, which encouraged the progressive establishment of a new minority of warrior, land-owning barbarians and, in the long term, the political hegemony of military over civilian power.

Bibliography

Primary Sources

Cassiodorus Senator, *Variae*, ed. Th. Mommsen [MGH, *Scriptores – Auct. ant.* XII] (Berlin, 1894) // Flavio Magno Aurelio Cassiodoro Senatore, *Varie*, (dir. A. Giardina), eds. A. Giardina – G.A. Cecconi – I. Tantillo – F. Oppedisano, 6 vols. (Roma 2014*ff.*)

Cassiodorus, *The Variae. The Complete Translation*, ed. M. Shane Bjornlie (Oakland 2019) (Engl. transl. only)

Codex Euricianus, ed. A. D'Ors, *El Código de Eurico* (Madrid 2014²)

Codex Justinianus, ed. P. Krüger in T. Mommsen – P. Krüger, *Corpus Iuris Civilis*, II *Codex* (Berlin, 1954¹¹) – Engl. transl. B.W. Frier, *The Codex of Justinian. A New Annotated Translation*, I–III, (Cambridge 2016)

Codex Theodosianus: *Theodosiani libri XVI cum constitutionibus Sirmondianis et leges Novellae ad Theodosianum pertinentes*, ed. Th. Mommsen – P.M. Meyer, 2 vols. (Berlin 1905) – Engl. transl. by C. Pharr *The Theodosian Code and Novels and the Sirmondian Constitutions. A Translation with commentary, glossary, and bibliography*, [Princeton 1952], 10th Print (Union N.J. 2001)

Chronica Gallica anni CCCCLII, ed. R. Burgess, 'The Gallic Chronicle of 452: A New Critical Edition with a Brief Introduction', in R.W. Mathisen – D. Shantzer, eds., *Society and Culture in Late Antique Gaul: revisiting the sources* (Aldershot 2001)

Epistolae Arelatenses genuinae, ed. W. Gundlach [MGH, *Epistolae*, III] (Berlin 1892)

Ferrandus Fulgentius, *Vita Fulgentii*, ed. A. Isola [CC s.l.] (Brepols 2016)

Fredegarius, ed. B. Krusch, *Chronicarum quae dicuntur Fredegarii Scholastici libri IV. Cum Continuationibus* [MGH, *Script. rer. Merov.*, 2 *Fredegarii et aliorum Chronica*] (Hannover 1888) // eds. H. Wolfram – A. Kusternig, *Chronicarum quae dicuntur Fredegarii libri quattuor, Quellen zur Geschichte des 7. und 8. Jahrhunderts* (Darmstadt 1982)

Hydatius, *Chronicon*, ed. R. Burgess, *The Chronicle of Hydatius and the Consularia Constantinopolitana: two contemporary accounts of the final years of the Roman Empire* (Oxford 1993)

(Pseudo-)Joshua the Stylite, *Chronicon*, ed. W. Wright, *The chronicle of Joshua the Stylite: composed in Syriac A.D. 507* (Cambridge 1882) – Engl. transl. F.R. Trombley – J.W. Watt, *The Chronicle of Pseudo-Joshua the Stylite* (Liverpool 2000)

Leges Burgundionum, ed. L.R. de Salis [MGH, *Leges nat. Germ.* II.1] (Hannoverae 1892)

Leges Visigothorum, ed. K. Zeumer [MGH, *Leges nat. Germ.* I] (Hannoverae et Lipsiae 1902)

Liber Constitutionum, in *Leges Burgundionum*; Engl. transl. K. Fisher Drew, *The Burgundian code: book of constitutions or law of Gundobad* (Philadelphia 1972)

Marius Aventicensis, *Chronicon*, ed. Th. Mommsen [MGH, *Auct. Ant.* XI] (Berlin 1894)

Mauricius, *Strategikon*, ed. and Engl. transl. G.T. Dennis, *Maurice's Strategikon. Handbook of Byzantine Military Strategy* (*Philadelphia* 1984)

Novellae Iustiniani, ed. R. Schöll – G. Kroll, in *Corpus Iuris Civilis*, III (Berlin 1963[8]) – Engl. transl. D.J.D. Miller – P. Sarris, *The novels of Justinian. A complete annotated English translation*, 2 vols. (Cambridge-New York 2018)

Novella Theodosii, v. *Codex Theodosianus*

Olympiodorus, ed. R.C. Blockley, *The Fragmentary Classicising Historians of the Later Roman Empire, Eunapius, Olympiodorus, Priscus and Malchus*, vol. 2 (Liverpool 1983)

Orosius, *Historia adversus paganos*, ed. M.-P. Arnaud-Lindet, *Orose, Histoires contre les Païens*, 3 vols. (Paris 2003)

Panegyrici Latini ed. R.A.B. Mynors, *XII Panegyrici Latini* (Oxford 1964)

Passio S. Vincenti Agenatensis, ed. B. de Gaiffier, 'La Passion de Saint Vincent d'Agen', in *Analecta Bollandiana*, 70 (1952), pp. 160–181

Paulinus Pellaeus, *Eucharistichus*, ed. H.G. Evelyn White, *Ausonius* II (Cambridge, Mass. 1921)

Philostorgius, *Historia Ecclesiastica*, ed. (and Engl. transl.) Ph.R. Amidon, *Philostorgius Church History* (Atlanta 2007).

Procopius Caesariensis, *De bellis – Tôn Polemôn Logoi*, ed. (and Engl. transl.) H.B. Dewing, *History of the Wars* (Cambridge, Mass. 1919)

Prosperus Tiro, *Epitoma Chronicon*, ed. Th. Mommsen [MGH, *Auct. Ant.* IX] (Berolini 1892)

Sidonius Apollinaris, *Epistulae et carmina*, ed. C. Luetjohann [MGH, *Auct. Ant.* VIII] (Berlin, 1887)

Victor Vitensis, *Historia persecutionis africanae provinciae sub Genserico et Hunrico regibus Wandalorum*, ed. K.F. von Halm [MGH, *Auct. ant.* III.1 (Berolini, 1879) (repr. 2018)

Secondary Literature

Anders, Friedrich. *Flavius Ricimer. Macht und Ohnmacht des weströmischen Heermeisters in der zweiten Hälfte des 5. Jahrhunderts* (Bern – Frankfurt am Main 2010)

Arcuri, Rosalba. 'Etnogenesi, «entelechia barbarica» e attuali orientamenti storiografici sulla Völkerwanderungszeit', in *Koinonia*, 37 (2013), pp. 107–141

Arnold, Jonathan J. 'Ostrogothic Provinces: Administration and Ideology', in *A Companion to Ostrogothic Italy*, eds. J.J. Arnold – M.S. Bjornlie – K. Sessa (Leiden-Boston 2016), pp. 47–72

Barbero, Alessandro. *Barbari. Immigrati, profughi, deportati nell'impero romano* (Roma 2006)

Barnish, Samuel J.B. 'Cuncta Italiae membra componere: Political Relations in Ostrogothic Italy', in *The Ostrogoths. From the Migration Period to the Sixth Century*, eds. S.J.B. Barnish – F. Marazzi (Woodbridge 2007), pp. 317–352

Basso, Patrizia – Zanini, Enrico (eds.), *Statio amoena. Sostare e vivere lungo le strade romane* (Oxford 2016)

Berardi, Caterina C. '«Barbari exsecrati gladios suos ad aratra conversi sunt» (Oros., adv. pag. 7, 41, 7): il tema della pacificazione dei barbari tra IV e V sec. d.C.', in *Auctores nostri: studi e testi di letteratura cristiana antica*, 5 (Bari, 2007), pp. 17–37

Bintliff, John. 'Aspects of settlement change in late Antiquity from regional survey evidence', in *New cities in Late Antiquity: documents and archaeology*, ed. E. Rizos (Turnhout 2017), pp. 13–17

Bjornlie, Michael S. 'Law, ethnicity and taxes in Ostrogothic Italy. A case for continuity, Adaption and Departure', in *Early Medieval Europe* 22 (2014), pp. 138–170

Bjornlie, Michael S. 'Governmental Administration', in *A Companion to Ostrogothic Italy*, eds. J.J. Arnold – M.S. Bjornlie – K. Sessa (Leiden-Boston 2016), pp. 73–97

Bransbourg, Gilles. 'Le grand écart. Les destins divergents de l'Orient et de l'Occident à travers le prisme fiscal de l'Empire tardif', in *Libera curiositas. Mélanges d'histoire romaine et d'Antiquité tardive offerts à J.-M. Carrié*, eds. C. Freu – S. Janniard – A. Ripoll (Turnhout 2016), pp. 67–81

Caliri, Elena, *Praecellentissimus rex. Odoacre tra storia e storiografia* (Messina 2017)

Carrié, Jean-Michel. 'Problèmes de la terre dans l'Antiquité Tardive: réalités archéologiques, structures agraires et normes juridiques", in *Atti dell'Accademia romanistica costantiniana* (AARC), XXII. *Questioni della terra. Società, economia, normazioni, prassi. In onore di M. Bianchini* (Napoli, 2017), pp. 5–36

Castritius, Helmut. 'Barbaren ante portas: die gentes zwischen Beutemachen und Ansiedlung am Beispiel von Bazas', in *Millennium*, 6 (2009), pp. 281–294

Cesa, Maria. *Impero tardoantico e barbari: la crisi militare da Adrianopoli al 418* (Como 1994)

Chauvot, Alain. 'Julien et les lètes. À propos d'Ammien, XX,8,13', in *Libera curiositas. Mélanges d'histoire romaine et d'Antiquité tardive offerts à J.-M. Carrié*, eds. C. Freu – S. Janniard – A. Ripoll (Turnhout 2016), pp. 141–150

Corsi, Cristina. 'Travelling first class ? Emperors, dignitaries and intellectuals on the roads of the Roman Empire', *L'Antiquité Classique*, 88 (2019), pp. 139–177

Cosentino, Salvatore. 'Fine della fiscalità, fine dello stato romano ?', in *Le trasformazioni del V secolo. L'Italia, i barbari e l'Occidente romano*, eds. P. Delogu – S. Gasparri (Turnhout 2010), pp. 17–35

Cosentino, Salvatore. 'Sul transito dei soldati (in merito a Iust. Nov. CXXX)', in *Bulgaria Mediaevalis*, 2 (2011), pp. 25–38

Delaplace, Christine. *La fin de l'Empire romain d'Occident. Rome et les Wisigoths de 382 à 531* (Rennes 2015)

Di Paola, Lucietta. *Viaggi, trasporti, e istituzioni. Studi sul «cursus publicus»* (Messina 1999)

Dumézil, Bruno. 'Le patrice Liberius: développment et redéploiement d'un réseau dans la première moitié du VIe siècle', in *Échanges, communications et réseaux dans le haut Moyen Âge*, eds. A. Gautier – C. Martin (Turnhout 2011), pp. 27–44

Dumézil, Bruno. *Servir l'état barbare dans la Gaule franque: du fonctionnariat antique à la noblesse médiévale, IVe–IXe siècle* (Paris 2013)

Eisenberg, Merle. 'A New Name for a New State: the Construction of the Burgundian Regio', in *The Fifth Century, age of transformation*, eds. J.W. Drijvers – N. Lenski (Bari 2019), pp. 157–167

Fascione, Lorenzo. 'Barbari e lavoro della terra in Occidente da Teodosio I (382) a Odoacre (476)', *Historia et ius*, 11 (2017) – paper 9: http://www.historiaetius.eu/uploads/5/9/4/8/5948821/fascione_11.pdf

Fasolino, Francesco. 'Forme di pubblicità degli atti di alienazione nel tardo antico: dall'intervento dei vicini all'*insinuatio apud acta*', in *Disciplina degli atti negoziali inter vivos nelle fonti di IV–VII secolo in Occidente* (Sant'Arcangelo di Romagna, 2019), pp. 131–143

Gatzka, Friederike. *Cassiodor, Variae 6. Einführung, Übersetzung und Kommentar* (Berlin 2019)

Gaupp, Ernst Th. *Die germanischen Ansiedlungen und Landtheilungen in den Provinzen des Römischen Westreiches in ihrer völkerrechtlichen Eigenthümlichkeit und mit rücksicht auf verwandte Erscheinungen der alten Welt und des späteren Mittelalters* (Breslau 1844)

Geary, Patrick. 'Political Identity, Ethnic Identity, Genetic Identity: the Dangers of Conceptual Confusion', in *Neue Wege der Frühmittelalterforschung. Bilanz und Perspektiven*, eds. W. Pohl – M. Diesenberger – B. Zeller (Wien 2018), pp. 35–41

Goffart, Walter. *Barbarians and Romans A.D. 418–584: the Techniques of Accommodation* (Princeton 1980)

Goffart, Walter. *Barbarian tides. The migration age and the later Roman Empire* (Philadelphia 2006)

Goffart, Walter. 'The Technique of Barbarian Settlement in the Fifth Century: A Personal, Streamlined Account with Ten Additional Comments', in *Journal of Late Antiquity*, 3/1 (2010), pp. 65–98

Goffart, Walter. 'Le début (et la fin) des «sortes Vandalorum»', in in *Expropriations et confiscations dans les royaumes barbares. Une approche régionale*, eds. P. Porena – Y. Rivière (Roma 2012), pp. 115–128

Goffart, Walter. 'Administrative Methods of Barbarian Settlement in the Fifth Century. The Definitive Account', in *Gallien in Spätantike und Frühmittelalter. Kulturgeschichte einer Region*, eds. S. Diefenbach – G.M. Müller (Berlin 2013), pp. 45–58

Grey, Cam. 'Revisiting the problem of agri deserti in the late Roman Empire', in *Journal of Roman Archaeology*, 20 (2007), pp. 362–376

Grey, Cam. 'The «ius colonatus» as a model for the settlement of barbarian prisoners-of-war in the late Roman Empire ?', in *Romans, barbarians, and the transformation of the Roman world. Cultural interaction and the creation of identity in Late Antiquity*, eds. R.W. Mathisen – D. Shanzer (Aldershot 2011), pp. 147–160

Halsall, Guy. *Barbarian migrations and the Roman West, 376–568* (Cambridge-New-York 2007)

Halsall, Guy. 'The Ostrogothic Military', in *A Companion to Ostrogothic Italy*, eds. J.J. Arnold – M.S. Bjornlie – K. Sessa (Leiden-Boston 2016), pp. 177-183

Heather, Peter. *Goths and Romans, 332–489* (Oxford 1991)

Heather, Peter. 'Foedera and Foederati of the Fourth century', in *Kingdoms of the Empire. The Integration of Barbarians in Late Antiquity* (Leiden-New York-Köln 1997), pp. 57–74

Heather, Peter. 'La diplomazia romana e il problema goto: 376–418', in *Romani e barbari. Incontro e scontro di culture* (*Atti del Conv. di Bra, 11–13 aprile 2003*), ed. Silvia Giorcelli Bersani (Torino 2004), 160–172

Heather, Peter. 'Merely an Ideology? – Gothic Identity in Ostrogothic Italy', in *The Ostrogoths. From the Migration Period to the Sixth Century*, eds. S.J.B. Barnish – F. Marazzi (Woodbridge 2007), pp. 31–79

Heather, Peter. *Empires and Barbarians: The Fall of Rome and the Birth of Europe* (Oxford 2009)

Heather, Peter. 'Elite militarisation and the post-roman West', in *Istituzioni, carismi ed esercizio del potere* (*IV–VI secolo d.C.*), eds. G. Bonamente – R. Lizzi Testa (Bari 2010), pp. 245–266

Heather, Peter. 'East & West in the Fifth Century', in *Governare e riformare l'impero al momento della sua divisione. Oriente, Occidente, Illirico*, eds. U. Roberto – L. Mecella (Roma 2015) http://books.openedition.org/efr/2819

Huntzinger, Hervé. *La captivité de guerre en Occident dans l'Antiquité tardive* (*378–507*) / *War captivity in Late Antique Occident* (*378–507*) (Thèse Univ. Strasbourg, École Doctorale des Sciences de l'Homme et des Sociétés, oct. 2009)

Jaillette, Pierre. 'Les dispositions du Code Théodosien sur les terres abandonnées', in J.-L. Fiches, ed., *Le IIIe siècle en Gaule Narbonnaise. Données régionales sur la crise de l'Empire* (Sophia Antipolis 1996), pp. 333–404

Jaillette, Pierre. 'La structure agraire dans l'Empire tardif au miroir du Code Théodosien: quelques observations', in *Antiquité Tardive*, 20 (2012), pp.123–135

Kerneis, Soazick. 'Rome et les barbares: aux origines de la personnalité des lois', in *Civitas, iura, arma: organizzazioni militari, istituzioni giuridiche e strutture sociali alle origini dell'Europa (secc. III–VIII)*, eds. F. Botta – L. Loschiavo (Lecce 2015), pp. 103–116

Kulikowski, Michael. 'Sundered Aristocracies, New Kingdoms, and the End of the Western Empire', in *Gallien in Spätantike und Frühmittelalter. Kulturgeschichte einer Region*, eds. S. Diefenbach – G. Müller (Berlin 2013), pp. 79–90

Laquerrière-Lacroix, Aude. 'La vicinitas à l'épreuve du droit: la mutation des iura vicinitatis dans l'Antiquité Tardive', in *Droit, religion et société dans le Code Théodosien*, eds. J.-J. Aubert – P. Blanchard (Neuchâtel 2009), pp. 245–257

Loschiavo, Luca. 'Oltre la milizia. Fisco e civilitas per i Goti di Teoderico', in *Cittadinanze medievali. Dinamiche di appartenenza a un corpo comunitario*, ed. Sara Menzinger (Roma, 2017), pp. 3–26

Lot, Ferdinand. 'Du régime de l'hospitalité', in *Revue Belge de Philologie et d'Histoire*, 7 (1928), pp. 975–1011

Maragno, Giorgia. *'Punire e sorvegliare'. Sanzioni in oro, imperatori, burocrazia* (Napoli 2020)

Marcone, Arnaldo. 'Dal contenimento all'insediamento: i Germani in Italia da Giuliano a Teodosio Magno', in *Germani in Italia*, eds. B. Scardigli – P. Scardigli (Roma 1994), pp. 239–252

Marcone, Arnaldo, *L'ultimo anno dell'impero. Roma: 476* (Roma 2021)

Marotta, Valerio, 'Il problema dei 'laeti'. Fonti e storiografia', in *Civitas, iura, arma: organizzazioni militari, istituzioni giuridiche e strutture sociali alle origini dell'Europa (secc. III–VIII)*, eds. F. Botta, L. Loschiavo (Lecce 2015), pp. 117–158

Mathisen, Ralph W., *Roman Aristocrats in barbarian Gaul. Strategies for survival in an age of transition* (Austin Tex. 1993)

Mathisen, Ralph W. 'The End of the Western Roman Empire in the Fifth Century CE: Barbarian Auxiliaries, Independent Military Contractors, and Civil Wars', in *The Fifth Century: Age of Transformation. Proceedings of the 12th Biennial Shifting Frontiers in Late Antiquity Conference*, eds. J.W. Drijvers – N. Lenski (Bari 2019), pp. 137–156

Mathisen, Ralph W. 'A Prosopography of Sidonius', in *The Edinburgh Companion to Sidonius Apollinaris*, eds. G. Kelly – J. van Waarden (Edinburgh 2020), pp. 76–154

Mathisen, Ralph W. and Hagith S. Sivan, 'Forging a New Identity: the Kingdom of Toulouse and the Frontiers of Visigothic Aquitania (418–507)', in *The Visigoths. Studies in Culture and Society*, ed. A. Ferreiro (Leiden-Boston-Köln 1999), pp. 1–62

Meier, Misha. *Geschichte der Völkerwanderung. Europa, Asien und Afrika vom 3. bis zum 8. Jahrhundert n.Chr.* (München 2019)

Modéran, Yves. 'L'établissement territorial des Vandales en Afrique', in *Antiquité Tardive*, 10 (2002), pp. 87–122

Modéran, Yves. 'L'établissement de barbares sur le territoire romain à l'époque impériale', in *La mobilité des personnes en Méditerranée da l'Antiquité à l'époque moderne. Procédures de contrôle et documentation d'identification*, ed. C. Moatti (Roma 2004), pp. 337–397

Oppedisano, Fabrizio. 'Il generale contro l'imperatore: la politica di Maioriano e il dissidio con Ricimero', in *Athenaeum*, 97 (2009), pp. 543–561

Palazzi, Massimo. *«Foedus cum natione». I trattati alle radici dell'Europa. Aspetti delle relazioni fra impero romano e popolazioni barbariche fra II e V secolo* (Gallarate 2014)

Papa, Giovanni. 'Multae gentes ... quibus terrae laeticae administrandae sunt. A proposito di CTh.13.11.10', in *Studia et documenta historiae et iuris*, 82 (2016), pp. 177–190

Perevalov, Sergej M. 'Bazas 414: la rupture de l'alliance alano-gothique' (transl. from Russian), in *Dialogues d'Histoire Ancienne*, 26 (2000), pp. 175–193

Pohl, Walter. 'Völkerwanderung', in *Migrationen im Mittelalter: ein Handbuch*, ed. M. Borgolte (Berlin 2014), pp. 231–238

Pohl, Walter. 'Von der Ethnogenese zur Identitätsforschung', in *Neue Wege der Frühmittelalterforschung. Bilanz und Perspektiven*, eds. W. Pohl – M. Diesenberger – B. Zeller (Wien 2018), pp. 9–34

Pohl, Walter. 'Dinamiche etniche nel corso delle migrazioni', in *Le migrazioni nell'Alto Medioevo* (Atti Settimane del CISAM, LXVI) (Spoleto 2019), pp. 1–21

Porena, Pierfrancesco and Yann Rivière, eds., *Expropriations et confiscations dans les royaumes barbares. Une approche régionale* (Roma 2012)

Porena, Pierfrancesco. *L'insediamento degli Ostrogoti in Italia* (Roma 2012)

Porena, Pierfrancesco. 'La posizione dell'elemento militare nell'Impero romano e i 'regni romano-barbarici'', in *Civitas, iura, arma: organizzazioni militari, istituzioni giuridiche e strutture sociali alle origini dell'Europa (secc. III–VIII)*, eds. F. Botta – L. Loschiavo (Lecce 2015), pp. 221–257

Rance, Philip. 'The farmer and the soldier should be friends: Justinian's legislation on the provisioning of soldiers (Novel 130)', in *Journal of Late Antiquity*, 12.2 (2019), pp. 380–421

Roberto, Umberto, *Il secolo dei Vandali. Storia di un'integrazione fallita* (Palermo 2020)

Rosafio, Pasquale. 'Inquilinus', in *Opus*, 3 (1984), pp. 121–131

Rosafio, Pasquale. '*Constitutio de Scyris* e l'insediamento dei barbari nell'impero', in *Figures d'Empire, fragments de mémoire: pouvoirs et identités dans le monde romain impérial, IIe s. av. n. è.-VIe s. de n. è.*, eds. S. Benoist – A. Daguet-Gagey – C. Hoët-Van Cauwenberghe (Villeneuve-d'Ascq 2011), pp. 407–423

Rosafio, Pasquale. 'Nuove ricerche sul colonato', in *Uomini, istituzioni, mercati. Studi di storia per Elio Lo Cascio*, eds. M. Maiuro – G.D. Merola – M. De Nardis – G. Soricelli (Bari 2019), pp. 363–371

Schmidt-Hofner, Sebastian. 'Barbarian migrations and socio-economic challenges to the Roman landholding elite in the fourth century CE', in *Journal of Late Antiquity*, 10.2 (2017), pp. 372–404

Slootjes, Daniëlle and Michael Peachin, eds., *Rome and the Worlds beyond its frontiers* (Leiden 2016)

Swain, Brian. 'Goths and Gothic Identity in the Ostrogothic Kingdom', in *A Companion to Ostrogothic Italy*, eds. J.J. Arnold – M.S. Bjornlie – K. Sessa (Leiden-Boston 2016), pp. 203–233

Tarozzi, Simona. 'Riforma dello Stato e gestione della terra. La questione degli agri deserti nella prospettiva dei Codici tardo antichi. Alcuni spunti di riflessione', *Mélanges de l'École Française de Rome. Antiquité* (MEFRA), 125 (2013), pp. 461–467

Vannesse, Michael. 'La militarizzazione dell'impero: il ruolo dei barbari (395–455 d.C.)', in *Potere e politica nell'età della famiglia teodosiana (395–455): i linguaggi dell'impero, le identità dei barbari*, eds. I. Baldini – S. Cosentino (Bari 2013), pp. 87–112

Vera, Domenico. 'Questioni di storia agraria tardoromana: schiavi, coloni, villae', in *Antiquité Tardive*, 20 (2012), pp. 115–122

Vera, Domenico. 'Stato, fisco e mercato nell'Italia gotica secondo le Variae di Cassiodoro: fra ideologia politica e realtà', in *Agoranomes et édiles. Institutions des marchés antiques*, eds. L. Capdetrey – C. Hasenohr (Bordeaux 2012), pp. 245–286

Wickham, Chris. *Framing the early middle ages. Europe and the Mediterranean, 400–800* (Oxford 2005)

Wiemer, Hans-Ulrich. 'Theoderich und seine Goten. Aufstieg und Niedergang einer Gewaltgemeinschaft', in *Gewaltgemeinschaften. Von der Spätantike bis ins 20. Jahrhundert*, ed. W. Speitkamp (Göttingen 2013), pp. 15–38

Wiemer, Hans-Ulrich. 'Die Goten in Italien. Wandlungen und Zerfall einer Gewaltgemeinschaft', in *Historische Zeitschrift*, 296 (2013), pp. 593–628

Wiemer, Hans-Ulrich. *Theoderich der Grosse. König der Goten, Herrscher der Römer* (München 2018)

Wood, Ian. 'Ethnicity and the Ethnogenesis of the Burgundians', in *Typen der Ethnogenese unter besonderer Berücksichtigung der Bayern*, eds. H. Wolfram – W. Pohl (Wien, 1990), pp. 53–69

Wood, Ian. 'L'installation des Burgondes dans l'Empire romain: histoire événementielle', in *Expropriations et confiscations dans les royaumes barbares. Une approche régionale*, eds. P. Porena – Y. Rivière (Roma 2012), pp. 69–90

Wood, Ian. *The Modern Origins of the Early Middle Ages* (Oxford 2013)

Wood, Ian. 'The Uses and Abuses of the Barbarian Invasions in the Nineteenth and Twentieth Centuries', in *Manufacturing Middle Ages: entangled history of medievalism in nineteenth-century Europe*, eds. P. Geary – G. Klaniczay (Leiden 2013), pp. 51–70

Wood, Ian. *The Transformation of the Roman West* (Leeds 2018)

Wood, Ian. 'Responses to Migration and Migrants in the Fifth- and Sixth-Century West', in *Le migrazioni nell'Alto Medioevo* (Atti Settimane del CISAM, LXVI) (Spoleto 2019), pp. 177–204

Young, Andrew T. 'Hospitalitas: Barbarian settlements and constitutional foundations of medieval Europe', in *Journal of Institutional Economics*, 14/4, pp. 715–737

CHAPTER 4

Warrior Names and Military Language of the Westgermanic Peoples: Franks and Langobards

Wolfgang Haubrichs

Abbreviations: CDL = *Codice Diplomatico Longobardo*; cop. = copy; DELI = *Dizionario Etimologico della Lingua Italiana*; Engl. = English; ER = *Edictus Rothari*; EWA = *Etymologisches Wörterbuch des Althochdeutschen*; Frk. = Frankish; Germ. = Germanic; Goth. = Gothic; It. = Italian; Lat. = Latin; LEI = *Lessico Etimologico Italiano*; Lgb. = Langobardic; MGH = *Monumenta Germaniae Historica*; MHG = Middle High German; MLat. = Medieval Latin; MLG = Middle Low German; MD = Middle Dutch; Norw. = Norwegian; OE = Old English; OF = Old French; OFr = Old Frisian; OHG = Old High German; OI = Old Italian; ON = Old Nordic; OS = Old Saxon; PD = *Paulus Diaconus*; Rom. = Romance.

1 Names of Nobles, Kings and Warriors

In the year 591 king Gunthram, ruler over the Merovingian *regnum* of Burgundy and over other extensive parts of the Frankish empire, stood as godfather in the baptism of his nephew, only seven years old, son of king Chilperich (561–584) and of his widow Fredegunde, ruling over the *regnum* of Neustria and his capital Paris. This was then an act of extreme political importance, and it took place in the *palatium* or *vicus* of Nanterre (near Paris): Gunthram made the young man to his *filius spiritualis* and *alumnus*, gave him the name of his *avus* Chlothar, father of Gunthram, and legitimated him herewith as future king and heir. According to Gregory of Tours, those who arranged this act, by a special formula, staged a double semantization of the given name. This formula was certainly arranged between the two royal parties, the boy bearing his name informally already long since:[1]

1 Gregor of Tours, *Histor. libri decem*, X,28. Cf. Wolfgang Haubrichs, 'Identität und Name. Akkulturationsvorgänge in Namen und die Traditionsgesellschaften des frühen Mittelalters', in *Die Suche nach den Ursprüngen. Von der Bedeutung des frühen Mittelalters*, ed. W. Pohl (Wien 2004), here pp. 85–91.

 | DOI:10.1163/9789004698017_005

> Post haec autem [Fredegundis] legatos ad Gunthchramnum regem mittit, dicens: 'Profisciscatur dominus meus rex usque Parisius, et arcessitu filio meo, nepote suo, iubeat eum baptismatis gratia confirmare; ipsumque de sancto lavacro exceptum, tamquam alumnum proprium habere dignetur'. Haec audiens rex, commotis episcopis, id est Aetherium Lugdonensim [Lyon], Siacrium Agustidunensim [Autun] Flavumque Cavillonensim [Chalon] vel reliquis, quos voluit, Parisius accedere iubet, indecans se postmodum secuturum. Fuerunt etiam ad hoc placitum multi de regno eius tam domestici quam comites ad praeparanda regalis expensae necessaria. Rex autem, deliberatione acta, ut ad haec deberet accedere, pedum est dolore prohibitus. Postquam autem convaluit, accessit Parisius, exinde ad Rotoialinsim villam [Rueil] ipsius urbis properans, evocato puero, iussit baptisterium praeparari in vico Nemptudoro [Nanterre] …
>
> Et haec dicens, legatis discedentibus, rex accedens ad lavacrum sanctum, obtulit puerum ad baptizandum. Quem excipiens, Chlotharium vocitari voluit, dicens: 'Crescat puer et huius sit nominis exsecutur ac tale potentia polleat, sicut ille quondam, cuius nomen indeptus est'. Quod misterium celebratum, invitatum ad epulum parvolum multis muneribus honoravit. Similiter et rex ab eodem invitatus, plerisque donis refertus abscessit et ad Cavillonensem urbem redire statuit.

The ritualism of the scene is all too clear: The king calls the Burgundian bishops of Lyon, Autun and Chalon and other *episcopi* from his kingdom together at Paris. They are accompanied by magnates of his *regnum*, officials of the royal court (*domestici*) and counts (*comites*). It is safe to assume that Fredegundis too mobilized all those of her royal court, magnates and the available high clergy. After the baptismal act the ceremony was corroborated by the mess and confirmed with mutual invitations and rich gifts.

The most important part of the act, however, was the address spoken by the royal godfather to the newly baptized and to the high-ranking public: 'This boy may prosper (*crescat*) and make true, what his name means; and he may flourish in such an aboundance of power (*potentia*), as once the one whose name he got'. The formula spoken by the Merovingian king reveals a double *veriloquium nominis*.[2] The name is treated like a charm-formula, as it refers,

2 Wolfgang Haubrichs, 'Veriloquium Nominis. Zur Namenexegese im frühen Mittelalter', in *Verbum et Signum. Festschrift für Friedrich Ohly*, 1, eds. H. Fromm – W. Harms – U. Ruberg (München 1975), pp. 231 *ff.*; Id., 'Namendeutung im europäischen Mittelalter', in *Namenforschung – Name Studies – Les noms propres. Ein internationales Handbuch zur Onomastik*, 1, eds. E. Eichler – G. Hilty *et al.* (Berlin/New York 1995), pp. 381 *ff.*

just as in the *historiola* ('example-story') of a charm, to the enormous *potentia* of the *avus*, the shining example of king Chlothar I, who unified (since 558) the Frankish empire. But even before that the semantic dimension of the name *Chlothar* < **Hluda-harja-*, which the newly baptized prince has to fulfill, was conjured up. This semantic dimension has to be understood against the background of the frequently interweaving early names of the Merovingian *stirps*, related to the semantic fields of war and glory:[3]

a. *Chlodio* < **Hludjôn*, derivation from Germ. **hlutha-,*hluda-* 'audible, famous' (cf. Engl. *loud*, German *laut* < Germ. **hlûda-*);
b. *Merovechus* < **Mêr(o)-wîhaz* 'famous fighter', compound of Germ. **mêr(j)a-* 'known, famous' + Germ. **wîhan, *wîgan* 'to fight' (Goth. *weihan*, OHG *wîhan, wîgan*);
c. *Childericus* < *Hild(i)-rîkaz* 'mighty in battle', compound of Germ. **hildjô* 'conflict, battle' + Germ. **rîka-* 'powerful, mighty, ruler';
d. *Chlodovechus* < **Hlud(o)-wîhaz* 'famous fighter', compound of Germ. **hluda-* 'audible, famous' + Germ. **wîhan* 'to fight';
e. *Chlodomeres* < **Hlud(o)-mêr(j)az* 'famous and known', compound of Germ. **hluda-* 'audible, famous' + **mêr(j)a-* 'known, famous';
f. *Childebercthus* < **Hild(i)-berhtaz* 'splendid in battle', compound of Germ. **hild(j)ô* 'conflict, battle' + Germ. **berhta-* 'brilliant, splendid'
g. *Chlotharius* < **Hluda-harjaz* 'famous warrior', compound of Germ. **hluda-* 'audible, famous' + **harja-* 'army, army-warrior'

In a similar sense, at the end of the ninth century, after the report of the semi-official *Annales Fuldenses*, the son of king Arnulf, born in 893, was baptized by the most important magnates of the East-Franconian *regnum*, archbishop Hatto of Mainz and bishop Adalbero of Augsburg, and was given the meaningful name of his great-grandfather *Ludwig* (Louis the German), the founder of the *regnum*. The East-Franconian *Hludowicus* (†876) himself stood in succession to the Merovingian royal names Chlodwig/Ludwig and Chlothar/Lothar, adopted by Charlemagne:[4]

> De qua ei non multum post filius nascebatur, quem Haddo Magonciacensis [Mainz] archiepiscopus et Adalpero Augustae Vindelicae [Augsburg] episcopus sacro fonte baptismatis chrismantes nomine avi sui Hludawicum appellaverunt.

3 Cf. for the names of the Merovingian *stirps* from a historical perspective Eugen Ewig, 'Die Namengebung bei den ältesten Frankenkönigen und im merowingischen Königshaus', in *Francia* 18,1 (1991) pp. 21*ff.*

4 *Annales Fuldenses* a. 893 (ed. Kurze, p. 122).

> Shortly after, she [the queen] gave birth to a son and the archbishop Hatto of Mainz and bishop Adalbero of Augsburg consecrated him by Holy Chrism in the holy font of baptism and they gave him the name of his great-grandfather Hludawic.

Already the Pseudo-Fredegar (about 642), in the legendary report (III, 12) of a prophetic dream, performed by the mother of Chlodwig, had connected his name with his future character as a great and excellent fighter:[5]

> ... peperit filium nomen Chlodoveum. Haec fuit magnus et pugnator egregius, ad instar leoni fortissimus cyteris regibus.

> ... she gave birth to a son named Chlodwig. He became a great and excellent fighter, like the lion [as king of the beasts] he was the strongest among all the kings.

Louis the Pious, whose father Charlemagne, with the names of Louis and his brother Lothar, deliberately established a relationship of his own family with *Chlodwig*, the founder of the Frankish empire, and his famous son *Chlothar*, by naming two of his own sons *Hludo-wig* and *Hlod-hari*. The court poet Ermoldus Nigellus (a. 826/28) connected the king's name with the Roman war-god Mars (v. 44). Correctly he interpreted the name-element *Hluto* as 'famous', and correctly *Wicgh* as 'war' (v. 49f.). The former courtier did not forget to underline that it was a *prodigium* (v. 43), a prophecy, when the parents gave him such a significant name in baptism:[6]

Partibus aequatis crebrescit fama per orbem,
   Et Hludowicus ovans credita regna petit.
Prodigium fuerat sic hunc vocitasse parentes,
   Quod foret insignis Marte, potensque, pius.
Nam Hludowicus enim ludi de nomine dictus,
   Ludere subiectos pacificando monet.
Seu quis Franciscam mavult reserare loquelam,
   Nominis ut possit noscere notitiam:
Nempe sonat Hluto praeclarum, Wicgh quoque Mars est,

5 *Chronicarum quae dicuntur Fredegarii libri quattuor* III,12 (ed. Wolfram – Kusternig, pp. 94–96).

6 Ermoldus Nigellus, *Carmen in honorem Hludowici christianissimi Caesaris Augusti* (ed. Dümmler, p. 6). Cf. Norbert Wagner, 'Das Erstglied von Ludwig', in *Beiträge zur Namenforschung* NF, 21 (1986), pp. 78 *ff.*

Unde suum nomen conposuisse patet.
Iam puer excelsus sacro spiramine plenus
Auxit honore locum Marte fideque suum.

Ermoldus also offers, in the style of Isidor's *Etymologiae*, a Latin interpretation, based on the phonetic similarity of Latin *ludus*, 'game' (v. 45f.). But priority takes the *Francisca … loquela* (v. 47), the military *veriloquium*, assigned to Mars (v. 51f.).

A *veriloquium nominis* is also accentuated in the case of another member of the Merovingian *stirps regia*, Chilperich, son of Chlothar I and father of Chlothar II. This ruler of Neustria bore a name, rather isolated in the onomastic fund of the early Merovingians: *Chilpe-ricus*, romanized with the Merovingian spelling <ch> for Germ. [h], Frk. **Hilpe-rîk*, which can be well translated with the German *hilf-reich*, i.e., English *help-ful*. The court poet Venantius Fortunatus, Italian by origin, translated the ruler's name in a *carmen panegyricum*, certainly in arrangement with the court, as *adiutor fortis*, 'powerful helper'. This semantic hypothesis is revealed by the Frankish translation of an *interpres barbarus* (v. 27):[7]

… nomen ut hoc resonet non impedit aura nec unda:
sic tibi cuncta simul, terra vel astra, favent.
rex bonitate placens, decus altum et nobile germen.
in quo tot procerum culmina culmen habent.
auxilium patriae, spes et tutamen in armis,
fida tuis virtus, inclitus atque vigor.
Chilperice potens: si interpres barbarus extet,
'adiutor fortis' hoc quoque nomen habes.
non fuit in vacuum sic te vocitare parentes:
praesagum hoc totum laudis et omen erat.
iam tunc indicium praebebant tempore nato,
dicta priora tamen dona secuta probant.

… neither winds nor waves are able to impede the sounding of this name: earth and heaven together will serve you. You are a king enjoying by goodness, a fine specimen of your high lineage, a noble scion, many highborn ancestors got in your person their perfection: help of the country, hope and protection in times of war, give your reliable force and glorious power to your folks! Mighty Chilperic: a vernacular interpreter would confirm,

7 Venantius Fortunatus, *Opera Poetica* IX,1 (ed. Leo, pp. 201–202). Cf. Haubrichs, 'Identität und Name', p. 88.

> that you bear the predicate 'strong helper' even in your name. It was not without reason, that your parents gave you that name: it was a true prediction of your fame and an omen. At that time already, they granted an announcement to the newborn, the following talents then confirmed the early spoken words.

The name of the Merovingian king Chilperich comes from the Burgundian legacy of the Frankish *stirps regia*, to compare with the name of his brother *Gunthram* < **Gunth-hrabn* 'battle-raven', transporting the characteristic first element **gunthô* 'fight, battle' of the Burgundian royal family. The models are here *Chilpericus* rex (455-after 474), brother of *Gundo-wech* (and probably son of a *Gunda-har*), and Chilperich II (†476/77 ?), whose daughter Chrodechild (†548) had married the Frankish king Chlodwig (†511). Chilperich of Neustria was her grandson. However, a name with the second element **rîk-* fitted well a *genealogia*, which knew with Chlodwig's father *Childe-ricus* (460–481) an analogous *nomen regis*.[8] But the poet of Romance origin mainly emphasized the military meaning (v. 25sq.): Chilperich is the 'helper of the country, its hope and protection in times of war', he gives his followers and subjects 'reliable force and glorious power'.

As onomastic 'leitmotiv' the Burgundian royal family of the Gibichungs had elected a war-word, i.e. the above-mentioned Germ. **gunthô,* and actualized it constantly for a century in varying names of kings:[9] the ancestor *Gunda-hari* 'battle-warrior' (a. 436/37), *Gundo-wech* 'battle-fighter' (a. 455-ca. 474), his son *Gundo-bad* < **-badwa-* 'battle-combatant' (ca. a. 476/77–516) and his great-grandson *Gundo-bad* II. As in many Germanic dithematic 'full names' (composed of two stems), the compound preserved one element, but the other was varied – and here the second element was taken from the same semantic field. So, the Burgundian onomastic creations could, so to speak, double the semantics of fighting. Even more narrowly connected were the *gunth*-names with other names of the *stirps regia* by a second Germanic style feature, normally used in heroic and magic poetry: allitteration (*stabreim*), here made using the letter G: The eponymos of the family bore the name *Gibicha*, other

8 Cf. Reinhold Kaiser, *Die Burgunder* (Stuttgart, 2004), p. 265.

9 See Wolfgang Haubrichs, 'Typen der anthroponymischen Indikation von Verwandtschaft bei den 'germanischen' *gentes*: Tradition – Innovationen – Differenzen', in *Verwandtschaft, Name und soziale Ordnung (300–1000)*, eds. S. Patzold – K. Ubl (Berlin/Boston 2014), here pp. 43–44; Id., 'Funktion und Performanz. Namen als Instrumente der Sicherung von Herrschaft und Identität im frühen Mittelalter Westeuropas', in *Machterhalt und Herrschaftssicherung. Namen als Legitimationsinstrument in transkultureller Perspektive*, eds. M. Becher – H. Hess (Bonn 2019), here pp. 250–252.

members repeatedly had the names *Godo-mar* 'good and famous', *Gisla-har* 'arrow-warrior', *Gisla-had* 'arrow-fighter' and *Godi-gisil* 'good arrow'.

This onomastic style is reminiscent of the name strings we know from German heroic poetry. So we find in the Old High German 'Song of Hildebrand' (*Hildebrandslied*, late eighth c.) the sequence: grandfather *Heri-brand* 'army-sword', father *Hilde-brand* 'battle-sword', son *Hadu-brand* 'fight-sword'. The preserved (second) element is here represented by Germ. **branda-* 'sword' (OE *brand*, ON *brandr*), the etymon of the Italian loanword *brando* and OF *brant*, meanwhile the first elements **harja-* 'army', **hild(j)ô* 'battle' and **hadu-* 'fight, conflict' (cf. German *Hader*) offer the semantic variation in the field of fighting. The protagonists are all warriors who, as we see in many other anthroponyms, identify themselves with their weapons.[10]

Lineages, related by allitteration and semantic variation, are also found in the Anglo-Saxon heroic poem *Beowulf* (v. 57sqq.), for example the Scyldings with the ancestor *Healf-dene* ('Half-Dane') and his sons *Heoro-gâr* ('sword and spear') with son *Heoro-veard* ('sword-guardian'), then *Hrôth-gâr* ('glorious spear') with the sons *Hrôth-rîc* ('glorious ruler') and *Hrôth-mund* ('glorious protector'), finally *Halga* ('the inviolable') with his son *Hrôth-ulf* ('glorious wolf').[11]

Similarly, semantic variation and alliteration change in the Langobard *stirps regia* of the Lethingi, ancestors and descendants of king Wacho (†540), who placed the center of his *regnum* in Bohemia.[12] In the beginning, the lexical bases of the names alternate in the same semantic field of fighting and beating with the grandfather *Claffo* < **Klapo* ('beater') and his sons *Tato* < **Dado* ('man of action') and *Zuch-ilo* < **Tuk-ilo* ('drawer of a weapon'). But Claffo's third son bears, we don't know why, a variant dithematic name, refering to a weapon, i.e. *Wini-gis* 'spear of friendship', a compound of Germ. **wini-* 'friend' and **gîsa-* 'spear, shaft of arrow'. He again gives to his son the monothematic name *Wacho* < **Wakôn* 'brave', which fits well the onomastic tradition of the family, but is also related to the father's name by alliteration and therefore establishes a new tradition. The new custom is actualized in the names of the next generation, in the names of the daughters *Wisi-garda* and *Walde-rada* and of the son *Walt-hari* (†546), who bears a 'talking name': Germ. **Walda-harja-* 'ruling warrior', expressing a *veriloquium* of kingdom and warriorship. The new and successful branch of the dynasty created a new onomastic custom.

10 Haubrichs, 'Typen', pp. 42–43; Id., 'Funktion und Performanz', pp. 249–250.

11 Haubrichs, 'Typen', p. 44; Id., 'Funktion und Performanz', pp. 147–148.

12 Haubrichs, 'Identität und Name', pp. 86–91; Id., 'Typen', pp. 44–46; Id., 'Funktion und Performanz', pp. 236–237, 240.

Other names of Langobard kings, nobles, warriors too show military semantics: We may for instance mention the kings *Authari* (584–590) < **Auda-harja-* 'the rich warrior'; *Rothari* (636–652) < **Hrôth-harja-* 'glorious warrior' with his son *Rodoald* (†653) < **Hrôth-walda-* 'glorious ruler'; *(H)Aripert* (653–661) < **Harja-berhta-* 'splendid in the army', the founder of the so called 'Bavarian' dynasty, which preferred exclusively the lexem **berhta-* 'brillant, splendid, illuster', and his son *Perct(h)ari* (671–688) < **Berhta-harja-* 'splendid warrior' with inversion of the father's name, finally the great-grandson *H(A)ripert* II (703–712).

It is evident that the meaning of names could preserve an identitary function only as long as the name elements and lexemes were clear and could be easily understood. For the times of Late Antiquity and Sub-Antiquity (fourth to eighth centuries) there is little doubt about this. But in his *Gesta Karoli* (II,12) (a. 883/84) Notker Balbulus shows that also in the ninth century (and later) a linguistic understanding of Germanic personal names preserving their archaic semantic values was possible.[13] He characterized a horrible alamannic warrior, born in the Thurgau (CH) and active in the wars of Charlemagne against the *Avari … iuxta nomen suum magna pars 'terribilis exercitus', vocabulo Eishere …*, derived from Germ. **Agis-harja-*, a compound of **agis-* 'horror, terror' (late OHG *eis-*) and **harja-* 'exercitus, army, warrior'. As late as in the late tenth century new 'talking names' conveying a special meaning could be created. About 980 with reference to Ottonian wars against the Ungarians, then identified with the Huns, we hear of a warrior bearing a 'nom de guerre', i.e. *Hunin-gêr* 'spear against (?) the Huns' with his sons *Hunin-wê* ('pain of the Huns'), *Hunnin-leit* ('grief of the Huns') and finally *Hunin-tot* ('death of the Huns').[14]

Among the semantic complexes used for Germanic name-giving the themes of 'war' and 'fight' had the largest part, beside of the neighbouring themes of glory, rulership, nobility and wealth. It follows a small selection of lexemes from the semantic field of 'battle' (extracted from Langobardic personal names before a. 774)[15]: **gunthô, gundô* 'struggle'; **hild(j)ô* 'combat'; **wala-* 'battlefield, death in battle'; **sintha-* 'route, journey, military expedition'; **segu-, *segiz*

13 Notker Balbulus, Gesta Karoli, II, 12 (ed. Rau, p. 404).

14 Wilhelm Störmer, 'Hoch- und spätmittelalterliche 'Familien'-Namengebungspraxis vornehmlich beim Niederadel in Altbayern und Franken', in *Personennamen und Identität. Namengebung und Namengebrauch als Anzeiger individueller Bestimmung und gruppenbezogener Zuordnung*, ed. R. *Härtel* (Graz 1997), here p. 218.

15 Cf. Wolfgang Haubrichs, 'Langobardic Personal Names: Given Names and Name-Giving among the Langobards', in *The Langobards before the Frankish Conquest: An Ethnographic Perspective*, eds. G. Ausenda – P. Delogu – C. Wickham (Woodbridge/San Marino 2009), here pp. 196–210.

'victory'; **druhti-* 'company, troop of followers'; **fara-* 'group of travellers, followers'; **harja-* 'army'; **branda-* 'sword'; **agi(s)-* 'fear, horror, terror'; **grasan-* 'to devour'; **warnôn-* 'to warn'; **gaidô* (cf. OE *gâd*) 'sting, point, arrow'; **gaiza-*, Westgerm. **gaira-* 'spear'; **gîs-*, **gîsal-* 'staff, arrow' etc.

Frequently wild and aggressive animals were evoked to be compared with warriors: above all **wulfa-* 'wolf', then **berôn* 'bear', **arôn*, **ar-n* 'eagle', **ebur-* 'boar' and *hrabn-* 'raven'. The raven was also the bird of the Germanic and Langobard war-god *Wodan/Godan*, who played a prominent role in Langobardic mythical history (*Origo gentis Langobardorum*, 7th c.). We already had to point out the importance of the wolf-names in the Anglo-Saxon heroic poem *Beowulf* with the lineage of the Scyldings, ending with *Hrôth-ulf* 'glorious wolf'. The eponym *Beo-wulf* 'bee-wolf' bears a tricky wolf-name, a cryptonym for the honey-loving 'bear'. In this regard, mention should be made of the Langobardic king *Agilulf* (592–616) < **Agila-wulfa-* 'terrible wolf', the Friulian dukes *Bill-ulf* ('sword-wolf'), *Gis-ulf* ('arrow-wolf'), *Gras-ulf* ('devouring wolf'), all of them appearing more than one time in the family; just as good in the family of the Gothic king Ermanarîk (†376) with *Achiulf* < **Agjô-wulfa-* ('sword-wolf'), *Odvulf* < **Ôda-wulfa-* ('heritage-wolf'), *Ediulf* < ? + **wulfa-* and *Vult-vulf* < **Wulthu-wulfa-* ('splendid wolf'). If we look among the ancestors of the Carolingian *stirps*, we have *Arnulf* < **Arn-wulfa-* 'eagle and wolf', first royal *domesticus*, then bishop of the Austrasian capital Metz († ca. 640) with his son *Chlodulf* < **Hluda-wulfa-* ('famous wolf'), who became his third successor on the episcopal sede upon the Moselle.[16]

In interpreting scrupulously the names of the ancestors of Jesus, a commentary on the bible of the 5th or sixth centuries – the *Opus imperfectum in*

16 See Wolfgang Haubrichs, 'Tierische Identitäten. Zur symbolischen Kommunikation in Namen des frühen Mittelalters', in *Tiere: Begleiter des Menschen in der Literatur des Mittelalters*, eds. J. Klinger – A. Kraß (Cologne/Weimar/Wien 2017), here 239–244. Cf. Gunter Müller, 'Germanische Tiersymbolik und Namengebung', in *Frühmittelalterliche Studien*, 2 (1968); Id., *Studien zu den theriophoren Personennamen der Germanen* (Cologne/Vienna 1970); Heinrich Beck, 'Das Problem der theriophoren Personennamen im Germanischen', in *Zum Problem der Deutung frühmittelalterlicher Bildinhalte*, ed. H. Roth (Sigmaringen, 1986); Karen Høiland Nielsen, 'The Wolf-Warrior-Animal Symbolism on Weaponry of the sixth and seventh centuries', in *Archäologisches Zellwerk. Beiträge zur Kulturgeschichte in Europa und Asien. Festschrift für Helmut Roth*, eds. E. Pohl – U. Recker – C. Theune (Rahden-Westphalia 2001); Chiara Staiti, 'Agilulf e altri. Il lupo nell'antroponimia germanica dei primi secoli', in *Simbolismo animale e letteratura*, ed. D. Faraci (Rome 2003); Wilhelm Heizmann, 'Tiernamen', in *Reallexikon der Germanischen Altertumskunde*, 35 (2007); Christa Jochum-Godglück, 'Wolf und Bär in germanischer und romanischer Personennamengebung', in *Interferenz-Onomastik. Namen in Grenz- und Begegnungsräumen in Geschichte und Gegenwart*, eds. W. Haubrichs – H. Tiefenbach (Saarbrücken 2011).

Matthaeum, written in the Danubian provinces of the Empire – highlighted similarities in the name-giving customs of Germanic peoples:[17]

> Sicut solent et barbarae gentes nomina filiis imponere ad devastationem respicientia bestiarum ferarum, vel rapacium volucrum, gloriosum putantes filios tales habere, ad bellum idoneos, et insanientes in sanguinem.
>
> In the same way the barbarian peoples follow the custom to give their sons names, focused on devastation, that is names of wild animals and rapacious birds. They pretend to win fame and glory by sons, which are suited for war and blood-curdling until ecstasy.

A taboo-name for the wolf, like the mentioned *Beo-wulf* for the bear, is undoubtedly the Runic *Widu-hu(n)dar* 'forest-dog' (fibula from Himlingøje, DK, Sjaelland). A formation of similar age is to be found in the name of the Gothic legendary hero, the *fortissimus Gothorum Vidi-goia*, mentioned by Priscus (a. 472) and also Jordanes (a. 556), and also in the name of the Alamannic king *Vidi-gabius*, completed by a Frankish *comes* named *Widi-gowo* (a. 799–823, Lobdengau on the lower Neckar), the OS personal name *Widu-gô* and MHG *Wite-gouwe*. Following Maurits Schönfeld, this dithematic name derives from the Germanic compound **Widu-gawja* 'inhabitant of the forest', in its turn coming from **gawjôn* 'inhabitant, settler', which can be understood as a cryptonym for the wolf.[18]

Also *Isengrim* < Germ. **Isan-grîma-* ('iron-mask'), the name of the wolf in the animal fables, is a cryptonym: In the sense of a 'mask grey like iron', the compound represents the snout of the wolf; in the same sense *Brûn*, 'the brown one', stands for the bear. *Isan-grîm* exists also as warrior-name and therefore leads over to the anthroponyms, which deal with the disguise of men and fighters as animals.[19] Onomastic formations with the element **grîma-* 'mask, mask-helmet' (OE *grîma*, ON *gríma*) like *Isan-grîm, Hathu-grîm* (formed with **hathu-* 'conflict, fight'), *Hildi-grîm* (formed with **hild(j)ô* 'battle') would suggest a mask-helmet for battle. Theriophoric names, combined with **grîma-*, are frequent, in Scandinavia, England and on the continent. Examples are: ON *Arn-grímr* 'eagle-mask', Frk. *Hrafan-grim* 'raven-mask' and *Ebur-grim* 'boar-mask', Bavarian *Pera-grim* ('bear-mask'), Frk., Bavarian, Alamannic *Wolf-grim*,

---

17 *Opus imperfectum in Matthaeum*, Homilia I, MPG 56, c. 626–627.

18 Maurits Schönfeld, *Wörterbuch der altgermanischen Personen- und Völkernamen* (Heidelberg 1910), p. 263. Cf. Haubrichs, 'Tierische Identitäten', p. 246.

19 Cf. Müller, *Studien*, pp. 212–223; Haubrichs, 'Tierische Identitäten', pp. 246–247.

OE *Wulf-grim*, ON *Ulf-grímr* 'wolf-mask' with the rather frequent inversion **Grîma-wulfa-* 'mask-wolf', completed finally by the name *Grimoald, Grimald* < **Grîm-walda-* 'ruler with the mask-helmet', characteristic for the early Carolingians and the Langobardic-Friulian ducal family (seventh-ninth c.). In his pioneering studies on theriophoric personal names Gunter Müller rightly considered that behind this type of name-giving there was a real 'theriomorphic disguise of the heads' of warriors or 'helmets showing pictures or figures of animals'.[20]

Also the personal names Frk. OHG *Wolf-hetan*, ON *Úlf-heðinnn*, Old Swedish (Runic) *Ulf-hithin* < Germ. **Wulfa-hedana-* 'coat made of wolf-skin' (cf. the West Frankish inversion *Heden-ulf*) point at a real wolf-dress. They correspond, as G. Müller puts it, 'undoubtedly with the appellative ON *úlfheðinn*, a Bahuvrihi-compound, whose second component can be identified with ON *heðin* 'short sleeveless dress with a hood' and OE *heden* 'fur dress, hood with collar''.[21] Such warriors, dressed with wolf-skins instead of a breastplate and going into battle with brutish howls, are documented since the ninth century in Scaldic poetry and sagas. The sources particularly stress their bloodthirst, their 'bloody shields' and their 'reddened spears', corresponding to the commentary of the late antique *Opus imperfectum* mentioned above.[22]

For the bear exists in the same way the onomastic compound ON *Biarnheðinn* 'dress made of bear-skin' and the inversion Old Swedish (Runic) *Hithin-biarn* (cf. the analogous appellative *mannbjörn* 'man-bear'). However, name parallels for 'the blood drinking' band of warriors, called *Berserkr* in the sources of the North do not exist. After all, we have evidence that the Scandinavian North preserved the theriomorphic warrior dress and the correspondent name-giving, while in that semantic field the continental onomastic proofs are limited to the Franks and the neighbouring *gentes* of *Alamanni* and *Baiuwarii*.[23]

But the concept of animal warriors, transforming their identity at least mentally in an aggressive fighting animal, must have spread much wider on the background of a warrior society. In the first place, this is proved by name formations like *Wolf-hroc* 'wolf-dress' (and the inversion *Hrocc-olf*), semantically analogous to just mentioned Germ. **Wulfa-hedena-* and to be found frequently in Frankish, Bavarian and Alamannic sources. Proofs for the name *Wolf-gang* < **Wulfa-ganga-* 'man going as wolf' and the inversion *Gang-olf*, refering to the

20 Müller, *Studien*, pp. 218–220.

21 Müller, *Studien*, p. 212. Cf. Gunter Müller, 'Zum Namen *Wolfhetan* und seinen Verwandten', in *Frühmittelalterliche Studien*, 1 (1967).

22 Cf. Høiland Nielsen, 'The Wolf-Warrior', p. 478.

23 See Haubrichs, 'Tierische Identitäten', p. 247.

'ecstatic […] fight-walk of the animal warriors' (G. Müller), lead us to the same semantic area.[24] Mainly in the region of Alemannia were spread the analogous formations *Wolf-dregi*, derivated from Goth. **thrag-jan*, OE *draegan* 'to run', and, provided with a l-suffixe, *Wolf-drigil* < **thregila-* 'runner'. Here we touch another semantic field of the warrior onomastics, i.e. the euphemistic designations of warfare and warpath as *gangan* 'to go' (cf. Engl. *gang* 'troop, band'), **wadan* 'to walk, to stride, to wade', **laikan* 'to jump, to dance, to play' and **faran* 'to travel, to drive, to ride' (cf. Germ. *fara* 'company of travelling warriors') etc. The military euphemism culminates in personal names like *Far(a)ulf* < **Fara-wulfa* 'travelling wolf', *Faroald* < Germ. **Fara-walda-* 'ruler of the *fara*' and *Fara-mund* 'protector of the *fara*', as was (according to the *Liber Historiae Francorum*) the first king of the Franks.[25]

But above all the semantics of theriophoric names correspond with the Alamannic and Frankish representations of wolf-warriors of the sixth and seventh centuries, found in graves at Gutenstein (D, Sigmaringen) and Obrigheim (D, Bad Dürkheim), finally also with the warrior (on a bronze 'pressblech'), dressed in wolf-skin, from Torslunda (S, Öland, 7th–8th c.).[26] Those representations have their roots in widely spread ideas over the possibility of real transformation of a human being into a wolf, already known by Plinius (*Naturalis Historia* VIII, 22: *homines in lupos verti*).[27] In the older parts of the *Vølsunga-Saga* the heroes *Sigmundr* and *Sinfjötli* (OHG *Sintarfizzilo*, OE *Fitila*) change into wolves, live in the forest to fulfill their obligation to revenge. For this transformation we find in OE the compound *wer(e)-wulf*, in OHG *wer-wolf* 'man-wolf, man taking the 'shape of a wolf', derived from Germ. **wira-* 'man' and **wulfa-*. As personal name this formation is documented in a Langobardic context as *Guer-olf* (romanized with the phonetic substitution of Germ. [w] by [gu]). Probably, also the Frankish names *Wer-olf, Wer-ulf* have to stand here. But certainly, the attestations of **Mana-wulfa-* (derived from Germ. **manôn* 'man, human'), reaching from the Visigothic realm over the Franks up to Scandinavia, and finally Frk. *Gomulf* < **Guma-wulfa-* (derived from Germ. **gumôn* 'man') represent the same concept.[28]

---

24 Müller, *Studien*, p. 213.

25 Wolfgang Haubrichs, '*Leudes, fara, faramanni* und *farones*. Zur Semantik der Bezeichnungen für einige am Konsenshandeln beteiligte Gruppen', in *Recht und Konsens im frühen Mittelalter*, eds. V. Epp – C.H.F. Meyer (Ostfildern 2017), here pp. 249–253.

26 See Høiland Nielsen, 'The Wolf-Warrior', pp. 476–477.

27 Gudrun Lange, 'Tierverwandlung', in *Reallexikon der Germanischen Altertumskunde*, 30 (2005).

28 Müller, *Studien*, p. 218. Cf. Friedrich Kluge and Elmar Seebold, *Etymologisches Wörterbuch der deutschen Sprache*, 25th ed. (Berlin/Boston 2011), p. 984.

The masculine (but also the feminine) name-world of the Germanic peoples, organized in many facets, reflects clearly the orientation of the society according to battle, fight and war, as to be seen also in the historic narratives and the archaeological findings of the time.

2 Names of Peoples (Ethnonyms), Groups and Families

Names are signa of personal and group identities of early societies. So we may expect to find in designations of the *gentes*, in ethnonyms, not only frequent references to ethnos, origin, settlement and customs, but also to qualities in fighting and institutions of war. Some examples:[29]

- *Varini* < Germ. **War-inôz* 'men fit for battle, men able to defend themselves', derived from Germ. **warija-* 'to defend';
- *Franci*, OHG *Frankôn*, OE *Francan*, ON *Frakkar*, derived from Germ. **fraka-* 'courageous, fast';
- *Thuringi* < Germ. **Thuringôz* 'the courageous, the bold';
- *Ouargiones* < Germ. **Warg-jônez*, derivated from Germ. **wargjaz* 'rapacious, wolfish';
- *Saxones*, OHG *Sahsôn* 'men with daggers or short swords', probably so called for their main weapon, OHG OS *sahs*, OFr *sax*, OE *seax* 'knife, dagger, short sword';
- *Harii* < **Harjaz* 'army-warriors', derived from Germ. **harja-* 'army, troop';
- *Alamanni* < 'the full men, perfect men, real men', compound of Germ. **ala-* 'full, whole, complete';
- the OE ethnonym *Wulf-ingas* ('Widsith', v. 29) refers to the well known beast of prey Germ. **wulfa-* 'wolf'.

---

29 Cf. for the following ethnonyms Ludwig Rübekeil, 'Völkernamen Europas', in *Namenforschung – Name Studies – Les noms propres. Ein internationales Handbuch zur Onomastik*, eds. E. Eichler – G. Hilty *et al.* (Berlin/New York, 1996); Wolfgang Haubrichs, 'Nomen gentis. Die Volksbezeichnung der Alamannen', in *Röllwagenbüchlein. Festschrift Walter Röll*, eds. J. Jaehrling et *al.* (Tübingen, 2002); Id., 'Il linguaggio della guerra presso i popoli germanici occidentali (Longobardi e Franchi)', in *L'esercito romano e l'alba dell'Europa. Modelli concettuali e sperimentazioni sociali tra linguaggi, istituzioni e diritto (secoli IV-VIII)*, ed. G. Bassanelli Sommariva (Santarcangelo di Romagna, 2020), here pp. 213–214; Stefan Zimmer, 'Germani und die Benennungsmotive für Völkernamen in der Antike', in *Zur Geschichte der Gleichung 'germanisch-deutsch'. Sprache und Namen, Geschichte und Institutionen*, eds. H. Beck *et al.* (Berlin 2004); Alexander Sitzmann and Friedrich E. Grünzweig, *Die altgermanischen Ethnonyme. Ein Handbuch zu ihrer Etymologie*, [Philologica Germanica, 29] (Vienna, 2008).

Also the names of noble and royal lineages, of *stirpes* or *genealogiae* offer similar concepts. Some examples again:[30]

- OE *Brond-ingas* (*Beowulf*, v. 521) 'sword-people', derived from OE *brond* < Germ. **branda-* 'sword';
- OE *Helm-ingas* (*Beowulf*, v. 620) 'helmet-people', derived from Germ. **helma-* 'helmet';
- *Amali*, the royal stirps of the Goths < Germ. **Amalôz* 'the ardent, the busy';
- the Thuringian *genus Anauuas* of the Langobard king Agilulf (590–616) < **Ana-hwatôz* 'the most courageous', derived from Germ. **hwat-* 'courageous, keen, bold';
- *Balthi*, the Wisigothic royal family, derived from Germ. **baltha-* 'bold, courageous';
- *H(u)osi*, a Bavarian *genealogia* < **Hôsiz* 'the grey ones', most likely a cryptonym for wolves;
- *Throzza, Draozza*, a Bavarian *genealogia* < **Thrautôz* 'the pushing, the pressing, the threatening ones', derived from Germ. **thrautan* 'to threaten, to push'.

In morphology, semantics and function the names of groups, of *stirpes* and *genealogiae* and of peoples (ethnonyms) look notably alike. They are authentically Germanic creations. The opinion recently disseminated in historical research that the names of many *gentes* and groups were given by the Romans, therefore, is quite unlikely. Certainly, though the chroniclers of the Imperium Romanum contributed to the spreading and generalization of some ethnonyms.

3 The Military Language of the Westgermanic Peoples: The Examples of the Franks and the Langobards

Very useful for the reconstruction of the military language of the Germanic peoples are loanwords, transferred from the Germanic languages into 'Vulgar'

30 Cf. Norbert Wagner, 'Gausus und Harodus. Odinsnamen oder Stammesnamen in germanischen Königsgenealogien?', in *Beiträge zur Namenforschung*, NF 13 (1978); Id., 'Der Name der Huosi', in *Münchener Studien zur Sprachwissenschaft*, 58 (1998); Wolfgang Haubrichs, '*Nomina stirpium*. Sippennamen und Ethnonyme: Probleme einer Typologie der Personengruppen-bezeichnungen', in *Language and Text. Current Perspectives on English and Germanic Historical Linguistics and Philology*, eds. A.J. Johnston – F. von Mengden – S. Thim (Heidelberg, 2006); Id., 'Il linguaggio della guerra', pp. 214–215.

or Medieval Latin, into the 'latin parlé' (Michel Banniard).[31] Early evidence is often found in chronicles and law texts. So, for instance, we have vernacular lexemes followed by Latin translations in the Langobardic *Edictus Rothari* (ER),[32] published a. 643 and in the *Historia Langobardorum* of Paulus Diaconus (end of eighth c.).[33] The Italoromance (or Old Italian = OI) as well as the Galloromance (or Old French = OF) are poorly documented before the 12th and 13th centuries. But there are two methods in dating the loanwords more precisely. First, by the presence or absence of the results of phonological developments of the source language, for instance the OHG sound shift of the sixth and seventh centuries, allowing to fix a *terminus post quem*, if present, a *terminus ante quem*, if not present. If the Germanic borrowings did not realize this sound change, we get an indication for the acquisition of the term from an older Germanic stratum, for instance from Gothic or from a Koinè, spoken by Germanic members of the Roman *exercitus*, or even from the language of the Eastgermanic *gentes*, like *Gepidi* or *(H)Eruli*, which invaded Italy in 568 together with the Langobards. A second criterion for dating loanwords follows from presence or absence of the results of phonological developments in the Romance target language, allowing to fix a *terminus ante quem*. Finally, there exists an absolute *terminus ante quem* depending on the silent death of the source language. For the Langobardic vernacular idiom, the gradual end of interferences and contacts with the Italoromance came in the late eighth century, for the (West)Frankish of the Northern and Eastern Gallia in the eighth and ninth centuries, except a small stripe round about the language border. For the Frankish of the Central French area surrounding Paris (Ile de France) the language death certainly came much earlier.

In the following paragraphs words of the military language, arranged after semantic fields, will mainly be treated if originating from Frankish or Langobardic. It is evident that there will be overlappings with neighbouring fields at the margins, for instance the fields of handicrafts, of the social strata and of

31 Michel Banniard, *Viva voce. Communication écrite et communication orale du IVe au IXe siècle en Occident latin* (Paris 1992).

32 *Edictum Rothari*. Cf. for the loanwords and Langobardic terms in the legal texts Florus van der Rhee, *Die germanischen Wörter in den langobardischen Gesetzen* (Amsterdam 1970); Id., 'Die germanischen Wörter in der Historia Langobardorum des Paulus Diaconus', in *Romanobarbarica* (1980); Nicoletta Francovich Onesti, *Vestigia longobarde in Italia (568–774)* (Rome 1999), pp. 51–138; Wolfgang Haubrichs, 'Sprache und Schriftlichkeit im langobardischen Italien – Das Zeugnis von Namen, Wörtern und Entlehnungen', in *Filologia Germanica – Germanic Philology*, 2 (2010), here 182–188.

33 Paulus Diaconus, *Historia Langobardorum*. Cf. van der Rhee, 'Die germanischen Wörter in der Historia Langobardorum'.

mentality. Such overlappings are unavoidable. Loanwords designating aggressive qualities or related to warlike mentalities have to be factored into the analysis: we may suppose that the borrowing took place in bilingual milieus of the army, especially if an analogous term already existed in Latin.

The following semantic fields can be analyzed and will be illustrated with a few examples:[34]

3.1 *The Army and Its Detachments*

– Lgb., Frk., OHG *hari-* < Germ. **harja-* 'army, troop':[35] exists in many military compounds, e.g. in the laws:[36] *ariman(nus)* Liut. 44, Rat. 1,2, 10; CDL, I, no, 19 a. 715 Siena; *ariman* CDL, I, no. 20 a. 715 Vico Vallari[37](cf. the synonym *homo exercitalis*: ER 23, 24), etymologically 'army-man', i.e. 'l'*homo liber* longobardo in quanto membro del *populus-exercitus*';[38] *hari-gawerc* 'military equipment' (ER 225 in an alliterating law-formula: *res suas proprias, id est [h] andegauuerc et harigauuerc*), etymologically 'equipment for the army (service)' < Germ. **harja-* + **ga-werki* 'equipment';[39] *harschild, arischild* 'armed attack, raid, devastation of a settlement', a law term (Liut. 134 a. 733; Liut 141 a. 734) < Lgb. **hari-skildu* 'part of an army, troop of warriors', compound of Germ. **harja-* + **skilthu*, derived from **skiljan* 'to split' (cf. Norw. *skilja* 'to separate, to cut', Norw. *fara herskildi* 'to devastate');[40] *haritra(h)ib* 'attack, raid, insidious damaging of an estate' < Lgb. **hari-traiba-* (ER 379: ... *si casa, ubi habitatur, disturbaverit, conponat, sicut in hoc edictum legitur, haritrahib*),

34 For more examples and more bibliographical references see Haubrichs, 'Il linguaggio della guerra', pp. 220–246.

35 Cf. Francesco Sabatini, 'Riflessi linguistici della dominazione longobarda nell'Italia Mediana e Meridionale', in *Atti dell'Accademia Toscana di Scienze e Lettere La Colombara*, 28 (1963/64), here 219–220; *EWA*, IV, c. 971–975; Kluge – Seebold, *Etymologisches Wörterbuch*, p. 402.

36 Liut. = Leges Liutprandi regis; Rat. = Leges Ratchis regis.

37 CDL, 1.

38 Francovich Onesti, 'Vestigia', pp. 62–63. Cf. Jan Frederik Niermeyer (in coll. with C. van de Kieft), *Mediae Latinitatis Lexicon Minus*, I (Leiden, 2002), p. 631; Sabatini, 'Riflessi linguistici', pp. 158–159; Florus van der Rhee, *Die germanischen Wörter in den langobardischen Gesetzen* (Amsterdam, 1970), pp. 83–85; Heinrich Tiefenbach, *Studien zu Wörtern volkssprachiger Herkunft in karolingischen Königsurkunden. Ein Beitrag zum Wortschatz der Diplome Lothars I. und Lothars II.* (Munich, 1973), pp. 66–68; *EWA*, IV, c. 978.

39 Cf. van der Rhee, *Die germanischen Wörter*, p. 81; Francovich Onesti, 'Vestigia', p. 96; Wolfgang Haubrichs, 'Sprache und Schriftlichkeit im langobardischen Italien – Das Zeugnis von Namen, Wörtern und Entlehnungen', in *Filologia Germanica – Germanic Philology*, 2 (2010), p. 151.

40 Cf. van der Rhee, *Die germanischen Wörter*, pp. 31–33; Karl-Friedrich Krieger, 'Heerschild, -ordnung', in *Lexikon des Mittelalters*, 4 (1989); Francovich Onesti, 'Vestigia', pp. 63–64.

compound of Germ. **harja-* + Germ. **draib-jan* 'to drive, to scatter, to push';[41] OF *arban* 'call up to army', loanword Mlat. *heribannus*, frequently in *capitularia*, but testified only since Carolingian times (partly with OHG 'umlaut' [a] > [e] of the late eighth century < Westgerm. Frk. **harja-banna-* (cf. OHG *heri-pan* 'call up to the army, conscription');[42] Northern French *heri-berge* 'lodging, quarters', verb *héberger*; Southern French *ar-, al-berge*, French *auberge*, It. *al-bergo* (also attested in other Romance languages); MLat. loanword, first documented in the 'Reichenau Glosses' (Corbie, late eighth c.) in *Castro*: *heribergo*,[43] later in capitularies and law documents, with 'umlaut' [a] > [e] < Frk. **hari-berga-* (cf. OHG OS *heri-berga*), compound of Germ. **harja-* + **berga-* 'protection'; here also MLat. *heribergare* 'to billet the army' in the *Capitulare Bononiense* (*§ 2*) *of Charlemagne, treating with military affairs* (*a. 811*) *as part of the bannus*, mentioned together with *wacta* 'guard', *scara* 'troop' and *warda* 'guarding': *Ut non per aliquam occasionem, nec de wacta nec de scara nec de warda nec pro heribergare neque pro alio banno, heribannum comes exactare praesumat …*;[44] *harizuht*, romanized *herizuph* 'raid of an armed band' (Latin gloss: *collecta* 'illegal company, band, gang'), documented in *capitularia* of Charlemagne and Charles the Bald, compound of Germ. **harja-* + Westgerm. **tuhti* fem. 'expedition, march' (cf. MHG *zuht* in the same sense), derivated from Germ. **teuhan* 'to lead, to draw, to pull' (OHG *ziohan* with sound shift [t] > [ts]);[45] *herisliz* 'desertion', a Frk. term appearing four times in *capitularia* of Charlemagne (*Cap. Bononiense*, a. 811: *de hoste reversus fuerit, quod factum Franci herisliz dicunt*) and in the *Annales regni Francorum*;[46] in toponyms *Herstal* near Liège (B): ca. a. 718 cop. eighth c. (*Vita S. Landberti*) … *Cheristalius* (with Merovingian spelling <ch> for the Romance phonetic substitution of Germ. [h]); a. 723 cop.

---

41 Cf. van der Rhee, *Die germanischen Wörter*, pp. 85–86; Francovich Onesti, 'Vestigia', p. 96.

42 Niermeyer, *Mediae Latinitatis Lexicon*, pp. 629–630; Tiefenbach, *Studien*, pp. 64–68; de Sousa Costa, *Studien*, pp. 110–120; *EWA*, IV, c. 975; Kluge – Seebold, *Etymologisches Wörterbuch*, p. 402. Cf. *bannire*: Haubrichs, 'Il linguaggio della guerra', p. 240 fn. 110.

43 Klein, *Die Reichenauer Glossen*, p. 158 no. 304.

44 *Capitolare Bononiense* (ed. Boretius, p. 166 no. 74). Cf. Niermeyer, *Mediae Latinitatis Lexicon*, p. 630; Sabatini, 'Riflessi linguistici', pp. 219–220; Annette de Sousa Costa, *Studien zu volkssprachigen Wörtern in karolingischen Kapitularien* (Göttingen, 1993), pp. 263–268; Bernd-Ulrich Hergemöller, 'Herberge', in *Lexikon des Mittelalters*, 4 (1989); Ruth Schmidt-Wiegand, 'Herberge', in *Reallexikon der germanischen Altertumskunde*, 14 (1999); Kluge – Seebold, *Etymologisches Wörterbuch*, p. 411.

45 De Sousa Costa, *Studien*, pp. 318–322. Cf. Kluge – Seebold, *Etymologisches Wörterbuch*, p. 1009.

46 De Sousa Costa, *Studien*, pp. 282–288. Cf. Kluge – Seebold, *Etymologisches Wörterbuch*, p. 810 (*schleißen, Schlitz*).

tenth c. … *actum Harastallio*; later *Herstelle* on the Weser south of Höxter (D, Niedersachsen): a. 797 (*Annales regni Francorum*): … [*Carolus rex*] *positis castris apud Wisoram fluvium locum castrorum Heristelli vocari iussit* …; a. 797 (*Ann. Laur.*) … [*locum*] *etiam Heristelli appellavit, eo quod ab exercitu suo fuerant constructae mansiones ubi habitabant* …; a. 797 (*Ann. Mosellani*): … *usque ad locum, cui nomen imposuit Niwi Haristalli* ('New Herstal') < Germ. **harja-* + **stalja-* 'stand, base, residence'; we may compare the Italian settlement name of Frankish provenance *Guastalla* (north of Reggio Emilia), a. 864 or. *Vuardistalla* < **Warda-stalja* 'place of guards, watchtower'.[47]

- Lgb. *fara* 'travelling company of warriors' (also in Marius of Avenches, Chronicon, about a. 581), later (Paulus Diaconus) in the *regnum Langobardorum* interpreted as *generatio, linea*; early and frequently documented in Italian toponyms, but also in Frankish and Bavarian regions; further in Germ. personal names like *Fara* 'travelling warrior' (see above); in Burgundian in compounds like *fara-manni* 'men of the *fara*', *Burgundo-faro* 'travelling companion of the Burgundians'; a similar formation is Lgb. **fara-mannia* 'company of *faramanni*'.[48]
- OF *guaite* fem. 'guard', MLat. loanword *wacta* 'guards, watch service' (with Romance phonetic substitution of Germ. [ht] by [ct]), documented since the *Pactus Childeberti et Chlotharii* (a. 511/58) in law texts like the Carolingian *Lex Francorum Chamavorum* and in the *Capitulare Bononiense* (a. 811) < Frk. **wahta-* (cf. OHG *heri-wahta*), derived from Germ. **wâken* 'to watch over, to be wake'; It. *guata, guatare* 'guard, to keep watch' probably loaned from OF. Cf. OF *eschar-gaite* fem. 'guards' < Frk. **skar(a)-wahta* 'guarding by a troop'.[49]

3.2 *The Warriors and Their Social Stratification*

- Lgb. *waregang* 'stranger, submitted to the law of the Langobards', also 'immigrant warrior asking for royal protection' (ER 367), etymologically

47 See Wolfgang Haubrichs, 'Langobardisch-fränkische Ortsnamen in Oberitalien: Zu den toponymischen Typen *Stuttgart, Gamundio* und *Herstall/Wardstall*', in *Namenkundliche Informationen*, 109/110 (2017), here 279–284.

48 Sabatini, 'Riflessi linguistici', pp. 146–153; Carla Falluomini, 'Sui toponimi longobardi dei territori di Arenzano delle Valli Stura e Orba (Genova)', in *Rivista Italiana di Onomastica*, 18 (2012), here p. 132; Wolfgang Haubrichs, 'Ein namhaftes Volk – Burgundische Namen und Sprache des 5. und 6. Jahrhunderts', in V. Gallée, ed., *Die Burgunder – Ethnogenese und Assimilation eines Volkes* (Worms, 2008), here 162–163; Id., 'Leudes', pp. 248–263; Id., 'Sprache und Schriftlichkeit', pp. 158, 167–169.

49 Niermeyer, *Mediae Latinitatis Lexicon*, pp. 629–630; *DELI*, p. 703; de Sousa Costa, *Studien*, pp. 256–260. Cf. *EWA*, IV, c. 981; Kluge – Seebold, *Etymologisches Wörterbuch*, p. 965.

< **wara-ganga-* 'who goes or serves under royal protection', compound of Westgerm. *wara* 'attention, protection'; cf. the *war-gengus* 'stranger in the service of the king' in the *Lex Chamavorum* (c. 9); OE *waer-ganga* 'man, asking for protection, stranger'.[50]

- OF *garçun* 'servant, lad, puer' < Westgerm. **wrekk-jôn* 'the persecutee, the exiled' (cf. OS *wrekkio*, OE *wrecca*, OHG *reccheo*), originally also designation of the stranger in royal service.[51]
- *an-druscio, an-trustio, an-trussio* 'vassal, follower, member of the royal *trustis*', MLat. loanword (*quod puer regis est*) in the *Lex Salica*, *Formulae Marculfi*, *Pactus legis Salicae* and *Lex Ribuaria* etc. (in Carolingian times we can establish a semantic development to 'armed troop') < Frk. **trusti-*, by apophony ('ablaut') connected with Germ. **trausta-* 'comfort, help' (cf. ON *traust*, OHG *trôst*, OS *helm-gitrôstea* 'armed follower') and Goth. *trausti* 'treaty, alliance'.[52]

3.3 *Horse Breeding and Riders*

- Lgb. *maris-scalco, mar-scalc* 'master of the horses and stables, strator' (CDL no. 253 a. 771 Chiusi; no. 257 a. 771 Brescia); OF *mareschal*, also MLat. loanword *mariscalcus* in the *Lex Salica* and the *Lex Alamannorum* < Westgerm. **marah-skalka-* 'servant and guardian of the stables and horses', compound of Germ. **marah-* 'horse' + **skalka-* 'servant'.[53]
- MLat. *stod-arius* 'guardian of the horses' (a. 774 Benevento),[54] *stotarius* in the *Pactus Alamannorum* (5, 5), derived from Germ. **stôda-* 'herd of horses' (cf. Engl. *stud farm*). Cf. the certainly Lgb. toponym, frequent north of the river Po **stôda-gardôn* 'stud farm', in the 13th c. also documented as loanword

50 *Lex Chamavorum*, § 9 (ed. Sohm, p. 272). Cf. van der Rhee, *Die germanischen Wörter*, p. 133; Francovich Onesti, 'Vestigia', pp. 129–130; Niermeyer, *Mediae Latinitatis Lexicon*, II, p. 1468–1469.

51 Cf. Heinrich Beck, 'Recke', in *Reallexikon der Germanischen Altertumskunde*, 24 (2003); Kluge – Seebold, *Etymologisches Wörterbuch*, p. 751.

52 Niermeyer, *Mediae Latinitatis Lexicon*, pp. 63–64. Cf. Siegmund Feist, *Vergleichendes Wörterbuch der gotischen Sprache* (Leiden, 1939), p. 479; Hans Kuhn and Rainer Wenskus, 'Antrustio', in *Reallexikon der Germanischen Altertumskunde*, 1 (1973); Erich Gabriel, 'Antrustio', *Lexikon des Mittelalters* 1 (1980); de Sousa Costa, *Studien*, pp. 330–333; Kluge – Seebold, *Etymologisches Wörterbuch*, p. 932.

53 *DELI*, p. 935; Francovich Onesti, 'Vestigia', p. 103. Cf. *EWA*, IV, c. 152–155; Kluge –Seebold, *Etymologisches Wörterbuch*, p. 603.

54 Sabatini, 'Riflessi linguistici', p. 228; Haubrichs, 'Langobardisch-fränkische Ortsnamen', p. 273 fn. 14.

at Brescia; likewise as appellative and toponym in Bavaria and Alamannia (cf. Stuttgart).[55]

- French *galoper* 'to galop' < Frk. **wala hlaupan* 'to run well', verbal compound of Germ. **welô* 'well' (cf. OS *wola, wela, wala*, OHG *wola*, OE *wel*) + Germ. **hlaupan* 'to run'.[56]
- Cf. below It. *sperone*, MLat. *sporuni* 'spurs'.

3.4 *Weapons, Armament, Equipment*

- It. *schiniere* 'part of the armour for protection of the shin', MLat. *schineria*, derived from Westgerm. **skinô* fem. 'splint, shin' (cf. OHG *skina*, OE *scinu*).[57]
- OF *hauberc* masc. 'cuirass' < Frk. **hals-berg-* 'protection for the neck', compound of Germ. **halsa-* 'neck' + **berga-* 'protection'.[58]
- It. *sperone*, var. *sprone*, also MLat. loanword *sporuni* plur. (CDL, II, no. 295 a. 768 Pisa etc.) < Germ. **spurôn* 'spur' (cf. OE *spora*, OHG *sporo*), derived from Germ. **spurnan* 'to spur'; exists as loanword too in OF.[59]
- OF *gunfanun*, It. *gonfalone* 'war flag' < Lgb. Frk. *gundfano* 'vexillum' (OHG *gund-fano*), compound of Germ. **gunthô* fem. 'war, fight' + **fanôn* 'flag' (OS OHG *fano*, OE *fana*).[60]
- French *estandart* masc. 'standard' < Frk. **stand-hardu* 'steady, steadfast', compound of Germ. **standan* 'to stand' + **hardu-* 'strong, firm, hard'.[61]

3.5 *Fortification*

- It. *spalto* 'palisade stake, parapet, balustrade of a wall', a. 1158 MLat loanword *spaldus* < Lgb. **spalto* 'splitted wood', derived from Westgerm. **spaldôn* 'to split, to crack' (MLG *spalden*, OHG *spaltan*).[62]

---

55 Francovich Onesti, 'Vestigia', p. 121. Cf. Wolfgang Haubrichs, 'Un termine longobardo nella toponimia dell'Italia settentrionale: germ. **stôda-gardôn*, long. **stode-garda* 'recinto per cavalli', in *L'Italia Dialettale. Rivista di Dialettologia Italiana*, 80 (2019); Id., 'Langobardisch-fränkische Ortsnamen', pp. 270–273.

56 Kluge – Seebold, *Etymologisches Wörterbuch*, p. 329.

57 *DELI*, p. 1469. Cf. Kluge – Seebold, *Etymologisches Wörterbuch*, p. 803.

58 Cf. Ortwin Gamber, 'Haubert', *Lexikon des Mittelalters* 4 (1989); *EWA*, IV, c. 780.

59 *DELI*, p. 1584. Cf. Kluge – Seebold, *Etymologisches Wörterbuch*, p. 870.

60 *DELI*, p. 678. Cf. Tiefenbach, *Studien*, pp. 32–33; de Sousa Costa, *Studien*, pp. 191–192; *EWA*, III, c. 47–49; Kluge – Seebold, *Etymologisches Wörterbuch*, p. 272.

61 Cf. Kluge – Seebold, *Etymologisches Wörterbuch*, p. 876.

62 *DELI*, pp. 1575–1576. Cf. Kluge – Seebold, *Etymologisches Wörterbuch*, p. 861.

3.6 *War Actions*

- OI *guerra*, OF *guerre* fem. 'turmoil, riot, war', MLat. *werra* (only in capitularies of Charles the Bald) < Frk. **werra* 'confusion, muddle, conflict' (cf. OHG *werra* 'discord, scandalum'; OHG *werran* 'to confuse, to devastate', OS *werran* 'to press hard, to attack'); MLat. loanword *werrire* 'to devastate' a. 860 in a capitulary of Charles the Bald. The meaning 'war' is certainly product of a secondary development, but the term was so widely spread, that it replaced the inherited Lat. *bellum* nearly in the whole Romania.[63]
- It. *guardare*, French *garder* < Germ. **wardên* 'to guard, to watch' (cf. OS *wardon, wardien*, OE *weardian*, OHG *wartên*, ON *varda*). The substantive Frk. MLat. *warda* is documented in the Carolingian *Lex Chamavorum* and in Charlemagne's *Capitulare Bononiense* (a. 811). Compare the *curtis regia* of *Guastalla* (north of Reggio) in Northern Italy, a. 864 *Vuardistalla* < **warda-stalja-* 'watch tower'.[64]
- OF *eschec, eschac* 'armed robbery, fight'; MLat. *scachum* 'rapacious attack'; as vernacular Frk. term in a capitulary of Charles the Bald a. 853 (*adsalitura, quod scach vocant*), much later also in diplomes of Otto I. and Konrad II., partly of Italian provenance (spreading from there in OI dialects) < Frk. **skâk* < Westgerm. **skaeka-* 'robbery' (cf. OHG *schâh* 'latrocinium', OFr *skâk*, OHG *skâchâri* 'robber' in the 'Ludwigslied' a. 881/82,[65] OFr *skâkere*, OE *scêacere*).[66]
- It. *zava* fem. 'armament, armed riot' (Rat. 10: ... *per singulas civitatis mali hominis zavas et adunationes contra iudicem suum agendum faciebant*), MLat. loanword from Lgb. **zawa*, derived (with shift of the Tenuis [t] > [ts]) from Germ. **taw-ijan* (cf. OHG *zowên* 'to prepare', *zawên* 'to be successful', OS *tôgian* 'to make', OE *tawian* 'to prepare'; further OE *ge-tawe* 'armament', Norw. *gotvar* < **ga-tâwjia*, OHG *gi-zawa* 'household effects'; cf. MHG *ge-zouwe*, German *Ge-zähe* 'equipment'.[67]

---

63 *DELI*, pp. 703–704; de Sousa Costa, *Studien*, pp. 293–297; Kluge – Seebold, *Etymologisches Wörterbuch*, pp. 379, 991 (*Wirren*).

64 *DELI*, pp. 700–701; Sabatini, 'Riflessi linguistici', pp. 164–165; de Sousa Costa, *Studien*, pp. 260–262; Falluomini, 'Sui toponimi', pp. 132–133; Kluge – Seebold, *Etymologisches Wörterbuch*, p. 972. Cf. Wolfgang Haubrichs, 'Gamondio bei Marengo. Westgermanische Ortsnamen in Italien und ihre europäischen Verwandten', in *Nomina. Studi di onomastica in onore di Maria Giovanna Arcamone*, eds. D. Bremer – D. De Camilli – B. Porcelli (Pisa, 2013), here 263–264; Id., 'Langobardisch-fränkische Ortsnamen', pp. 279–284.

65 *Ludwigslied*, v. 17 (ed. von Steinmeyer, pp. 85–86 no. XVI). Cf. Kluge – Seebold, *Etymologisches Wörterbuch*, p. 791.

66 Niermeyer, *Mediae Latinitatis Lexicon*, II, p. 1228. Cf. de Sousa Costa, *Studien*, pp. 309–315.

67 van der Rhee, *Die germanischen Wörter*, p. 143; Francovich Onesti, 'Vestigia', p. 134; Kluge – Seebold, *Etymologisches Wörterbuch*, p. 358 (*Gezähe*).

- Frk. *scaftlegi* 'laying down of arms, armistice', documented in the *Capitulare missorum Wormatiense* of Louis the Pious (*id est armorum depositio*) and in the *Edictum Pistense* of Charles the Bald, MHG *schaft-lege*, compound of Germ. **skafta-* masc. 'shaft, rod, spear' (OHG *scaft* translates the Lat. lemmata *hasta, telum, iaculum*).[68]

3.7 *Warrior Mentality and Emotions*

- a. 643 MLat. *asto animo* 'with hostile intentions', also *asto* adv., important law terms (ER 201: ... *si asto animo, id est volontariae, occiserit*; cf. ER 146, 149, 201, 229, 248, 264 294, 342; Liut. 71, 151), with Romance phonetic substitution of Germ. [ai] by [a] < Lgb. **hai(f)st-* 'angry, aggressive' < Germ. **haifsti-* 'conflict, effort' (cf. Goth. *haifsts*; OHG *heifti* 'violent, furious'); from the same stem also It. *astio* 'grudge, anger, hostility'. By way of contrast, the law term *haistan* (acc. sing.) – with Latin translation and preservation of Germ. [h] – represents a genuine Lgb. word with identic meaning (ER 277 titulus: *De haistan, id est furorem*; text: *Si quis in curtem alienam haistan, id est irato animo, ingressus fuerit, ...*).[69] Cf. with slightly changed meaning also OF *haste* 'violence' < Frk. **haifsti-*.[70]
- OF *balt, baud* 'joyful, cheerful, merry'; It. *baldo* 'who shows dextery and competence' < Germ. **baltha-* 'bold, courageous' (cf. OS OHG *bald*, OE *beald*).[71]
- OF *estout* 'bold, proud, arrogant' < Westgerm. Frk. **stulta-* (cf. OFr *stult*, MLG *stolt*, MNL *stout*, OHG *stolz* 'superbus').[72]
- French *hai:r* 'odiare, to hate', MLat. loanword *hatire* (*O[disti] : hadisti*) in the 'Reichenau Glosses' (Corbie, eighth c.) < Germ. **hat-jan*, derived from Germ. **hataz-* 'odium, inimicitia, hate' (cf. Goth. *hatis*; ON *hatr*; OE *heti*, OS *hete*, OHG *haz*).[73]

---

68 De Sousa Costa, *Studien*, pp. 288–290; Kluge – Seebold, *Etymologisches Wörterbuch*, p. 793 (*Schaft*).

69 *DELI*, p. 140 (*astio*). Cf. van der Rhee, *Die germanischen Wörter*, p. 20, 79–80; Max Pfister, 'I prestiti linguistici di origine germanica fra Tardo Antico e Alto Medioevo', in *La cultura in Italia fra Tardo Antico e Alto Medioevo*, I (Rome, 1981), here pp. 273–275; Kluge – Seebold, *Etymologisches Wörterbuch*, p. 398 (*Hast*), p. 402 (*heftig*).

70 Cf. Haubrichs, 'Il linguaggio della guerra', p. 241, fn. 120.

71 *DELI*, p. 169; *LEI*, I, pp. 106–127. Cf. *EWA*, I, pp. 434–436; Kluge – Seebold, *Etymologisches Wörterbuch*, p. 84.

72 Kluge – Seebold, *Etymologisches Wörterbuch*, p. 887. Cf. Rudolf Schützeichel, *Althochdeutscher und altsächsischer Glossenwortschatz*, IX (Tübingen 2004), p. 246.

73 Hans-Wilhelm Klein, *Die Reichenauer Glossen*, pars I: *Einleitung, Text, vollständiger Index und Konkordanzen* (Munich, 1968), p. 180 no. 1118. Cf. *EWA*, IV, pp. 871–873; Kluge –Seebold, *Etymologisches Wörterbuch*, p. 421.

- French *honte* fem. 'shame, disgrace, dishonour'; MLat loanword **haunitas* (*Ignominia : hauntes*) in the 'Reichenau Glosses' (Corbie, eighth c.)[74] < Frk. **haun-itha-*, derived from Germ. **hauna-* 'inferior, contemptible' (cf. Goth. *hauns*; OHG *hôna* 'scorn, derision'). It is noteworthy that the term includes aspects of public and social valuation.[75]

3.8 *The Military Language of the Langobards*[76]

In the present state of research approximately 280 borrowings from Langobardic can be traced for Italian and its dialects or regional languages. All of them must have been transferred before the vernacular of the *gens Langobardorum* was extinguished (end of eighth c.). Just a quarter of the loanwords (ca. 75) represent the semantic fields of war and fighting, while the rest belongs to the spheres of law, agriculture, handicrafts and even of everyday vocabulary (e.g. parts of the body), so suggesting an intensive long-lived bilingualism. A small quantity of words could also be adopted from Frankish, that, as language of administration and negotiation was also influential in Northern and Central Italy after 774. In some cases we may recognize double borrowings, like for example *mari-scalc-* 'strator', *sperone* 'spur', *gonfalone* 'war flag', *guardare* 'to guard', *guerra* 'confusion, turmoil, war', *baldo* 'bold courageous' etc.

Going through the semantic fields, in which the military language settled, we can find a very variable distribution. Only a rather small quantity of lemmata (10%) can be placed in the section of the army and its detachments. Perhaps not fortuitously, many words in this field are 'supragentile' terms like *hari* 'army', *fulca* 'troop of warriors', *scara* 'army detachment', that is, they appear in the languages of different *gentes*. The same term *fara*, documented for units of organization of the *exercitus Langobardorum* in the times of the migration and the first battles in Italy (late sixth c.), is also attested for Franks, Alamanni, Bavarians and Burgundians.

A considerable quantity of lemmata refers to the terminology of warriors and their social stratification (24%). This group of words is strongly oriented towards the vocabulary of royal power, of court and administration, that is, to the vocabulary of the leading classes, which fulfils military functions at the same time. The following terms seem to be especially Langobardic: (*h*)*ari-man* '(homo) exercitalis', *scario* 'leader of an army detachment', *scilde-por* 'bearer of the shield, armiger', *bane-sagius* (a kind of herald), *ga-staldius* 'administrator of an estate, head of a district' and *aldio* 'semiliberus', finally *mar-pais*

74 Klein, *Die Reichenauer Glossen*, p. 173 no. 825.

75 *EWA*, IV, c. 1125, 1130–1131. Cf. Kluge – Seebold, *Etymologisches Wörterbuch*, p. 421.

76 For more details see Haubrichs, 'Il linguaggio della guerra', pp. 220–234.

'administrator of a stud farm, of the stables' (in contrast to Frankish **marah-skalk* for the same function, attested early).[77] In general, the breeding and training of horses was very important for the Germanic *gentes*; here we have to place also the Germano-Latin hybrid term *studarius* 'guardian of the horses'.

The most represented semantic field of the military language is that referring to weapons, armament and equipment (about 30%). Many of the terms belonging to this field immediately concern weapons: for instance, **branda-* 'sword'; *scild* 'shield'; **strala* 'arrow'; *gaida* 'arrow point' and *gis(il)* 'arrow shaft, spear shaft'; **gaira-* 'spear' and the synonyms **stanga* and **stollo*. To the special equipment of horse warriors refer **spurôn* 'spur' (It. *sperone*) and *staffa* 'stirrup'. The following words seem to be of particular Langobardic origin: **rauba-* 'armour' (It. *roba*); *hari-gawerki* 'equipment for the army (service)'; It. *schiniere*, MLat. *schineria* 'part of the armour for protection of the shin', then *gaida, gis(il), *stollo, *stôzza* 'hammer', *zippa* 'tip' < Germ. **tip(p)* and **zaino* 'quiver'. In many cases we find lexical parallels to other Westgermanic languages, especially to Frankish and Old High German.

The terminology of fortification is scantily represented in Langobardic. Only two terms, referring to earth ramparts and similar constructions, are to find: It. *spalto* 'wooden palisade of a protecting wall' < Lgb. *spalto* 'slit, splitting, splitted wood' and It. *rampa* < Germ. Lgb. **(h)rampa-* 'ramp'. The latter term appeared also in Frankish. The weak presence of fortification terms does not surprise, since *castra* etc. required complex techniques, available only with difficulties to Germanic peoples.

The sphere of military actions, of attack and defense, is much better represented. Within it, there is a clear priority for aggressive actions, above all in specifically Langobardic terms, concentrated nearly exclusively on raids, robbery and conspiration: Lgb. *hobe-ros* 'illegal entry of someones house or courtyard'; *hari-treib* 'raid, holdup'; *hari-schild* 'armed holdup'; *walo-paus* 'attack of masked men'; *zuffa* 'short furious fight'; finally **zawa* 'armed riot'. Very well represented in Germanic languages are also verbs, belonging to this semantic field like It. *rattrapare* 'to catch', derived from Lgb. Frk. **trappa-* 'trap', already documented together in the *Lex Salica* with the *mallobergo*-synonym *falla* < Westgerm. *fallôn* 'trap'; It. *ar-raffare* 'to snatch with violence' < Lgb. **(h)raffôn* 'to obtain'; It. *bluttare, de-bluttare* 'to bare, to rob' < Lgb. **blutt-* 'naked, poor'; It. *guarire* < Germ. **war(i)jan* 'to defend himself'; It. *guarnire* < Westgerm. Lgb. **warnôn* 'to be careful, to watch out, to equip'; It. *schermire* < Lgb. **skerm-jan* 'to protect, to fight' (with Frk. and OHG parallels).

---

77 Cf. Haubrichs, 'Il linguaggio della guerra', p. 225 fn. 29.

Only a few lemmata (10%) settle in the field of warrior mentality and emotions. This result does not surprise, if we consider that this semantic area was certainly densely covered with terms of Latin provenance. Most of the borrowings seem to be specifically Langobardic, like It. *snello* 'quick, agile' < Germ. **snella-* 'fast, agile', It. *baldo* < Germ. **baltha-* 'bold, courageous' or *It. stracco* < Westgerm. **straka-* 'hard, tight', are also documented in Frankish and Old High German. Only Lgb. are *haistan* 'irato animo, angry' (cf. OI *asto*; It. *astio* 'grudge') and OI *riffo* 'strong', at least the vocables of flight like *a-stalin* 'flight, treason' and *fraida* 'illegitime flight'.

3.9 *The Military Language of the Franks*[78]

Although many specific studies of Langobardic vocabulary have been carried out,[79] a complete list and exhaustive linguistic analysis of the borrowings from Germanic languages in the area of the Italoromania has firstly appeared in the commendable first volume of the 'Germanismi' of the 'Lessico Etimologico Italiano (LEI)' (edited by Elda Morlicchio),[80] which in fact records only the lemmata beginning with the letters A, B/P. On the contrary, the early documented linguistic relics of the Langobards, above all the legal terminology, are well explored.[81] For Frankish, as 'superstratum' of the Galloromania, the situation seems to be precisely the opposite. We are well informed about the loanwords of Germanic provenance by some special volumes of the 'Französisches Etymologisches Wörterbuch (FEW)'[82] by Walther von Wartburg, by the first volume of the 'Romania Germanica' of Ernst Gamillscheg, first edited in 1933 and subsequently deeply revised in 1970,[83] and thanks to an overview sketched by Max Pfister and Wolfgang Haubrichs, published in 2008 in the 'Wieser

78 For more details see Haubrichs, 'Il linguaggio della guerra', pp. 234–246.

79 One of the initial works concerning the linguistic heritage of the Langobards in Italy is: Ernst Gamillscheg, *Romania Germanica. Sprach- und Siedlungsgeschichte der Germanen auf dem Boden des alten Römerreichs*, I, 2nd ed. (Berlin/Leipzig, 1935), pp. 57–229.

80 *LEI – Germanismi*, I. Cf. Elda Morlicchio, 'Die sprachliche Hinterlassenschaft der Germanen auf der italienischen Halbinsel', in *Akkulturation – Probleme einer germanisch-romanischen Kultursynthese in Spätantike und frühem Mittelalter*, eds. D. Hägermann – W. Haubrichs – J. Jarnut (Berlin, 2004).

81 Surveys are: Max Pfister, 'Langobardische Superstratwörter im Italienischen', in *Jahrbuch für Internationale Germanistik*, 11 (1979); Francovich Onesti, 'Vestigia', pp. 51–138; Haubrichs, 'Sprache und Schriftlichkeit', pp. 138–144.

82 Walther von Wartburg, *Französisches Etymologisches Wörterbuch* (*FEW*), vol. 15–17 (Basel, 1959–1969). Cf. Gamillscheg, *Etymologisches Wörterbuch*; Wilhelm Meyer-Lübke, *Romanisches Etymologisches Wörterbuch* (Heidelberg, 1992).

83 Gamillscheg, *Romania Germanica*, I, pp. 35–425 (with a special chapter about military language: § 117–126).

Enciclopedy – Western European Languages'.[84] The early documented Germanic borrowings in the Galloromania, however, are rarely studied, above all because of the unsolved problems of text tradition of the Old Frankish 'Malberg Glosses' in the _Lex Salica,_[85] often corrupted and obscured by romanization and misunderstandings, which prevented precise etymological interpretations.

Nonetheless, we can make some preliminary remarks for the Frankish words belonging to the field of military language. We can note approximately 500 loanwords in Old French – mainly in the Northern and Eastern dialects, but also by a lot of Middle Latin borrowings – as showing the strong and widespread influence of the language of the new masters of the Gallia in Merovingian times. From the beginning, borrowings from Frankish expanded into wide areas of the everyday life of a society mostly formed by agriculture and handicrafts: farming, stockbreeding, beekeeping, animals, parasites, insects, plants, toolmaking, vessels etc. We have to add words referring to aspects of house construction and settlement of the Franks. Considerable quantities of loanwords were developed in semantic fields dealing with activities of warriors and nobles like hunting, weapons and military techniques, riding and horse breeding, but also administration, public language, law and official functions. Certainly, especially for the military language, the wordlist is actually incomplete. But it seems evident, that the about 60 Galloromance (including Old French) and Middle Latin loanwords, to be found in the semantic field of war and battle, comprise a much smaller share (about 12%) of the totality of borrowings, compared with the terminology of the Langobards, where they amount to a quarter of the whole.

Regarding the Franks, also the vocabulary refering to the army and its subdivisions has just the same modest dimensions as compared to the Langobards (10%). There is however a difference: apart from generic terms, the dominance of the technical terminology of a well-organized army stands out. This is testified by technical army-words like MLat _hari-_, _heri-bannus_ < Frk. _*harja-banna-_ 'call up to army, conscription'; Northern French _heri-berge_ < Frk. _*hari-berga-_ 'army lodging, quarter' and by technical terms touching the organization of the service like MLat _wacta_ < Frk. _*wahta-_ 'guard, watch service' or _*scar-wahta-_ and the connected Romance hybrid OF _eschariere_ < _*scar-aria_ 'army detachment', derived from Frk. _*skara-_ 'troop, group of warriors'.

84 Wolfgang Haubrichs and Max Pfister, 'Fränkisch (Frankish)', in _Wieser Enzyklopädie: Sprachen des europäischen Westens_, 1, eds. U. Ammon – H. Haarmann (Klagenfurt/Vienna/Ljubljana, 2008), pp. 249–274 (with a bibliography).

85 Ruth Schmidt-Wiegand, 'Malbergische Glossen', in _Lexikon des Mittelalters_, 6 (1992).

In the semantic area of warriors and their social stratification we find not more than 9% of the attested terms, focusing (surely not by chance) on the king and his court. As examples we may mention **marah-skalk-* 'servant and guardian of the horses and the stables' and **sene-skalk-* 'the responsible for the household and court' (etymologically 'the oldest of the servants'); to join is the latinized *antrustio* 'vassal, member of the royal *comeatus*', derived from Frk. **trusti* as designation of the *comitatus*, etymologically rooted in the sphere of helping and consolation in difficulties and distress, a term which underlines the emotional aspects of the royal or seigneurial environment. Also, the Old French term *garçun*, developed from Frk. **wrekjo* 'exsul', originally designated most likely the *puer regis* of foreign origin, living under the king's protection: in the majority of the cases the *puer* was certainly a warrior. Finally, we have to emphasize the venerable institution of the *leudes* < Germ. **leudi-*, plur. 'people', designating the free men belonging to the *populus-exercitus* of Franks and Burgundians. In the times of the new Carolingians, Charles Martell (†741) and Pippin (†768) the *leudes*, perhaps joined together in a sort of guild, mainly acting as warriors and protectors of the king.

The words for weapons, armament and military equipment (included horse breeding and keeping) fill the most extensive semantic area (39%); as amongst the Langobards (30%) they are strongly represented. Technical terms prevail; not less than 14 lemmata refer to offensive weapons: **speuta-* 'spear, spit'; **speru-* 'spear, lance'; **fleukka-* 'arrow'; **bultjôn* 'missile, bolt'; **helt-* 'sword handle'; **habjôn* 'billhook, war-scythe' (OF *hache* 'a bent knife'), also used in the agrarian sense of 'sickle'; **staka-* 'pointed staff, rod, made for sticking'. Here we find too weapons, which are considered as typically Frankish: *ango* 'lance with barbs'; **daroth-* 'javelin, dart'; *scrama-sax* 'short sword' and *barda* 'axe', may be the axe for throwing, called in early Frankish sources *francisca* 'the Frankish one'. The following terms also seem to be specifically Frankish: **targô* 'shield' (instead of *skild* used in Langobardic and Old High German); furthermore Galloromance **guanta* < Frk. **wanta* for a glove used in sword fighting, who was so important that it could serve as 'supernomen' for warriors; finally French *garrot* 'javelin', derivated from Frk. **wrokkôn*. Only two terms refer to the protection of the warriors: first **brunjôn* 'breast plate, cuirass', second **hals-berg* 'protection for the neck, collarium', whereas for OF *heaume*, It. *elmo* 'helmet' borrowing from Gothic (?) is supposed. In contrast to the Langobards, apart from the generic **gund-fano* 'battle-flag', the Franks also had a second special term, i.e. **stand-hardu-* 'standard', perhaps marking originally the flag of the *castrum* or the army camp.

Concerning the equipment of horse warriors, beside the Westgermanic term **hangista-* the Franks also had two further words for the stallion: first **stallôn*,

OF *estalon*, and then **wrain-jôn*, Anglonorman *gareignun* (cf. OHG *reinno*), the latter originally designating the young male horse. We have to add the term for the gallop, important for the attack with horses: **walu hlaupan* (OF *waloper, galoper* with the Galloromance phonetic substitution [g] for Germ. [w]), litterally 'to run well'. Apart from the normal term for the spur, OF *esperon* < Germ. *spurôn*, the Franks had a special term for the stirrup, i.e. **streupa-*, MLat. *strepus* (instead of Lgb. *staffa*).

The second semantic field, richly represented in Frankish with 19 lemmata (32% – clearly more than the 22% of the Langobards) is that of war actions. In Langobardic, the aggressive and generally illegal actions (agression, robbery, conspiration) dominated, while in the Frankish vocabulary technical aspects of army-organization prevailed: MLat loanword *bannire* 'to call to the arms, praeparare ad hostem' (*Pactus Childeberti et Chlotharii*, sixth c.); MLat. *adfannire* 'to call to the colours', derivated from Germ. *fanôn* 'banner'; *land-wêri* 'defensio patriae'; MLat. *warnire* 'to equip' (OF *garnir*) < Westgerm. **warnôn* 'to take care, to equip'; furthermore, the Frankish technical terms *spehôn* 'to spy'; **têri* 'forming up for battle' (Jewish French *teire*; cf. OHG *zieri* 'decoration'); OF *estur* 'attack, fight' < Germ. **sturma-* 'storm'. But the Franks too did not miss the designations of aggressive, not institutionalized activities like **werra* 'turmoil, riot' (MLat. *werra* about 877; OF *guerre*); **strîda* 'fight, conflict' (OF *estrit*); **haifsti* 'violence, furor' (OF *haste*); MLat. *marrire* 'to create disorder, to mess up' (French *marrer*) < Westgerm. **marr-jan* 'to disturb, to devastate'; **harijôn* 'to devastate' (OF *harigoter* 'to tear up'); MLat. *scachum*, a. 853 OHG *scach* 'armed robbery', OF *eschec, eschac* 'robbery, battle' < Frk. **skâk-*; MLat. *texaga, tesceia* 'violent robbery', already documented in the 'Lex Salica', derived from Germ. **takan* 'to take'. However, in this semantic area appear also evidently younger terms of Rhenano-Frankish origin like *hari-zuht* 'robbery raid of a band'; *herisliz* 'desertion' and again as technical term *scaft-legi* 'armistice'. A much earlier borrowing, documented since the sixth century, is represented by MLat *fredum* 'pacification' < Germ. **frithu-* 'peace, treaty', referring to containment and closing of war and fighting.

Corresponding with the situation in Langobardic, also in Frankish the semantic area of loanwords referring to warrior mentality and emotions comprises only a small quantity of terms (10%). Once again, such a scantiness is not unexpected, for that very field was already abundantly planted with Latin terms. By contrast, it is surprising that we can find any borrowings at all from the Frankish superstratum-language. These are above all positive qualities of fighters and warriors, which are provided with borrowings from the Germanic vernacular: **baltha-* 'bold, courageous' (OF *balt*); **stulta* 'proud, bold'; **ur-gôl* 'glorious' (OF *orgolz*) and the specifically Frankish word **man-haidu* 'nature of man, courage, strength' (OF *manaie* 'power'), compound of Germ. **man-* 'man, male' and

**haidu-* 'nature, essence, form'. On the opposite pole of the warrior qualities, we have a verbal term describing the basis of *inimicitia,* that is MLat. *hatire* (French *hai:r*) < Frk. **hat-jan* 'to hate', replacing the Latin verb *odire*, derivated from *odium*. On the negative side as well, the term **haun-itha* 'shame, disgrace' (OF *honte*) belongs to the area of dis-honour, surprisingly substituting in French the Latin linguistic heritage, meanwhile Latin *honor* ('honour') maintains its place.

4 Conclusions

Personal names, ethnonyms, group designations, early documented terms and loanwords in neighboring Romance languages from the fields of military semantics (army detachments, social stratification of warriors, horse breeding, weapons, armament, fortification, war actions, mentality and emotions) show how deeply Germanic peoples (here the Franks and the Langobards) were influenced by war and fight. The process of substitution of the Latin lexical heritage by Langobardic and Frankish terms shows us how intensive must have been the linguistic contacts and exchanges between the (quantitatively marginal) Germanic elites and the Roman-Romance upper classes. The intensity of contacts can also be observed on the basis of borrowings from semantic fields not analysed here, which necessarily hosted a Latin or Romance lexical *patrimonium*, for instance the terminology of the human body or the vocabulary of everyday communication. Such intimate linguistic interferences can only be explained by a daily mixing of languages, by a code-switching practised in the administrative and military elites first in the late Roman *exercitus* and then in the armies of the post-imperial *regna*. Beside new weapons and new warfare techniques, it is the partial bilinguality of the elites that explains the high percentage of military terms of Germanic provenance found in the languages of the Franks and the Langobards as well as in the Romance successor-languages, where they are borrowed from the language of the conquerors.

Bibliography

Primary Sources

Annales Fuldenses, ed., F. Kurze [MGH, *Scrip. Rer. Germ. in usum scholarum*, 7] (Hannoverae, 1891)

Capitolare Bononiense, ed. A. Boretius, in *Capitularia regum Francorum*, 1, [MGH, *Leges* II.1] (Hannoverae 1883/1897) p. 166 no. 74

CDL = *Codice Diplomatico Longobardo*, I, ed. L. Schiaparelli [Fonti per la storia d'Italia, 62] (Roma, 1929)

Chronicarum quae dicuntur Fredegarii libri quattuor, ed. B. Krusch [MGH, *Script. rerum Merov.* 2] (Hannoverae, 1888), pp. 1–193 // ed. H. Wolfram – A. Kusternig, *Quellen zur Geschichte des 7. und 8. Jahrhunderts* (Darmstadt 1982)

Ermoldus Nigellus, *Carmen in honorem Hludowici christianissimi Caesaris Augusti*, ed. E. Dümmler [MGH, *Poetae Latini*, 2] (Berlin, 1884)

Gregorius Turonensis, *Libri Historiarum X*, ed. B. Krusch – W. Levison [MGH, *Script. Rer. Meroving.* I.1] (Hannover, 1951) – Engl. transl. by L. Thorpe, *Gregory of Tours, The history of Franks* (London, 1974)

Leges Liutprandi, ed. C. Azzara and S. Gasparri, *Le leggi dei Longobardi. Storia, memorie e diritto di un popolo germanico* (Rome, 2005), pp. 137–244 // ed. F. Bluhme, in *Leges Langobardorum* [MGH *Leges* IV] (Hannoverae, 1868), pp. 96–175

Leges Ratchis, ed. C. Azzara and S. Gasparri, *Le leggi dei Longobardi. Storia, memorie e diritto di un popolo germanico* (Rome, 2005), pp. 259–277 // ed. F. Bluhme, in *Leges Langobardorum* [MGH *Leges* IV] (Hannoverae 1868), pp. 183–193

Lex Chamavorum, ed. R. Sohm [MGH *Leges* V] (Hanoverae 1875/79), pp. 269–276

Ludwigslied, ed. E. von Steinmeyer, *Die kleineren althochdeutschen Sprachdenkmäler* (Berlin 1916), pp. 85–88

Notkerus Balbulus, *Gesta Karoli Magni*, ed. R. Rau, *Quellen zur karolingischen Reichsgeschichte*, 3 (Darmstadt, 1962), pp. 322–426 – Engl. transl. by D. Ganz, *Einhard and Notker the Stammerer: Two Lives of Charlemagne* (London, 2008), pp. 55–116

Opus imperfectum in Matthaeum, ed. J.-P. Migne, *Patrologia Graeca*, vol. 56 (Paris 1862), pp. 601–946

Paulus Diaconus, *Historia Langobardorum*, ed. G. Waitz [MGH *Scriptores rerum Germanicarum in usum scholarum separatim editi*, 48] (Hanoverae 1878)

Venantius Fortunatus, *Opera poetica, Carminum epistularum expositionum libri XI*, ed. F. Leo [MGH, *Auct. Ant.*, 4,1] (Berolini 1881) – Engl. transl. by J. George, *Venantius Fortunatus, Personal and Political Poems* (Norwich, 1995)

Secondary Literature

Banniard, Michel. *Viva voce. Communication écrite et communication orale du IVe au IXe siècle en Occident latin* (Paris1992)

Beck, Heinrich. 'Das Problem der theriophoren Personennamen im Germanischen', in *Zum Problem der Deutung frühmittelalterlicher Bildinhalte*, ed. H. Roth (Sigmaringen 1986), pp. 305–315

Beck, Heinrich, 'Recke', in *Reallexikon der Germanischen Altertumskunde*, 24 (2003) pp. 295–297

DELI = Manlio Cortelazzo and Michele A. Cortelazzo eds., *Dizionario Etimologico della lingua Italiana*, 2th ed. (Bologna 1999)

de Sousa Costa, Annette. *Studien zu volkssprachigen Wörtern in karolingischen Kapitularien* (Göttingen 1993)

EWA = A.L. Lloyd – R. Lühr, *Etymologisches Wörterbuch des Althochdeutschen*, I–VII (Göttingen 1998–2021)

Ewig, Eugen, 'Die Namengebung bei den ältesten Frankenkönigen und im merowingischen Königshaus', in *Francia* 18,1 (1991) pp. 21–69

Falluomini, Carla. 'Sui toponimi longobardi dei territori di Arenzano delle Valli Stura e Orba (Genova)', in *Rivista Italiana di Onomastica*, 18 (2012) pp. 130–142

Feist, Siegmund. *Vergleichendes Wörterbuch der gotischen Sprache* (Leiden 1939)

Francovich Onesti, Nicoletta, *Vestigia longobarde in Italia (568–774)* (Rome 1999)

Gabriel, Erich. 'Antrustio', *Lexikon des Mittelalters* 1 (1980) 735–736

Gamber, Ortwin, 'Haubert', *Lexikon des Mittelalters* 4 (1989) 1959

Gamillscheg, Ernst. *Etymologisches Wörterbuch der französischen Sprache* (Heidelberg 1969)

Gamillscheg, Ernst. *Romania Germanica. Sprach- und Siedlungsgeschichte der Germanen auf dem Boden des alten Römerreichs*, I, 2nd ed. (Berlin, 1970); II (Berlin/Leipzig 1935)

Haubrichs, Wolfgang, 'Veriloquium Nominis. Zur Namenexegese im frühen Mittelalter', in *Verbum et Signum. Festschrift für Friedrich Ohly*, 1, eds. H. Fromm – W. Harms – U. Ruberg (Munich 1975), pp. 231–266

Haubrichs, Wolfgang. 'Namendeutung im europäischen Mittelalter', in *Namenforschung – Name Studies – Les noms propres. Ein internationales Handbuch zur Onomastik*, 1, eds. E. Eichler – G. Hilty *et al.* (Berlin/New York 1995), pp. 381–390

Haubrichs, Wolfgang. 'Nomen gentis. Die Volksbezeichnung der Alamannen', in *Röllwagenbüchlein. Festschrift Walter Röll*, eds. J. Jaehrling et *al.* (Tübingen 2002) pp. 20–42

Haubrichs, Wolfgang, 'Identität und Name. Akkulturationsvorgänge in Namen und die Traditionsgesellschaften des frühen Mittelalters', in *Die Suche nach den Ursprüngen. Von der Bedeutung des frühen Mittelalters*, ed. W. Pohl (Wien 2004), pp. 85–105

Haubrichs, Wolfgang, '*Nomina stirpium.* Sippennamen und Ethnonyme: Probleme einer Typologie der Personengruppenbezeichnungen', in *Language and Text. Current Perspectives on English and Germanic Historical Linguistics and Philology*, eds. A.J. Johnston – F. von Mengden – S. Thim (Heidelberg 2006), pp. 57–78

Haubrichs, Wolfgang – Pfister, Max, 'Fränkisch (Frankish)', in *Wieser Enzyklopädie: Sprachen des europäischen Westens*, I, eds. U. Ammon – H. Haarmann (Klagenfurt/Vienna/Ljubljana 2008) pp. 249–274

Haubrichs, Wolfgang. 'Ein namhaftes Volk – Burgundische Namen und Sprache des 5. und 6. Jahrhunderts', in V. Gallée, ed., *Die Burgunder – Ethnogenese und Assimilation eines Volkes* (Worms 2008), pp. 135–184

Haubrichs, Wolfgang, 'Langobardic Personal Names: Given Names and Name-Giving among the Langobards', in *The Langobards before the Frankish Conquest: An Ethnographic Perspective*, eds. G. Ausenda – P. Delogu – C. Wickham (Woodbridge/San Marino 2009) pp. 195–236

Haubrichs, Wolfgang, 'Sprache und Schriftlichkeit im langobardischen Italien – Das Zeugnis von Namen, Wörtern und Entlehnungen', in *Filologia Germanica – Germanic Philology*, 2 (2010) pp. 133–201

Haubrichs, Wolfgang, 'Gamondio bei Marengo. Westgermanische Ortsnamen in Italien und ihre europäischen Verwandten', in *Nomina. Studi di onomastica in onore di Maria Giovanna Arcamone*, eds. D. Bremer – D. De Camilli – B. Porcelli (Pisa 2013), pp. 263–279

Haubrichs, Wolfgang. 'Typen der anthroponymischen Indikation von Verwandtschaft bei den 'germanischen' *gentes*: Tradition – Innovationen – Differenzen', in *Verwandtschaft, Name und soziale Ordnung (300–1000)*, eds. S. Patzold – K. Ubl (Berlin/Boston 2014), pp. 29–71

Haubrichs, Wolfgang. 'Langobardisch-fränkische Ortsnamen in Oberitalien: Zu den toponymischen Typen *Stuttgart, Gamundio* und *Herstall/Wardstall*', in *Namenkundliche Informationen*, 109/110 (2017) pp. 269–290

Haubrichs, Wolfgang. '*Leudes, fara, faramanni* und *farones*. Zur Semantik der Bezeichnungen für einige am Konsenshandeln beteiligte Gruppen', in *Recht und Konsens im frühen Mittelalter*, eds. V. Epp – C.H.F. Meyer (Ostfildern 2017), pp. 235–263

Haubrichs, Wolfgang. 'Tierische Identitäten. Zur symbolischen Kommunikation in Namen des frühen Mittelalters', in *Tiere: Begleiter des Menschen in der Literatur des Mittelalters*, eds. J. Klinger – A. Kraß (Cologne/Weimar/Wien 2017), pp. 229–254, 281–284

Haubrichs, Wolfgang. 'Funktion und Performanz. Namen als Instrumente der Sicherung von Herrschaft und Identität im frühen Mittelalter Westeuropas', in *Machterhalt und Herrschaftssicherung. Namen als Legitimationsinstrument in transkultureller Perspektive*, eds. M. Becher – H. Hess (Bonn 2019), pp. 235–264

Haubrichs, Wolfgang, 'Un termine longobardo nella toponimia dell'Italia settentrionale: germ. **stôda-gardôn*, long. **stode-garda* 'recinto per cavalli', in *L'Italia Dialettale. Rivista di Dialettologia Italiana*, 80 (2019) pp. 211–220

Haubrichs, Wolfgang. 'Il linguaggio della guerra presso i popoli germanici occidentali (Longobardi e Franchi)', in *L'esercito romano e l'alba dell'Europa. Modelli concettuali e sperimentazioni sociali tra linguaggi, istituzioni e diritto (secoli IV-VIII)*, ed. G. Bassanelli Sommariva (Santarcangelo di Romagna 2020), pp. 213–247

Heizmann, Wilhelm, 'Tiernamen', in *Reallexikon der Germanischen Altertumskunde*, 35 (2007), pp. 151–160.

Hergemöller, Bernd-Ulrich, 'Herberge', in *Lexikon des Mittelalters*, 4 (1989) p. 2148

Høiland Nielsen, Karen, 'The Wolf-Warrior-Animal Symbolism on Weaponry of the sixth and seventh centuries', in E. Pohl – U. Recker – C. Theune, eds., *Archäologisches Zellwerk. Beiträge zur Kulturgeschichte in Europa und Asien. Festschrift für Helmut Roth*, eds. E. Pohl – U. Recker – C. Theune (Rahden-Westphalia 2001), pp. 471–481

Jochum-Godglück, Christa, 'Wolf und Bär in germanischer und romanischer Personennamengebung', in *Interferenz-Onomastik. Namen in Grenz- und Begegnungsräumen in Geschichte und Gegenwart*, eds. W. Haubrichs – H. Tiefenbach (Saarbrücken 2011), pp. 447–477

Kaiser, Reinhold. *Die Burgunder* (Stuttgart 2004)

Klein, Hans-Wilhelm, ed., *Die Reichenauer Glossen*, pars I: *Einleitung, Text, vollständiger Index und Konkordanzen* (Munich 1968)

Kluge, Friedrich – Seebold, Elmar. *Etymologisches Wörterbuch der deutschen Sprache*, 25th ed. (Berlin/Boston 2011)

Kuhn, Hans – Wenskus, Rainer, 'Antrustio', in *Reallexikon der Germanischen Altertumskunde*, 1 (1973) pp. 360–361

Krieger, Karl-Friedrich, 'Heerschild, -ordnung', in *Lexikon des Mittelalters*, 4 (1989), pp. 2007–2008

Lange, Gudrun. 'Tierverwandlung', in *Reallexikon der Germanischen Altertumskunde*, 30 (2005), pp. 608–610

LEI = *Lessico etimologico italiano – Germanismi*, ed. E. Morlicchio (Wiesbaden 2015)

Meyer-Lübke, Wilhelm, *Romanisches Etymologisches Wörterbuch* (Heidelberg 1992)

Morlicchio, Elda. 'Die sprachliche Hinterlassenschaft der Germanen auf der italienischen Halbinsel', in *Akkulturation – Probleme einer germanisch-romanischen Kultursynthese in Spätantike und frühem Mittelalter*, eds. D. Hägermann – W. Haubrichs – J. Jarnut (Berlin 2004), pp. 99–110

Müller, Gunter. 'Germanische Tiersymbolik und Namengebung', in *Frühmittelalterliche Studien*, 2 (1968), pp. 202–217

Müller, Gunter. *Studien zu den theriophoren Personennamen der Germanen* (Cologne/Vienna 1970)

Müller, Gunter. 'Zum Namen *Wolfhetan* und seinen Verwandten', in *Frühmittelalterliche Studien*, 1 (1967), pp. 200–212

Niermeyer, Jan Frederik (in coll. with C. van de Kieft), *Mediae Latinitatis Lexicon Minus*, I (Leiden 2002)

Pfister, Max. 'Langobardische Superstratwörter im Italienischen', in *Jahrbuch für Internationale Germanistik*, 11 (1979), pp. 100–110

Pfister, Max, 'I prestiti linguistici di origine germanica fra Tardo Antico e Alto Medioevo', in *La cultura in Italia fra Tardo Antico e Alto Medioevo*, I (Rome 1981) pp. 261–283

Rübekeil, Ludwig. 'Völkernamen Europas', in *Namenforschung – Name Studies – Les noms propres. Ein internationales Handbuch zur Onomastik*, eds. E. Eichler – G. Hilty *et al.* (Berlin/New York 1996) pp. 1330–1343

Sabatini, Francesco. 'Riflessi linguistici della dominazione longobarda nell'Italia Mediana e Meridionale', in *Atti dell'Accademia Toscana di Scienze e Lettere La Colombara*, 28 (1963/64) pp. 123–249

Schmidt-Wiegand, Ruth, 'Herberge', in *Reallexikon der germanischen Altertumskunde*, 14 (1999) pp. 394–396

Schmidt-Wiegand, Ruth, 'Malbergische Glossen', in *Lexikon des Mittelalters*, 6 (1992), pp. 166–167

Schönfeld, Maurits. *Wörterbuch der altgermanischen Personen- und Völkernamen* (Heidelberg 1910)

Schützeichel, Rudolf. *Althochdeutscher und altsächsischer Glossenwortschatz* (Tübingen 2004)

Sitzmann, Alexander – Grünzweig, Friedrich E. *Die altgermanischen Ethnonyme. Ein Handbuch zu ihrer Etymologie*, [Philologica Germanica, 29] (Vienna, 2008)

Staiti, Chiara. 'Agilulf e altri. Il lupo nell'antroponimia germanica dei primi secoli', in *Simbolismo animale e letteratura*, ed. D. Faraci (Rome 2003), pp. 197–220

Störmer, Wilhelm, 'Hoch- und spätmittelalterliche 'Familien'-Namengebungspraxis vornehmlich beim Niederadel in Altbayern und Franken', in *Personennamen und Identität. Namengebung und Namengebrauch als Anzeiger individueller Bestimmung und gruppenbezogener Zuordnung*, ed. R. Härtel (Graz 1997), pp. 215–235

Tiefenbach, Heinrich, *Studien zu Wörtern volkssprachiger Herkunft in karolingischen Königsurkunden. Ein Beitrag zum Wortschatz der Diplome Lothars I. und Lothars II.* (Munich 1973)

van der Rhee, Florus. *Die germanischen Wörter in den langobardischen Gesetzen* (Amsterdam 1970)

van der Rhee, Florus. 'Die germanischen Wörter in der Historia Langobardorum des Paulus Diaconus', in *Romanobarbarica* (1980) pp. 271–296

von Wartburg, Walther. *Französisches Etymologisches Wörterbuch* (FEW), vol. 15–17 (Basel 1959–1969)

Wagner, Norbert. 'Das Erstglied von Ludwig', in *Beiträge zur Namenforschung* NF 21 (1986) pp. 78–84

Wagner, Norbert. 'Gausus und Harodus. Odinsnamen oder Stammesnamen in germanischen Königsgenealogien?', in *Beiträge zur Namenforschung*, NF 13 (1978) pp. 241–260

Wagner, Norbert. 'Der Name der Huosi', in *Münchener Studien zur Sprachwissenschaft*, 58 (1998), pp. 163–170

Zimmer, Stefan. 'Germani und die Benennungsmotive für Völkernamen in der Antike', in *Zur Geschichte der Gleichung 'germanisch-deutsch'. Sprache und Namen, Geschichte und Institutionen*, eds. H. Beck *et al.* (Berlin 2004) pp. 1–23

CHAPTER 5

The Gothic Language of Warfare

Carla Falluomini

1 Introduction

During the Late Antiquity, the presence of soldiers of Germanic origin in the Roman army increased progressively, as it is well known. Their cultural and linguistic Romanisation, subsequently extended to their respective communities, represents one of the most relevant consequences of these contacts.[1] There is indeed no doubt that Germanic mercenaries and auxiliaries learned Latin, the language of command and military administration.[2] On the other hand, Germanic words were incorporated in the Latin language first,[3] and then in the Romance languages. The period and modality of the introduction of the

1 There is no doubt that the Roman army represented a typical 'community of practice', which, in the course of its history, have developed a specific vocabulary, enriched also by the contribution of several languages, spoken by soldiers of different provenience. The rule of the army as promoter of linguistic exchanges and of the creation of a *Sondersprache* has long been recognised, see, e.g., Maria Grazia Mosci Sassi, *Sermo castrensis* (Bologna 1983); Roxana-Gabriela Curcă, '*Sermo castrensis* in Moesia Inferior', in *Roman Army and Local Society in the Limes Provinces of the Roman Empire: Papers of an International conference, Iași, June 4th–6th, 2018*, eds. L. Mihailescu-Bîrliba and W. Spickermann (Rahden/Westf. 2019), pp. 85–93; Marius Alexianu-Florian Matei-Popescu, 'The Linguistic Impact of the Roman Army in the Province of *Moesia Inferior*', in *Roman Army and Local Society in the Limes Provinces of the Roman Empire: Papers of an International conference, Iași, June 4th–6th, 2018*, eds. L. Mihailescu-Bîrliba and W. Spickermann (Rahden/Westf. 2019), pp. 63–84. The literature on the dynamics of the communities of practice, analysed from linguistic, anthropological, and sociological points of views, is vast, see, e.g., Adrian Goldsworthy and Ian Haynes (eds.), *The Roman Army as a Community* (Portsmouth, RI 1999); Miriam Meyerhoff, 'Communities of Practice', in *The Handbook of Language Variation and Change*, eds. by J. K. Chambers, Peter Trudgill, and Natalie Schilling-Estes (Oxford 2002), pp. 525–548.

2 '[A]lthough Latin may not have been spoken, the military terminology of the Romans must have penetrated', as Dennis H. Green (*Language and History in the Early Germanic World* [Cambridge 1998], p. 232) pointed out. See also Hans Kuhn, 'Das römische Kriegswesen im germanischen Wortschatz', in *Zeitschrift für deutsches Altertum und deutsche Literatur*, 101 (1972), pp. 13–53. On the role of Latin, used to stress the authority and rituals of the Roman army, see James N. Adams, *Bilingualism and the Latin Language* (Oxford 2004; on Arminius' code-switching between his Germanic language and Latin, see at p. 20).

3 See, e.g., Green, *Language and History*, pp. 230–235; Philip Rance, '*quam gentilitate appellant.* The Philological Evidence for *Germani* in the Late Roman Army: Germanic Loanwords in

 | DOI:10.1163/9789004698017_006

earlier Germanic borrowings into Latin are, however, often obscure, as is the exact language from which they derive. There is a general consensus among scholars that most of the Germanisms common to Romance languages were introduced in the period before the collapse of the Roman Empire and the dismission of the army, and were already present in the Vulgar Latin vocabulary, albeit attested in a later period.[4]

A warfare term which had been introduced by a particular Germanic group and become part of the specific military jargon could probably survive and be used also by other soldiers, without interruption. Some divergences among Germanic dialects already existed in Late Antiquity, but pragmatic contexts and common purposes may have facilitated the mutual intelligibility and reinforced the linguistic exchange among soldiers speaking different Germanic languages.[5]

Among the Germanic groups which served as mercenaries and auxiliaries (but also as commanders) in the Roman army and took part in the process of Romanisation, the Goths are the best known.[6] The cultural consequences of their relationship with Romans were already evident in antiquity, as John Chrysostom's speech to Gainas, the commander of the Gothic contingent of the Roman army, shows. According to Theodoret,[7] Gainas asked Arcadius for a

---

Roman Military Vocabulary', in Romania Gothica II. *The Frontier World: Romans, Barbarians and Military Culture*, ed. T. Vida (Budapest 2015), pp. 51–93.

4 See, e.g., the survey by Arrigo Castellani, *Grammatica storica della lingua italiana*. I. *Introduzione* (Bologna 2000), pp. 29–67 and the literature cited.

5 See William Moulton, 'Mutual Intelligibility among Speakers of Early Germanic Dialects,' in Germania: *Comparative Studies in the Old Germanic Languages and Literatures*, eds. Daniel G. Calder and T. Craig Christy (Wolfeboro NY 1988), pp. 9–28, here at pp. 15–27; Verio Santoro, 'L'intercomprensione orale tra parlanti lingue germaniche antiche', in *Studies on Late Antique and Medieval Glossography and Lexicography in Honour of Patrizia* Lendinara, eds. Claudia Di Sciacca – Concetta Giliberto *et al.* (Pisa 2018), pp. 663–686, here at p. 674.

6 The literature on the history of the Goths is vast: for a general background, see, e.g., Herwig Wolfram, *Die Goten. Von den Anfängen bis zur Mitte des sechsten Jahrhunderts. Entwurf einer historischen Ethnographie* (Munich 1990); Peter Heather, *The Goths* (Oxford 1996). See also Wolfgang J.H. Liebeschuetz, *Barbarians and Bishops: Army, Church, and State in the Age of Arcadius and Chrysostom* (Oxford 1990); Ralph Mathisen, 'Barbarian Invasions or Civil Wars? Goths as Auxiliary Forces in the Roman Army', in *Empire in Crisis: Gothic Invasions and Roman Historiography. Beiträge einer internationalen Tagung zu den Wiener Dexipp-Fragmenten* (Dexippus Vindobonensis), eds. F. Mitthof – G. Martin – J. Grusková (Vienna 2020), pp. 263–286. On the Ostrogothic military organisation, see also Guy Halsall, 'The Ostrogothic Military', in *A Companion to Ostrogothic Italy*, eds. J. Arnold – S. Bjornlie – K. Sessa (Leiden 2016), pp. 173–199.

7 Theodoret, *Kirchengeschichte*, eds. L. Parmentier – F. Scheidweiler – G.C. Hansen, 3rd rev. ed., GCS 19 (Berlin 1998), V.32.4–11.

church in which to worship in the Arian cult together with his troops, and on this occasion, he emphasized that he had undergone 'so many toils in war for Romans'. Then, John Chrysostom replied to the Goth:

> You have greater rewards for your labours, you are a general; you are vested in the consular robe, and you must consider what you were formerly and what you are now – your indigence in the past and your present prosperity; what kind of raiment you wore before you crossed the Ister, and what you are robed in now. Consider, I say, the littleness of your labours and the greatness of your rewards and be not unthankful to them who have shewn you honour.[8]

This passage testifies well to the privileges gained by Goths serving in the Roman army and puts in evidence the impact on their economic and social life. Further information is provided by the analysis of their military terminology, which is attested more anciently than that of other Germanic *gentes*.

The fourth-century version of the Greek Bible in Gothic, of which long parts of the Gospels and Pauline Epistles together with a short part of the Old Testament survive,[9] transmits several military terms as translations of the corresponding Greek words.[10] This version, realised by the Visigothic bishop Wulfila († 383), testifies to the vocabulary in use among the *Gothi minores*, the group of which he was the religious and political leader in Dacia and Moesia Inferior. This source is therefore significant because chronologically, geographically and semantically well-defined.

Further sources (onomastic material, Gothic borrowings in Latin and Romance languages) contribute to reconstructing warfare terminology. Personal names, transmitted mostly by Latin texts, testify to numerous stems that often find parallels in other Germanic languages but are not attested in direct tradition.[11] As previously mentioned, the investigation of the Gothic loanwords

---

8 Transl. by J. Blomfield in Philip Schaff – Henry Wace, eds., *A Selected Library of Nicene and Post-Nicene Fathers of the Christian Church*, 3rd vol. (New York *et al.* 1892).

9 It is preserved in seven fragmentary manuscripts, likely copied in the Ostrogothic kingdom of Italy; for an overview, see Carla Falluomini, *The Gothic Version of the Gospels and Pauline Epistles: Cultural Background, Transmission and Character* (Berlin-Boston 2015).

10 The Greek *Vorlage*, however, is lost; see the reconstruction in Wilhelm Streitberg, *Der gotische Text und seine griechische Vorlage. Mit Einleitung, Lesarten und Quellennachweisen sowie den kleineren Denkmalern als Anhang*, mit einem Nachtrag von Piergiuseppe Scardigli, 7. Aufl. (Heidelberg 2000).

11 See, e.g., Wolfgang Haubrichs, 'Krieg, Volk und Verwandtschaft. Zur Struktur und kulturellen Signifikanz ostgotischer Frauennamen', in *Archiv für Kulturgeschichte*, 99 (2017),

in Latin and in Romance languages is difficult, as these loans often cannot be separated from those derived from other Germanic languages.[12] Apart from early Gothic borrowings in Vulgar Latin, other terms entered the Romance languages from Visigothic, spoken in Gaul from the beginning of the fifth century and later in Spain, and Ostrogothic, spoken in Italy from the end of the fifth century. Such late Gothic loans are hence the results of close contacts and cultural amalgamation with Latin-speaking peoples in the West; a process concluded with the obsolescence of the Gothic language (in both Visigothic and Ostrogotic variants), between the second half of the sixth and the first half of the seventh century.

Direct and indirect sources, therefore, indicate or offer a possibility for reconstructing the Gothic terminology of warfare, referred to organisation, personnel, equipment, technique and military attitude:[13]

---

pp. 297–339 and his contribution in this volume. I wish to thank Wolfgang Haubrichs for his generous comments on the onomastic forms discussed in my text.

12 Linguistic criteria are insufficient to ascertain the Gothic origin, except in a few cases. Gothic, like other East Germanic languages, does not present rhotacism and Umlaut phenomena. Furthermore, it offers a particular evolution of Germ. *-*ww*- (> Goth. -*ggw*-), while it preserves Germ. **ē* (against West and North Germ. > *ā*, but OE *ǣ*). The presence of the same form (*mutatis mutandis*) in Italy and/or southern France, Spain and Portugal may speak for a Gothic origin, although terms once spoken in wider territories might have disappeared. Terms of Germanic origin that are attested only in Spanish and Portuguese very probably derive from Gothic (see Steven N. Dworkin, *A History of the Spanish Lexicon: A Linguistic Perspective* [Oxford 2012], p. 70). Finally, multiple introductions of a certain word, from different Germanic languages, may be taken into consideration. In several cases, however, a clear attribution of a certain loan to a donor language is impossible (see Castellani, *Grammatica storica*, pp. 29–67, with literature).

13 The following analysis is principally based on Joseph M. Piel – Dieter Kremer, *Hispano-gotisches Namenbuch. Der Niederschlag des Westgotischen in den alten und heutigen Personen und Ortsnamen der Iberischen Halbinsel* (Heidelberg 1976); Winfred P. Lehmann, *A Gothic Etymological Dictionary: based on the Third Edition of* Vergleichendes Wörterbuch der Gotischen Sprache *by Sigmund Feist* (Leiden 1986); Hermann Reichert, *Lexikon der altgermanischen Namen*, 2 vols.: I. *Text*; II. *Register*, with the cooperation of R. Nedoma (Vienna 1987–1990) – to be noted: the names are occasionally cited as reported in the sources; Vladimir E. Orel, *A Handbook of Germanic Etymology* (Leiden 2003); Antje Casaretto, *Nominale Wortbildung der gotischen Sprache. Die Derivation der Substantive* (Heidelberg 2004); Nicoletta Francovich Onesti. *I nomi degli Ostrogoti* (Florence 2007). See also Giuseppe Restelli, 'I Goti nella «Romania». Influssi della loro lingua sul latino', in *Aevum* 51 (1977), pp. 207–238. The Greek correspondence(s), with the Vulgate readings in brackets, is/are followed by the biblical passages in which the Gothic term is/are attested. 'Sk' indicates the commentary on John's Gospel known as *Skeireins*, of uncertain dating (but composed after Wulfila's translation) and source. For Romance languages, see Walther von Wartburg, *Französisches Etymologisches Wörterbuch. Eine Darstellung des galloromanischen Sprachschatzes* (Bonn *et al.* 1922–2002), with the cooperation of Oscar

2 Organisation and Personnel

Gothic terms for military structures and persons are mainly known through Wulfila's translation of the Bible.

Military organisation:

hansa 'cohort', 'band of men', 'crowd', 'troop' translates Gr. σπεῖρα (Lat. *cohors*) in Jn 18:3, Jn 18:12 and Mk 15:16, and πλῆθος in Lk 6:17 (Lat. *multitudo*). The term is attested in other Germanic languages (OE *hōs* 'company', 'band', OHG *hansa* 'cohort', 'company').

harjis 'army', 'legion', 'armed host' translates Gr. λεγεών (Lat. *legio*) in Lk 8:30, where the sense is 'large number' (in this case, the name of the unclean spirit). In Lk 2:13, this word renders Gr. στρατιά (Lat. *militia*), referred to the heavenly host. There is no doubt that *harjis*, which is also attested in personal names (e.g., Visigothic *Moda-harius* ca. y. 470; Ostrogothic *Ufita-hari* y. 551),[14] belonged to the Gothic traditional military language (with parallels in other Germanic languages: ON *herr*, OE *here*, OS *heri*, and OHG *heri* 'army', 'people'). It is attested in the first part of the compound ****hari-bairgo*** 'accommodation for the army', which entered in Late Latin from a Germanic (Gothic?) language and continues in It. *albergo*, Cast. *albergo*, Cat. *albergue*, OPort. *alibergo*, Prov. *auberga*, *auberc*.[15] Very probably, the word reflects the military custom of accommodating the army in provisory form during its movements and campaigns. *Harjis*

Jänicke *et al.* [= *FEW*]; Wilhelm Meyer-Lübke, *Romanisches etymologisches Wörterbuch*, 3. vols. (Heidelberg 1935) [= *REW*]; Salvatore Battaglia (ed.), *Grande dizionario della lingua italiana* (Turin 1967–2002); Carlo Battisti – Giovanni Alessio, *Dizionario etimologico italiano* (Florence 1975); Joan Corominas, *Diccionario crítico etimológico castellano e hispánico*, with the collaboration of José A. Pascual (Madrid 1980–1983); Tullio De Mauro (ed.), *Grande dizionario italiano dell'uso* (Turin 1999–2004); Elda Morlicchio (ed.). *LEI. Lessico etimologico italiano. Germanismi* (Wiesbaden 2000-). The asterisk before a form indicates that it has been reconstructed on the base of indirect sources. Abbreviations: Cat.: Catalan; Cast.: Castilian; Engl.: English; Fr.: French; Frank.: Frankish; Germ.: Germanic; Gr.: Greek; It.: Italian; Lang.: Langobardic; Lat.: Latin; MHG: Middle High German; O: Old; OE: Old English; OHG: Old High German; ON: Old Norse; OS: Old Saxon; Port.: Portuguese; Prov.: Provencal. If a borrowing continues in modern Romance languages, the sole modern form will be given.

14 Reichert, *Lexikon*, I, p. 733.

15 The exact origin and spread of this word are controversial, and it is not possible to draw a firm conclusion, see Castellani, *Grammatica storica*, p. 57. It has been suggested that the forms with Umlaut, i.e. OFr. *herberge, herberc*, derive from Frank. **heriberga*, see von Wartburg, *FEW*, XVI, pp. 160–161; Ernst Gamillscheg, Romania Germanica. *Sprach- und Siedlungsgeschichte der Germanen auf dem boden des alten Römerreiches*, 3 vols. (Berlin 1934–1936; 2nd ed. of the vol. I in 1970), I^2, p. 378; Meyer-Lübke, *FEW*, s.n. 4045; Corominas – Pascual, *Diccionario*, I, p. 118.

is also part of the compound ***hari-hrings** 'circle of the army', 'reunion of the army', attested in Cast., Cat., Port. and Prov. *arenga* 'line', 'rank' and It. *arengo* 'place of the assembly'. The Gothic origin of these terms is not certain, because a parallel form was probably in use also in the languages of the Franks and Langobards.[16]

It is worth mentioning that, in Mk 5:9 and 5:15, a group of unclean spirits is named *laigaion* in Gothic, a loanword from Gr. λεγεών, in its turn from Lat. *legio* 'legion'. As in Lk 8:30, the Greek word is used as a personal name. The different translation in Gothic (*laigaion* and *harjis*) is not surprising, because a Greek term is often rendered with two or more Gothic forms, a characteristic that may be imputed to the work of more translators or/and different period of the translation. It seems odd that the borrowing is not from Lat. *legio*, as expected – and usual – in case of military vocabulary (see below).

Army personnel:

The word(s) for 'leader', 'chief of the army' 'leader of the warband/army' is/are not attested in biblical Gothic, a fact which depends on the nature of the source. The reconstruction of **skarja* 'captain', based on Prov. *escara* 'bandit', is uncertain, because the Provencal word is a *hapax*, and the author (Peire Vidal) is referring to an Italian context.[17]

gadrauhts 'soldier', 'warrior', 'serviceman' translates Gr. στρατιώτης (Lat. *miles*) in Mt 8:9; Jh 19:2; Lk 7:8; Mk 15:16 and 2Tim 2:3. The term, formed with the cumulative prefix *ga-* 'together' and a derivative from **drauhts* 'army', 'warband', '*comitatus*' (see below), suggests a conception of the soldier and warrior as a part of a group. Similar formations are Gr. σύμμαχος, συστρατιώτης and Lat. *commilito*, *commiles* 'fighting along with'; and OE *gedryht* 'troop'. All these compounds with cumulative prefix may depart from similar conceptual bases, and their formal agreement may be accidental. *Gadrauhts* is the sole term used for translating στρατιώτης in Wulfila's Gothic.

The term ***gahlaiba*** 'companion', 'comrade' deserves particular attention. It translates Gr. συμμαθητής 'fellow disciple' (Lat. *condiscipulus*) in Jn 11:16 and συστρατιώτης 'fellow soldier' (Lat. *commilito*) in Ph 2:25. In the Ostrogothic deed of Naples (y. 551), the term, attested in a slightly different form, indicates a companion in an ecclesiastic context.[18] The word is a compound, formed again with the cumulative prefix *ga-* 'together' and a derivative of *hlaifs* 'bread'; its significance stems from the meaning 'someone who has the bread together',

16 See Corominas-Pascual, *Diccionario*, I p. 322; von Wartburg, *FEW*, XVI, pp. 240–247; Castellani, *Grammatica storica*, p. 58.

17 Reconstruction in Meyer-Lübke, *FEW*, s.n. 7980; against von Wartburg, *FEW*, XVII, p. 96.

18 See Piergiuseppe Scardigli, *Die Goten. Sprache und Kultur* (Munich 1973), pp. 283–284.

i.e. 'companion' (see OHG *gileibo* 'companion', 'comrade'). Other formations with *ga-* are attested in Gothic, like *gadrauhts* 'soldier' (see above) and *gasinþa* 'fellow traveller' (see below). Scholars do not agree about the origin of *gahlaiba*: some[19] claim that the Gothic word is an old, pre-Wulfilian loan translation of Lat. *companio* (derived from *cum* 'with' + *panis* 'bread'), analogous formation to the Gothic term; more correctly, other[20] argue a Germanic (Gothic?) origin of the Latin term because this is not attested before the sixth century.[21] There is however a consensus on the hypothesis that *gahlaiba/companio* originate in a military context and reveal close contacts and linguistic exchanges between Goths and Romans. At the same time, the rendering of Gothic *ga-* with Latin *cum-* > *com-* (or *vice versa*) indicates a linguistic awareness of the semantic correspondence of these prefixes, a fact that reflects a bilingual milieu.

gasinþa 'fellow traveller' translates Gr. συνέκδημος (Lat. *comes*) in 2Cor 8:19; ***gasinþja*** (or ***gasinþi***) 'fellow traveller' renders Gr. συνοδία (Lat. *comitatus*) in Lk 2:44.[22] The terms are not attested in military contexts, but this fact may depend on the character of the source. In other Germanic languages, the parallel words respectively indicate a 'member of the *comitatus*' (Frank. and Lang.-Lat. *gasindus*, *gasindius*) and the '*comitatus*' (Lang.-Lat. *gasindio*).[23]

militondans, translation of Gr. στρατευόμενοι (Lat. *milites*) in Lk 3:14, referred to Roman soldiers, is the participle form (plur.) of ***militon*** 'to serve in the army', 'to be a soldier'. The Gothic word derives from Lat. *militare* 'to serve as a soldier', 'to perform military service'. The participial form is transmitted only in Gothic,

---

19 Harry Velten, 'Studies in the Gothic Vocabulary with Special Reference to Greek and Latin Models and Analogues', in *The Journal of English and German Philology*, 29 (1930), p. 345; Günter Reichenkron, *Historische latein-altromanische Grammatik. I. Einleitung. Das sogenannte Vulgärlatein und das Wesen der Romanisierung* (Wiesbaden 1965), pp. 170–171.

20 Antoine Thomas, 'L'étymologie du mot français *compagnon*', *Comptes rendus des séances de l'Académie des Inscriptions et Belles-Lettres*, 1931 (75), pp. 79–86; Lehmann, *A Gothic Etymological Dictionary*, p. 139; see also Angela Della Volpe, 'On Gothic *gahlaiba* and Latin *companion*: An Excursus in Historical Linguistics Methodology', in *LACUS Forum XXX: Language, Thought and Reality*, eds. G. Fulton – W.J. Sullivan – A.R. Lommel (Houston, TX 2004), pp. 5–28.

21 The first attestation is in the Frankish *Pactus Legis Salicae* LXIII.1, ed. K.A. Eckhardt, MGH LL Nat. Germ. IV.1 (Hannover 1962).

22 The formation of *gasinþjam* is not clear, see Casaretto, *Nominale Wortbildung*, pp. 242 and 263–264; see further Elmar Seebold, 'Gt. *gasinþa** ‚Reisegefährte' und *gasinþi**, Reisegesellschaft'', in *Beiträge zur Geschichte der deutschen Sprache und Literatur* (Tübingen), 96 (1974), pp. 1–11.

23 See Nicoletta Francovich Onesti, *Vestigia longobarde in Italia (568–774). Lessico e antroponimia* (Rome 1999), p. 91.

whereas OE *milite* pl. 'soldiers' derives from Lat. *miles*, a fact that points to their independent borrowing.

The Greek verb στρατεύεσθαι (Lat. *militare*) 'to do military service', 'to serve in the army' 'to make a military expedition', 'to do military duty', 'to fight' is rendered with ***drauhtinon*** 'to serve under a warlord', 'to gather as a warrior' in 1Cor 9:7, 2Cor 10:3, 2Tim 2:4. This term is a secondary formation from ****drauhts*** 'army', 'warband', '*comitatus*' (ON *drótt*, OE *dryht*, OHG *truth*), which is not attested in direct sources but is part of the Gothic name *Droctarius* (y. 525–545).[24]

One passage of Wulfila's version deserves particular attention, because the rendering of Gr. στρατεύεσθαι (Lat. *militare*) diverges from the usual *drauhtinon*:

> 1Tim 1:18
> þo anabusn anafilha þus, barnilo Teimauþaiu, bi þaim faura faursniwandam ana þuk praufetjam, ei **driugais** in þaim þata godo **drauhtiwitoþ**
>
> ταύτην τὴν παραγγελίαν παρατίθεμαί σοι, τέκνον τιμόθεε, κατὰ τὰς προαγούσας ἐπὶ σὲ προφητείας, ἵνα στρατεύῃ ἐν αὐταῖς τὴν καλὴν στρατείαν
>
> This charge I commit to you, Timothy, my son, in accordance with the prophetic utterances which pointed to you, that inspired by them you may wage the good warfare. (RSCVE)

The term ***driugan*** is a *hapax legomenon* and only in this context translates Gr. στρατεύεσθαι. Its exact meaning is not entirely sure. Cognates in other Germanic languages had the more general value of 'to perform', 'to exert', 'to accomplish', 'to carry out', which could also be assumed for the Gothic word. The translation of Gr. στρατεία (Lat. *militia*) 'expedition', 'campaign', 'military service', 'warfare' is also worthy of note: ***drauhtiwitoþ***, literarily 'military law', perhaps 'military discipline'. The compound, which is formed by **drauhts* 'army', 'warband', '*comitatus*' and *witoþ* 'law',[25] does not have parallels in other Germanic languages and is linguistically independent from Gr. στρατεία and Lat. *militia*. The general meaning of the Gothic passage seems to be 'you may carry out the good military law (or discipline)'. In 2Cor 10:4, instead, Gr. στρατεία is translated with ***drauhtinassus*** 'warfare', 'service under a warlord', 'military service', an abstract formation attested only in Gothic, from the verb *drauhtinon* 'to serve under a warlord'. The reason of the different translation of the Greek word is not evident. The context refers to the *militia Christi*, modelled – as known – on

24 Reichert, *Lexikon*, I, p. 239.

25 In the biblical passages, *witoþ* corresponds to Gr. νόμος (Lat. *lex*) 'law', while Gr. συνήθεια (Lat. *consuetudo*) *'custom, usage, practice'* corresponds to Goth. *biuhti*.

the well-hierarchised Roman army. Wulfila likely needed to underline that the warfare of the *miles Christi* should be conducted with rules and discipline, and probably the indirect allusion in this passage is to the rules and discipline of the Roman army, well-known by the Goths. It is therefore possible that *drauhtiwitoþ* represents a (pre-Wulfilian?) rendering of Lat. *disciplina exercitus*.

wahtwo or ***wahtwa*** 'guard' derives from *wakan* 'to watch', 'to keep awake'; it translates Gr. φυλακή (= Lat. *vigilia*) in Lk 2:8. The word has parallels in other Germanic languages (OS *wahta*, OHG *wahta* 'guard').[26]

wardja 'watchman' is attested in Mt 27:65 (plur.), where corresponds to *custodes* of some *Vetus Latina* manuscripts (against Gr. κουστωδία and Lat. *custodia* of part of the NT tradition); **wardja* 'guard service' is attested in the *Lex Visigothorum*[27] (IX.2.9). It. *guardia* and Cast. *guardia* 'watchman' and 'guard' are considered borrowings from Gothic.[28] Etymologically connected with this term is **wards* 'custodian', attested in the compound *daurawards* 'custodian of the door', 'porter' (Gr. θυρωρός) in Neh 7:1; 7:45 and Jn 10:3.[29]

Of Latin – or less likely Greek – origin, ***spaikulatur*** 'sentinel' translates Gr. σπεκουλάτωρ (Lat. *speculator*) in Mk 6:27. The Gothic form, with *-u-* for -ω- or *-o-*, seems to reflect an oral transmission of the borrowing and, for this reason, its pre-Wulfilian origin seems to be plausible.[30] In this passage, the loanword indicates a specific figure of the Roman army.

Wulfila's version transmits two loan translations of Greek terms that indicate figures of the Roman army:

hundafaþs 'centurion', 'commander of a hundred-man military unit' translates Gr. ἑκατόνταρχος/ἑκατοντάρχης (Lat. *centurio*) in Mt 8:5, Mt 8:8, Mt 8:13,

26 On OIt. and Cat. *guaita* 'watchman', from the corresponding Frankish word, see Gamillscheg, Romania Germanica, I^2, p. 279; Castellani, *Grammatica storica*, p. 94; Joan María Jaime Moya, 'La aportación del léxico militar de origen germánico en la formación de la lengua catalana', in *Literature, Science & Religion: Textual Transmission and Translation in Medieval and Early Modern Europe*, eds. M. Bellmunt Serrano – J. Mahiques Climent (Kassel 2020), p. 259.

27 *Lex Visigothorum*, ed. K. Zeumer, MGH LL Nat. Germ. I.1 (Hannover/Leipzig 1902).

28 On the Italian form, see Castellani, *Grammatica storica*, p. 57; on the Spanish word, see Corominas – Pascual, *Diccionario*, III, p. 246–247. It. *guardare*, Fr. *garder*, Prov. Cat. Cast. Port. *guardar* 'to watch', 'to protect' are Germanisms, see Meyer-Lübke, *REW*, s.n. 9502; von Wartburg, *FEW*, XVII, p. 523; Gamillscheg, Romania Germanica, I^2, p. 279. The exact donor language(s) is/are not clear.

29 The Gothic origin of **sculca* 'patrol-duty', 'guard', 'place of guard' is instead not fully accepted, see Rance, *quam gentilitate appellant*, pp. 79–81; see also Norbert Wagner, 'Sculca und skalks', in *Sprachwissenschaft*, 45 (2020), pp. 275–278.

30 Vittoria Corazza, 'Le parole latine in gotico', in *Accademia Nazionale dei Lincei. Memorie. Classe di Scienze morali, storiche e filologiche*, XIV (Rome 1969), pp. 55–56.

Mt 27:54, Lk 7:2, Lk 7:6; and Gr. κεντυρίων (Lat. *centurio*) in Mk 15:39, Mk 15:44 and Mk 15:45.

þusundifaþs 'commander of a thousand men' translates Gr. χιλίαρχος (Lat. *tribunus*) in Jh 18:12 and Mk 6:21.

Both terms are compounds, respectively, with *hunda-* 'hundred' and *þusundi-* 'thousand' in union with *-faþs* 'chief', 'commander'. The last term, of Indo-European origin, survives only in Gothic among the Germanic languages. It is also attested in the compounds *bruþ-faþs* 'bridegroom' and *swnagoga-faþs* 'chief of the synagogue' (the latter is undoubtedly a Wulfilian formation). The models of *hundafaþs* and *þusundifaþs* are instead not clear. They may have been created by Wulfila or be older loan translations of the respective Greek terms ἑκατόντ-αρχος and χιλί-αρχος. It is also possible that these terms have an independent history, and that only one of these is a Wulfilian creation.

According to Dahn,[31] *þusundifaþs* is the form from which the term ***thiufadus*** derives. This latter is attested only in Latin sources from Spain (*Lex Visigothorum* and other legal documents) and poses several semantic and formal difficulties.[32] The sources indicate that the *thiufadus* (variants: *teuphadus, thiuphadus, thifadus, tiufadus, tiumphadus, thiufandus, tiupadus*) was an officer, apparently of a lower level than a *dux* and a *comes*. He may govern peoples (*populos regere* in the *Lex Visigothorum* IX.2.9) and absolve fiscal duties within the *thiufadia* (variants: *tiufa, thiufa, tiupha, thiupha, teupa, tiufada, thiuphada, tiufadia*), a word attested in the *Lex Visigothorum* IX.2.1, IX.2.4, and IX.2.6, indicating both the persons and the territory controlled by the *thiufadus*.

The etymology of *thiufadus*, undoubtedly Gothic, is only partially clear: the second part (*-fadus*) is a regular Latin rendering of the above-mentioned Goth. *-faþs* 'chief', 'commander'. Denominal abstract from *-faþs* is *-*fadja*[33] 'pertaining on/ruled by the *-faþs*', attested in *thiufadia*: it is latinised in *-fadia/-phadia* and later reduced in *-fada/-phada* > *-fa/-pha*. The first part, *thiu-*, is less clear, although Latin *th-* indubitably reflects Goth. *þ-* at the beginning of a word.

31 Felix Dahn, *Die Könige der Germanen. Das Wesen des ältesten Königthums der germanischen Stämme und seine Geschichte bis auf die Feudalzeit*, 6 (Würzburg 1871), p. 374.

32 The literature on the value and origin of this term is vast, see, e.g., Dietrich Claude, '*Millenarius* und *thiufadus*', in *Zeitsch. der Sav. Stift. – Germ. Abt.*, 88 (1971), pp. 181–190; Walter Goffart, *Barbarians and Romans A.D. 418–584. The Techniques of Accommodation* (Princeton 1980), esp. pp. 80–88; Wolfram, *Die Goten*, pp. 222–223; Gideon Maier, *Amtsträger und Herrscher in der* Romania Gothica. *Vergleichende Untersuchungen zu den Institutionen der ostgermanischen Völkerwanderungsreiche* (Stuttgart 2005), pp. 228–230. See also Esperanza Osaba's contribution in this volume (ch. 11).

33 See Casaretto, *Nominale Wortbildung*, p. 587.

Dahn argued that *þusundifaþs* and its supposed Latin derivative *thiufadus* would be a loan translation of Lat. *millenarius*, a word attested in contexts related to Vandalic, Visigothic and Ostrogothic kingdoms.[34] The relationship between the *thiufadus* and the *millenarius*, both cited in the *Lex Visigothorum* II.1.27 (*dux, comes, vicarius, pacis adsertor, thiuphadus, millenarius, quingentenarius, centenarius*), is however problematic. It is not clear whether they indicated the same figure or not. In addition, the derivation of *thiu-fadus* from *þusundi-faþs* is difficult to justify from a formal point of view: the evolution from *-u-* (*þu-*) to *-iu-* (*thiu-*) and the fall of two post-tonic syllables (*-sundi-*) is unlikely. In other words, the derivation of *thiu-fadus* from *þusundi-faþs* is linguistically implausible.

According to Gamillscheg,[35] *thiu-* goes back to Goth. *þius* 'servant', also attested in *þiu-magus* 'servant boy'. This etymology, formally possible, was accepted by Claude[36] and subsequently by Wolfram,[37] who defined the *thiufadus* as 'master of military servants', considering that the younger or less heavily armed warriors, who supported the elite troops, were looked upon as servants, regardless of their social standing. Claude[38] argued that, initially, the *thiufadus* and the *millenarius* represented two different figures. Conversely, Wolfram[39] claimed that the term *thiufadus* replaced Lat. *millenarius* and that both forms indicate the same figure. The substitution of *millenarius* with *thiufadus* would have occurred before the mid-sixth century when the Gothic language was still spoken in the Visigothic kingdom. Finally, Martin's hypothesis[40] sees a juxtaposition and fusion of these two figures, initially independent, because they had similar competences.

A third etymological explanation of *thiufadus*, also proposed by Dahn,[41] sees in *thiu-* a derivation from the well-attested Goth. *þiuda* 'people'. The syncope of the post-tonic syllable (**þiuda- > thiu-*) is well-attested in the transmission of several Gothic names: e.g. Ostrogothic *Theo-rico* (y. 526) and *Theo-hadus* (y. 536),[42] Visigothic *Tue-rigo* (y. 989) and *Te-miro* (y. 1034).[43]

34 Claude, *Millenarius*, p. 182.
35 Gamillscheg, Romania Germanica, I, p. 357.
36 Claude, *Millenarius*, p. 183.
37 Wolfram, *Die Goten*, 222.
38 Claude, *Millenarius*, p. 186.
39 Wolfram, *Die Goten*, p. 223.
40 Céline Martin, *La géographie du pouvoir dans l'Espagne visigothique* (Lille 2003), pp. 151–156.
41 Dahn, *Die Könige der Germanen*, p. 345.
42 Reichert, *Lexikon*, I, pp. 671 and 660.
43 Piel – Kremer, *Hispano-gotisches Namenbuch*, pp. 261–266.

According to this explanation, which seems the best by far, *thiufadus* would derive from Goth. **þiudafaþs*, with the original semantic value of 'commander of the people' ('people in arms'?). Indeed, the compound *thiufadus* would perfectly reflect the duty – *populos regere* – expressed in the *Lex Visigothorum* IX.2.9. It is not to exclude that this word predates the Gothic migration period westwards, as it indicated a leader of a subgroup or warband. The antiquity of the formation would justify its phonetic transformation and the loss of its semantic transparency. The **þiudafaþs* 'commander of the 'people', would have been subsequently equate with the *millenarius* 'commander of a thousand men', on the basis of the common semantic value of 'commander of a great number of men'. Only accidentally, the word would have been preserved in the sole Visigothic milieu.[44] It is worth of note that quasi-parallel formations are attested among other Germanic *gentes*, namely Saxons and Anglo-Saxons: OS *thiod-kuning* and *þéod-cyning* 'king of the people'.

In any case, independently from its origin, the term *thiufadus* represents a relic of the Visigothic military and juridical language. It survives a long time after the integration of the Visigoths with the Ibero-Romans, being attested in Catalan juridical documents from the ninth to the eleventh century.[45]

3 Weapons and Military Technique

Weapons played an important social role in the Gothic society. They were among the most evident symbols of power and helped to define the role of the leader within a group of warriors. They were also present in warriors' graves and used in important rituals such as the *adoptio per arma*, known through literary sources.[46]

---

44 Completely to exclude the claim that the word originates from the Gothicisation of the Latin word *devotus* 'devoted', i.e. 'the honorific title of the executive officer who [...] was generally a Goth' (Leo Wiener, *Commentary to the Germanic Laws and Mediaeval Documents* [Cambridge 1915], p. 31).

45 See *Glossarium Mediae Latinitatis Cataloniae*, <http://gmlc.imf.csic.es/codolcat/>, s.v. *thiufadus*.

46 See Alexander K. Nefedkin, 'Weaponry of the Goths of the mid-3rd to the 7th Century AD', in *Journal of Roman Military Equipment Studies*, 16 (2008), pp. 147–155; Hans U. Wiemer – Guido M. Berndt, 'Instrumente der Gewalt: Bewaffnung und Kampfesweise gotischer Kriegergruppen', in *Millennium. Jahrbuch zu Kultur und Geschichte des ersten Jahrtausends n. Chr.*, 13 (2016), pp. 197–198; see also Attila. P. Kiss, '*Per arma adoptio*. Eine gotische Sitte in den frühmittelalterlichen schriftlichen Quellen', in Romania Gothica II. *The Frontier World: Romans, Barbarians and Military Culture*, ed. T. Vida (Budapest 2015), pp. 95–108.

Gothic terms for weapons are mostly of Germanic origin and have cognates in other Germanic languages:

arƕazna 'arrow' translates Gr. βέλος (Lat. *telum*) in Eph 6:16. Gothic exhibits a form slightly different from that of other Germanic languages (ON *ǫr*, OE *earh* 'arrow').

The sword has several designations, some of them metonymical, and different forms, as the archaeological finds show: ***hairus*** 'sword' translates Gr. μάχαιρα (Lat. *gladius*) in several passages, and ῥομφαία (Lat. *gladius*), which indicates a long and wide sword, in Lk 2:35. The Gothic word has parallels in other Germanic languages (ON *hjǫrr*, OE *heoru*, OS *heru-*). Gr. μάχαιρα (Lat. *gladius*) is also translated with ***meki*** 'sword' in Eph 6:17. This word, of unclear etymology, is spread in Germanic (ON *mækir*, OE *méce* and OS *māki*), Slavic, Celtic and Finnish, and it is considered a cross-cultural term of great antiquity.[47] The semantic divergence between *hairu* and *meki* is not evident, because both translate μάχαιρα in Wulfila's version.

Forms indicating the 'sword' are preserved also in indirect sources: **agja-* 'sharpness', 'sword', attested e.g. in the Ostrogothic name *Agiulf* (written *Achiulf*),[48] with parallels in all Germanic languages (ON *egg*, OE *ecg*, OS *eggia*, OHG *ecke* 'edge', 'blade', 'sword'); ****brands*** 'sword', attested e.g. in the Ostrogothic and Visigothic name *Brandila* (y. 523/526; y. 683),[49] with parallels in other Germanic languages (ON *brandr*, OE *brand*, OHG *brant*); it may originally have been a metonymical form for 'the blade gleaming like fire'.[50] Derivatives from this Germanic word are attested in some Romance languages (It. *brando*, OFr. *brant*, Prov. *bran* 'sword'), but the donor language(s) is (are) unclear.[51] The 'short sword', well-attested in archaeological sources, is denominated **scrama* in the *Lex Visigothorum* IX.2.9 (Goth. ****skrama***, from *skramasax*).[52]

brunjo 'armor', 'breastplate' translates Gr. θώραξ (Lat. *Lorica*) in Eph 6,14 and 1Th 5,8. The word has cognates in other Germanic languages (ON *brynja*, OE *byrne*, OS *brunnia*, and OHG *brunna* 'corslet', 'coat of mail') and derives ultimately from Celtic.[53]

hilms 'helm' renders Gr. περικεφαλαία (Lat. *galea*) in Eph 6:17 and 1Thess 5:8. The word has parallels in other Germanic languages (ON *hjálmr*, OE *helm*,

---

47 Lehmann, *A Gothic Etymological Dictionary*, p. 250.

48 Jordanes, *Romana et Getica*, ed. Th. Mommsen, MGH Auct. ant. V, 1 (Berolini 1882), XIV.79. See Reichert, *Lexikon*, I, p. 9.

49 Reichert, *Lexikon*, I, p. 148.

50 Green, *Language and History*, p. 70.

51 Morlicchio, LEI *Germ.* I, coll. 1235–1247.

52 See also *scramasaxus* in Gregorius Turonensis, *Libri Historiarum X, editionem alteram* cur. B. Krusch – W. Levison, MGH SS rer. Merov. I.1 (Hannover 1951), IV.46.

53 Orel, *A Handbook*, p. 58, s.v. **ƀrunjōn*.

OS *-helm*, OHG *helm*). This word seems to be the basis of Cast., It. and Port. *elmo* 'helm', with Goth. [i] > Romance [e]. On the other hand, OFr. *helme* > Fr. *heaume*, Cast. *yelmo* and Prov. *elm* presuppose an initial form with *-e-* (**helm-*) and suggest a derivation from Frank. **helm*, Latinised in *helmus* in the Reichenau glosses (8th c.).[54]

sarwa 'arms', 'armament', 'weapons' translates Gr. ὅπλα (Lat. *arma*) in Rom 13:12 and πανοπλία (Lat. *armatura*) in Eph 6:11 and Eph 6:13. Parallels of this word are attested in other Germanic languages (OE *searo* 'craft', OS *saro* and OHG *saro* 'arms', 'armament') and in personal names, e.g., Visigothic *Sarus* (y. 407). Another translation of Gr. ὅπλα (Lat. *arma*) is ***wepn*** 'weapon', always in the plural form, in Jn 18:3, 2Cor 6:7 and 2Cor 10:4, with parallels in other Germanic languages (ON *vápn*, OE *wǣpen*, OS *wāpan*, OHG *wāfan*).

skildus 'shield' translates Gr. θυρεός (Lat. *scutum*) in Eph 6:16; it has parallels in other Germanic languages (ON *skjǫldr*, OE *scield*, OS *skild*, OHG *skilt*). Another Gothic word for 'shield' derives from ***lind*-** 'linden-tree' because – evidently – the shield was made of wood derived from this tree. It is attested only in onomastic sources, e.g. in the Ostrogothic name *Lendarit* (y. 557).[55] The generic term for 'shelter', 'protection', derived from *****mund*-**, with parallels in all Germanic languages (ON, OE, OS *mund* and OHG *munt*), a common stem in the onomastic sources, see e.g. Ostrogothic *Mundo* (y. 505).[56]

The passages of the Gothic version that contained the term(s) for 'spear' are lost. The onomastic sources offer *****gais*-** 'spear', 'pike', transmitted e.g. in the Ostrogothic name *Gesimund* (y. 526)[57] and also attested in other Germanic languages (ON *geirr*, OE *gār*, OS, OHG *gēr*).

Sources on the military technique and its vocabulary are scanty. Describing a Gothic warfare technique, Ammianus Marcellinus[58] cites the term ***carrago***. It consisted of a wagon-train used as defensive barricade or laager. The word, attested in subsequent Latin and Greek authors in connection with warfare practices in the Balkans and later adopted by the East Roman army, has been interpreted as a compound: the first part as a borrowing from Lat. *carrus* 'wagon', the second part as a derivation from Goth. **hago-* 'hedge'.[59] If this etymological explanation is correct, the word represents a sign of the language hybridisation among Gothic warbands.

---

54 Gamillscheg, Romania Germanica, I², p. 286; Meyer-Lübke, *REW*, s.n. 4130.

55 Reichert, *Lexikon*, I, p. 461.

56 Reichert, *Lexikon*, I, p. 513.

57 Reichert, *Lexikon*, I, p. 353.

58 Ammianus Marcellinus, *Rerum gestarum libri*, ed. J.C. Rolfe, 3 vols. (Cambridge, Mass. 1935–9), XXXI.7, 7.

59 Rance, *quam gentilitate appellant*, pp. 56–57.

The Gothic presence in the Roman army is reflected through terms that have been passed down to Latin and, subsequently, to Romance languages. Lat. *bandum* and Gr. βάνδον originally indicated the 'military banner' and later, metonymically, a regiment, i.e. the body of troops that served under the same banner. Attested in the Latin-based jargon of the East Roman army, it has been considered a borrowing from Goth. ***bandwo*** 'sign',[60] attested in Mk 14:44 as translation of Gr. σύσσημον (Lat. *signum*) and in 2Th 3:17 for Greek σημεῖον (Lat. *signum*), rendered also with ***bandwa*** in 1Cor 14:22, without a clear semantic distinction between the two forms. It probably passes to Cast. and Port. *bando* 'annoncement', 'group', 'faction' and to It. *bando* 'ban', 'decree', It. *banda* 'group'. The influence of the corresponding Langobardic word may have reinforced the use of this term in Italian.[61] The words have cognates in other Germanic languages, with a slight different meaning.

garedan 'to provide', well-attested in the Gothic texts, passes into Vulgar Latin as **conredare* (with substitution of Goth. *ga-* with Lat. *cum-* > *con-*, see above), possibly through the military language, and then to Romance languages: Cast. *conrear* and Cat. *conrear* 'to cultivate', It. *corredare* 'to equip', OFr. *conreer* 'to fit out', 'to equip', Prov. *correzar, correar*; with the Latin prefix *ad-*: Cast. *arrear* and Cat. *arrear* 'to equip', 'to prepare', It. *arredare* 'to equip', OFr. *areer*.[62] The Gothic origin is very likely because of the vocalism *-ē-* (against *-ǣ-/-ā-* of the other Germanic languages, see ON *-ráða*, OE *gerǣdan*, OS *girādan*, OHG *girātan* 'to advise', 'to support', 'to intend').

manwjan 'to prepare' 'to make ready', well-attested in Gothic, entered in Late Latin as *admanuire*, from which several Romance forms derive (It. *ammannire* 'to prepare', OFr. *(en)manevir*, OProv. *marvir*, Cat. *amanir* 'to prepare [often food]').[63]

---

60 Rance, *quam gentilitate appellant*, pp. 63–64.

61 See Johannes Kramer, 'Ein Gräzismus gotischer Herkunft im Italienischen: bando', in *Balkan-Archiv NF*, 12 (1987), pp. 197–207; Morlicchio, LEI *Germ.* I, coll. 282–359; see also Corominas – Pascual, *Diccionario*, I, p. 487; Gerold Hilty, 'La herencia visigótica en el léxico de la Península Ibérica', in *Latin et langues romanes : Études de linguistique offertes à József Herman à l'occasion de son 80ème anniversaire*, eds. S. Kiss – L. Mondin – G. Salvi (Tübingen 2005), p. 475; Dworkin, *A History of the Spanish Lexicon*, p. 71.

62 See von Wartburg, *FEW*, XVI, pp. 698–701; Corominas – Pascual, *Diccionario*, I, p. 350–351; Castellani, *Grammatica storica*, p. 57.

63 See von Wartburg, *FEW*, XVI, pp. 515–516; Castellani, *Grammatica storica*, pp. 58–59.

4 Relationships with the Enemy

Several words indicate the relationships with the enemy in their different typologies. Some of these terms are peculiar to the sole Gothic and have no cognates in other Germanic languages. Indirect sources offer some forms that are not attested in Wulfila's version.

fijands 'enemy', 'foe' translates Gr. ἐχθρός (Lat. *inimicus*) in several passages and ἔχθρα (here Lat. *inimicus*) in Rom 8:7, but ἐχθρός (Lat. *inimicus*) is also rendered as ***hatands*** 'enemy' in Lk 1:71 and 6:27. Both are present participles respectively from the well-attested verbs *fijan* 'to be inimical towards', 'to feel hostile towards', from which also ***fijaþwa*** 'enmity', 'hostility' derives (rendering of Gr. ἔχθρα : Lat. *inimicitia* in Gal 5:20; Eph 2:15 and 2:16), and *hatan* 'to hate', 'to be inimical towards', with parallels in the Germanic languages (respectively, ON *fjá*, OE *féon*, OHG *fījēn* 'to be hostile', and ON *hata*, OS *haton*, OHG *hazzen* 'to hate').

Cast. *espía*, Cat. It. *spia*, OFr. *espie*, Port. and Prov. *espia* 'spy' have occasionally been considered derivatives from Goth. ****spaiha*** 'spy',[64] in its turn from the Gothic verb **spaihon* 'to spy' (ON *spá* 'to prophetise'; OHG *spehon* 'to explore', 'to inquire'). However, the Germanic donor language(s) of these Romance terms is/are difficult to ascertain.

brakja 'wrestling', 'struggle' translates Gr. πάλη (Lat. *colluctatio*) in Eph 6:12. The term has no perfect parallels in other Germanic languages.

haifsts 'contestation', 'competition', 'conflict' translates Gr. ἀγών (Lat. *certamen*) in 1Tim 6:12, 2Tim 4:7, Php 1:30; ἐριθεία (Lat. *contentio*) in 2Cor 12:20, Php 1:17, Php 2:3; ἔρις in Rom 13:13, Gal 5:20, 1Tim 6:4, Php 1:15 (Lat. *contentio*). The term finds parallels in other Germanic languages (ON *heifst* 'conflict'?, OE *hǣst* 'violence') and entered in Italian as *astio* 'grudge', 'anger', 'hostility' (if it is not of Langobardic origin).[65] A derivative of this form is the verb ***haifstjan*** 'to compete in a contest', 'to contend', 'to fight', which translates Gr. ἀγωνίζεσθαι (Lat. *contendere*) in Jn 18:36, 1Cor 9:25, 1Tim 6:12, 2Tim 4:7, and ἀθλεῖν (Lat. *certare*) in 2Tim 2:5.

þrasa- 'quarrel' is attested in Sk 5:11 and in East Germanic onomastic sources, e.g. in the Ostrogothic name *Trasaric* (y. 589).[66] Its etymology and the connection with other Germanic languages are doubtful.

---

64 Gamillscheg, Romania Germanica, I², p. 279; Corominas – Pascual, *Diccionario*, II, pp. 742–743, against von Wartburg, *FEW*, XVII, p. 175.

65 Battisti – Alessio, *Dizionario etimologico italiano*, I, p. 339.

66 Reichert, *Lexikon*, I, p. 713.

waihjo 'battle', 'quarrel', 'wrangling' translates Gr. μάχη (Lat. *pugna*) in 2Cor 7:5. This formation is attested only in Gothic. Etymologically connected with this word is ***wigan*** 'armed conflict', 'war', rendering of Gr. πόλεμος (Lat. *bellum*) in Lk 14:31; ***weihan*** 'to battle', 'to combat', 'to dispute', which translates Gr. θηριομαχεῖν (Lat. *ad bestias pugnare*) in 1Cor 15:32 (= *du diuzam weihan*) and λογομαχεῖν (Lat. *contendere verbis*) in 2Tim 2:14 (= *waurdam weihan*); and ***andweihan*** (or ***andwaihan***?) 'to fight in opposition to',[67] rendering of gr. ἀντιστρατεύεσθαι (Lat. *repugnare*) in Rom 7:23. The root is well-attested in other Germanic languages (see ON *vega*, OE, OHG *wīgan* 'to fight').

Indirect sources provide other stems indicating the 'fight':

****bega*** 'quarrel', 'squabble', reconstructed on the basis of It. *bega*[68] 'quarrel', derived from a Gothic form (given the maintenance of Germ. **ē*, against ON *bágr* 'contest', OE *bāg* 'bragging', OS *bāg* 'praise', OHG *bāg* 'quarrel'), possibly attested as first part of the Visigothic name *Bigesvindus* (y. 693).[69] Other Gothic stems are transmitted by personal names:[70] ****badu-*** 'fight', 'battle', e.g. in the name *Badvila* (Ostrogothic king, y. 541–552), with Germanic parallels (ON *bǫð*, OE *beadu*, OS *badu-*, OHG *batu-*'battle'); ****gunþ-*** 'fight', 'struggle', e.g. in the Ostrogothic name *Theodagund* (y. 507–511), with Germanic parallels (ON *gunnr*, *gúðr*, OE *gūð* and OS *gūthea* 'war', 'battle'); ****haþu-*** 'fight', 'struggle', 'battle', e.g. in *Theodahad*, *Theodahatus* (Ostrogothic king, y. 534–536), with Germanic parallels (ON *hǫð*, OE *heatu-*, OS **hathu-*, OHG *hadu-*); ****hildi-*** 'fight', 'struggle', 'battle', e.g. in the Ostrogothic name *Hildevar* (y. 523), with Germanic parallels (ON *hildr*, OE, OS *hild*, OHG *hilt*(*i*)*a*). Their semantic distinction is not evident.

Some stems/terms related to the semantic field of the victory and peace are known:

sigis 'victory' translates Gr. νῖκος (Lat. *victoria*) in 1Cor 15:54, 1Cor 15:55 and 1Cor 15:57. It is often attested as part of personal names, e.g. in the Ostrogothic *Sigismer* (y. 526),[71] and finds parallels in other Germanic languages (ON *sigr*, OE *sige*, OS *sigi-*, OHG *sigu* 'victory').

triggws 'trustworthy', 'faithful' renders Gr. πιστός (Lat. *fidelis*) in several passages of the Bible translation. This term is etymologically connected with ***triggwa*** 'covenant', 'arrangement', 'testament', 'entrustment', translation of Gr. διαθήκη (Lat. *testamentum*) in several passages. The word passes into the Romance languages, where it is attested in a slightly different meaning: It.

67 See Casaretto, *Nominale Wortbildung*, pp. 254–255.

68 See Morlicchio, LEI *Germ.* I, coll. 705–711.

69 Reichert, *Lexikon*, I, p. 141.

70 See Reichert, *Lexikon*, I, pp. 111, 427, 660–664.

71 Reichert, *Lexikon*, I, p. 607.

tregua, Prov. *tregua*, Cast. *tregua*, Port. *tregoa* 'truce'.[72] The presence and geographical distribution of the consonant cluster *-gu-* (< Goth. *-ggw-*) suggests a Gothic origin of these forms, while OFr. *trève* seems to derive from Frankish **triuwa*.[73] This lexeme is also attested in personal names, e.g. in the Ostrogothic *Triggva* (y. 510–521).[74]

In addition, onomastic sources transmit ****friþu***- 'protection', 'peace', e.g. in the Visigothic name *Fredigern* (4th c.),[75] with parallels in all Germanic languages (ON *friðr*, OE *friþ*, OS *frethu*, OHG *frithu* 'peace'). This term is etymologically connected with ***gafriþon*** 'establish peace between', 'reconcile', translation of Gr. ἀποκαταλλάσσειν in Col 1:20, Col 1:21 and Eph 2:16 and καταλλάσσειν in 2Cor 5:18 and 5:19 (Lat. *reconciliare*), and ***gafriþons*** 'establishment of peace', 'reconciliation', rendering of Gr. καταλλαγή (Lat. *reconciliatio*) in 2Cor 5:18 and 5:19; ****hroþa***- 'victory', 'fame', attested in the Ostrogothic name *Ruderic*[76] (y. 542–546), with Germanic parallels (ON *hróðr*, OE *hrēþ*, OS *hrōth*, OHG *ruot-* 'fame').

5 Warfare Attitude

Terms related to 'fear' and 'boldness' are attested both in direct and indirect transmission:

****grimms*** 'furious', 'wrathful' (see Cast. and Cat. *grima* 'shiver', 'horror', cf. Neapolitan *grimma* 'wrinkled', 'old',[77] and Prov. *grima* 'sadness'). OIt. *grimo* 'old', 'sickly', 'miserable', attested in northern Italian dialects, is instead considered of Langobardic origin.[78] Parallels in other Germanic languages are ON *grimmr*, OE *grimm*, OS and OHG *grim* 'grim'.

balþei 'boldness', 'confidence', 'frankness' translates Gr. παρρησία (Lat. *fiducia*) in Eph 3:12, Eph 6:19, 1Tim 3:13 and Sk 8:5 and 6. It is an abstract substantive

---

72 Meyer-Lübke, *REW*, s.n. 8927; Corominas – Pascual, *Diccionario*, V, 648.

73 von Wartburg, *FEW*, XVII, p. 361. The Gothic origin of the Italian form is not fully accepted by Castellani, *Grammatica storica*, p. 87, who preferred a Langobardic origin, see also Max Pfister, 'Longob. **baug-*, **trewwa*, **waiss-*. Fonti e metodologia per lo studio del superstrato longobardo', in *Studi linguistici e filologici per Carlo Alberto Mastrelli*, eds. L. Agostinelli *et al.* (Pisa 1985), pp. 361–371. However, the Langobardic influence on an older Gothic borrowing is possible.

74 Reichert, *Lexikon*, I, p. 722.

75 Reichert, *Lexikon*, I, p. 288.

76 Reichert, *Lexikon*, I, p. 576.

77 See Francesco D'Ascoli, *Lingua spagnuola e dialetto napoletano* (Naples 1972), p. 65.

78 Battisti – Alessio, *Dizionario etimologico italiano*, III, p. 1872.

from the Gothic adjective ***balþs*** 'bold' (from which the adverb *balþaba* 'boldly', 'openly' derives), attested also in personal names, e.g. Visigothic *Balthi* and Ostrogothic (?) *Heldebaldus* (y. 496–507).[79] Regarding *Balthi*, the name of Alaric's family, Jordanes indicates its origin, from 'Baltha, id est audax'.[80] The adjective is attested in other Germanic languages (OE *bielde*, OS *bald*, OHG *bald* 'bold', 'strong'). The donor language of It. *baldo* 'bold' is unclear: it seems to derive from the corresponding Frankish word via OFr. *balt*, *bald*, *baud* 'bold', like Prov. *baut* 'bold' (> OCat. *baudos* 'happy').[81] A derivative verb is ***balþjan*** 'to be bold', 'to dare', attested in Sk 2:1, with parallels in other Germanic languages (ON *bella* 'to deal with', OE *bieldan*, OS *beldian*, OHG *belden* 'to make bold').

ananan þjan 'to dare', 'to take courage', 'to be audacious' translates ἀποτολμᾶν (Lat. *audere*) in Rom 10:20 and τολμᾶν (Lat. *audacter* [esse], *audere*) in Mk 15:43, 2Cor 11:21 and in Sk 4:27. The word is a compound, based on ***nanþ-*** 'bold', also attested in the Ostrogothic name *Nanderit*[82] (y. 575). *-nanþjan* finds parallels in other Germanic languages (ON *nenna* 'to strive', OE *nēþan*, OS *nāthian*, OHG *nenden* 'to dare').

gadaursan 'to venture', 'to dare', 'to be bold', 'to be courageous' translates Gr. θαρρεῖν (Lat. *audere*) in 2Cor 10:2, παρρησιάζεσθαι (Lat. *audere*) in 1Cor 6:1, Eph 6:20, and τολμᾶν (Lat. *audere*) in several passages. The Gothic word has parallels in other Germanic languages (OE *durran*, OS (*gi*)*durran*, OHG (*ki*)*turran* 'to dare').

reiran 'to tremble', 'to quiver', translation of Gr. τρέμειν (Lat. *tremere*) in Lk 8:47 and Mk 5:33. The word, isolated among the Germanic languages, passes in Spanish as *rehilar* 'to tremble', 'to quiver'.[83]

The Visigothic onomastic sources also offer ***funs*** 'ready', 'prepared', attested e.g. in *Fons*[84] (y. 589), with parallels in other Germanic languages (ON *fúss* 'willing', 'eager', OE *fūs*, OS *fūs*, OHG *funs* 'ready').

79 Francovich Onesti, *I nomi degli Ostrogoti*, pp. 58–59.

80 Jordanes, *Romana et Getica*, XXIX.146. See also Reichert, *Lexikon*, I, p. 115.

81 See von Wartburg, *FEW*, XV, pp. 29–30; Morlicchio, *LEI Germ.* I, coll. 106–127. It is not completely to rule out that the Italian word derives from Langobardic *bald*, very common in the onomastic sources, see *Garibaldi* (Haubrichs, personal communication).

82 Reichert, *Lexikon*, I, p. 515.

83 See Corominas – Pascual, *Diccionario*, IV, 856; Hilty, *La herencia visigótica*, p. 475; Dworkin, *A History of the Spanish Lexicon*, p. 72.

84 Reichert, *Lexikon*, I, p. 272.

6 Latin Loanwords

Latin military terms had already entered Gothic, together with other Latinisms belonging to the everyday life and commerce, before the mid-fourth century, as their attestation in Wulfila's version indicates.[85] Beside *militon* (see above) and, perhaps, if not of Greek provenience, *spaikulator* (see above), this source transmits a few other Latin terms indicating specific referents of the Roman army and military culture:

anno 'soldier's pay', 'wages' translates Gr. ὀψώνιον (Lat. *stipendium*) in Lk 3:14 and 1Cor 9:7. This is a loanword from Lat. *annona* 'soldier's pay', 'wages', a term that entered in the military language of the Goths as result of their mercenary service in the Roman army. The form *anno* revels an oral acquisition of the Latin word.

kapillon 'to get a haircut', 'to be shorn' translates Gr. κείρασθαι (Lat. *tonderi*) in 1Cor 11:6. The term is an independent formation from the Latin substantive *capillus* 'hair'. There is no doubt that the reason of the borrowing reflects the usage of Roman soldiers to wear short haircut, compared to the Gothic/Germanic custom of having long hair.[86]

karkara 'prison', 'jail' translates Gr. δεσμωτήριον (Lat. *vincula*) in Mt 11:2 and φυλακή (Lat. *carcer*) in several passages. The term derives from Vulgar Latin **carcara*, also attested in other Old Germanic languages (OE *carcern*, OS *karkari*, OHG *karkari*) to indicate a closed space for punishing, a custom absent among Germanic peoples.

7 Final Remarks

The richness of the vocabulary that emerges from the analysis is not surprising, considering the importance of warfare in the Gothic society. Of course, given the nature of the sources, it is not possible either to trace a history of the military lexicon in Gothic or to assess whether all the Gothic *gentes* used the same terminology. However, one feature is indisputable: most of the terms transmitted by Wulfila's Bible translation have perfect linguistical parallels in other Germanic languages and in stems attested in Visigothic and Ostrogothic

85 See Corazza, *Le parole latine*, *passim*.
86 See Wolfram, *Die Goten*, pp. 111 and 122.

personal names, a fact that indicates the antiquity and wider use of these words among Gothic armies and warbands.[87]

Some terms are peculiar only to Wulfilian Gothic (*brakja* 'wrestling', *drauhtiwitoþ* 'military law', 'military discipline', *gadrauhts* 'soldier', *þrasa-* 'quarrel', *waihjo* 'battle', and the loan translations *hundafaþs* 'centurion' and *þusundifaþs*[88] 'commander of a thousand men'), but their unicity may be accidental and due to the type of extant sources. The diffusion of the Bible version suggests their understanding and probable use among other Gothic groups.

The military vocabulary of the Greek model, which principally refers to the Roman army described in the New Testament, is mainly rendered by Wulfila with Gothic terms, through a process of loan translation and semantic borrowing, probably in order to offer lexical clarity and cultural closeness to his addresses. The influence of the Greek military vocabulary on the Gothic version is almost absent, while the Latin influence is limited to some terms, which indicate customs of the Roman world: the payment of the soldiers (Goth. *anno*), the short haircut (Goth. *kapillon*), the existence of a closed place for prisoners (Goth. *karkara*) and the organised military service (Goth. *militon*) and guards (Goth. *spaikulatur*, if of Latin origin).[89] It is worthy of note that the words *anno, kapillon*, and *spaikulatur* are not attested in other Germanic languages; a particularity that suggests that they probably entered the language of the Goths after their migrations from northern Europe and their separation from other Germanic *gentes*. The phonetic and morphologic integration of these borrowings speaks for their antiquity, and their use in the Gothic Bible indicates that they were understood also in non-military contexts and belonged to the common language. For these reasons, their introduction may be dated back to the period between the mid-third and the beginning of the fourth century.

The degree of knowledge of Latin, the command language and military *lingua franca*, varied certainly among Gothic soldiers. Undoubtedly, Latin competence represented a way to reach the highest levels of the military career, serving as linking between the Roman commander(s) and the Gothic troops.

---

87 On Gothic warbands, see Guido M. Berndt, 'Aktionsradien gotischer Kriegergruppen', in *Frühmittelalterliche Studien*, 47 (2013), pp. 7–52.

88 The linguistic analysis has led to exclude that *þusundifaþs* is the form from which *thiufadus* derives. Its origin may instead be seen in **þiudafaþs* 'commander of the people (people in arms?)', see above.

89 See the scenario depicted by Edward A. Thompson, *The Visigoths in the Time of Ulfila* (Oxford 1966), p. 39: 'Roman traders together with the hundreds of Visigoths who returned home from service with the Imperial forces planted an interesting list of Latin words in the Gothic language'.

Cross-lingual influences, code-switching, language transfers and mutual borrowings probably took place more frequently among the lowest layers of the army, where the dominance of the Latin language was less necessary and under control. In these contexts, it is easy to imagine the formation and the spread of Gothic-Latin words like *admanuire* 'to prepare', 'to make ready', *carrago* 'wagon-train', **conredare* and **adredare* 'to provide', 'to equip', *companio* 'companion', *kapillon* 'to get a haircut'. The introduction of new customs had the potential to influence the languages: Gothic terms – *bandwo* and *bandwa* 'sign' (later 'military banner'), *hilm* 'helm', *triggwa* 'entrustment', 'truce', and possibly *wardja* 'watchmann', **haribairgo* 'accommodation for the army', **spaiha* 'spy' – entered Latin, and Latin terms – *annona* 'soldier's pay', 'wages', **carcara* 'prison' and perhaps *spaikulatur* 'sentinell', together with *militare* 'to serve as a soldier', 'to perform military service' and *capillus* 'hair' – passed into Gothic. Very probably, the Gothic terms that entered the language of the Late Roman army formed a kind of *koine*, i.e. a jargon continued in their use (and enlarged) also by other Germanic-speaking soldiers, given the mutual intelligibility of their languages in the late antique period. Some of these words continued in Romance languages.

In certain periods and contexts, Gothic contingents of the Roman army probably formed separated unities that continued to use their own religion and language. This scenario clearly emerges from Theodoret's account about Gaina[90] and is reflected in the epigram 285 of the *Anthologia Latina* (*De conviviis barbaris*),[91] where the anonymous Latin author complains about the Gothic (or Vandalic)[92] words that he heard, probably spoken from soldiers in a tavern: *Inter 'eils' Goticum 'scapia matzia ia drincan | non audet quisquam dignos edicere versus* 'Among the Gothic 'eils scapia matzia ia drincan' ['Hail! Let's prepare (something to) eat and drink] no one ventures to recite decent verses'.

90 See above.

91 *Anthologia Latina sive poesis Latinae supplementum I: Carmina in codicibus scripta, I. Libri Salmasiani aliorumque carmina, editio altera*, ed. A. Riese (Leipzig 1894), n° 285; see Magnús Snædal, 'The 'Vandal' Epigram', in *Filologia Germanica/Germanic Philology*. 1 (2009), pp. 181–213 (with literature).

92 The term 'Goticum' might indicate both these East Germanic languages. On the perception of their likeness, see Procopius of Caesarea (Procopius Caesariensis, *Bellum Vandalicum*, in *Opera omnia*, ed. J. Haury, 2 vols. [Leipzig 1905; repr. Leipzig 1962] I.2.5). On the language of the Vandals, see Nicoletta Francovich Onesti, *I Vandali. Lingua e storia* (Rome 2002); Frederik Hartmann, *The Vandalic Language: Origins and Relationships* (Heidelberg 2020).

Practical reasons and spontaneous inclination in using the proper mother language may have had reflexes in constructing a feeling of union and community among Gothic soldiers; a behaviour that – in some cases – might have slowed down their process of linguistic and cultural integration and their acceptance as part of the Roman army by the civilian population. However, on the other hand, the act of following the *drauhtiwitoþ*, i.e. the 'military law', the 'military discipline' of the Roman army, which includes the carrying out of duties and the access to the *annona militaris*, equated Gothic soldiers to the Roman ones. *Obtorto collo*, the presence of the Goths in the Roman world could no longer be ignored, as emerges from Latinus Pacatus Drepanius' words in 388:

> Shall I not speak of the Goths received into your service, as a soldier in your camps and a farmer in your fields? [...] You grant the status of fellow soldier to barbarian peoples voluntarily vowing to serve you, both so that a suspect force might depart from the frontier and so that an auxiliary might arrive in military service; enticed by your benevolence, all of the Scythian nations congregated in such great swarms that you appeared to have requisitioned a levy from barbarians from which you excused your own people [...]. As a Roman, a one-time enemy marched under Roman generals and Roman standards and followed Roman banners that he once had opposed [...]. The Goth, the Hun, and the Alan answered the roll call and took their turns standing watch [...] if ever there was a shortage of rations, each bore the short-age patiently and relieved the amount of the reduction by conserving his *annona*.[93]

Glossary

agja*- 'sharpness', 'sword' → 3; ***ananan̄þjan* 'to dare', 'to take courage', 'to be audacious' → 5; ***andweihan*** (or ***andwaihan***?) 'to fight in opposition to' → 4; ***anno*** 'soldier's pay', 'wages' → 6; ***arƕazna*** 'arrow' → 3; ****badu***- 'fight', 'battle' → 4 ; ***balþei*** 'boldness', 'confidence', 'frankness' → 5; ***balþjan*** 'to be bold', 'to dare' → 5; ****balþs*** 'bold' → 5; ***bandum*** 'military banner' → 3; ****bega*** 'quarrel', 'squabble' → 4; ***brakja*** 'wrestling', 'struggle' → 4; ****brands*** 'sword' → 3; ***brunjo*** 'armor', 'breastplate' → 3; ***carrago*** → 3; ***drauhtinassus*** 'warfare', 'service under

93 *Pacati panegyricus*, eds. C.E.V. Nixon – B. Saylor Rodgers – R.A.B. Mynors, *In Praise of Later Roman Emperors: The panegyrici latini* (Berkeley 1994), 2(12).22.3–4; engl. transl. by Mathisen, *Barbarian Invasions*, p. 278.

a warlord', 'military service' → 2; ***drauhtinon*** 'to serve under a warlord', 'to gather as a warrior' → 2; ***drauhtiwitoþ*** 'military law', 'military discipline' → 2; ****drauhts*** 'army', 'warband', '*comitatus*' → 2; ***driugan*** 'to perform', 'to exert', 'to accomplish', 'to carry out' → 2; ***fijands*** 'enemy', 'foe' → 4; ***fijaþwa*** 'enmity', 'hostility' → 4; ****friþu-*** 'protection', 'peace' → 4; ****funs*** 'ready', 'prepared' → 5; ***gadaursan*** 'to venture', 'to dare', 'to be bold', 'to be courageous' → 5; ***gadrauhts*** 'soldier', 'warrior', 'serviceman' → 2; ***gafriþon*** 'establish peace between', 'reconcile' → 4; ***gafriþons*** 'establishment of peace', 'reconciliation' → 4; ***gahlaiba*** 'companion', 'comrade', 'co-breader' → 2; ****gais-*** 'spear', 'pike' → 3; ***garedan*** 'to provide' → 3; ***gasinþa*** 'fellow traveller' → 2 ***gasinþja/gasinþi*** 'fellow traveller' → 2; ****grimms*** 'furious', 'wrathful' → 5; ****gunþ-*** 'fight', 'struggle' → 4; ***haifstjan*** 'to compete in a contest', 'to contend', 'to fight' → 4; ***haifsts*** 'contestation', 'competition', 'conflict' → 4; ***hairus*** 'sword' → 3; ***hansa*** 'cohort', 'band of men', 'crowd', 'troop' → 2; ***harjis*** 'army', 'legion', 'armed host' → 2; ****hari-bairgo*** 'accommodation for the army' → 2; ***hatands*** 'enemy' → 4; ****haþu-*** 'fight', 'struggle', 'battle' → 4; ****hildi-*** 'fight', 'struggle', 'battle' → 4; ***hilms*** 'helm' → 3; ****hroþa-*** 'victory', 'fame' → 4; ***hundafaþs*** 'centurion', 'commander of a hundred-man military unit' → 2; ***kapillon*** 'to get a haircut', 'to be shorn' → 6; ***karkara*** 'prison', 'jail' → 6; ****lind-*** 'shield', 'lime-tree' → 3; ***manwjan*** 'to prepare' 'to make ready' → 3; ***meki*** 'sword' → 3; ***militon*** 'to serve in the army', 'to be a soldier' → 2; ****mund-*** 'shelter', 'protection' → 3; ****nanþ-*** 'bold' → 5; ***reiran*** 'to tremble', 'to quiver' → 5; ***sarwa*** 'arms', 'armament', 'weapons' → 3; ***sigis*** 'victory' → 4; ***skildus*** 'shield' → 3; ****skrama*** 'short sword' → 3; ****spaiha*** 'spy' → 4; ***spaikulatur*** 'sentinel' → 2; ***thiufadus*** → 2; ***þrasa-*** 'quarrel' → 4; ***þusundifaþs*** 'commander of a thousand men' → 2; ***triggwa*** 'covenant', 'arrangement', 'testament', 'entrustment' → 4; ***triggws*** 'trustworthy', 'faithful' → 4; ***wahtwo/wahtwa*** 'watch' → 2; ***waihjo*** 'battle', 'quarrel', 'wrangling' → 4; ***wardja*** 'watchman' → 2; ****wardja*** 'guard service' → 2; ****wards*** 'custodian' → 2; ***weihan*** 'to battle', 'to combat', 'to dispute' → 4; ***wepn*** 'weapon' → 3; ***wigan*** 'armed conflict', 'war' → 4.

Bibliography

Primary Sources

Ammianus Marcellinus, *Rerum gestarum libri*, ed. J.C. Rolfe, 3 vols. (Cambridge, Mass. 1935–9)

Anthologia Latina sive poesis Latinae supplementum I: Carmina in codicibus scripta, I. Libri Salmasiani aliorumque carmina, editio altera, ed. A. Riese (Leipzig 1894)

Glossarium Mediae Latinitatis Cataloniae <http://gmlc.imf.csic.es/codolcat/>

Gregorius Turonensis, *Libri Historiarum X, editionem alteram* cur. B. Krusch – W. Levison, MGH SS rer. Merov. I.1 (Hannover 1951)
Jordanes, *Romana et Getica*, ed. Th. Mommsen, MGH Auct. ant. V,1 (Berolini 1882)
Pacati panegyricus, eds. C.E.V. Nixon – B. Saylor Rodgers – R.A.B. Mynors, *In Praise of Later Roman Emperors: The panegyrici Latini* (Berkeley 1994)
Lex Visigothorum, ed. K. Zeumer, MGH LL Nat. Germ. I.1 (Hannover/Leipzig 1902)
Pactus Legis Salicae, ed. K.A. Eckhardt, MGH LL Nat. Germ. IV.1 (Hannover 1962)
Procopius Caesariensis, *Bellum Vandalicum*, in *Opera omnia*, ed. J. Haury, 2 vols. (Leipzig 1905; repr. Leipzig 1962)
Theodoret, *Kirchengeschichte*, eds. L. Parmentier – F. Scheidweiler – G.C. Hansen, 3rd rev. ed., GCS 19 (Berlin 1998)

Secondary Literature

Adams, James N., *Bilingualism and the Latin Language* (Oxford 2004)
Alexianu, Marius – Matei-Popescu, Florian, 'The Linguistic Impact of the Roman Army in the Province of *Moesia Inferior*', in *Roman Army and Local Society in the Limes Provinces of the Roman Empire: Papers of an International conference, Iași, June 4th-6th, 2018*, eds. L. Mihailescu-Bîrliba and W. Spickermann (Rahden/Westf. 2019), pp. 63–84.
Battaglia, Salvatore (ed.), *Grande dizionario della lingua italiana* (Turin 1967–2002)
Battisti, Carlo – Alessio, Giovanni, *Dizionario etimologico italiano* (Florence 1975)
Berndt, Guido M., 'Aktionsradien gotischer Kriegergruppen', in *Frühmittelalterliche Studien*, 47 (2013), pp. 7–52
Casaretto, Antje, *Nominale Wortbildung der gotischen Sprache. Die Derivation der Substantive* (Heidelberg 2004)
Castellani, Arrigo, *Grammatica storica della lingua italiana. I. Introduzione* (Bologna 2000)
Claude, Dietrich, '*Millenarius* und *thiufadus*', in *Zeitsch. der Sav. Stift. – Germ. Abt.*, 88 (1971), pp. 181–190
Corazza, Vittoria, 'Le parole latine in gotico', in *Accademia Nazionale dei Lincei. Memorie. Classe di Scienze morali, storiche e filologiche*, XIV (Rome 1969)
Corominas, Joan, *Diccionario crítico etimológico castellano e hispánico*, with the collaboration of José A. Pascual (Madrid 1980–1983)
Curcă, Roxana-Gabriela, '*Sermo castrensis* in Moesia Inferior', in *Roman Army and Local Society in the Limes Provinces of the Roman Empire: Papers of an International conference, Iași, June 4th-6th, 2018*, eds. L. Mihailescu-Bîrliba and W. Spickermann (Rahden/Westf. 2019), pp. 85–93
D'Ascoli, Francesco, *Lingua spagnuola e dialetto napoletano* (Naples 1972)
Dahn, Felix, *Die Könige der Germanen. Das Wesen des ältesten Königthums der germanischen Stämme und seine Geschichte bis auf die Feudalzeit*, 6 (Würzburg 1871)

De Mauro, Tullio (ed.), *Grande dizionario italiano dell'uso* (Turin 1999–2004)

Della Volpe, Angela, 'On Gothic *gahlaiba* and Latin *companion*: An Excursus in Historical Linguistics Methodology', in *LACUS Forum XXX: Language, Thought and Reality*, eds. G. Fulton – W.J. Sullivan – A.R. Lommel (Houston, TX 2004), pp. 5–28

Dworkin, Steven N., *A History of the Spanish Lexicon: A Linguistic Perspective* (Oxford 2012)

Falluomini, Carla, *The Gothic Version of the Gospels and Pauline Epistles: Cultural Background, Transmission and Character* (Berlin-Boston 2015)

Francovich Onesti, Nicoletta, *Vestigia longobarde in Italia (568–774). Lessico e antroponimia* (Rome 1999)

Francovich Onesti, Nicoletta, *I Vandawahtli. Lingua e storia* (Rome 2002)

Francovich Onesti, Nicoletta, *I nomi degli Ostrogoti* (Florence 2007)

Gamillscheg, Ernst, Romania Germanica. *Sprach- und Siedlungsgeschichte der Germanen auf dem Boden des alten Römerreiches*, 3 vols. (Berlin 1934–1936; 2nd ed. of the vol. I in 1970)

Goffart, Walter, *Barbarians and Romans A.D. 418–584. The Techniques of Accommodation* (Princeton 1980)

Goldsworthy, Adrian – Haynes, Ian (eds.), *The Roman Army as a Community* (Portsmouth, RI 1999)

Green, Dennis H, *Language and History in the Early Germanic World* (Cambridge 1998)

Hartmann, Frederik, *The Vandalic Language: Origins and Relationships* (Heidelberg 2020)

Haubrichs, Wolfgang, 'Krieg, Volk und Verwandtschaft. Zur Struktur und kulturellen Signifikanz ostgotischer Frauennamen', in *Archiv für Kulturgeschichte*, 99 (2017), pp. 297–339

Haubrichs, Wolfgang, 'Il linguaggio della guerra presso i popoli germanici occidentali (Longobardi e Franchi)', in *Ravenna Capitale. L'esercito romano all'alba dell'Europa. Modelli concettuali e sperimentazioni sociali tra linguaggi, istituzioni e diritto (secoli IV–VIII)*, ed. G. Bassanelli Sommariva (Santarcangelo di Romagna 2020), pp. 213–247

Heather, Peter, *The Goths* (Oxford 1996)

Hilty, Gerold, 'La herencia visigótica en el léxico de la Península Ibérica', in *Latin et langues romanes : Études de linguistique offertes à József Herman à l'occasion de son 80ème anniversaire*, eds. S. Kiss – L. Mondin – G. Salvi (Tübingen 2005), pp. 473–480

Jaime Moya, Joan María, 'La aportación del léxico militar de origen germánico en la formación de la lengua catalana', in *Literature, Science & Religion: Textual Transmission and Translation in Medieval and Early Modern Europe*, eds. M. Bellmunt Serrano – J. Mahiques Climent (Kassel 2020), pp. 257–268

Kiss, Attila. P., '*Per arma adoptio*. Eine gotische Sitte in den frühmittelalterlichen schriftlichen Quellen', in Romania Gothica II. *The Frontier World: Romans, Barbarians and Military Culture*, ed. T. Vida (Budapest 2015), pp. 95–108

Kramer, Johannes, 'Ein Gräzismus gotischer Herkunft im Italienischen: bando', in *Balkan-Archiv* NF, 12 (1987), pp. 197–207

Kuhn, Hans, 'Das römische Kriegswesen im germanischen Wortschatz', in *Zeitschrift für deutsches Altertum und deutsche Literatur*, 101 (1972), pp. 13–53

Lehmann, Winfred P., *A Gothic Etymological Dictionary: based on the Third Edition of* Vergleichendes Wörterbuch der Gotischen Sprache *by Sigmund Feist* (Leiden 1986)

Liebeschuetz, J.H. Wolfgang, *Barbarians and Bishops: Army, Church, and State in the Age of Arcadius and Chrysostom* (Oxford 1990)

Maier, Gideon, *Amtsträger und Herrscher in der* Romania Gothica. *Vergleichende Untersuchungen zu den Institutionen der ostgermanischen Völkerwanderungsreiche* (Stuttgart 2005)

Meyerhoff, Miriam, 'Communities of Practice', in *The Handbook of Language Variation and Change*, eds. J. K. Chambers, Peter Trudgill, and Natalie Schilling-Estes (Oxford 2002), pp. 525–548

Martin, Céline, *La géographie du pouvoir dans l'Espagne visigothique* (Lille 2003)

Mathisen, Ralph, 'Barbarian Invasions or Civil Wars? Goths as Auxiliary Forces in the Roman Army', in *Empire in Crisis: Gothic Invasions and Roman Historiography. Beiträge einer internationalen Tagung zu den Wiener Dexipp-Fragmenten* (Dexippus Vindobonensis), eds. F. Mitthof – G. Martin – J. Grusková (Vienna 2020), pp. 263–286

Meyer-Lübke, Wilhelm. *Romanisches etymologisches Wörterbuch*, 3. vols. (Heidelberg 1935) [= *REW*]

Morlicchio, Elda (ed.), *LEI. Lessico etimologico italiano. Germanismi* (Wiesbaden 2000-)

Mosci Sassi, Maria Grazia. *Sermo castrensis* (Bologna 1983)

Moulton William, 'Mutual Intelligibility among Speakers of Early Germanic Dialects,' in Germania: *Comparative Studies in the Old Germanic Languages and Literatures*, eds. Daniel G. Calder and T. Craig Christy (Wolfeboro, NY 1988), pp. 9–28

Nefedkin, Alexander K., 'Weaponry of the Goths of the mid-3rd to the 7th Century AD', in *Journal of Roman Military Equipment Studies*, 16 (2008), pp. 147–155

Orel, Vladimir E. *A Handbook of Germanic Etymology* (Leiden 2003)

Osaba, Esperanza, 'Answering the Call to Arms. *Lex Visigothorum* 9, 2', in this volume Ch. 11

Pfister, Max, 'Longob. **baug-*, **trewwa*, **waiss-*. Fonti e metodologia per lo studio del superstrato longobardo', in *Studi linguistici e filologici per Carlo Alberto Mastrelli*, eds. L. Agostinelli *et al.* (Pisa 1985), pp. 361–371

Piel, Joseph M. – Kremer, Dieter, *Hispano-gotisches Namenbuch. Der Niederschlag des Westgotischen in den alten und heutigen Personen und Ortsnamen der Iberischen _Halbinsel* (Heidelberg 1976)

Rance, Philip, '*quam gentilitate appellant*. The Philological Evidence for *Germani* in the Late Roman Army: Germanic Loanwords in Roman Military Vocabulary', in

Romania Gothica II. *The Frontier World: Romans, Barbarians and Military Culture*, ed. T. Vida (Budapest 2015), pp. 51–93

Reichenkron, Günter, *Historische latein-altromanische Grammatik. I. Einleitung. Das sogenannte Vulgärlatein und das Wesen der Romanisierung* (Wiesbaden 1965)

Reichert, Hermann. *Lexikon der altgermanischen Namen*, 2 vols.: I. *Text*; II. *Register*, with the cooperation of R. Nedoma (Vienna 1987–1990)

Restelli, Giuseppe, 'I Goti nella «Romania». Influssi della loro lingua sul latino', in *Aevum* 51 (1977), pp. 207–238

Santoro, Verio, 'L'intercomprensione orale tra parlanti lingue germaniche antiche', in *Studies on Late Antique and Medieval Glossography and Lexicography in Honour of Patrizia* Lendinara, eds. Claudia Di Sciacca – Concetta Giliberto *et al.* (Pisa 2018), pp. 663–686

Scardigli, Piergiuseppe, *Die Goten. Sprache und Kultur* (Munich 1973)

Schaff, Philip – Wace, Henry, eds., *A selected Library of Nicene and Post-Nicene Fathers of the Christian Church*, 3rd vol. (New York *et al.*, 1892)

Seebold, Elmar, 'Gt. *gasinþa** ‚Reisegefährte' und *gasinþi** ‚Reisegesellschaft'', in *Beiträge zur Geschichte der deutschen Sprache und Literatur* (Tübingen), 96 (1974), pp. 1–11

Snædal, Magnús, 'The 'Vandal' Epigram', in *Filologia Germanica/Germanic Philology*. 1 (2009), pp. 181–213

Streitberg, Wilhelm, *Der gotische Text und seine griechische Vorlage. Mit Einleitung, Lesarten und Quellennachweisen sowie den kleineren Denkmalern als Anhang*, mit einem Nachtrag von Piergiuseppe Scardigli, 7. ed. (Heidelberg 2000)

Thomas, Antoine, 'L'étymologie du mot français *compagnon*', *Comptes rendus des séances de l'Académie des Inscriptions et Belles-Lettres*, 1931 (75), pp. 79–86

Thompson, Edward A., *The Visigoths in the Time of Ulfila* (Oxford 1966)

Velten, Harry, 'Studies in the Gothic Vocabulary with Special Reference to Greek and Latin Models and Analogues', in *The Journal of English and German Philology*, 29 (1930), pp. 332–351

Wagner, Norbert, 'Sculca und skalks', in *Sprachwissenschaft*, 45 (2020), pp. 275–278

von Wartburg, Walther, *Französisches Etymologisches Wörterbuch. Eine Darstellung des galloromanischen Sprachschatzes* (Bonn *et al.*, 1922–2002), with the cooperation of Oscar Jänicke *et al.* [= *FEW*]

Wiemer, Hans U. – Berndt, Guido M., 'Instrumente der Gewalt: Bewaffnung und Kampfesweise gotischer Kriegergruppen', in *Millennium. Jahrbuch zu Kultur und Geschichte des ersten Jahrtausends n. Chr.*, 13 (2016), pp. 141–210

Wiener, Leo, *Commentary to the Germanic Laws and Mediaeval Documents* (Cambridge 1915)

Wolfram, Herwig, *Die Goten. Von den Anfängen bis zur Mitte des sechsten Jahrhunderts. Entwurf einer historischen Ethnographie* (Munich 1990)

PART 2

Social and Juridical Structures

CHAPTER 6

Militia and *Civitas* between Third and Sixth Century CE

Valerio Marotta

1 *Barbari Dediticii* and Citizenship

A famous epigraph at Aquincum (*Pannonia inferior*[1]), of uncertain dating,[2] commemorates a *miles* of Frankish origin:

> Francus Ego Cives (sic!) Romanus miles in armis / egregia virtute tuli bello mea dextera sem<p>er.
>
> I am Francus, I am a Roman citizen in the army, I have always wielded, with excellent virtue, my right hand in war.

It is usually interpreted as if the person speaking maintained, in celebrating the memory of his military virtues, that he was a *civis Francus* and a *miles Romanus in armis*.[3] However, if we think about it, its real meaning is very different.[4] This inscription honours someone deceased who served in the army, obtaining *civitas* either during the prescribed service or when he was discharged. Hence, this is a witness that, even more explicitly than the tombstones of the cemetery of Concordia,[5] perpetuates the memory of a successful circumstance of integration into Roman citizenship.[6]

The controversial content of the Aquincum epigraph summarises an exemplary story that intersects at least part of the subjects I would like to address in this essay. These subjects are certainly related to each other, indeed closely

1 *CIL* III 3576: cf. Paolo Cugusi, *Per un nuovo corpus dei Carmina Latina Epigraphica. Materiali e discussion* (Roma, 2007), pp. 108, 172.
2 It varies between the end of the second and the beginning of the fifth century.
3 Thus Soazick Kerneis, "Francus civis, miles Romanus': les barbares de l'Empire dans le code Théodosien', in *Droit, religion et société dans le Code Théodosien*, eds. J.-J. Aubert – Ph. Blanchard (Genève 2009), esp. p. 390.
4 Kent Rigsby, 'Two Danubian Epitaphs', in Zeitschr. für *Papyrologie und Epigraphik* 126 (1999).
5 *Infra*, notes 114 and 126.
6 An overall picture *infra*, pp. 000 *ff.*

 | DOI:10.1163/9789004698017_007

interconnected: on the one hand, the extent of the army's contribution to the diffusion of Roman citizenship in Late Antiquity; on the other – looking at the Constantinian period[7] as a whole – the various legal provisions that governed access to *civitas* between the fourth and fifth centuries.

An obligatory starting point is the *constitutio Antoniniana*. Indeed, depending on whether one or another of the many interpretations of its provision is accepted, one can propose markedly divergent reconstructions of the history of Roman citizenship in the third and fourth centuries:

> D. 1.5.17 (Ulp. 22 *ad ed.*) In orbe Romano qui sunt ex constitutione imperatoris Antonini cives Romani effecti sunt.
>
> According to a Constitution of the Emperor Antoninus, all those who were living in the Roman world were made Roman citizens.
>
> P. Giss. 40 col. I, ll. 7–9 (...) δίδωμι τοῖς συνάπα- [σιν -ca.?- κατὰ τ]ὴν οἰκουμένην π[ολιτ]είαν Ῥωμαίων, [μ]ένοντος [τοῦ δικαίου τῶν πολιτευμ]άτων, χωρ[ὶς] τῶν [[δε]δειτικίων (...).
>
> I give everyone in the Roman World the Roman Citizenship, preserving the law of the politeumata, except the dediticii. (...).
>
> Dio *Hist.* 77(78).9.4–5 (...) τῶν τε τελῶν τῶν τε ἄλλων ἃ καινὰ προσκατέδειξεν, καὶ τοῦ τῆς δεκάτης ἣν ἀντὶ τῆς εἰκοστῆς ὑπέρ τε τῶν ἀπελευθερουμένων καὶ ὑπὲρ τῶν καταλειπομένων τισὶ κλήρων καὶ δωρεᾶς ἐποίησε πάσης, τάς τε διαδοχὰς καὶ τὰς ἀτελείας τὰς ἐπὶ τούτοις τὰς δεδομένας τοῖς πάνυ προσήκουσι τῶν τελευτώντων καταλύσας (οὗ ἕνεκα καὶ Ῥωμαίους πάντας τοὺς ἐν τῇ ἀρχῇ αὐτοῦ, λόγῳ μὲν τιμῶν, ἔργῳ δὲ ὅπως πλείω αὐτῷ καὶ ἐκ τοῦ τοιούτου προσίῃ διὰ τὸ τοὺς ξένους τὰ πολλὰ αὐτῶν μὴ συντελεῖν, ἀπέδειξεν)
>
> The taxes, both the new ones which he (Caracalla) promulgated and the ten per cent, tax that he instituted in place of the five per cent, tax applying to the emancipation of slaves, to bequests, and to all legacies; for he abolished the right of succession and exemption from taxes which had been granted in such cases to those who were closely related to the deceased. This was the reason why he made all the people in his empire Roman citizens; nominally he was honouring them, but his real purpose

7 I refer, of course, to the famous periodization of Santo Mazzarino, *Antico, tardoantico ed èra costantiniana*, 2 vols. (Bari, 1974–1980).

was to increase his revenues by this means, inasmuch as aliens did not have to pay most of these taxes.[8]

Did the *edictum* of 212 involve the entirety of the free inhabitants of the Empire – as these three different witnesses would seem to attest (even more so if coordinated among each other) – or, to the contrary, did it exclude a certain number of provincials or certain groups from the benefit of citizenship?

I do not think it is possible to answer this question.[9] Although almost all scholars, today as in the past, rely on examination of the text transmitted to us by the Giessen Papyrus to resolve this question, I am doubtful that it can provide us with decisive evidence or even relevant clues. Even if we were to integrate line 9 by reading, in accordance with what is still today the *communis opinio*,[10] the words χωρ[ὶς] τῶν [δε]δειτικίων, we could not identify them with a clause of exclusion,[11] consequently reconnecting them to the main verb δίδωμι in line 7.[12] In short, this document can tell us very little about the results and limits of the process of extension of *civitas*. Moreover, even while ignoring the rules of Greek syntax,[13] we should nonetheless recognize – in connecting these words to the verb δίδωμι – that the term *dediticius*, unlike a skilled quick-change artist on the stage, struggles to interpret several parts "in a play": in fact, it is unsuitable to label, at the same time, the disparate categories of individuals excluded for various reasons,

8 Engl. Trad. Cary (Loeb).

9 In this regard, I can repeat what I have already expressed in Valerio Marotta, 'Inclusione o esclusione? La 'constitutio Antoniniana' e i limiti del cosmopolitismo universale romano', in *Strumenti, percorsi e strategie dell'integrazione nelle società multiculturali*, eds. G. Cerrina Feroni – V. Federico (Napoli 2018), esp. pp. 138 *ff.*

10 Lastly see Peter van Minnen, 'Three Edicts of Caracalla? A New Reading of P.Giss. 40', *Chiron* 46 (2016), pp. 217 *ff.* see also p. 219 and note 66, in which the author of this nonetheless excellent article focuses on a secondary detail (the presence of transliterations of Latin words in Greek inscriptions or papyri), but unfortunately does not dwell on the main objection formulated against the conjecture [δε]δειτικίων: Valerio Marotta, *La cittadinanza romana in età imperiale* [*secoli I–III d.C.*]. *Una sintesi* (Torino 2009), p. 114. Indeed, this term does not seem able to assume all the categories of individuals who – if the words χωρ[ὶς] τῶν [δε]δειτικίων had really proposed an exclusion clause – would not have been included among the beneficiaries of the *Theía Dōreá*, first of all the Latin freedmen (see what I write here at fn. 83).

11 For the reasons I provide, *infra*, in the text and, *supra*, in fn. 10.

12 If we accept the conclusions of Gino Segré, 'L'editto di Caracalla relativo alla concessione della cittadinanza romana e il papiro Giessen 40, I., in Id., *Scritti giuridici* II., (Roma, 1938) 97–271, 126 note 1, 129 note 1, 230 *ff.*, with other references to the philological debate following the discovery and publication of the papyrus.

13 It seems necessary to Joseph Mélèze Modrzejewski, *Droit et Justice dans le monde grec et hellénistique*, (Warszawa 2011) esp. 486 f. and note 48, with literature.

according to historians, from the *Theía Dōreá*: in order, *liberti dediticii Aeliani, deportati,* those condemned to *opus publicum perpetuum*, tributary peasant masses extraneous to the civic ranks of a *políteuma* (primarily Egyptians[14] of the χώρα and λαοί of Asia minor and *Syria*), and finally *barbari dediti in fidem*.

It would be sufficient to observe that by straining its literal sense, perhaps beyond what might be legitimate, we could relate it to the notion of *dediticius* – together with prisoners *dediti in fidem* and with *liberti dediticii Aeliani* – the *deportati* and those condemned *in opus publicum*. Yet, I would not really be able to count the *liberti Latini* – a category abolished only by Justinian – among them. Including them among the beneficiaries of the *edictum de civitate* would imply the deliberate choice of the emperor to expropriate their former masters (as well as their heirs and legatees) without any form of compensation. We could theoretically assume the existence of a clause that safeguarded the *iura patronatus*, on which basis these *liberti*, although becoming *cives Romani*, nonetheless died *tamquam Latini*. But there is no trace of its presence in the text of the Papyrus.[15]

Indeed, reading the provision of the edict *iuxta propria principia*, i.e. as a legal document,[16] without projecting on it suggestions borrowed from social history[17] or political history,[18] it seems very difficult to conclude that line 9 of the Giessen Papyrus and in particular the words χωρ[ὶς] τῶν [δε]δειτικίων define a clause of exclusion. It is no coincidence that the integration [ἀδ]δειτικίων, a transliteration of the neutral *additicia*,[19] is gradually gaining the consensus of

14 Or better among Egyptian peasants subjected in full to payment of the *tributum capitis* (*laographouménoi*) and *dediticii*: see Ulrich Wilcken, *Grundzüge und Chrestomatie der Papyruskunde* I.1. *Erster Band. Historischer Teil. Erste Hälfte: Grundzüge* (Leipzig 1912), pp. 59–60. Michael Rostowtzew, *Studien zur Geschichte des römischen Kolonates* (Leipzig – Berlin 1910), pp. 407 *ff.*, adopted these conclusions, referring *verbatim* to the text of a letter by Ulrich Wilcken. Therefore, also in his opinion, the *dediticii* of *P. Giss.* 40 col. I would coincide – as Paul Martin Meyer maintained – with the *laographouménoi* and thus on a strictly objective level with the ὁμόλογοι. In other words, the latter would be identified with the entire population of Egypt subject to poll tax, more precisely with men between 14 and 60 years of age required to pay the *tributum capitis*. However, the editors of *P. Ryl.* II 209 (see *Documents of the Ptolemaic and Roman Periods*) contested the basis of the equivalence, conjectured by Wilcken, between ὁμόλογοι and *dediticii*.

15 Gai. 3.72–73.

16 Which, as such, should be aimed at defining a regulatory provision consistent first of all with itself as well as with the existing order.

17 Santo Mazzarino, *L'Impero romano*, II (Roma – Bari, 1986) esp. pp. 609 *ff.*, 689 *ff.* But see, with other bibliography, Marotta, *Inclusione o esclusione?*, nt. 9, pp. 137 *ff.*

18 Claudia Moatti, *'Res publica'. Histoire romaine de chose publique* (Paris 2018), pp. 361 *ff.*, esp. 369 *ff.*

19 On this point, with a detailed review of opinions, see Gianfranco Purpura, 'Constitutio Antoniniana de civitate (212 d.C.). (P.Giff. 40 I = FIRA I, 88)', in *Revisione ed integrazione*

scholars.[20] Nevertheless, unlike what James Oliver thought,[21] it is impossible to identify a perfect symmetry between the clause *et sine deminutione tributorum et vectigalium populi et fisci* of the *Tabula Banasitana*[22] and the words χωρ[ὶς] τῶν [ἀδ]δειτικίων.[23] The former, in the extract from the *Commentarius civitate Romana donatorum*, cannot be configured as an addition (*additicium*) to *salvo iure gentis*; at most it is a further clarification, probably essential to explain unequivocally that the *princeps gentis Iulianus* and his family, in addition to remaining subject to all obligations imposed on their tribe, had not been exempt from paying taxes and perhaps even from the *tributum capitis*. Vice versa, the sub-clause χωρ[ὶς] τῶν [ἀδ]δ̣ειτ̣ικίων – devised with some flexibility if we accept this hypothesis – would seem to underline that all personal statuses additional to the *ius*[24] of each community of the Empire,[25] primarily the fiscal privileges granted to certain individuals, had to be safeguarded, even though wholly or partly inconsistent with the specific condition granted to the other members of the *políteuma* of which they were *originarii*. Hence the safeguard clause [μ]ένοντος |9. [τοῦ δικαίου τῶν πολιτευμ]άτων[26] and the

---

dei Fontes Iuris Romani Anteiustiniani (FIRA). *Studi preparatori*, I, *Leges*, ed. G. Purpura (Torino 2012), esp. pp. 703*f.*

20 Lastly Alex Imrie, *The Antonine Constitution. An Edict for the Caracallan Empire* (Leiden-Boston 2018), pp. 139*ff.*, esp. 146–150, 149*f.*

21 James H. Oliver, 'Text of the Tabula Banasitana, A.D. 177', *American Journal of Philology* 93 (1972) and Id., *Greek Constitutions of Early Roman Emperors from Inscriptions and Papyri* (Philadelphia 1989), no. 260, pp. 496–505, esp. 504. Of fundamental importance is the accurate palaeographic and philological analysis of Peter Alois Kuhlmann, *Die Giessener literarischen Papyri und die Caracalla-Erlasse. Edition, Übersetzung und Kommentar* (Giessen 1994), pp. 215*ff.*, 221, 234*ff.*

22 IAM II. 94 = *ILMaroc* 94 ll. 35–37: ... *We grant Roman citizenship to them, preserving the law of tribe* (*salvo iure gentis*) *and without harm to the taxes and duties of the People or the Imperial Purse.*

23 In this regard, the arguments of J. Mélèze Modrzejewski, *Loi et coutume dans l'Égypte grecque et romaine* (Warszawa, 2014), p. 322, certainly appear persuasive.

24 Here *ius* (perhaps in *P. Giff.* 40 col. I, l. 9, δίκαιον or other similar term, *ius* in the *tabula Banasitana*) has the meaning of the order of things by virtue of which each person, as *polítēs* of a city or member of a tribe, has his own specific *status* which results in rights, but also obligations and duties.

25 See Valerio Marotta, 'Doppia cittadinanza e pluralità degli ordinamenti. La tabula Banasitana e il Papiro di Giessen 40 col. I', in *Liber amicorum. Mélanges en l'honneur de Jean-Pierre Coriat*, eds. E. Chevreau – C. Masi Doria – J.M. Rainer (Paris 2019), pp. 568*ff.* Regarding the problem of payment of the *tributum capitis*, which was perpetuated even after 212, see Lutz Neesen, *Untersuchungen zu den direkten Staatsabgaben der römischen Kaiserzeit* (*27 v. Chr. - 284 n. Chr.*) (Bonn 1980), pp. 129, 132.

26 The status (*ius*: *supra*, note 24) of each community of the Empire remained valid. In other words, with regard to the problem of individual statuses, the provincial population did

words χωρ[ὶς] τῶν [ἀδ]δειτικίων, in coordination, would be limited to sanctioning the existing obligations towards the *civitates* and the other communities of the Roman oecumene (aimed, in turn, at providing certain services to the provincial administrations), at the same time preserving[27] also the *addeitíkia*, i.e. those additional regulations – at times more favourable, at others more unfavourable – assigned to or imposed on certain categories of people.

It is well established that the issuing, in 212, of the *edictum de civitate* (with extension of the tax base of the *decima* [*vicesima*] *hereditatium* and the *decima* [*vicesima*] *manumissionum*) conformed above all to the need to increase the flow of financial resources destined to maintaining the army and paying the *praemia militiae* to veterans.[28] It is certainly probable, as Marco Rocco recently argued,[29] that the *constitutio* of Caracalla attempted also to obviate the problem of reintegration of the complements of the legions. Since – as the sources seem to attest – the *cives Romani* preferred to enlist in the *auxilia*, the emperor decided to extend the legionary recruitment basin to virtually all the free inhabitants of the Empire. This would also explain the massive enlistment of Thracians and Pannonians into the Roman legions starting from the early third century.

In contrast, at least in my opinion, as we cannot identify in the Giessen Papyrus the presence of an exclusion clause that discriminates the so-called *dediticii*, those conjectures aiming to connect it to the need – explicitly felt by Caracalla – to safeguard the long-established practice of recruiting units of specialists drawn from the less Romanized populations of the Empire, and even from certain tribes based outside the borders of the *provinciae*,[30] are

---

not appear, after 212, to be less heterogeneous than before. On the interpretation of the clause μένοντος [τοῦ δικαίου τῶν πολιτευμ]άτων (ll. 8–9), the comments by Jean-Michel Carrié, 'Les distributions alimentaires dans les cités de l'Empire romain tardif', in *Mélanges de l'École française de Rome*, 87.2 (1975), esp. pp. 1026–1029, are, in my opinion, decisive.

27 Χωρ[ὶς], in the language of the papyri, is sometimes equivalent to our 'apart from': not so much, however, in the sense of 'except' as instead in that of 'in addition to': see Kuhlmann, *Die Giessener literarischen Papyri*, p. 232.

28 Valerio Marotta, 'Tre riflessioni sulla cittadinanza: da Roma antica al mondo attuale'. *Iuris Antiqui Historia*, 5 (2013), pp. 60 *ff.*, with further information.

29 See Marco Rocco, *L'esercito romano tardoantico. Persistenze e cesure dai Severi a Teodosio I.* (Limena 2012), pp. 29 *ff.*, esp. 35 *ff.* However, this problem had already been lucidly understood by Mazzarino, *L'impero romano*, II, pp. 691 *ff.* On this point, see also Imrie, *The Antonine Constitution*, pp. 81 *ff.*, esp. 93 *ff.*

30 It is a hypothesis shared by Méléze Modrzejewski, *Droit et Justice*, (*supra*, fn. 13), who, in this regard, explicitly recalled, adopting them himself, the conclusions of the monograph of Soazick Kerneis, *Les celtiques: servitude et grandeur des auxiliaires bretons dans l'Empire Romain* (Clermont-Ferrand 1998) (*non vidi*), as well as by van Minnen, 'Three Edicts of

unfounded. In other words – although it is true, as Marco Rocco pointed out,[31] that ethnic units (a long-lasting phenomenon) played a role that was anything but marginal in the armies of the Severan and post-Severan period – it is preferable to follow the prudent approach of Michael Speidel,[32] limiting ourselves to acknowledging its existence without venturing into legal-historical conjectures on the *edictum de civitate* and its supposed exclusion clause. After having ascertained that in the Severan period the term *numerus* still indicated military units of various kinds,[33] we must bear in mind, on the basis of the findings of Patrick Le Roux,[34] that although the *Mauri gentiles* and the other troops recruited from among the *gentes* did not acquire Roman citizenship at the end of the prescribed service, this does not mean that they were included among the *dediticii*, although – as the *Tabula Banasitana* underlines (ll. 4–5) – Roman citizenship was not granted to them *nisi maximis meritis provocata*. On other occasions – as confirmed by the events of the *Palmyreni sagittarii*[35] and perhaps of the *Mauri equites*[36] – *civitas Romana* was conferred without *ius conubii*, likely on the basis of the provisions of the *foedus* that this caravan city and some Mauri tribes had concluded with Rome. Yet all that, at least in the case of Palmyra,[37] did not prevent them from passing, in a period of about eighty years, from a condition of relative marginality – attested by the tenor of the *diplomata* concerning its *milites*, all dated between 120 and 126, and by the almost coeval *De munitione castrorum* of Pseudo-Hyginus[38] – to full

Caracalla?' and by Rocco, *L'esercito romano*, and, obviously, by Kerneis herself (*infra*, fn. 41 and 63).

31 Rocco, *L'esercito romano*, pp. 49 *ff.*

32 Michael Speidel, 'The Rise of Ethnic Units in the Roman Imperial Army', in *Aufstieg und Niedergang der römischen Welt* (ANRW), II.3 (1975), esp. pp. 213–223, 228.

33 See D. 49.16.4.10 (Menander 1 *de re militari*).

34 Patrick Le Roux, 'Les diplômes militaires et l'évolution de l'armée romaine de Claude à Septime Sévère: 'auxilia', 'numeri' et 'nationes'', in *Heer und Integrationspolitik. Die römischen Militärdiplome als historische Quelle*, eds. W. Eck – H. Wolff (Köln – Wien 1986), pp. 367–370.

35 *RMD* I 17 = *ZPE* 133, 279 = *AÉ* 1958, 30 = *AÉ* 1959, 31 = *AÉ* 1960, 336 = *IDR* I 5 = *ZPE* 208, 257 (a. 120); *RMD* I 28 = *IDR* I 9 = *AÉ* 1977, 696; *IDR* I 6a «… (Hadrianus) granted citizenship to the Palmireni *sagittarii* who are in *Syria* under the command of Julius Severus …» (a. 120); *RMD* I 20 = *AÉ* 1975, 714 = *AÉ* 1980, 763 = *ZPE* 208, 259 (a. 123); *RMD* IV 237 (a. 120–126).

36 *CIL* XVI 144. It is necessary to differentiate the *Mauri equites* from the *Mauri gentiles*: Le Roux, 'Les diplômes militaires', p. 370.

37 In this regard, see the easy-to-read, but captivating, book by Michael Sommer, *Palmyra. A History* (London - New York 2018), pp. 112 *ff.*

38 § 2; § 29.

integration into the *civitas*, as demonstrated by its status as a *colonia* with *ius Italicum* already in a period prior to the issuing of the *constitutio Antoniniana.*[39]

Therefore, I do not doubt the decisive importance of recruitment of the *gentiles* into the military formations of the Empire. There were certainly many men of barbarian stock who, coming from the outside, were without any *ius originis* in the Roman *provinciae*, but who nonetheless already constituted an important factor for social stability, for the economy[40] and in particular for the defence of the third-century imperial oecumene.[41] For now, I would merely underline that – since the urgency to clarify who were the *dediticii* of line 9 of the Giessen Papyrus has, in my opinion, disappeared – it is misleading to identify exclusively in this term the common thread that should connect together the sources on the history of the army and the recruitment of so-called ethnic units from the Severan age to the end of the fourth century and beyond.

Obviously, we do not know if other clauses of the edict – whether illegible (because of the large gaps in the Giessen Papyrus) or not passed down to us – might have included a list of those excluded from the concession of *civitas*. In this case, the edict undoubtedly should have indicated, together with the *liberti dediticii Aeliani,*[42] the *liberti Latini* and perhaps also the true *dediticii*, that is – as Gaius writes[43] – those who took up arms to fight against the Roman people and, once defeated, surrendered.[44]

Therefore, the question of the *dediticii* and especially the *barbari* reduced to the status of *dediticii* – albeit in my view considerably curtailed – still remains open. This is because – as underlined by some scholars[45] – *gentes devictae* and *deditae in fidem* were often settled in depopulated regions of the Empire between the third and fourth centuries. At the same time, others observe[46] that, already in the Severan age, the *numeri* were more often than not recruited from these populations. Nevertheless, we know of only one unit characterized,

39 D. 50.15.1.4–5 (Ulp. 1 *de cens.*).

40 Miroslava Mirkovic, *The Later Roman Colonate and Freedom*, [Transactions of the American Philosophical Society. New Series, 87.2] (Philadelphia 1997), pp. 85 *ff.*

41 Soazick Kerneis, 'Les numeri ethniques de l'armée romaine au IIe et IIIe siècles', in *Rivista storica dell'antichità*, 26 (1996), esp. pp. 81–92.

42 However, it should be borne in mind that, on the basis of the *lex Aelia Sentia*, the *liberti dediticii Aeliani* would have already been excluded, in principle and by virtue of their specific status, from any possibility of gaining *civitas Romana*.

43 Gai. 1.14: *Those enemies are called dediticii who, having formerly taken up arms and fought against the Roman people afterwards have been conquered and have surrendered at discretion.*

44 For the *formula deditionis*, see Liv. 1.38.

45 Mirkovic, *The Later Roman Colonate*, pp. 87 *ff.*

46 Kerneis, *Les numeri ethniques*, pp. 92–94.

in its own nomenclature, as *dediticius*: the *Brittones dediticii Alexandriani* (?) of the Walldürn inscription[47] quartered, in 232, in *Germania superior* (in the Odenwald). It is assumed that the *deditio* of the group, from which they had been recruited, occurred during the British campaign of Septimius Severus between 209 and 211.[48]

However, can this witness alone sustain the weight of the conjectures that relate – almost as if were dealing with a routine certified by extensive documentation – the recruitment in the ethnic *numeri* to the condition of *barbarus dediticius*?

I do not think that this question can be answered at the moment. In my opinion – if we move to the level of the history of law – we can only wonder whether the *status* of *dediticius* was temporary or – like that of the so-called *liberti dediticii Aeliani* on which there is no need to linger here – hereditary and thus tendentially perpetual.[49]

With respect to Gaius, Isidore[50] makes a different, albeit obscure, point. Like the Antonine jurist, he recalls how, originally, the *dediticii* took their name from the act of surrender: "we define surrender (*deditio*) when the enemies, now defeated or on the verge of being so, give themselves up to the victors. This was, in fact, the origin of the term *dediticius*". Yet, immediately afterwards, he specifies that some *servi*, having risen up in arms against the Roman people and finally being defeated, surrendered to the victors. Once imprisoned, they were subjected to various horrendous treatments. Later, some, freed by their

47 Cf. CIL XIII 6592 = ILS 9184 (12 August 232): *Dedicated to the Goddess Fortuna Sancta. The exploratores [Sturi] and the officiales of the Brittones gentiles and the dediticii Alexandriani restored, with their own money and on the initiative of the centurion Titus Flavius Romanus of the XXII legio Primigenia Pia Fidelis, the balneum that had collapsed due to its age. On the Ides of August while Lupus and Maximus were consuls* (12 August 232). However, as emerges from the integration [*et*] in line 4, significantly different conjectures could also be advanced.

48 Kerneis, *Les numeri ethniques*, pp. 73–76.

49 In the sources, the term *dediticius* can assume different meanings. A broader examination of the set of texts regarding the lemma *dediticius* can be found in Christoph Sasse, *Die Constitutio Antoniniana. Eine Untersuchung über den Umfang der Bürgerrechtsverleihung auf Grund des Papyrus Giss. 40 I*. (Wiesbaden 1958), pp. 70–124; Hartmut Wolff, *Die Constitutio Antoniniana und Papyrus Gissensis 40 I*, (Köln 1976), I. pp. 210–271; II. pp. 443–502; Imrie, *The Antonine Constitution*, pp. 66–60.

50 Max Laistner, *The Source of Isidore* (*Etym*. 9, 4, 49–50), in *The Journal of Roman Studies*, 11 (1921). The source of Isidore can be identified as the Gaian *Institutiones* and, in particular, §§ 14 and 15. In the *Liber glossarum*, to be compared with this passage from Isidore's *Etymologiae*, use was also made of § 13: cf. *Liber glossarum* (DE 284, *si barbarus tradat se Romanis*).

masters, did not however acquire the status of Roman citizens on account of the scars attesting to the tortures inflicted upon them.

It would seem that, in the *Etymologiae*, the Bishop of Seville wished in his own way[51] to explain why the *legislator* of the *Aelia Sentia* had termed *dediticii* the *liberti* on whose body was clearly imprinted the stigma of the turpitudes suffered as a penalty for the crimes committed while they were slaves:[52]

> Isid. *Etymol.* 9.4.49–50: Dediticii primum a deditione sunt nuncupati. Deditio enim dicitur quando se victi aut vincendi hostes victoribus tradunt: quibus haec origo nominis fuit. Dum quondam adversus populum Romanum servi armis sumptis dimicassent, victi se dederunt, conprehensique varia turpitudine affecti sunt. [50] Ex his quidam postea a dominis manumissi, propter suppliciorum notas, quas manifeste perpessi sunt, ad dignitatem civium Romanorum non pervenerunt.
>
> The dediticii were initially named after the deditio, the act of surrender. In fact, one speaks of surrender when enemies, now defeated or about to be defeated, hand themselves over to the victors. This was the origin of the name dediticius of the slaves, who had risen up in arms against the Roman people, once defeated, gave themselves up to the victors, were imprisoned and made the object of various iniquities. Later on, some of them, freed from their masters, did not attain the dignity of Roman citizens because of the scars, clear evidence of the tortures they had suffered.

---

51 According to Alvaro d'Ors, 'Estudios sobre la "Constitutio Antoniniana", II. Los dediticios y el edicto de Caracala', in *Anuario de historia del derecho español* (AHDE), 15 (1944), esp. p. 180, this is a rather 'curious' definition, which proposes a strange and, in some ways, incomprehensible contamination between the category of the *hostes dediti in fidem* and that of the *liberti peregrini Aeliani*: on the one hand the *hostes* who offered themselves *in fidem*, on the other hand the *servi vincti*.

52 The *lex Aelia Sentia* took into account a series of dishonest behaviours by the slaves before their eventual liberation. The *servi* receiving penalties so severe as to mark their behaviour as reprehensible (the *damnatio ad bestias, ad ludum* or *ad gladium*: see Gai. 1.13 and *Tituli ex corpore Ulpiani* 1.11), i.e. chained (*vincti*), thrown into the *latumiae*, subjected to custody, branded on their body and in particular on their face (cf. Mart. *Epigr.* 3.21; Iuv. *Sat.* 12), interrogated under torture for a crime of which they were adjudged guilty, became, after the *manumissio*, *liberti* in the same condition as *peregrini dediticii*. They were barred from any possibility of gaining *civitas Romana* (Gai. 1.27). This is fully summarized in *Epit. Gai* 1.1.3. The stigma of branding regarded those who had been *damnati ad metallum*.

However, I believe that I can only detect in Leo the Great the use of this term, a reference to individuals or better to groups of individuals destined, like their descendants, to remain perpetually in a condition of lowly submission because of their specific status. In three passages of the *Tractatus septem et nonaginta*, there is juridical content over and above the *status* of *deditus in fidem*, regulated by *ius belli* and thus by *ius gentium*, or *libertas dediticia ex lege Aelia Sentia*. He speaks of *originalis dediticii generis servitus*, of *ius ferreum dediticii seminis*, of *captiva progenies*, as well as of *captivitatis vincla* and *propago dediticia*. Obviously, Leo the Great, using the notion of *ius originis*[53] as it was usually defined in Late Antiquity,[54] utilizes these legal metaphors for the sole purpose of better explaining the notion of original sin, which was then still theologically controversial. Nevertheless, his remarks are extremely interesting because they characterize *propago dediticia* as *ius ferreum dediticii seminis*, that is to say a *status* destined to be perpetuated from generation to generation, as a degraded condition of subjection of the prisoner[55] and the vanquished from which it is impossible to escape without the intervention of a saviour.[56]

However, for some, the legal provisions described by the texts of Gaius, Isidore and Leo the Great are not confirmed by the data emerging from scrutiny of the witnesses from the third and fourth centuries. Examination of the sources strengthens the sensation that they did not operate within the practices of the coeval *ius gentium*. In this regard, Gerhard Wirth[57] observed that the *dediticii*, mentioned by the sources, were not characterized by their common and uniform legal status. To the contrary, each defeated group would have conformed with peculiar obligations and conditions according to the circumstances. The

53 Charles Saumagne, 'Du role de l'origo et du census dans la formation du colonat romain', in *Byzantion* 12 (1937), pp. 487*ff.*

54 A legal provision essential to understand how the imperial power attempted, in practice, to limit the freedom of movement of certain individuals: the *collegiati* or *corporati*, among whom I would also include the *laeti* (see Valerio Marotta, 'Il problema dei 'laeti'. Fonti e storiografia', in *'Civitas, Arma, Iura'. Organizzazioni militari, istituzioni giuridiche e strutture sociali alle origini dell'Europa* [*secc. III–VIII*], eds. F. Botta - L. Loschiavo (Lecce 2015), pp. 120, 134*ff.*), as well as, above all, the *coloni*.

55 It is one of those texts in which there is a sort of equivalence between *dediticius* and *captivus*: d'Ors, *Estudios sobre la "Constitutio Antoniniana"*, p. 164.

56 *Tractatus septem et nonaginta* 22.107; *ibid.* 22.107; *ibid.* 30.136; *ibid.* 69.60. Hence, the *dediticius* cannot be identified *tout court* as the *tributarius*, although the latter pays the tax, which he is obliged to pay almost as a penalty (*poena belli*), in exchange for a freedom recognized to him as such by the Roman power.

57 Gerhard Wirth, 'Rome and its Germanic Partners in the Fourth Century', in *Kingdoms of the Empire. The Integration of Barbarians in Late Antiquity*, ed. W. Pohl (Leiden – New York – Köln 1997), esp. pp. 26*ff.*

deditio, it seems, would not necessarily have involved deprivation, destined to be prolonged from generation to generation, of the right to pass on one's goods and to receive them from others on the basis of the specific rules of each political community.[58] From this point of view, the *dediti in fidem*, as such because they surrendered before or after the defeat, could not be assimilated *tout court* with those who had been reduced to the condition of stateless persons or of *peregrini nullius civitatis* according to the rules of the *lex Aelia Sentia* or a sentence of *deportatio* or *opus publicum perpetuum*.[59] Moreover, for *barbari*, *deditio* could also open the doors to the possibility of a prompt and effective participation in the advantages of Roman civilization, at the expense of (in the best case) an independence undoubtedly more insecure in the long term.[60] Indeed, the cases in which Rome, having obtained *deditio*, did not reconstitute the subjected community in this way – as already seen, thanks to the Alcantara *tabula*, in the story of the *populus Seanocorum* (104 BCE)[61] – seem to have been very few up until most of the third century, even though an episode like that recorded by Suetonius must be taken into account.[62]

Nevertheless, on closer inspection, the picture that emerges from the sources – as put together by Gerhard Wirth – can be reconciled with what appears in the *Tractatus* of Leo the Great. In his sermons, the bishop of Rome meant simply to underline that the *propago dediticia*, in the absence of a specific act of reintegration (which he implicitly compared with the ablution of Christian baptism), would never have been able to be freed, by its merits alone, from its degrading condition. The fact that the imperial power most often restored – immediately, after a few months or a few years – freedom to the subjugated communities does not *per se* imply that, at its discretion, *captivitatis vincula* could not endure forever or for a long period of time.

It was the intention of the Roman military authorities to break up the ethnic homogeneity of a certain region, for example by settling tribal populations

58 Therefore, different from the *liberti dediticii Aeliani*: cf. *Tituli ex corpore Ulpiani* 20.14; Gai. 1.25.

59 Cf. D. 32.1.2 (Ulp. 1 *fideicomm*); D. 48.19.17.1 (Marcian 1 *inst.*).

60 Wirth, 'Rome and its Germanic Partners', p. 30*f.*

61 *Tabula Alcantarensis*: D. Nörr, *Aspekte des römischen Völkerrechts. Die Bronzetafel von Alcántara*, (München, 1989) 19–20, 23, ll. 1–13. The words '*se dedit*' indicate the surrender, while the phrase '*esse iussit agros et aedificia leges cetera omina quae sua fuissent pridie quam se dediderunt*' proves the *restitutio*. The fundamental premise of this institute (*deditio in fidem*) consisted in recognition of the superiority (*maiestas*) of Rome. However, included in it was a norm that guaranteed the life and, as a rule, the freedom of the *dediti*.

62 Suet. *Tib.* 9.1–2. See Dio 72.11.4–5.

defeated in *Britannia* or elsewhere in the areas of the Rhenish *limes*.[63] As the *Notitia Dignitatum* attests – at the end of the fourth century – their communities, administered by *praefecti* (and thus by military officers), were exempt by statute from the usual mechanisms of government of each *provincia*, although the recruitment in these communities of immigrants and deportees settled on public lands conformed to the usual discipline.[64] Moreover, at least in the border regions of the Gauls, these communities were almost always situated within the *tractus*, i.e. in military regions created between the third and fourth centuries.[65] However, these *praefecti*[66] did not act as commanders of fighting units. Instead, as military officials they dealt with organizing the settlement of immigrants or deportees, while also overseeing their surveillance and – if need be – the first sorting of recruits or defence of the territory. In other words, these *praefecturae* constituted administrative frameworks, not military units quartering in the same zone under the command of the *praefectus*.[67] However, although these remarks identify exactly the discipline that presided over the deportation of immigrant populations from one region to another and their settlement in the *tractus*, this does not mean – as Gerhard Wirth observed – that, despite this initial condition of *gentes devictae* and *dediticiae*, they were excluded in perpetuity, by virtue of the administrative practices and rules defined by the imperial power, from any possibility of being integrated into the frameworks of citizenship.[68]

---

63 Soazick Kerneis, *La Bretagne rhénane. Note sur les établissements bretons dans les Champs Décumates*, in *Latomus* 58 (1999), pp. 357 *ff.*

64 C.Th. 7.20.12. A reading of Amm. 21.13.16, which does not contextualize this witness within Julian's story, would seem to leave room for other hypotheses.

65 Thus, in my opinion rightly, Jean-Pierre Poly, 'La corde au cou. Les Francs, la France et la loi salique', in *Genèse de l'État moderne en Méditerranée* (Rome – Paris 1993), pp. 287 *ff.*; Id., "Leges Barbarorum'. La création des lois des nations', in *Une histoire juridique de l'Occident. Le droit et la coutume* (*IIIe–IXe siècle*), ed. S. Kerneis (Paris 2018), pp. 198 *ff.*, with other bibliography. A different point of view in Karl Ubl, 'L'origine contestée de la loi salique. Une mise au point', in *Revue de l'Institut français d'histoire en Allemagne*, 1 (2009). See, in this volume, the essay by Ian Wood.

66 *Not. Dign. Occ.* (ed. Seeck) 42.33–44.

67 See Joachim Szidat, 'Laetensiedlungen in Gallien im 4. und 5. Jahrhundert', in *La politique édilitaire dans les provinces de l'Empire romain: IIe-IVe siècles après J.-C.*, eds. R. Frei-Stolba – H.E. Herzig (Berne 1995), pp. 290–291; Id., '"Terrae leticae" (Cod. Theod. 13,11,10)', in *Historische Interpretationen. Gerold Walser zum 75. Geburstag dargebracht von Freunden Kollegen und Schuldern*, eds. H. Heinen – F. Paschoud – K. Raaflaub – H. Temporini (Stuttgart 1995); Marotta, 'Il problema dei 'laeti'', pp. 145 *ff.*; Giovanni Papa, "Multae gentes … Quibus Terrae laeticae administrandae sunt'. A proposito di CTh. 13.11.10', in *Studia et documenta historiae et iuris* (*SDHI*), 82 (2016), pp. 177 *ff.*, with other literature.

68 Wirth, *Rome and its Germanic Partners*, pp. 29 *f.*

2 Access to *Civitas* in the Fourth and Fifth Century

After 212, *civitas Romana* lost much of its prestige and power of attraction.[69] In this regard, the conclusions of almost all Late Antiquity historians converge. Yet, there is no consensus on the provisions that would have governed access to citizenship after the *edictum* of Caracalla.[70]

Because of the extremely incomplete documentation, lines of research have emerged which, if evaluated according to the rules of legal history, seem extravagant to say the least. This is the case, for example, of the history of the hypotheses of Ralph Mathisen,[71] which nonetheless have met with a certain success and not only among English-language scholars.[72] In his opinion, also due to the effect of the *constitutio Antoniniana*, *barbari* would have acquired the *status* of *cives Romani* – a *status* which, like any other, attributes rights and duties – exclusively by virtue of conclusive behaviour or, at most, of a mere declaration of will, according to the individual desire of each one and the necessities of the moment. In the fourth and fifth centuries, those who, like citizens, used the Roman negotiation schemes did so not because they had formally received citizenship, but pursuant to a clause of the *edictum de civitate* (of which obviously there is no trace in the text of P. Giss 40 col. I or elsewhere) which would have perpetuated the effects of its provision. Not for this, however, were all the *barbari* settled in the territory of the *provinciae* automatically counted among the *cives*, according to Mathisen. Only those who made use of *ius civitatis* would have been considered such. Consequently, on

69 Therefore, in this regard I agree with what has been written by, among others, Wolfgang Liebeschuetz, 'Citizen Status and Law in the Roman Empire and the Visigothic Kingdom', in *Strategies of Distinction, the Construction of Ethnic Communities, 300–800*, eds. W. Pohl – H. Reimitz (Leiden 1998) and Claudia Moatti, 'Mobilità nel Mediterraneo: un progetto di ricerca', in *Storica*, 27 (2003), pp. 107*ff.*

70 On the *constitutio*, see finally, with bibl., Imrie, *The Antonine Constitution*, pp. 50*ff.*, 139*ff.*; Arnaud Besson, Constitutio Antoniniana. *L'universalisation de la citoyenneté romaine au 3ᵉ siècle* (Basel 2020), esp. pp. 23*ff.*

71 Ralph Mathisen, "Peregrini', 'Barbari', and 'cives Romani': Concepts of Citizenship and the Legal Identity of Barbarians in the Later Roman Empire', in *American Historical Review* (*AHR*), 111/4 (2006); Id., 'Concepts of Citizenship', in *The Oxford Handbook of Late Antiquity*, ed. S.F. Johnson (Oxford 2012); Id., '"Becoming Roman, Becoming Barbarian": Roman Citizenship and the Assimilation of Barbarians into the Late Roman World', in *Migration and Membership Regimes in Global and Historical Perspective*, eds. U. Bosma – G. Kessler – L. Lucassen (Leiden 2010), esp. pp. 199*ff.*

72 Recently Javier Martínez Jiménez and Robert Flierman reiterad Mathisen's Theses: 'The Uses of Citizenship in the Post-Roman West', in *Citizenship in Antiquity. Civic Communities in the Ancient Mediterranean*, eds. J. Filonik, C. Plastow, R. Zelnick-Abramovitz (London-New York 2023), pp. 672*f.*

the assumption of the existence of such a legal provision, *civitas* and *status civitatis* would consist only in a question of participation, integration and above all self-identification. Everyone could consider himself and present himself either as *civis Francus* or as *civis Romanus*. Indeed, according to the notion of dual citizenship or dual identity (to which the formula *Roma communis nostra patria est* would also allude), he would have had both *civitates* at the same time. In short, after Constantine, *barbari* would have acquired citizenship by common and tacit consent.[73]

Mathisen's theory rests on three pillars: a) the *constitutio Antoniniana* was self-perpetuated in time. One of its clauses would have explicitly envisaged this, providing for a procedure to allow the naturalization of all new immigrants; b) many *barbari* held positions in the imperial apparatuses, sometimes ascending even to the consulate; they prepared wills according to the rules of the *ius civile*, they petitioned the courts of the Empire, they made use of the institutes of Roman law, even without a formal grant of citizenship; c) all the *peregrini*, who settled in the *provinciae*, assuming the identity and obligations of the citizens of a *municipium*, became "potential *cives Romani*". According to Mathisen, this does not mean that every barbarian immigrant was thus registered among the *cives*: citizenship, in his opinion, merely constituted a matter of self-identification. What made a barbarian *civis de iure*, as well as *de facto*, was primarily his will to make use of the *ius Romanorum*.

Of the three presuppositions of this complex theory, only one withstands scrutiny of the sources.[74] Indeed, it is true that many individuals of Germanic origin – it would be sufficient to recall the names of Bauto[75] and Arbogast[76] – had such fortunate careers that they managed to hold high positions: in the army as *magistri militum*, in the *res publica* even as consuls. After all, this is attested to by several prosopographic studies.[77] Moreover, it is well known that numerous high officials of Frankish or Vandal origin were awarded the imperial *nomen* Flavius, a *Rangsbezeichnung* typical of Late Antiquity.[78]

73 Mathisen, "Peregrini', 'Barbari', and 'cives Romani'; p. 1037.

74 Also Kerneis, 'Francus civis', critically discusses some uses of the sources in these works by Mathisen and in particular the reckless use that the American scholar made of Pan. Lat. 4(8), 21, 1.

75 PLRE, I (260–395) 159*f.*

76 PLRE, I (260–395) 95–97.

77 See, for an overview, which should probably be updated, Manfred Waas, *Germanen im römischen Dienst im 4. Jahrhundert nach Christus* (Bonn 1971²), pp. 5–13.

78 James G. Keenan, 'The Names Flavius and Aurelius as Status Designations in Later Roman Egypt', I–II, in *Zeitschr. für Papyrologie und Epigraphik* (ZPE), 11 (1973) and 13 (1974).

Nor has the recourse to negotiation schemes and institutes of Roman law by individuals of barbarian origin – not unlike their full integration at the top of the imperial apparatuses – ever really been questioned. In reality (and we will verify it shortly[79]), throughout the fourth century, many *peregrini* continued to obtain citizenship, *viritim* or at the end of their military service.

Finally, the use in Mathisen's contributions of the controversial notion of dual identity raises many perplexities, especially because he relates it – in effect confusing the *patria loci* of the *municeps* with the *patria iuris* of the citizen of a *civitas libera* or, in any case, not subject to the Empire – to the formula *Roma communis nostra patria est.*[80] The opposite has often been maintained;[81] however, on a strictly juridical level, the latter does not assume, after the *constitutio Antoniniana*, an entirely different meaning from the past. The Romanization of the Empire between the first and third centuries CE, like that of Italy earlier, was accomplished in the light of the principle of two homelands: *Roma communis patria* and local *patria.*[82] As it is known, the controversial problem of dual citizenship[83] – i.e. the set of practices and rules that allowed, as early as the triumviral period, to reconcile possession of *civitas Romana* with the exercise of honours and priesthoods (*Heimathsrecht*) in the community – should not be confused with that of 'municipal' citizenship which the jurists of the Principate, starting with Neratius, would indicate by the term *origo,*[84] a provision governed, in the male line, by the same mechanism as legitimate filiation and thus of paternal authority. Hence, the formula *Roma communis patria* and the technical device it conceals cannot be invoked to argue that someone could, according to his conveniences or inclinations of the moment, behave or pass himself off as a *civis Romanus* or a *civis Francus*.

These considerations allow us to address another crucial point of Ralph Mathisen's argumentation. He maintains, if I have understood his point correctly, that the *peregrini* who settled after 212 in the territory of the Empire,

---

79 *Infra*, p. 182 and, for an exemplification, note 112.

80 D. 50.1.33 (Mod. *l.s. de manumissionibus*).

81 Francesco Casavola, 'Il concetto di 'urbs Roma': giuristi e imperatori romani', and Id., 'Lo spazio nell'esperienza giuridica del mondo antico', both in Id., *Sententia legum tra antico e moderno*, 1. *Diritto romano* (Napoli 2000), esp. p. 360*f.* and pp. 527*f.* respectively.

82 Cf. Cic. *De leg.* 2.5: see Yan Thomas, *"Origine" et "commune patrie". Étude de droit public romain (89 av. J.-C.-212 ap. J.-C.)* (Paris – Rome 1996), pp. 3*ff.*

83 The notion of dual citizenship should be reconsidered more carefully also in relation to the imperial period. Some remarks, with further bibliographical references, in Valerio Marotta, 'Una nota su D. 49.14.32 (Marcian. 14 inst.) e sulla condizione giuridica degli 'obsides'', in *Annali dell'Università di Palermo* (AUPA), 61 (2018), esp. pp. 234*f.*

84 D. 50.1.9 (Ner. 3 *membr.*).

assuming the identity and obligations of the citizens of a *municipium*, became "potential *cives Romani*" by virtue of a provision of the *constitutio Antoniniana*, which – as has already been underlined – would be 'self-perpetuated'. Perhaps – although his reasoning unfolds without the use of the usual technical vocabulary of Roman law scholars – Mathisen is thinking of the so-called *adlectio*.[85] Yet, Yan Thomas[86] demonstrated that, after the *leges de civitate* of 90–89 BCE which granted Roman citizenship as a whole to the Italic communities, for a *peregrinus* it would have been fraudulent, a crime punishable by severe sanctions, to have himself included in the lists of citizens of a *municipium* or *colonia* without being a *civis Romanus*. Therefore, as a result of their incorporation into the frameworks of *civitas Romana*, these cities (now having become *municipia*) lost the right to register new citizens who were not already *cives Romani*. Regulating the access to Roman citizenship during the Principate, regardless of the authority of the *populus* (which remained only in theory) was a matter of exclusive jurisdiction of the *princeps*. Therefore, the *municipia* perpetuated their civic bodies according to four traditional juridical forms: filiation (and, consequently, the provision of *ius originis*), adoption, liberation of slaves and, more sporadically, *adlectio* of Roman citizens. If Mathisen meant to argue that this picture radically changed during the third century after the *constitutio Antoniniana*, he should have recalled how the leading scholar on these subjects, Yan Thomas, thought only of the *adlectio*, in the positions of a *municipium*, of *cives* already *Romani*. Otherwise he would have at least had the onus of advancing some concrete evidence in support of his conjectures. However, the only testimony[87] in support of his theory refers to a Diocletian provision which was limited, within the new framework defined by the *edictum de civitate* of 212 and by the general extension of *civitas*, to setting

85 Mathisen, 'Peregrini', p. 1034: «The opportunity to choose one's own *patria* would have been most applicable to those who did not already have one, a criterion that would have fit barbarian veterans. Citizens of a municipality were *ipso facto* Roman citizens». It is pointless to reiterate that the veteran could have settled in a *civitas*, assuming there the status of *municeps*, only if the *princeps*, previously or at the time of his discharge, had already granted him Roman citizenship.

86 Thomas, *"Origine" et "commune patrie"*, pp. 83 *ff.*, esp. 101 *f.*

87 I refer to a *constitutio* of Diocletian, in which, for the first time, there is explicit mention, in the pre-Justinian and Justinian compilations, of *adlectio* as a source of local citizenship, alongside *origo*, *manumissio* and *adoptio*: C. 10.40.7pr. «*Origo*, adoption and *adlectio* make citizens, while, as the divine Hadrian openly declared by his own edict, domicile makes *incolae*». However, in this regard it should be considered that Ettore De Ruggiero, *La* patria *nel diritto pubblico romano* (Roma 1921), pp. 121 *f.*, reconnected, not without well-founded reasons (in my opinion), also the first part of C. 10.40.7pr. (*cives-adoptio*) to Hadrian's edict.

adlectio alongside *origo*, *adoptio* and *manumissio*. Ralph Mathisen, perhaps realizing this,[88] presumes precisely for this reason the existence of a clause of the *constitutio Antoniniana* that would have envisaged the 'self-perpetuation' of its effects, granting new arrivals to the territory of the Empire the possibility of requesting integration into Roman citizenship. Now, as I underlined earlier, there is no trace of such a clause in the sources. Therefore, in the present state of our knowledge, Mathisen's conclusions must be considered unfounded.

By other means, in his fine book on the *Barbari*,[89] Alessandro Barbero, utilizing some premises put forward by Pierangelo Catalano,[90] arrives at conclusions that coincide substantially with those of Mathisen, although he proposes and argues them better and with greater caution. Drawing on the presumed lack of evidence and thus solely on arguments *e silentio*, he maintains that the notion of *peregrinitas* gradually disappeared during the fifth century CE. In his opinion, the *constitutio Antoniniana*, from the fifth century onwards, would have been perceived "as an extension of citizenship to anyone who came to live in the Empire and submitted to imperial authority". Now this affirmation does not consider the fact that the dichotomy between *liberi* and *servi*, at least up to Justinian, does not coincide *tout court* (apart from the case of the *barbari*) with that between *cives Romani* and *servi*.[91] And not all *liberi ingenui*, at least in the provinces of the *pars Occidentis*, were considered *cives Romani*. In this way, the existence of different categories of *liberi* and primarily of the descendants of Junian Latins would be forgotten. The latter, unlike the *liberti dediticii Aeliani*, were not entirely precluded acquiring *civitas Romana*.[92] The Breviary of Alaric II transmitted a *sententia* of the fourth book of the homonymous work attributed to Paulus, which makes explicit reference to the case of the *Latina ingenua*, i.e. the daughter of *liberti Latini* born after the manumission of her parents:[93]

88 But he does not explicitly account for it.

89 Alessandro Barbero, *Barbari. Immigrati, profughi, deportati nell'impero romano* (Roma – Bari 2006), pp. 47, 67*f.*

90 Pierangelo Catalano, "Ius Romanum'. Note sulla formazione del concetto', in *Da Roma alla terza Roma. Documenti e studi. La nozione di "Romano" tra cittadinanza e universalità* (Napoli 1984), p. 544.

91 See Sidon. *Ep*. 1.6.

92 Salvianus *Ad ecclesiam* 3.7.34 also seems to allude to Latin freedmen: *quasi Latinos iubes esse libertos, ut vivant scilicet quasi ingenui et moriantur ut servi*. See Gai. 3.56. However, an overall evaluation of chapters 6 and 7 of the third book *Ad ecclesiam* forces us to refer the phrase *moriuntur ut servi* not to the *Latini Iuniani* but to children who belong to a religious order.

93 See Jolanda Ruggiero, 'Una breve nota sulla condizione dei liberti latini e dei loro discendenti in età tardoantica', in *ΚΟΙΝΩΝΙΑ*, 41 (2017).

PS. 4.9.8 (*Ad senatus consultum Tertullianum*) = PV 4.9.1 = Liebs 4.8.8: Latina ingenua ius Quiritium consecuta si ter peperit, ad legitimam filii hereditatem admittitur: non est enim manumissa.

Free-born Latina obtains Roman citizenship if she has given birth three times: in fact she has not been enfranchised.
INTERPRETATIONE non eget.
No explanation necessary.

This case[94] attests to the existence of families in which the lowly condition of Latin was perpetuated, sometimes for more than one generation. Thus, it is not true, as Catalano and Barbero maintain, that the Edict of Caracalla, "with the assumption of a 'spatial' criterion, supplementing the 'personal' one", would have upset "the relationship between 'citizenship' and 'non-citizenship'". The assertion that the registers in which to record the names of new citizens had disappeared also raises some perplexity. Liberations of slaves were certainly numerous in Late Antiquity, and *manumissio vindicta* and *manumissio testamento* survived for centuries.[95] Constantine, in establishing *manumissio in ecclesia,*[96] conferred *civitas Romana* on those who had been freed in this way; evidently great importance was attributed to this, especially in the case of former slaves, primarily for patrimonial reasons. Indeed, *caput* 3 of the *formula* of Alvernia[97] assumes – although it can be dated to the beginning of

94 Which is paired with PS. 2.21a.1 on the subject of *senatusconsultum Claudianum*. This text comes from the *codices aucti*. They contain *sententiae* not included in the official version of the *lex Romana Visigothorum* but added later: Jolanda Ruggiero, *Ricerche sulle* Pauli Sententiae (Milano 2017), pp. 88–89, fn. 30.

95 Kyle Harper, *Slavery in the Late Roman World AD 275–425* (Cambridge 2011), pp. 463*ff.*

96 To the book by Fabrizio Fabbrini, *La manumissio in ecclesia* (Milano 1965) it is necessary to add the important contribution by Giorgio Barone-Adesi, 'Istanze servili alle 'libertates': alle origini della disciplina costantiniana "de his qui in ecclesiis manumittuntur"', in *Iuris Antiqui Historia* (*IAH*), 5 (2013); see also *infra*, fn. 142.

97 *Formulae Merowingici et Karolini aevi, Formulae Arvernenses,* 3, p. 30, l. 13 (…) «Whoever (secular or religious), once granted freedom, wanted to confer it on his slaves according to Roman law, could do so, i.e. (he could confer the status) of *Latina*, of *dolitia* [*dediticia*], of Roman citizen (…)». Cf. also, although they allude only to *civitas Romana* and *ingenuitas, ibid.*, 4, p. 30; *Formulae Turonenses* 12, p. 141; *Formulae Bituricenses*, 9, p. 172; *Formulae Salicae Merkelianae*, 13, p. 246; *ibid.* 44, p. 257; *Formulae codicis Laudunensis*, 14, p. 518. Also see *Formulae Imperiales* 33, 35; *Formulae Augienses* 42; *Formulae Sangallanses miscellaneae* 6; *Formulae Visigothicae*, 3, 5. In this regard, see Alice Rio, *Legal Practice and the Written Word in the Early Middle Ages: Frankish Formulae, c. 500–1000* (Cambridge 2009), pp. 80*f.*; Alexandre Jeannin, "Vigor actorum'. La mise en forme romanisante de la pratique', in *Une histoire juridique de l'Occident. Le droit et la coutume* (*IIIe–IXe siècle*), ed. S. Kerneis (Paris 2018), pp. 290–294.

the eight century – the existence of an older archetype, perhaps developed during the 6th century[98] when the condition of the *liberti Latini* and *dediticii* was still clearly differentiated from that of the *cives Romani* freedmen, above all because *manumissiones*, in the first place those conferring Roman citizenship, had to be registered in the municipal *acta*.[99] No silence from the sources concerning the specific problem of *civitas Romana* can be advanced as proof or support of the fact that the very notion of *peregrinitas* had disappeared in the fifth century CE. In other words, in my opinion, to conclude that in Late Antiquity (in the fourth or fifth century) the *constitutio Antoniniana* was perceived "as an extension of citizenship to anyone who came to live in the Empire and submitted to the imperial authority"[100] is rash.[101]

3 The Justinian Period

If there was a turning point, it must be identified in the Justinian period, as I believe can be inferred – following the conclusions of a splendid article by Fausto Goria[102] – from *caput* V of the Novella LXXVIII dated to 539. While Ulpianus 22 *ad edictum* D. 1.5.17 and its use in the compilation do not allow any conjecture, the Novella of 539 does not seem to assume that the *constitutio Antoniniana* is an act limited only to those who, in due course, benefited from it and to their descendants, but rather a provision that forever eliminated any difference between Roman citizens and *peregrini*, with the implicit consequence that, from then onward, when someone entered the category of *subiecti* of the Empire, he also obtained *civitas Romana*. According to the Justinian chancellery, the *constitutio Antoniniana* would have provided for the future, so that all inhabitants of the territories subjected *ex novo* could enjoy Roman citizenship and the rights due to them.

98 Friedrich Karl von Savigny, *Geschichte des römischen Rechts im Mittelalter*, II (Heidelberg 1816), pp. 128–132, maintains that the authors of this formulary would have been aware, in this case, of *Epitome Gai* 1.1, as well as *Epitome Iuliani* const. 72, cap. 258.

99 Indeed, in Late Antiquity, *manumissio in ecclesia* was the only act of attribution of *civitas Romana*, whose realization was not under to the control of the normally responsible authorities, i.e. the municipal curias in the provinces: cf. Roland Ganghoffer, *L'évolution des institutions municipales en Occident et en Orient au Bas-Empire* (Paris 1963), pp. 143*f.*

100 Barbero, *Barbari*, p. 47.

101 Catalano, 'Ius Romanum', p, 544.

102 Fausto Goria, 'Romani, cittadinanza ed estensione della legislazione imperiale nelle costituzioni di Giustiniano', (Napoli 1984) now in *Diritto romano d'Oriente. Scritti scelti di Fausto Goria*, eds. P. Garbarino – A. Trisciuoglio – E. Sciandrello (Alessandria 2016), esp. pp. 45*ff.*

Indeed, examination of Justinian and post-Justinian legislation reveals that, with regard to marriage, the distinction between Romans and foreigners existed and remained important: unions with *peregrini* were not protected by law.[103] In the light of Nov. 36.6, addressed to the prefect of the *praetorium* of *Africa*, even more explicitly in the Greek version of Theodorus, it was announced that all *Afri* would have to follow the Roman laws. And this would seem to have been true primarily for individuals of Vandalic stock. A clue in this regard comes from the fact that many Roman soldiers married the daughters or widows of such individuals and were convinced by them to claim as a dowry the lands previously belonging to them and which had been confiscated in favour of the treasury or the imperial patrimony. If, as it seems, these were real marriages, it follows that the Vandal women living in the re-conquered lands were considered Roman citizens.[104] Moreover, we know that the Goths resident in Ravenna, after the Justinian reconquest of Italy, followed the rules of Roman law.[105]

There is a legislative text from the period of Justinian which makes explicit reference to barbarians honoured with some office or rank.[106] Justinian[107] had imposed the use of dotal deeds on the highest dignitaries, including senators and all those who used the title *illustris*. Those of lower rank could be limited to a certificate of marriage written by a *defensor ecclesiae* and sealed by at least three witnesses. The lower plebs, peasants and infantry soldiers were still allowed to marry without making use of written documents. Although Justinian later[108] simplified these formalities, the provisions on marriages of dignitaries remained in force, with two exceptions: marriage contracted prior to the granting of a position remained valid even in the absence of dotal deeds; observance of this provision was spared to *barbari* subject to the *res publica* when they held high positions.

---

103 I. 1.10pr. «Roman citizens unite in legal marriage when they are joined according to the precepts of the law, and the males have attained the age of puberty and the females are capable of child-birth, whether they are the heads of families or the children of families ...». Also see I. 1.10.12.

104 Proc. *BV*. 2.14.8–10.

105 In this regard, it is worth recalling that, after the conquest of Sicily by Belisarius, during the negotiations with Theodatus he considered for some time the plan to fully incorporate all the Goths in the Empire, thus making them citizens: see Proc. *BG* 1.7.23–24.

106 Cf. also Auth. 117 c. 4.

107 Nov. 74.4.

108 Nov. 89.1.1.

As Fausto Goria opportunely pointed out,[109] the most relevant datum concerning this provision concerns its *interpretatio*. Already in the 6th century, in their *summae*, two epitomators, Anastasius of Emesa and Theodorus Scholasticus, assessed it, in normative terms, in a significantly different way. The former spoke in this regard of οἱ ὑπόσπονδοι βαρβαροί, thus showing that he referred the provision to the barbarian leaders of the allied peoples. The latter, vice versa, applied it to ὑποτελεῖς βαρβαροί, an expression indicating that many were fully incorporated into the Empire and thus subject to the tax obligations imposed on the *subiecti*. Fausto Goria demonstrated that this is without doubt the meaning to be attributed to that rule. In interpreting c. 4 of Nov. 117, it should be borne in mind that it can assume regulatory significance only if coordinated with Nov. 74.4, which as a general law should have been observed also by the *barbari*, accepted in the πολιτεία, who resided in the territories of the Empire and primarily in Constantinople. If this reconstruction is exact, it follows that high dignitaries of foreign origin, targeted by Nov. 117.4, were considered Roman citizens, since in the absence of express dispensations they should have applied the common Roman law. Yet, *nuptias facere nuda affectione* was granted to these new Romans by reason of their *simplicitas*.[110] Nonetheless, on the historical level, the interpretation of Anastasius of Emesa is just as interesting as the other one: it presupposes that the *barbari* incorporated into the structures of the Empire became Romans, a consequence of the new interpretation of the *constitutio Antoniniana* defined, as emerges from Nov. 78.5, in the Justinian period or immediately prior to it.

In conclusion, Goria's conjectures on the Justinian period[111] are often supported by the sources. However, in my opinion, it cannot be supposed – as did Alessandro Barbero – that identical provisions were already in operation in the fifth and, even more so, the fourth or third century.

4 *Viritim* Concessions and Veterans of Barbarian Origin

The contextual presence of foreign and Roman *milites* in the imperial army of Late Antiquity is a fact attested by a host of witnesses and above all, because of its incisiveness, by a passage from the *epistula* of Julian the Apostate to

109 Goria, 'Romani', pp. 61*ff.*

110 *Epitome Iuliani* const. 108, cap. 381.

111 Goria, 'Romani', p. 86.

the Athenians.[112] For *barbari*, military service remained the principal, if not the only, way to gain Roman citizenship even in the fourth century especially when performed in certain units of the imperial army.

Regarding the expansion of *ius civitatis* after the end of the tetrarchic period, the position adopted by Émilienne Demougeot[113] still appears to be, on the whole, the most persuasive one. Although recruitment no longer represented, as in the middle imperial period, the *pépinière* of citizenship, barbarian officers, primarily those of the palatine *auxilia*, were still included in the ranks of the *civitas*. This can also be verified from information emerging from the funerary epitaphs of the military cemetery of Concordia: the *Flavii* are above all *senatores, biarchi, centenarii, ducenarii* and *praepositi*. Among the simple soldiers only Fl. Abruna of the *auxilium* of the *Bataves Seniores* and Fl. Dassiolus of the *auxilium* of the *Mattiaci Seniores* seem to be citizens.[114]

Moreover, the palatine legionaries at the top of the list of the *Notitia Dignitatum* were *cives*;[115] instead, those in the *auxilia palatina* were *peregrini*. Finally, there is, at least for the fourth century, evidence that allows us to presume that, in the Roman army, service in certain units led to the acquisition of *civitas*, if recruits or veterans were not already citizens.[116]

However, before dealing with the subject of the attribution of citizenship following the completion of military service, I would first like to dwell for a moment upon the so-called *viritim* concessions. Indeed, examination of the sources reserves some surprises in this regard.

An exceptional document on the teaching of the law in the West, *interpretatio Gai Institutionum Augustodunensis*,[117] preserves – despite its broad gaps – an interesting digression on the law of citizenship between the fourth and fifth centuries. It is a paraphrase rooted only in the basic Gaian text. However,

112 Iul. *Ep. ad Athenienses* 285: cf., in particular, the words "Ανδρες στρατιῶται καὶ ξένοι καὶ πολῖται, μὴ προδῶτε τὸν αὐτοκράτορα 'soldiers and foreigners and citizens, do not betray the emperor'.

113 Émilienne Demougeot, 'Restrictions à l'expansion du droit de cité dans la seconde moitié du IVe siècle', in *Ktema*, 6 (1981).

114 CIL V 8744, CIL V 8750; CIL V 8754. Also see CIL V 4369 (Fl. Higgo); ILS 2777 (Fl. Baudio); CIL III 7494 (Fl. Stercorico): see Alain Chauvot, *Opinions romaines face aux Barbares au IVe siècle apr. J.-C.* (Paris 1998), pp. 156, 159, 171; Barbero, *Barbari*, p. 154; *infra*, fn. 126.

115 *Not. Dign. Occ.* [Seeck] 5, 144–156; 5, 157–222: for the first, cf. in particular *Ioviani* and *Herculiani*.

116 Jean-Pierre Poly, 'Le premier roi des Francs: la loi salique et le pouvoir royal à la fin de l'Empire', in *Auctoritas. Mélanges Olivier Guillot*, eds. G. Constable – M. Rouche (Paris 2006), p. 113.

117 José-Domingo Rodríguez Martín, *Fragmenta Augustodunensia* (Granada 1998), pp. 123–135.

this fragment, in giving us the *vox magistra*,[118] the voice of an unknown Gallo-Roman teacher, allows us to understand that the law developed in the second second century was still applied in the Late Antique Empire. Probably composed before the great upheavals of the following century (*Frag.Gai* 1–5:[119] see Gai. *Inst.* 1, 93–94), it provides, in my opinion, sure indications against the conjectures of Mathisen and Barbero. The Master dwelt in particular on the technical problems sometimes raised by the application to acquire citizenship. To our eyes, his clarifications assume a definite meaning only if we assume that, at the end of the fourth century, the *peregrini* still requested, as in the past, the 'gift' of *civitas*. In the first place, the problem of the children was raised. The unborn were distinguished from those already born. The latter, if adolescents, could attach their request to that of the father. Instead, pre-pubertal children had to present a specific request. Furthermore, when the unborn was the child of a *mulier* who had already made a request, a further one would undoubtedly have appeared superfluous. But the *pater*, to obtain *potestas* over his child, still had to take precautions, petitioning the emperor to grant a special benefit in this regard. Unfortunately, as we do not know the regulatory content of the Hadrianic period *senatus consultum*[120] mentioned by the Master of Autun, his reasoning escapes us, at least in part.

Although it is legitimate to doubt that the practice attested by the *Fragmenta* of Augustodunum concerned, in the fourth and fifth centuries, only officers of barbarian origin of the *auxilia* of the *scolae palatinae*, it is nevertheless certain, as opportunely observed by Soazick Kerneis,[121] that here (*Frag. Gai* 1–5), unlike the brief section dedicated to *ius Latii* (*Frag.Gai* 6–8[122]), the *interpretatio* does not rest on a purely historical digression. To the contrary, the regulatory provision described in this paraphrase still seems to have been in force in the *Galliae* of Late Antiquity, before the great upheavals caused by the breakup of the Rhenish *limes* in 406 CE.

118 See *C. Omnem* 1.

119 *Frag.Gai* 1–5.

120 Probably to be identified as the one mentioned on several occasions by Gaius: 1.30, 1.77, 1.80, 1.81; see also *Tituli ex corpore Ulpiani* 3.3.

121 Soazick Kerneis, 'Dans une école de Gaule. La leçon du maître d'Autun', in *Mélanges en l'honneur du professeur Nicole Dockès, La Mémoire du droit*, eds. Chr. Lauranson Rosaz – D. Deroussin (Paris 2014). Furthermore, in this regard, we can use a passage from the *Itinerarium Alexandri*, dated around 340 AD, which explicitly refers to the future possibility of conferring Roman citizenship on the inhabitants of the regions conquered after the end of the war against Persia: see *Itinerarium Alexandri* 2.5 (ed. Tabacco): ... *civitate Romana donati*, ... '... grantees of the Roman citizenship, ...'.

122 See Gai. 1.95–96.

Regarding the mechanisms for assigning *civitas* to veterans of barbarian origin, Émilienne Demougeot[123] – proceeding from the correct assumption that, up to 380, the recruitment into the Roman military forces of Franks, Alemanni, Alans, Sarmatians and other populations of the Transdanubian plains[124] did not take place, if not sporadically, on the basis of a *foedus*[125] – clarified that it was the army of *comitatenses*, largely favoured over *ripenses* (or *limitanei*), that allowed at least a certain number of these *milites peregrini* to become citizens. After Constantine, *civitas* was granted to Franks, Alemanni and other barbarians belonging both to the *scolae palatinae* of the new imperial guard and to the elite of officers of the *auxilia* and *vexillationes*. Moreover, it seems obvious to me that all those who held a military position of a certain importance (*tribunus armaturarum*, *comes domesticorum*, as well as, and even more so, *magistri equitum* or *peditum*) had previously obtained inclusion in the ranks of *civitas*.[126]

The French scholar wondered if the *barbari peregrini* who fought in the *auxilia* and in the legions received, like the *peregrini* of the *auxilia* and the *numeri* of the Principate, citizenship at the time of their *honesta missio*.[127] As is known, the practice of issuing *diplomata*[128] to the veterans of the praetorian cohorts ended perhaps in 306,[129] whereas this occurred further back in time in the case of *milites* of the other units: the last *diploma* concerning sailors of one of the *classes praetorie* (that of Ravenna and that of Misenum) dates to 250;[130] for the *auxilia*, instead, this goes back as far as 206.[131]

Nonetheless, we can use a different criterion of analysis. If we pay attention to the content of the *constitutio* of Constantine passed down in C.Th. 7.20.4

---

123 'Restrictions à l'expansion', p. 383.

124 Particularly intense, in the case of the first ones, in the infantry *auxilia*, and in the case of the second ones in the *vexillationes* of *equites*.

125 As instead would occur after the Adrianople catastrophe, following the decisions of Gratian and Theodosius in a situation of extreme emergency.

126 See, for example, what transpires from examination of the epigraphic materials of the cemetery of Concordia: see, by way of example (in addition to the reported inscriptions, *supra*, note 114), Giovanni Lettich, *Le iscrizioni sepolcrali tardoantiche di Concordia* (Trieste 1983), esp. no. 22 and no. 57.

127 Émilienne Demougeot, 'Le 'conubium' et la citoyenneté conféreé aux soldats barbares du Bas-empire', in *Sodalitas. Scritti in onore di Antonio Guarino*, IV (Napoli 1984), esp. pp. 1636 *ff.*

128 On this subject, consult Francesco Castagnino, in this volume.

129 *RMD* I 78.

130 *CIL* XVI 154.

131 Werner Eck, 'Das letzte Diplom für einen Auxiliarsoldaten aus dem Jahr 206 n.Chr.: der Text der Innenseite', in *Zeitschr. für Papyrologie und Epigraphik*, 208 (2018).

(17 June 325)[132] (a provision to be compared with the *sacrae litterae* preserved in the *tabulae* of Brigetio and Durostorum[133]), we realize that the veterans of the *comitatenses* and *ripenses* were exempted, after their *honesta missio*, for only one *caput*: their own.[134] Émilienne Demougeot maintains that this unique and personal *caput* – extending neither to the wife, nor to the parents, nor to the children – was perhaps an echo of the same restrictions envisaged, in the second century, for the concessions of the right of citizenship for veterans of certain bodies. As is known, in the case of the *auxilia*, after 139/140 the children of *milites* born during the period of service of their parents were no longer granted, as previously, *civitas*, while the veterans of *numeri*, consisting of *gentiles* enlisted (if we rely on the documents) among the *Mauri equites* (120 / 138–161 ca.) and the *Palmyreni* (120 / 126), although given citizenship were not granted *ius conubii* with the *peregrinae* women they had previously taken as wives.[135] Perhaps – if we give credit to this analogy – in the fourth century the personal *caput*, after twenty years of service, implicitly indicated that the

132 The fiscal privileges granted to the *milites* were more or less generous depending on whether they served among the *comitatenses*, *ripenses*, *protectores*, or among the *alares* and *cohortales*. Historiography has often related this constitution to the military organization defined by Constantine, based on the assumption that, at least initially, it had to coincide with the partitioning of the troops described by C.Th. 7.20.4. Only in a second phase, on the basis of the *comitantenses* / *limitanei* dichotomy attested a century later by the *Notitia Dignitatum*, would a further simplification be defined. However, in reality, as Jean-Michel Carrié demonstrated, the corps recalled in C.Th. 7.20.4 do not suggest any relationship to the new statutory definition of the Constantinian army, created after the establishment of the *magistri militum*, i.e. a few years after the issuing of C.Th. 7.20.4. The *comitatenses* mentioned in this text were in fact the tetrarchic *comitatus*, that is expeditionary corps consisting of *vexillationes* detached from other mother-units by the emperors on the occasion of this or that *expeditio*. The expression *ripenses* would instead have indicated those legions, or parts of legions, not grafted into the imperial *comitatus* and left to garrison the provincial cantonments. As for the veterans, the discipline established by the *constitutio* is not unequivocal. At least in an early phase, all those discharged with *emerita missio* would be granted exemption from a single *caput*: their own. See Jean-Michel Carrié, 'Constantin continuateur et liquidateur de l'expérience tétrarchique', in *Fra Costantino e i Vandali. Atti del Conv. Internaz. di studi per Enzo Aiello (1957–2013)* (Bari 2016), esp. pp. 82 *ff.*

133 Francesco Castagnino, 'Le 'Tabulae' di Brigetio e di Durostorum. Una 'constitutio' sui privilegi di 'milites' e veterani: lo 'status quaestionis'', in *Diritto ed esercito. Profili dell'epoca tardoantica*, ed. S. Puliatti, *Cultura giuridica e diritto vivente*, 7 (2020), pp. 10 *ff.*

134 Demougeot, 'Le 'conubium'', p. 1637.

135 Le Roux, 'Les diplômes', esp. pp. 357 *ff.*

barbarian veterans naturalized at the time of their discharge could not, in turn, father citizens: unless of course they married Roman women.[136]

5 Conclusions

I have entered a field (the law of the Justinian *Novellae* and the history of the Late Antique Roman army) very far from my usual interests and thus from my limited expertise. The overall picture of the sources allows the conclusion that the same legal provisions that had regulated the access of *barbari* – primarily those enlisted in the military forces of the Empire – to the ranks of citizenship remained substantially operative also in the 6th century.

It would probably make little sense to continue further, examining periods in which the imperial army and administration had already ceased to exist for a long time in the West (if we exclude part of *Italia*, *Sicilia*, *Sardinia* and *Corsica*, *Africa* and finally *Baetica* for a short period). In any case – regardless of the resilience of certain institutes – the documentation available to us not would not allow us to advance any conjectures.

After Christmas day in 800 CE, the notion of citizenship perhaps regained a semblance of vitality within the framework of the renewed idea of *Imperium* and *res publica*. This expression, belonging to the Roman legal and political vocabulary – whose use, after the fall of the Western Empire, almost completely vanished in official documents as well as in literary works – was rediscovered at the time of the Carolingians.[137] The protagonists of a page of the more recent *Chronicon* of Regino of Prüm[138] would seem to presuppose its existence when they sentence the guilty man to exclusion from the *res publica*. On 18 May 872, the Lombard prince Adelchis of Benevento, who had imprisoned

136 However, I cannot understand what Demougeot, 'Le 'conubium'', p. 1638, means when she speaks of citizenship granted without *conubium*. The *ius conubii*, as a specific privilege (attested by *diplomata militaria*), was necessary to marry a *peregrina* or a *Latina Iuniana*. Between two *cives Romani* it existed by statute, unless a *lex* (*lex Iulia de maritandis ordinibus*), *mandata principum* or *disciplina militum* expressly forbade certain unions or withheld, as to soldiers before 197 CE, the right to marry; see Sara Phang, *The Marriage of Roman Soldiers (13 B.C.–A.D. 235). Law and Family in Imperial Army* (Leiden – Boston – Köln 2001), pp. 22 *ff.*

137 In this regard, see Yves Sassier, 'L'utilisation d'un concept romain aux temps carolingiens: la 'res publica' aux IXe et Xe siècles', in *Mediévales* 15 (1988).

138 Bibl. West, 'Knowledge of the past and the judgement of history in tenth-century Trier: Regino of Prum and the lost manuscript of Bishop Adventius of Metz', in *Early Medieval Europe*, 24.2 (2016).

Louis II (*Hlodowicus*), was declared by the Roman senate[139] *tyrannus* e *hostis reipublicae*.[140] If we attribute real historical substance to this episode, we could perhaps assume that the protagonists of this process were aware of applying a rule that once would have entailed, for those branded as *hostes* by the *senatus*, also the further penalty of *amissio civitatis*.[141]

In contrast, considering the institutes of law that operated in the daily dimension of European societies between the seventh and 10th centuries, we must recognize that – except for the specific field of manumissions and especially *manumissio in ecclesia*[142] – Roman citizenship had vanished from the sky of history.

---

139 Obviously, the institute to which the chronicler alludes can in no way be considered the heir of the ancient senate, which disappeared between the sixth and seventh centuries, as is proved, if nothing else, by the absolute absence of any senator in the *adventus* of Constans II (5–17 July 663). However, well into the eight century, a senate re-emerged, in significant concomitance with the loosening of ties with Byzantium and the Eastern Empire. See Filippo Burgarella, 'Il Senato', in *Roma nell'Alto Medioevo* (Settimane del Centro Italiano di Studi sull'Alto Medioevo, XLVIII) (Spoleto 2001), esp. pp. 172 *ff.*, with other literature.

140 Reginonis *Chronicon*, p. 584.

141 See D. 4.5.5.1 (Paul. 11 *ad ed.*). We can only make conjectures on the condition and number of libraries in early medieval Rome. In 649, the archives of the Roman Church, and perhaps together with them also the Lateran library, were transferred from their previous location to San Lorenzo in Presina (later named San Lorenzo in Damaso in honour of the homonymous pope who had put them there). It is likely that in the eight century the papal library was also housed in the Lateran. Yet, we know precious little about it. Indeed, the oldest surviving catalogue dates to the pontificate of Boniface VIII. Any attempt to trace the titles present in the papal library between the eighth and ninth centuries would probably be in vain; see Pierre Toubert, "Scrinium' et 'Palatium'. La formation de la bureaucratie romano-pontificale aux VIIIe–IXe siècles', in *Roma nell'Alto Medioevo* (Settimane del Centro Italiano di Studi sull'Alto Medioevo, XLVIII) (Spoleto 2001) and Chris Wickham, *Roma medievale. Crisi e stabilità di una città 900–1150* (Roma 2013), pp. 420 *ff.*, esp. 422 and 425. Nevertheless, it is not impossible that it also possessed a copy of the Digest and the other parts of the compilation; see Luca Loschiavo, 'Was Rome still a Centre of Legal Culture between the 6th and 8th Centuries? Chasing the Manuscripts', in *Rechtsgeschichte – Legal History*, 23 (2015), esp. pp. 98 *ff.*

142 Indeed, many documents from the early Middle Ages attest that, following the *manumissio in ecclesia*, the registration of freedom and Roman citizenship was carried out on the basis of a precise procedure. Louis the Pious (814–840) set some guidelines for the drafting of the *libellus manumissionis* in order to protect the freed slaves, preventing the former masters from making claims against them: cf. *Capitulare ecclesiasticum*, no. 138 (*c.* 818/19), I, pp. 276–277. Cf. also *Lex Ripuar.* tit. *De libertis secundum legem Romanam* 64, 1. Lastly, with precise information, Stefan Esders, 'Early Medieval Use of Late Antique Legal Texts: The Case of the 'manumissio in ecclesia'', in *Configuration du texte en histoire. Proceedings of the 12th Intern. Conf. on Studies for Integrated Text Science*, ed. O. Kano (Nagoya 2012), esp. pp. 60 *ff.*; Alice Rio, *Slavery after Rome* (Oxford 2017), pp. 75 *ff.*

Bibliography

Primary Sources

Ammianus Marcellinus, *Rerum gestarum libri*, ed., J.C. Rolfe, 3 vols. (Cambridge, Mass., 1935–9)

Capitularia ecclesiastica, Hludowici capitularia intra annos 816 et 819 data - Capitulare ecclesiasticum (MGH Boretius 1883, 275–280)

Cicero (Marcus Tullius), *De legibus*, ed. C.F.W. Müller, *Bibliotheca scriptorum Graecorum et Latinorum Teubneriana*, Leipzig, *Cic.*[1], IV/2 (1878) 380–450.

CIL = *Corpus Inscriptionum Latinarum*, consilio et auctoritate Academiae litterarum Borussicae editum, Berlin, 1863-

Codex Justinianus, v. *Iustinianus Codex*

Codex Theodosianus: *Theodosiani libri XVI cum constitutionibus Sirmondianis et leges Novellae ad Theodosianum pertinentes*, ed. Th. Mommsen – P.M. Meyer, 2 vols. (Berlin, 1905) – Engl. transl. by C. Pharr *The Theodosian Code and Novels and the Sirmondian Constitutions. A Translation with commentary, glossary, and bibliography*, [Princeton, 1952], 10th Print (Union, N.J., 2001)

Digestum see Iustinianus

Dio (Cassius Cocceianus), *Historiae Romanae*, ed E. Cary, *Dio's Roman History*, 9 vols. with an Engl. transl. (London – New York, 1914–1927)

Epitome Gai institutionum, ed. Kübler, in Ph.G. Huschke, *Iurisprudentiae anteiustinianae reliquiae*, 6a edd., E. Seckel and B. Kübler, in *Bibliotheca Teubneriana*, Leipzig, 1908–1927, II/2, 398–431 = *Fontes iuris Romani antejustiniani*, vol. II, *Auctores*, edd. G. Baviera, C. Ferrini, G. Furlani, II ed., Barbera, Firenze 1968, 232–257.

Formulae Andecavenses, in *Formulae Merowingici et Karolini aevi*

Formulae Arvernenses, in *Formulae Merowingici et Karolini aevi*

Formulae Augienses, in *Formulae Merowingici et Karolini aevi*

Formulae Bituricenses, in *Formulae Merowingici et Karolini aevi*

Formulae codicis Laudunensis, in *Formulae Merowingici et Karolini aevi*

Formulae Imperiales, in *Formulae Merowingici et Karolini aevi*

Formulae Turonenses, in *Formulae Merowingici et Karolini aevi*

Formulae Salicae Merkelianae, in *Formulae Merowingici et Karolini aevi*

Formulae Sangallanses miscellaneae, in *Formulae Merowingici et Karolini aevi*

Formulae Visigothicae, in *Formulae Merowingici et Karolini aevi*

Formulae Merowingici et Karolini aevi, ed. K. Zeumer [MGH, *Formulae*] (Hannover 1886)

Fragmenta interpretationis Gai Institutionum Augustudunensia, in *Fontes iuris Romani antejustiniani*, vol. II, *Auctores*, edd. G. Baviera, C. Ferrini, G. Furlani, II ed., Barbera, Firenze 1968, 205–228.

Gaius, *Institutiones*, ed. (and Engl. transl.) W.M. Gordon – O.F. Robinson, *The Institutes of Gaius*, London 1988

Hyginus (Pseudo), *Liber de munitione castrorum*, in M. Lenoir, *Pseudo-Hygin: des fortifications du camp*, Les Belles Lettres, Paris 1979.

Isidorus Hispalensis, *Etymologiae*, ed. M. Reydellet (Paris, 1984) [ed. Lindsay] = 2 voll. in *Scriptorum classicorum bibiotheca Oxoniensis*, Clarendon Press, Oxford 1911.

Inscriptiones Latinae selectae (*ILS*), ed. H. Dessau. Weidmann, Berlin 1892–1916

Inscriptions Antiques du Maroc (*IAM*). 2. Inscriptions latines, éd. M. Euzennat, J. Marion, J. Gascou, Éditions du Centre National de la Recherche Scientifique, Paris 1982

Itinerarium Alexandri, a c. di R. Tabacco, *Per una nuova edizione critica dell'*Itinerarium Alexandri, Pàtron Editore, Bologna 1992.

Iuliani epitome Latina novellarum Iustiniani, secondo l'edizione di G. Hänel e col glossario d'Antonio Agustín. Testo, Indici, Concordanze a c. di P. Fiorelli, A.M. Bartoletti Colombo, Dipartimento di Teoria e Storia del Diritto, Firenze 1996.

Iulianus, *Epistula ad Senatum Populumque Atheniensium*, in J. Bidez, G. Rochefort, Chr. Lacombrade (édd.), *L'empereur Julien. Œuvres complètes*, I/1, *Orationes*, in *Collection des Universités de France publiée sous le patronage de l'Association Guillaume Bude*, Les Belles Lettres, Paris 1932.

Iustinianus, *Codex*, ed. P. Krüger in T. Mommsen – P. Krüger, *Corpus Iuris Civilis*, II (Berlin, 1954[11]) – Engl. transl. B.W. Frier, *The Codex of Justinian. A New Annotated Translation*, I–III, (Cambridge, 2016)

Iustinianus, *Digesta*, ed. T. Mommsen – P. Krüger, *Corpus Iuris Civilis*, I (Berlin, 1963[17]) - From: *The Digest of Justinian*, English translation edited by Alan Watson, 4 vols. (Philadelphia, Pennsylvania, 1985)

Iustinianus, *Institutiones*, ed. P. Krüger, in *Corpus Iuris Civilis*, I (Berlin, 1963[17]) – Engl. transl. S. Scott, *The civil law*, vol. 2 (Cincinnati, 1932)

Iustinianus, *Novellae*, ed. R. Schöll – G. Kroll, in *Corpus Iuris Civilis*, III (Berlin, 1963[8]) –Engl. transl. D.J.D. Miller – P. Sarris, *The novels of Justinian. A complete annotated English translation* (Cambridge-New York, 2018)

Iuvenalis, *Satirae*, ed. C.F. Hartmann, in *Bibliotheca scriptorum Graecorum et Latinorum Teubneriana*, Leipzig 1914.

Leonis Magni Romani pontificis, *Tractatus septem et nonaginta*: Recensuit Antonius Chavasse : I–II, vol. 138 of *Corpus Christianorum. Series Latina*, Brepols, Turnhout 1973

Lex Ribvaria, ed. F. Beyerle – R. Buchner [MGH, *Leges nat. Germ.* III.2] (Hannoverae, 1954)

Liber glossarum = Libgloss (huma-num.fr)

Livius, *Ab Urbe condita, Lib. I–X*, ed. R.S. Conway, CF. Walters, S.K. Johnson, in *Scriptorum classicorum bibiotheca Oxoniensis*, Clarendon Press, Oxford 1914.

Malalas (Johannes), *Chronicon*, ed. L. Dindorf, 2 vols. (Bonnae 1831)

Martialis, *Epigrammata*, ed. W. Heraeus, in *Bibliotheca scriptorum Graecorum et Latinorum Teubneriana*, Leipzig 1925 (riv. L. Borovskij, 1976).

Notitia Dignitatum: accedunt Notitia urbis Constantinopolitanae et laterculi prouinciarum, ed. O. Seeck (Berlin 1876 = Frankfurt, 1983); ed., C. Neira Faleiro, *La Notitia Dignitatum. Nueva edicion critica y comentario historico* (Madrid, 2005); ed. in progress, I. Maier, *The compilation 'notitia dignitatum', Cnd* (//www.notitiadignitatum.org). Units: Hoffmann, 1969–1970)

Novellae Iustiniani, v. *Iustiniani Novellae*

Novella Theodosii, v. *Codex Theodosianus*

Papyri Giessen (*P.Giss.* 40), *Griechische Papyri im Museum des oberhessischen Geschichtsverein zu Giessen*, I, Leipzig-Berlin 1910.1912 (Fasc. I, ed. O. Eger, E. Kornemann, P.M. Meyer)

Papyri Rylands (*PRyl.*), *Catalogue of the Greek Papyri in the John Rylands Library, Manchester*, ed. A.S. Hunt et alii, Manchester 1911.

Pauli Sententiae, ed. M. Bianchi Fossati Vanzetti, *Pauli Sententiae. Testo e interpretatio* (Padova 1995)

Πεῖρα (*Practica ex actis Eustathii Romani*), ed. C.E. Zachariae von Lingenthal, in J. Zepi – P. Zepi, *Jus graecoromanum*, IV (Athenai, 1931 = Aalen 1962)

Procopius Caesariensis, *De bellis – Tôn Polemôn Logoi*, ed. (and Engl. transl.) H.B. Dewing, *History of the Wars* (Cambridge, Mass., 1919)

Reginonis *Chronicon*, ed. H.H. Pertz in *MGH Scriptores* I, Hannover 1829 = *Quellen zur karolingischen Reichsgeschichte*, 3. Teil, herausgegeben von R. Rau (FSGA 7), Wissenschaftliche Buchgesellschaft, Darmstadt 1960, 179–319

Roman Military Diplomas (*RMD*), ed. M.M. Roxan, P.A. Holder, Institute of Classical Studies, London 1954-

Salvianus Ad ecclesiam 3.7.34, ed. G. Lagarrigue, (Oeuvres / Salvien de Marseille, *introduction, texte critique, traduction et notes par* G. Lagarrigue, Sources Chrétiennes 220, Paris 1975)

Sidonius Apollinaris, *Epistulae et carmina*, ed. C. Luetjohann [MGH, *Auct. Ant.*, 8] (Berlin, 1887)

Sidonius Apollinaris, *Epistulae*, ed. A. Loyen, *Sidoine Apollinaire, Œuvres*, 3 vols. (Paris, 1960–70)

Suetonius, *De vita Caesarum*, éd. H. Ailloud, 3 voll., in *Collection des Universités de France publiée sous le patronage de l'Association Guillaume Bude*, Les Belles Lettres, Paris 1931–1932.

Tabula Alcantarensis, in D. Nörr, Aspekte des römischen Völkerrechts. Die Bronzetafel von Alcántara, Verlag des bayerischen Akademie der Wissenschaften, München 1989, 19–23.

Tacitus (Publius Cornelius), *Annales* I–III and IV–VI, ed. P. Wuillemier (1923, 1949; Paris 2013, 2019)

Tituli ex corpore Ulpiani, ed. F. Schulz, Di Epitome Ulpiani des Codex Vaticanus Reginae 1128, Bonn 1926 = *Fontes iuris Romani antejustiniani*, vol. 0, *Auctores*, edd. G. Baviera, C. Ferrini, G. Furlani, II ed., Barbera, Firenze 1968, 262–301.

Secondary Sources

Barbero, Alessandro. *Barbari. Immigrati, profughi, deportati nell'impero romano* (Roma – Bari 2006)

Barone Adesi, Giorgio. 'Istanze servili alle 'libertates': alle origini della disciplina costantiniana "de his qui in ecclesiis manumittuntur"', in *IAH*, 5 (2013), pp. 89–120

Besson, Arnaud. Constitutio Antoniniana. *L'universalisation de la citoyenneté romaine au 3ᵉ siècle* (Basel, 2020)

Burgarella, Filippo. 'Il Senato', in *Roma nell'Alto Medioevo.* (Settimane di Studio del Centro Italiano di Studi sull'Alto Medioevo, XLVIII) (Spoleto, 2001), pp. 121–178

Carrié, Jean-Michel. 'Les distributions alimentaires dans les cités de l'Empire romain tardif', in *MEFRA* 87.2 (1975), pp. 995–1101

Carrié, Jean-Michel. 'Constantin continuateur et et liquidateur de l'expérience tétrarchique', in *Fra Costantino e i Vandali. Atti del Convegno internazionale di studi per Enzo Aiello (1957–2013) (Messina, 29–30 ottobre 2014)* (Bari, 2016), pp. 73–+94

Casavola, Francesco. 'Il concetto di 'urbs Roma': giuristi e imperatori romani', in Id., *Sententia legum tra antico e moderno*, con una *nota* di F. d'Ippolito e una *postfazione* dell'autore, I. *Diritto romano* (Napoli, 2000), pp. 355–364

Casavola, Francesco. 'Lo spazio nell'esperienza giuridica del mondo antico', in Id., *Sententia legum tra antico e moderno*, con una *nota* di F. d'Ippolito e una *postfazione* dell'autore, I. *Diritto romano* (Napoli, 2000), pp. 523–528

Castagnino, Francesco. 'Le 'Tabulae' di Brigetio e di Durostorum. Una 'constitutio' sui privilegi di 'milites' e veterani: lo 'status quaestionis'', in S. Puliatti (ed.), *Diritto ed esercito. Profili dell'epoca tardoantica*, in *Cultura giuridica e diritto vivente*, 7 (2020), pp. 1–16

Catalano, Pierangelo. ''Ius Romanum'. Note sulla formazione del concetto', in *Da Roma alla terza Roma. Documenti e studi. La nozione di "Romano" tra cittadinanza e universalità* (Napoli, 1984), pp. 531–558

Chauvot, Alain. *Opinions romaines face aux Barbares au IVᵉ siècle apr. J.-C.* (Paris, 1998)

Cugusi, Paolo. *Per un nuovo corpus dei Carmina Latina Epigraphica. Materiali e discussioni* (Roma, 2007)

Demougeot, Émilienne. 'Restrictions à l'expansion du droit de cité dans la seconde moitié du IVᵉ siècle', in *Ktema*, 6 (1981), pp. 381–393

Demougeot, Émilienne. 'Le 'conubium' et la citoyenneté conféreé aux soldats barbares du Bas-empire', in *Sodalitas. Scritti in onore di Antonio Guarino*, IV (Napoli 1984), pp. 1633–1643

De Ruggiero, Ettore.. *La* patria *nel diritto pubblico romano* (Roma 1921)

D'Ors, Álvaro. 'Estudios sobre la "Constitutio Antoniniana", II. Los dediticios y el edicto de Caracala', in *AHDE* 15 (1944) pp. 162–204.

Eck, Werner. 'Das letzte Diplom für einen Auxiliarsoldaten aus dem Jahr 206 n. Chr.: der Text der Innenseite', in *ZPE* 208 (2018), pp. 237–244.

Esders, Stefan. 'Early Medieval Use of Late Antique Legal Texts: The Case of the 'manumissio in ecclesia', in *Configuration du texte en histoire. Proceedings of the Twelfth International Conference on Studies for Integrated Text Science*, ed. O. Kano (Nagoya 2012), pp. 54–66

Fabbrini, Fabrizio. *La manumissio in ecclesia* (Milano 1965).

Ganghoffer, Roland. *L'évolution des institutions municipales en Occident et en Orient au Bas-Empire* (Paris 1963).

Goria, Fausto. 'Romani, cittadinanza ed estensione della legislazione imperiale nelle costituzioni di Giustiniano', in *Da Roma alla terza Roma*. II. *La nozione di «Romano» tra cittadinanza e universalità* (Napoli 1984), pp. 277–342

Harper, Kyle. *Slavery in the Late Roman World AD 275–425* (Cambridge 2011).

Imrie, Alex. *The Antonine Constitution. An Edict for the Caracallan Empire* (Leiden – Boston 2018).

Jeannin, Alexandre. '*Vigor actorum* La mise en forme romanisante de la pratique', in *Une histoire juridique de l'Occident. Le droit et la coutume (IIIe–IXe siècle)*, ed. S. Kerneis, (Paris 2018), pp. 249–300

Keenan, James G. 'The Names Flavius and Aurelius as Status Designations in Later Roman Egypt', in *ZPE* 11 (1973), pp. 33–63

Keenan James G. 'The Names Flavius and Aurelius as Status Designations in Later Roman Egypt', in *ZPE* 13 (1974), pp. 283–304

Kerneis, Soazick. 'Les numeri ethniques de l'armée romaine au IIe et IIIe siècles', in *RSA* 26 (1996), pp. 69-94

Kerneis, Soazick. *Les celtiques: servitude et grandeur des auxiliaires bretons dans l'Empire Romain*, (Clermont-Ferrand 1998)

Kerneis, Soazick. 'La Bretagne rhénane. Note sur les établissements bretons dans les Champs Décumates', in *Latomus* 58 (1999), pp. 357-390

Kerneis, Soazick. "Francus civis, miles Romanus': les barbares de l'Empire dans le code Théodosien', in *Droit, religion et sociétè dans le Code Théodosien*, eds. J.-J. Aubert, Ph. Blanchard (Genève 2009), pp.. 377-399

Kerneis, Soazick. 'Dans une école de Gaule. La leçon du maître d'Autun', in *Mélanges en l'honneur du professeur Nicole Dockès, La Mémoire du droit*, eds. Chr. Lauranson Rosaz, D. Deroussin (Paris 2014) pp. 397–416

Kuhlmann, Peter Alois. *Die Giessener literarischen Papyri und die Caracalla-Erlasse. Edition, Übersetzung und Kommentar* (Giessen 1994)

Laistner, Max Ludwig Wolfram. 'The Source of Isidore (Etym. 9, 4, 49–50), in *JRS* 11 (1921), pp. 267-268

Le Roux, Patrick. 'Les diplômes militaires et l'évolution de l'armée romaine de Claude à Septime Sévère: 'auxilia', 'numeri' et 'nationes'', in *Heer und Integrationspolitik. Die römischen Militärdiplome als historische Quelle*, eds. W. Eck, H. Wolff (Köln – Wien 1986), pp. 347-374

Lettich, Giovanni. *Le iscrizioni sepolcrali tardoantiche di Concordia* (Trieste 1983)

Liebeschuetz, Wolfgang. 'Citizen Status and Law in the Roman Empire and the Visigothic Kingdom', in *Strategies of Distinction, the Construction of Ethnic Communities, 300–800*, eds. W. Pohl, H. Reimitz, (Leiden 1998), pp. 131–152

Loschiavo, Luca. 'Was Rome still a Centre of Legal Culture between the 6th and eight Centuries? Chasing the Manuscripts', in *Rechtsgeschichte – Legal History*, 23 (2015), pp, 83–108.

Marotta, Valerio. *La cittadinanza romana in età imperiale [secoli I–III d.C.]. Una sintesi* (Torino 2009)

Marotta, Valerio. 'Tre riflessioni sulla cittadinanza: da Roma antica al mondo attuale' in *IAH* 5 (2013) pp. 53–72

Marotta, Valerio. 'Il problema dei 'laeti'. Fonti e storiografia', in *'Civitas, Arma, Iura'. Organizzazioni militari, istituzioni giuridiche e strutture sociali alle origini dell'Europa [secc. III–VIII]*, eds. F. Botta - L. Loschiavo (Lecce 2015) pp. 117-157

Marotta, Valerio. 'Inclusione o esclusione? La 'constitutio Antoniniana' e i limiti del cosmopolitismo universale romano', in *Strumenti, percorsi e strategie dell'integrazione nelle società multiculturali*, eds. G. Cerrina Feroni, V. Federico, (Napoli 2018), pp. 143–163.

Marotta, Valerio. 'Una nota su D. 49.14.32 (Marcian. 14 inst.) e sulla condizione giuridica degli *obsides'* , in *AUPA* 61 (2018), pp. 213–238

Marotta, Valerio. 'Doppia cittadinanza e pluralità degli ordinamenti. La tabula Banasitana e il Papiro di Giessen 40 col. I', in *Liber amicorum. Mélanges en l'honneur de Jean-Pierre Coriat*, eds.E. Chevreau, C. Masi Doria, J.M. Rainer (Paris 2019), pp. 551 ff.

Martínez Jiménez, Javier – Flierman, Robert. 'The Uses of Citizenship in the Post-Roman West', in *Citizenship in Antiquity. Civic Communities in the Ancient Mediterranean*, eds. J. Filonik, C. Plastow, R. Zelnick-Abramovitz (London-New York 2023) pp. 669–690.

Mathisen, Ralph W. 'Peregrini', 'Barbari', and 'cives Romani': Concepts of Citizenship and the Legal Identity of Barbarians in the Later Roman Empire', in *AHR* 111/4 (2006), pp. 1024–1028.

Mathisen, Ralph W. "Provinciales', 'gentiles', and Marriages between Romans and Barbarians in the Late Roman Empire', in *JRS* 99 (2009), pp. 140–155.

Mathisen, Ralph W. 'Concepts of Citizenship', in *The Oxford Handbook of Late Antiquity*, ed. S.F. Johnson (Oxford 2012), pp. 744–763

Mathisen, Ralph W. '"Becoming Roman, Becoming Barbarian": Roman Citizenship and the Assimilation of Barbarians into the Late Roman World' in *Migration and Membership Regimes in Global and Historical Perspective*, eds. U. Bosma, G. Kessler, L. Lucassen (Leiden 2013), pp. 191-217

Mazzarino, Santo. *Antico, tardoantico ed èra costantiniana*, Voll. I e II (Bari 1974–1980)

Mazzarino, Santo. *L'Impero romano*. Vol. II. Roma – Bari 1986²)

Mélèze Modrzejewski, Joseph. *Droit et Justice dans le monde grec et hellénistique* (Warszawa 2011)

Mélèze Modrzejewski, Joseph. *Loi et coutume dans l'Égypte grecque et romaine* (Warszawa 2014)

Mirkovic, Miroslava. *The Later Roman Colonate and Freedom* (Philadelphia 1997)

Moatti, Claudia. 'Mobilità nel Mediterraneo: un progetto di ricerca', in *Storica* 27 (2003), pp. 107-129

Moatti, Claudia. *'Res publica'. Histoire romaine de chose publique* (Paris 2018)

Neesen, Lutz. *Untersuchungen zu den direkten Staatsabgaben der römischen Kaiserzeit (27 v. Chr. - 284 n. Chr.)* (Bonn 1980)

Nörr, Dieter. *Aspekte des römischen Völkerrechts. Die Bronzetafel von Alcántara* (München 1989)

Oliver, James H. 'Text of the Tabula Banasitana, A.D. 177', in *AJPh* 93 (1972), pp. 336–340.

Oliver, James H. *Greek Constitutions of Early Roman Emperors from Inscriptions and Papyri* (Philadelphia 1989)

Papa, Giovanni. 'Multae gentes ... Quibus Terrae laeticae administrandae sunt'. A proposito di CTh. 13.11.10', in *SDHI* 82 (2016), pp. 177-190

Phang, Sara Elise. *The Marriage of Roman Soldiers (13 B.C. - A.D. 235). Law and Family in Imperial Army* (Leiden – Boston – Köln 2001)

Poly, Jean-Pierre. 'La corde au cou. Les Francs, la France et la loi salique', in *Genèse de l'État moderne en Méditerranée* (Rome – Paris 1993), pp. 287-320

Poly, Jean-Pierre. '*Leges Barbarorum*'. La création des lois des nations', in *Une histoire juridique de l'Occident. Le droit et la coutume (IIIe–IXe siècle)*, ed. S. Kerneis (Paris 2018), pp. 187–247.

Poole, Reginald Lane. *Lectures on the History of the Papal Chancery down the Time of Innocent III* (Cambridge 1915)

Purpura, Gianfranco. 'Constitutio Antoniniana de civitate (212 d.C.). (P.Giff. 40 I = FIRA I, 88)', in *Revisione ed integrazione dei Fontes Iuris Romani Anteiustiniani (FIRA). Studi preparatori*, I, *Leges*, ed. G. Purpura (Torino 2012), pp. 695–715.

Rigsby, Kent J. 'Two Danubian Epitaphs' in *ZPE* 126 (1999), pp. 175-176

Rio, Alice. *Legal Practice and the Written Word in the Early Middle Ages: Frankish Formulae, c.500–1000* (Cambridge 2009).

Rio, Alice. *Slavery after Rome 500–1100* (Oxford 2017)

Rocco, Marco. *L'esercito romano tardoantico. Persistenze e cesure dai Severi a Teodosio I* (Limena 2012)

Rodríguez Martín, José-Domingo. *Fragmenta Augustodunensia* (Granada 1998)

Rostowtzew, Michael. *Studien zur Geschichte des römischen Kolonates* (Leipzig u. Berlin 1910)

Ruggiero, Iolanda. 'Una breve nota sulla condizione dei liberti latini e dei loro discendenti in età tardoantica', in ΚΟΙΝΩΝΙΑ 41 (2017), pp. 461-473

Ruggiero, Iolanda. *Ricerche sulle* Pauli Sententiae (Milano 2017)

Sasse, Christoph. *Die Constitutio Antoniniana. Eine Untersuchung über den Umfang der Bürgerrechtsverleihung auf Grund des Papyrus Giss. 40 I.* (Wiesbaden 1958)

Sassier, Yves. 'L'utilisation d'un concept romain aux temps carolingiens: la 'res publica' aux IXe et Xe siècles', in *Mediévales*, 15 (1988), pp. 17-29

Saumagne, Charles. 'Du role de l'origo et du census dans la formation du colonat romain', in *Byzantion* 12 (1937), pp. 487-581

Segré, Gino. 'L'editto di Caracalla relativo alla concessione della cittadinanza romana e il papiro Giessen 40, I', in *Scritti giuridici* II (Roma 1938), pp. 97–271

Sommer, Michael. *Palmyra. A History* (London - New York 2018)

Speidel, Michael. 'The Rise of Ethnic Units in the Roman Imperial Army', in *ANRW* II.3 (1975), pp. 202–231

Szidat, Joachim. 'Laetensiedlungen in Gallien im 4. und 5. Jahrhundert', in *La politique édilitaire dans les provinces de l'Empire romain: IIe–IVème siècles après J.-C.* Actes du IIe colloque roumano-suisse, eds. R. Frei-Stolba, H.E. Herzig (Berne 1995), pp. 283–293

Szidat, Joachim. '*Terrae leticae* (Cod. Theod. 13,11,10)', in *Historische Interpretationen. Gerold Walser zum 75. Geburstag dargebracht von Freunden Kollegen und Schuldern*, eds. H. Heinen, F. Paschoud, K. Raaflaub, H. Temporini (Stuttgart 1995), pp. 151–159

Thomas, Yan. *"Origine" et "commune patrie". Étude de droit public romain* (*89 av. J.-C.-212 ap. J.-C.*) (Rome 1996)

Toubert, Pierre. "Scrinium' et 'Palatium'. La formation de la bureaucratie romano-pontificale aux VIIIe–IXe siècles', in *Roma nell'Alto Medioevo*. (Settimane di Studio del Centro Italiano di Studi sull'Alto Medioevo), XLVIII.1 (Spoleto 2001), pp. 57-119

Ubl, Karl. 'L'origine contestée de la loi salique. Une mise au point', in *Revue de l'IFHA* 1 (2013), pp. 1–18.

Van Minnen, Peter. 'Three Edicts of Caracalla? A New Reading of P.Giss. 40', in *Chiron* 46 (2016), pp. 205-221

Von Savigny, Friedrich Karl. *Geschichte des römischen Rechts im Mittelalter*, II, (Heidelberg 1816).

Waas, Manfred. *Germanen im römischen Dienst im 4. Jahrhundert nach Christ* (Bonn 1971^{2})

West, Charles. 'Knowledge of the past and the judgement of history in tenth-century Trier: Regino of Prum and the lost manuscript of Bishop Adventius of Metz', in *Early Medieval Europe* 24.2 (2016), pp. 137–159

Wickham, Chris. *Roma medievale. Crisi e stabilità di una città 900–1150* (Roma 2013)

Wilcken, Ulrich. 1912. *Grundzüge und Chrestomathie der Papyruskunde* I.1. *Erster Band. Historischer Teil. Erste Hälfte: Grundzüge* (Leipzig 1912)

Wirth, Gerhard. 'Rome and its Germanic Partners in the Fourth Century', in *Kingdoms of the Empire. The Integration of Barbarians in Late Antiquity*, ed. W. Pohl (Leiden – New York – Köln 1997), pp. 13-55

Wolff, Hartmut. *Die Constitutio Antoniniana und Papyrus Gissensis 40 I, Erster Band: Text* (Köln 1976)

Wolff Hartmut. *Die Constitutio Antoniniana und Papyrus Gissensis 40 I. Zweiter Band: Anmerkungen und Indizes* (Köln 1976)

CHAPTER 7

Persecuting *Latrones*, Maintaining *Disciplina*, Enforcing the *Velox Supplicium*: The Frankish *Centena* According to Childebert II's Decree

Stefan Esders

1 Introduction

The late Roman army provided a full repertoire of organizational structures, operative models and conceptual designs that could play a crucial role in the establishment of post-Roman kingdoms on former Roman provincial soil. We find its post-Roman impact most visibly echoed in official titles such as *dux*[1] or *comes*,[2] in the royal power to command (e.g. the *bannus*),[3] in the presence of overarching ideas such as fidelity[4] and discipline,[5] but also in the obligation to military service both in personal and fiscal terms and in the formulation and application of military law.[6] On a more basic level, we witness the maintenance of Roman roads and resources as part of a military infrastructure that had ceased to function on an empire-wide scale, but now came to operate

1 Stefan Esders, 'Spätantike und frühmittelalterliche Dukate. Überlegungen zum Problem historischer Kontinuität und Diskontinuität', in *Die Anfänge Bayerns. Von Raetien und Noricum zur frühmittelalterlichen Baiovaria*, eds. H. Fehr – I. Heitmeier, 2nd edition (St. Ottilien 2014), pp. 425 *ff.*; Michael Zerjadtke, *Das Amt dux in Spätantike und frühem Mittelalter. Der Ducatus im Spannungsfeld zwischen römischem Einfluss und eigener Entwicklung* (Berlin and Boston 2018).

2 Alexander Callander Murray, 'The position of the *grafio* in the constitutional history of Merovingian Gaul', in *Speculum* 61 (1986), pp. 787 *ff.*

3 Wulf Eckart Voss, 'Vom römischen Provinzialprozess der Spätantike zum Rechtsgang des frühen Mittelalters', in H. Siems – K. Nehlsen-von Stryk – D. Strauch (eds.), *Recht im frühmittelalterlichen Gallien. Spätantike Tradition und germanische Wertvorstellungen* (Köln u. a. 1995), pp. 73 *ff.*

4 Stefan Esders, 'Les implications militaires du serment dans les royaumes barbares (V^e^–VIIe siècles)', in *Oralité et lien social au Moyen Âge (Occident, Byzance, Islam): Parole donnée, foi jurée, serment*, eds. M.-F. Auzepy – G. Saint-Guillain (Paris 2008), pp. 17 *ff.*; Id., 'Rechtliche Grundlagen frühmittelalterlicher Staatlichkeit', in *Der Mittelalterliche Staat – Europäische Perspektiven*, eds. W. Pohl – V. Wieser (Vienna 2009), pp. 423 *ff.*

5 Alexander Callandar Murray, '*Pax et disciplina*. Roman public law and the Merovingian state', in *From Roman Provinces to Medieval Kingdoms*, ed. T.F.X. Noble (London 2006), pp. 376 *ff.*

6 Fabio Botta and Luca Loschiavo, eds., *Civitas, iura, arma. Organizzazioni militari, istituzioni giuridiche e strutture sociali alle origini dell'Europa* (*sec. III–VIII*) (Lecce 2015).

 | DOI:10.1163/9789004698017_008

within the more limited scope of a post-Roman kingdom.[7] Still, one should refrain from tracing the Roman army's impact on the formation of post-Roman polities as a question of continuity and discontinuity alone. Rather, one has to carefully differentiate according to kingdoms, regions, sectors and societies, whose degrees of 'militarization' could vary enormously. Moreover, while several post-Roman kingdoms appear to have grown out of the late-Roman military, it seems obvious that they underwent profound processes of transformation once they were firmly established on former provincial soil and began to exercise important political functions over a sub-Roman population largely composed of civilians. But even if we were able to put all the relevant pieces together, it would still be challenging to paint a coherent picture. Much detailed research remains necessary to understand how the individual 'building blocks' of the late Roman military came to fulfil essential functions within these new post-Roman polities and how we can imagine the long-term processes by which elements of the late Roman military were transformed.

2 The Frankish *Centena:* A Personal and Spatial Unit

This holds particularly true for those implications of the late Roman military that played an important part in the performance of 'public functions' such as the maintenance of peace and order, which is at the centre of the following contribution. The *centena* or 'hundred' appears to be a case in point, although some caution is necessary since it was not relevant in all barbarian kingdoms. We know, for instance, that office-holders called *centuriones* or *centenarii* turn up in sources from several barbarian kingdoms. But, apparently, it was only in the Frankish kingdom, and later on in the Anglo-Saxon kingdoms,[8] where we find the *centena* or *hundred* as a spatial unit, that is a district with military and

7 Ernst Baltrusch *et al.*, 'Infrastructures, Mobility and Water Management. The Iberian Peninsula in the Imperial and Post-Imperial Context', in *eTopoi. Journal for Ancient Studies* 6 (2016), pp. 220 *ff.*; Stefan Esders, 'The "Staffelsee inventory". Carolingian manorial economy, mobility of peasants, and 'pockets of functional continuity' in the transition from Antiquity to the Middle Ages', in *Journal of European Economic History* 49 (2020), pp. 206 *ff.*

8 Helen Maud Cam, *Local Government in Francia and England. A Comparison of the Local Administration and Jurisdiction of the Carolingian Empire with that of the West Saxon Kingdom* (London, 1912); Ead., 'The "Private" Hundred in England before the Norman Conquest', in Ead., *Law-Finders and Law-Makers in Medieval England* (London 1962), pp. 59 *ff.*; Ead., '*Manerium cum hundredo*: The Hundred and the Hundredal Manor' (1932), in Ead., *Liberties and Communities in Medieval England. Collected Studies in Local Administration and Topography* (London, 1963), pp. 64 *ff.*; Henry R. Loyn, 'The Hundred in England in the Tenth and Early Eleventh Centuries', in *British Government and Administration. Studies presented to*

administrative functions. This difference merits our attention, as it suggests that among the former Western provinces, the situation in Gaul and Germany could have differed significantly from that in Spain or Italy.

The difference between the *centurio* or *centenarius* as a functionary on the one hand, and the *centena* as a spatial administrative unit under the command of this functionary on the other, appears to have been crucial. Research on the topic of the 'hundred' has, at least in Germany, been dominated for a long time by attempts to trace back the origin of the Frankish *centenae* to the *centeni singulis ex plebe comites* mentioned by the Roman historian Tacitus in his *Germania* written around 120 AD.[9] An alternative, and much more plausible, explanation of how the Franks may have become familiar with the *centena* organisation is offered by a late fourth-century Christian burial inscription from Cologne that mentions a *centenarius* with the Roman name Emeterius, who was *ex numero gentilium*, which probably means that he commanded a unit of non-Roman troops.[10]

It was only in 1988 that Alexander Callandar Murray published a groundbreaking article on the Frankish *centenarii* and *centenae*.[11] Convincingly refuting any claim of Germanic continuity between Tacitus and the Franks, Murray detected many functional aspects in the late Roman army that led him to conclude that the Frankish *centena* should foremost be regarded as a continuation of a late Roman military institution. Moreover, he made an important methodological point in emphasizing that historians should not focus on a single office alone when trying to find links between late antique and early medieval positions: 'The term *centenarius* was not an isolated borrowing from the late Empire. The Merovingian hierarchy of *dux*, *comes*, *tribunus*, and *centenarius* was a system of ranks and offices adopted as a whole from the Roman military by the Franks and adapted to the conditions of the Merovingian kingdom

---

S. B. Chrimes, eds. H. Hearder – H.R. Loyn (Cardiff, 1974), pp. 1 *ff.*; George Molyneaux, *The Formation Formation of the English Kingdom in the Tenth Century* (Oxford, 2015), pp. 141–152.

9 Tacitus, *Germania*, c. 12 (ed. Lund, pp. 78 and 80). See in this vein, e.g., Claudius von Schwerin, *Die altgermanische Hundertschaft* (Breslau, 1907).

10 *Hic iacit Emeterius c(e)nt(enarius) ex numer(o) gentil(ium) qu/i v<i>xit ann(os) quinquaginta, militavit p(lus) m(inus)/ XXV. D(eo) d(omino) d(evotus)*. Winfried Schmitz, 'Die spätantiken und frühmittelalterlichen Grabinschriften in Köln (4.–7. Jahrhundert)', in *Kölner Jahrbuch* 28 (1995), here no. 4, late fourth century.

11 Alexander Callander Murray, 'From Roman to Frankish Gaul: *Centenarii* and *centenae* in the administration of the Merovingian kingdom', in *Traditio* 44 (1988), pp. 59 *ff.*

in Gaul'.[12] This is supported by a provision of the seventh-century Bavarian law-code:[13]

> If anyone in the army wishes through hostility to plunder the province without his duke's order, or to take hay or grain or burn buildings, we forbid this totally, and it is not to occur. And hereafter let a count be careful [of its happening] in his retinue; in fact, let him give his order to a hundred-man and a ten-man, and let each one watch over the troops that he commands so they do not act contrary to law. And if anyone does this audaciously, it [the act] is to be examined by the count whose man does it. And if the count neglects to enquire, who does it, let him restore all things from his own property; nevertheless, let him have time to investigate. And if such a powerful man does this that a count cannot restrain him, then let him tell his duke, and let the duke restrain him according to law. If he is a freeman, let him owe forty solidi, and let him restore all things with equal value. If a slave does this, let him be sentenced to capital punishment. Let his master, however, restore all things equally, since he did not forbid his slave to do such things. If you devour one another, you will perish quickly. Nevertheless, let a count not neglect to watch over his army, so that it does not act contrary to law within his province.

In short, we see here a military hierarchy at work which includes the duke (*dux*), the count (*comes*), the 'hundredman' (*centurio*) and finally the 'ten-man' (*decanus*) – a hierarchy to which we may add the Frankish king, in whose name this statute was issued – although in Bavaria the duke appears to have largely replaced the king in terms of military and administrative functions. As it seems, this hierarchy was introduced into Bavaria only in the context of the foundation of the Frankish duchy of Bavaria in the later 530s. Apparently, the Franks had taken over a model of a military hierarchy from the Roman army which they adapted to organize Bavaria as the south-eastern frontier area of their kingdom.[14]

In the passage quoted, the *centurio* is dealt with as part of a military hierarchy within the army, but the extent of its spatial implications is not clear from this text. While a spatial responsibility is likely to be assumed for the Bavarian

12 Murray, 'From Roman to Frankish Gaul', p. 73.

13 *Lex Baiuv.* II (*De ducibus et eius causis quae ad eum pertinent*), 5 (ed. Schwind, pp. 297–8; Transl. Rivers, pp. 125–126).

14 Stefan Esders, 'Late Roman Military Law in the Bavarian Code', in *La forge du droit. Naissance des identités juridiques en Europe* (*IVe–XIIIe siècles*), Clio@Themis, 10 (March 2016).

duke, it seems as if the *centurio* was only responsible for a certain number of *homines* performing military service, but not for a district. This impression may be supported by the observation that *centenarii* or *centuriones* only very rarely appear in Bavarian charters.

This differs strongly from the situation in other parts of the Frankish kingdoms where we have strong evidence suggesting that the *centena* was a spatial unit as early as the sixth century and continued to be so well into the ninth century and beyond.[15] While it is difficult to say whether the *centena* existed in all parts of the Frankish kingdom, Carolingian evidence seems indeed to suggest that by the later eight century, the *centena* was a sub-comital district in many, if not most parts of the Frankish realm. In a capitulary issued in 789 to mobilize troops for the imminent Avar war, Charlemagne obliged all adult males to swear an oath of fidelity to the king. It was the king's *missi* who would be travelling around in the kingdom for this business and accept the oath on behalf of the king:[16]

> … all shall swear. And the *missi* shall carry the names and number of those who have sworn with them in a breviary; and the counts likewise separate for each *centena*, [with the names of] those who were born in the *pagus* and thus are *pagensales*, as of those [who have come] from elsewhere and committed themselves into a bond of vasallage. And if someone, by maliciously fleeing from one county to another because of this oath, seeks to delay this or out of arrogance declines to swear, they [the officials] should report this separately by a breviary.

Judging from the capitularies, in the Carolingian Empire, the practice of oath-swearing according to *centenae* appears to have been a well-orchestrated and administered business. The list-keeping is indeed remarkable, and despite

15 For an analysis, see Stefan Esders, 'Amt und Bann: Weltliche Funktionsträger (*centenarii, vicarii*) als Teil ländlicher Gesellschaften im Karolingerreich', in *Kleine Welten. Ländliche Gesellschaften im Karolingerreich*, eds. T. Kohl – S. Patzold – B. Zeller (Ostfildern, 2019), pp. 255*ff*.

16 *Capitulare missorum* a. 789, c. 4: ' … omnes iurent. Et nomina vel numerum de ipsis qui iuraverunt ipsi missi in brebem secum adportent; et comites similiter de singulis centenis semoti, tam de illos qui infra pago nati sunt et pagensales fuerint, quamque et de illis qui aliunde in bassallatico commendati sunt. Et si fuerint aliquis qui per ingenio fugitando de comitatu ad aliud comitatu se propter ipsum sacramentum distulerit aut per superbia iurare noluerint, semoti per brebem renuntiare sciant, et tales aut per fideiussores mittant aut si ipsi fideiussores non habuerint qui in praesentia domini regis illos abducant, sub custodia servent; aut si in illo vicinio habitare voluerint, sicut caeteri iurent' (ed. Boretius, in *Capit. reg. Franc.* 1, no. 25, p. 67; my translation).

their temporally limited relevance, two such lists have come down to us from this period.[17] One comes from Reims from the time of Charles the Bald and can be linked to the activities of Archbishop Hincmar of Reims. In 854 Hincmar asked the count of the *pagus* of Tardenois, situated west of Reims, to prepare the local people for swearing fidelity so that Hincmar could accept the oath as the king's *missus*.[18] As an appendix to the capitulary of Attigny, which was issued in the same year 854, we have a list of those people who swore the oath at an assembly held in the *pagus* of Reims.[19] These documents illustrate the business of oathtaking within the administrative units of the *pagus* for the same year. The well-ordered list references 64 names composed in three columns, accurately distinguishing those 17 men who had already sworn the oath a longer time ago (*antiquitus*), from those 47 persons who had done so only recently, obviously because they had reached adulthood since the last oath of fidelity had been sworn. Some of these persons held the official function of a *decanus*, which seems to indicate that these groups were part of a *centena*. The list thus gives us a rare glimpse into the top-down processes by which local districts such as the *centenae* were used to bind the local population to their ruler. Taking such a perspective is also possible through the famous fiscal polyptych of Chur written in the mid ninth century, from which we learn how a Carolingian *centenarius* or *schuldais* extracted weapons, taxes, and certain services from the inhabitants of his district.[20]

There are certain dangers inherent in looking back from a much more developed system to the supposed 'origins' of this system. Yet, as Murray has shown, already the Merovingian evidence on the *centena* is quite remarkable, as it includes various important legal texts that sought to regulate its functioning on a more general level. Their wording makes clear that the *centena* as a district existed already in sixth-century Gaul. But how can we explain that *centena* was a personal as well as a territorial unit? And is there more behind the traditional top-down approach? This question is still open, even after Murray's important

17 For an Italian example of such a list, see Stefan Esders, Massimiliano Bassetti and Wolfgang Haubrichs, *Verwaltete Treue. Ein oberitalienisches Originalverzeichnis mit den Namen von 174 vereidigten Männern aus der Zeit Lothars I. und Ludwigs II.* (Roma 2024, in press)

18 Flodoardus, *Historia Remensis ecclesiae* III, 26: '[Hincmarus] Bertrano illustri comiti Tardunensis pagi, propinquo suo, pro sacramento regi agendo, qualiter regi fidelitatem iurare deberent, qui in ipsius comitatu consistebant' (ed. Stratmann, p. 342).

19 *Capitulare Attigniacense* a. 854 (ed. in *Capit. reg. Franc.* 2, no. 261, p. 278).

20 *Bündner Urkundenbuch* (ed. Meyer-Marthaler – Perret, 1, pp. 380–1 and 388–9); Esders, 'Amt und Bann', pp. 278–9; Id., 'The Church as a Governance Actor in a Period of Post-Imperial Transition: Delegation of Fiscal Rights and Legal Change in tenth-century Churraetia', in *Medieval Worlds* 10 (2019), pp. 17 *ff.*

article, though any answer must remain speculative. Part of an explanation seems to lie in regarding the Frankish top-down military hierarchy as a system that was imposed on pre-existing spatial structures, encountered by the Franks when they established their rule over Gaul. As the new kings had formerly been part of the Roman army, it may thus be helpful to imagine the foundation of the new kingdoms in Gaul as a dual process with the military as the backbone. As it seems, *civitates* and *pagi* that had existed in Gaul for centuries, under Frankish rule became subject to a military office-holder called *comes*, a count.[21] Already under the Burgundian kings of Southern Gaul reference is made to a functionary called *comes pagi*, a military official whose district was a *pagus*, a well-established spatial unit of Roman administration.[22] That there were *comites civitatis* and *comites pagi* in Burgundy seems to suggest that the traditional Roman administrative subordination of *pagi* under the *civitates* had ceased to exist, when both came to be governed by royally appointed counts.[23] Following the Frankish takeover of the Burgundian kingdom in 534, these *comites* would exercise military and certain judicial functions and were directly subordinated to the Frankish kings, while the late Roman provincial administration as a superstructure ceased to exist.[24]

Civitates and *pagi* were by no means the only districts to have existed in Gaul before the Burgundians, Visigoths and Franks established their kingdoms in the course of the fifth century. There were numerous sub-districts, some of which were called *condita, ager* or *ministerium* since the Roman period. And it is here where the Franks appear to have linked in the *centenarii* as functionaries subordinated to the counts. As it seems, the Merovingian rulers placed military office-holders at the top of these pre-existing districts.[25] As these districts were now governed by *centenarii* and *vicarii*, they also came to be called *centena* or *vicaria* in official documents such as capitularies that addressed

---

21 See Murray, 'The position of the grafio'.

22 *Lex Burgundionum, Prima Constitutio* 5: 'Burgundiones quoque et Romani civitatum aut pagorum comites vel iudices deputati, omnes etiam et militantes …' (ed. de Salis, *Leges Bugundionum*, p. 31).

23 Stefan Esders, 'Zur Entwicklung der politischen Raumgliederung im Übergang von der Antike zum Mittelalter: Das Beispiel des *pagus*', in *Politische Räume in vormodernen Gesellschaften. Gestaltung – Wahrnehmung – Funktion*, eds. O. Dally – F. Fless – R. Hänsch – F. Pirson – S. Sievers (Berlin, 2013), pp. 195*ff*.

24 However, it continued to exist in the spatial organisation of the church with its metropolitan sees and ecclesiastical provinces.

25 As already observed by Richard Schröder and Eberhard von Künßberg, *Lehrbuch der deutschen Rechtsgeschichte*, 7th ed. (Berlin, 1932), p. 131 fn. 17: 'Die *vicariae* [und *centenae*] lehnten sich meistenteils an altgallische Untergaue aus der Zeit der römischen Provinzialverfassung (*condita, aicis, ager, finis, ministerium*) an.'

these office-holders. This hypothesis on the genesis of the Frankish *centena* as a district may sound somewhat speculative, but it helps to explain why *centena* could become conceptualized as designating a personal as well as a territorial unit, and why apart from being the object of a top-down hierarchy with the king as ultimate reference, the *centena* also had certain communitarian features.

3 *Centenae* and the Persecution of Robbers in Sixth-Century Gaul

As a territorial unit, the *centena* had important administrative functions. While it becomes especially clear from Carolingian sources that it was first and foremost a unit for recruiting men to form groups exercising relevant military and policing functions, one can hardly doubt that this basic role was already characteristic of the sixth-century *centena*. Among the Merovingian laws dealing with *centenae*, most important is the *Pactus pro tenore pacis* concluded by the Merovingian kings Childebert I (511–558) and Chlothar I (511–561) and already dealt with at length by Murray,[26] who drew our attention to the first provision of Chlothar's decree:[27]

> Because those persons deployed as night watches would not catch thieves, as for various [reasons] they would in collusion [with the thieves] exercise their watch duty defaultingly, it has been decreed that hundreds shall be constituted. In whose *centena* anyone has lost something [by theft] he should receive the value of what he lost, and the robber shall be pursued even if he appears to have gone into the *centena* of another. Anyone summoned to this [search] who is negligent upon admonition shall be condemned to pay five *solidi*. He who had lost something, shall nonetheless without question receive its value from that *centena* [to which the thief had fled], that is the second or third [*centena*]. If the trail of the robber is brought to proof, he must be punished either now or later.

26 Murray, 'From Roman to Frankish Gaul', pp. 75–93.

27 *Decretio Chlotarii regis* (ed. in *Capit. reg. Franc. 1*, Nr. 3, pp. 5–6): 'Decretum est, ut qui ad vigilias constitutas nocturnas fures non caperent, eo quod per diversa intercedente conludio scelera sua pretermissas custodias exercerent, centenas fierent. In cuius centena aliquid deperierit, capitale qui perdiderit recipiat, et latro [insequatur] vel si in alterius centenam appareat deduxisse et ad hoc admonitus si neglexerit, quinos solidos condempnetur; capitale tamen qui perdiderat, ad centena illa accipiat absque dubio, hoc est de secunda vel tertia. Si vestigius conprobatur latronis, tamen presencia aut longe multandus' (my translation).

Most likely referring to regions of the northern parts of the Frankish kingdoms,[28] the *Pactus* echoes the division of the kingdom under the sons of King Clovis, and was issued in response to an encroachment upon legal security, possibly in 541.[29] Certain inadequacies or corruptive practices in protecting local communities against thieves or robbers (*fures*, *latrones*) through night watches are used in the text to justify the convocation of *centenae* (*centenas fierent*). This suggests that *centenae* were regarded as a more complex, but also more effective measure to cope with the problems posed by prevailing theft and robbery. The complexity of the measures taken also becomes apparent from the fact that the search for robbers was allowed to extend geographically, with efforts taken in both kingdoms requiring careful coordination.

In distinguishing night watches and *centenae* as two different institutional modes of handling robbery, the regulation appears to presuppose more complex mechanisms of local administration. In his important book of 1937, Julius Goebel pointed out how the robber posed a new and previously unheard problem to traditional Frankish law in which most regulations presume that a culprit was known and easily apprehendible. As a *latro* was less easy to get hold of, this caused extraordinary measures for persecution of secret offenders to be initiated by royal legislation.[30] That this could not happen by coincidence was demonstrated by Alexander Murray, who was first to point out that the measures taken by the two Merovingian kings Childebert I and Chlothar I followed Roman models of handling brigandage.[31] Performing services such as night watch had indeed been a Roman *munus publicum* attested in papyri from Egypt, where *nuktophylakes* often appear as liturgists in the fourth century and later, whose function was to protect villages and important places, and also to arrest thieves whom they were also permitted to chastise.[32] While their duties were regarded as dangerous and thus not very popular (the thieves often tried to escape), it is even attested that they occasionally appealed to the *riparii* or

28 See Murray, 'From Roman to Frankish Gaul', p. 76 on the frontier between Chlothar's and Childeric's kingdoms running between Paris and Soissons.

29 On the decree's date (541?) see Michael Glatthaar, 'Der Edictus Chilperichs *I.* und die Reichsversammlung von Paris (577)', in *Deutsches Archiv für Erforschung des Mittelalters* 73 (2017), at pp. 64–74.

30 Julius Goebel, *Felony and Misdemeanour. A Study in the History of English Criminal Procedure* (New York, 1937), pp. 66–76.

31 Murray, 'From Roman to Frankish Gaul', pp. 80–83.

32 Dieter Hennig, 'Nyktophylakes, Nyktostrategen und die παραφυλακή της πόλεως', in *Chiron*, 32 (2002), pp. 281 *ff.*; Clemens Homoth-Kuhs, *Phylakes und Phylakon-Steuer im griechisch-römischen Ägypten: Ein Beitrag zur Geschichte des antiken Sicherheitswesens* (München – Leipzig, 2005), pp. 77–79 and 135–138.

to more official militaries, to help them out.[33] By contrast, chasing deserters called for more complex measures that often involved military officials and staff.[34]

While night-watches point to the Roman system of compulsory public service, the *munera publica* or *liturgeiai*, Roman models may have also been relevant for the more militarized mode of coping with brigandage. Already in colonies and *municipia* in the early Roman imperial period, a functionary called *praefectus arcendis latrociniis* is often mentioned in inscriptions from Roman Gaul and Germany.[35] He could be appointed by the city's Duumvir upon the consent of the *decuriones* to lead a group of armed retainers who had to be recruited from among *coloni* and *incolae*, in order to protect the colony's borders. For this kind of self-defence, which seems to be paralleled in the peace-keeping functions exercised by *eirenarchi* often attested in Egypt,[36] according to the colonial statute of Urso in Spain the *praefectus* was entitled to use the power of command (*ius et animadversio*) as it was attributed to a military tribune in Rome.[37]

It seems that this public burden, whose efficiency was already being questioned in some late Roman contexts, as many other *munera publica*, did not cease to be enforced in Merovingian Gaul, as the Pactus's wording *qui ad vigilias constitutas nocturnas* would suggest. By contrast, the convocation of *centenae* as ordered by Chlothar's decree, because of the night watchmen's insufficiency and corruption,[38] points to some sort of military or police-keeping service, as the problem of robbers, thieves, and brigands, who were not based in local societies, apparently got out of hand. This presupposes a spatial definition

33 Homoth-Kuhs, *Phylakes*, pp. 72–73.

34 Anna Maria Kaiser, 'Die Fahndung nach Deserteuren im spätantiken Ägypten', in *Actes du 26e Congrès international de papyrologie*, ed. P. Schubert (Génève, 2012), pp. 381*ff.*

35 Arrigo Manfredini, '*Municipii* e città nella lotta ai *latrones*', in *Annali dell'Università di Ferrara, Sezione V: Scienze giuridiche*, NS 6 (1992), pp. 23*ff.*

36 Murray, 'From Roman to Frankish Gaul', p. 78.

37 *Lex coloniae Iuliae genetivae* c. 103: 'Quicumque in colonia Genetiva IIvir praefectusve iure dicundo praerit, eum colonos incolasque contributos quocumque tempore coloniae finium defendendorum causa armatos educere decuriones censuerint, quot maior pars qui tum aderunt decreverint, id ei sine fraude sua facere liceto. Eique IIviro aut quem IIvir armatis praefecerit idem ius eademque animadversio esto, uti tribuno militum populi Romani in exercitu populi Romani est, itque ei sine fraude sua facere liceto ius potestasque esto, dum it, quot maior pars decurionum decreverit, qui tum aderunt, fiat.' (ed. Riccobono, pp. 190–191).

38 On corruption connected to night watches in the Roman period see Werner Riess, *Apuleius und die Räuber. Ein Beitrag zur historischen Kriminalitätsforschung* (Stuttgart, 2001), pp. 207 and 214.

of *centenae*, which allowed for the more systematic persecution of robbers beyond an individual local community, as it also points to the possibility to convoke a posse on special order from among the local population, and to inflict monetary fines if the obligations of 'military service' were not met. In contrast to the night watches, the *centena* thus appears to have been a more *militia*-type organization that was either created here or, to my mind more likely, summoned to cope with a situation that had got out out of control.

The summoning of free men to serve in the *centena* as a local militia thus appears as a measure that could be taken in a militarized local society, in which some sort of military service was performed if the extended military administration required it. But it also becomes clear from the provision that the *centena* had financial liabilities and that there was an element of collective responsibility not only to catch a robber, but also to compensate for stolen property if the *centenae* did not manage to do so. The regulation thus seems to imply some sort of communitarian structure,[39] which is most aptly explained in the context of Roman traditions of public obligations. The communitarian structure of *centenae* is also reflected in the fact that *centenarii*, as attested in our charter evidence since the late seventh century, appear as officials witnessing certain property transactions (which could impact on the division of responsibilities) and exercizing jurisdiction at a lower level. It is not easy to identify places where *centenarii* were actually based, although one might assume that *centenae* had a certain central location that served as a type of base for the *centenarius*. To this one may add that among the vast number of places where the Merovingian rulers minted gold coins in the sixth and seventh centuries we find several *castra* and other fortified places, which suggests that there was often a connection between *centena* and the exaction and commutation of taxes and the minting of coins on a local level.

4 *The Centenae* in Childebert II's Decree of 596: The Historical Context

This line of inquiry, that is tracing the transformation of late Roman legal ideas and institutions in the context of the Frankish *centena*, can be pursued further by focusing on a second royal decree dealing with *centenae*, the so-called

39 On this important aspect, visible in a number of medieval texts, see Götz Landwehr, 'Genossenschaftliche Rechtsverfolgung im Mittelalter', in H. Möller, ed., *Studien zur Rechtsschutzversicherung in europäischen Ländern und in den Vereinigten Staaten* (Karlsruhe, 1975), pp. 45*ff.*

'Decretio' of the Frankish King Childebert II, issued in 596.[40] Although the law is often transmitted in manuscripts alongside the *Lex Salica*, it was certainly not conceived as an addition to Salic law, but as a decree designed for Austrasia and constituted as a law of its own.[41] As is stated in the text, its provisions had been agreed upon at three military assemblies subsequently held in Andernach, Maastricht and Cologne – three important late Roman fortifications that continued to be used under Frankish rule.[42] Andernach, according to the *Notitia dignitatum* was the place where a legion (*legio Acincensis*) was subordinated to the duke of Mainz around 400. It had a granary, a *horreum*, and its harbour continued to be used well into the 7th century under Frankish rule. It was a minting-place of coins and had a royal palace with two churches dedicated to St Genovefa of Paris and St Martin of Tours, both *patrocinia* closely associated with Merovingian kingship. As one would expect from these facts, there was also a large fiscal property located near Andernach. It belonged to Austrasia, the East Frankish kingdom. We even have a funeral inscription from the seventh century which attests to a notary with the Roman name *Amicatus*. Maastricht, the second place referred to in Childebert's law of 596, was a fortified *castrum* in late antiquity. It was part of the Roman system of infrastructure and supply of the army on the lower Rhine, where barbarian *laeti* are also attested. As in Andernach, we also have fiscal property attested for Maastricht, and it functioned as a minting-place into the seventh century. And finally Cologne, already briefly mentioned above as base of a *centenarius* in the late fourth century,[43] had been the provincial capital of *Germania inferior*. Again, fiscal property and a minting-place are attested, as are numerous stays of Merovingian kings, who seem to have resided in the former *praetorium*, now called *aula regia*. Moreover, Cologne was the centre for the formation of the *Ripuarii*, a late Roman unit of *limitanei* stationed along the Rhine. These *Ripuarii* became a Frankish sub-group in the course of the sixth and seventh centuries.[44]

---

40 *Decretio Childeberti II*. a. 596 (ed. in *Capit. reg. Franc. 1*, no. 7, pp. 15–17).

41 Wilhelm Alfred Eckhardt, 'Die *Decretio Childeberti* und ihre Überlieferung', in *Zeitschr. der Savigny-Stiftung für Rechtsgeschichte, Germ. Abt.* 84 (1967), at p. 1.

42 For what follows, see with further references Stefan Esders, 'Man-like discipline and loyalty against the "enemies of God": Some observations on the militarized frontier society of Eastern Francia around 600', in *Early Medieval Militarisation*, eds. E. Bennett – G.M. Berndt – S. Esders – L. Sarti (Manchester, 2021), pp. 181 *ff.*

43 See above n. 10.

44 Matthias Springer, '*Riparii* – Ribuarier – Rheinfranken nebst einigen Bemerkungen zum Geographen von Ravenna', in *Die Franken und die Alemannen bis zur 'Schlacht bei Zülpich' (496/97)*, ed. D. Geuenich (Berlin – New York, 1998), pp. 200 *ff.*; K. Ubl, *Köln im Frühmittelalter. Die Entstehung einer heiligen Stadt, 400–1100* (Köln 2022), pp. 52 *ff.*, 108 *ff.*

It was in these three places that the Austrasian King Childebert II met with his *leudes*, being his army, to take counsel about legislative measures that needed to be taken. The three places combined territories that formerly belonged to the Roman provinces of *Germania superior* and *Germania inferior* and which were settled by both Salian and Ripuarian Franks along with the Roman provincial population in the sixth-century kingdom of Austrasia. Of the three meetings held at Andernach, Maastricht and Cologne, each took place on the first of March, which also seems relevant. For it was on the 1st of March that in the late Roman period new rulers were appointed and new taxes announced, and this practice is still attested into the later sixth century.[45] In 579, for example, Marcus, referendary of King Chilperic, announced new taxes in the city of Limoges on the 1st of March which caused unrest among the people who burned the tax registers and almost assassinated Marcus, were it not for the local bishop who allowed him to seek asylum in the episcopal church.[46] Three years earlier one of Chilperic's dukes called Roccolen was about to announce new taxes in the city of Poitiers on the 1st of March, but died the night before, since, according to the remote diagnosis of our episcopal chronicler, he used to eat rabbit during lent.[47] These anecdotes illustrate the point that these were public and military assemblies that sought to establish *consensus* between the Frankish king and his soldiers and military retinue. It thus was only natural that King Childebert's decree of 596, one of the most important pieces of extant Merovingian legislation, deals quite extensively with public order and with the function of the *centenae*.[48]

5 The Robber's Execution 'without law' and the Late-Roman *Velox Supplicium*

The purpose of the remaining part of this paper is to argue that the Frankish *centena* structure not only suggests an adaption of models of late Roman military organization, and its implementation in local societies, but also points to the reception of ideas of Roman military law. For Childebert's decree allows us not only to get acquainted with some of the underlying military and

45 Matthias Springer, 'Jährliche Wiederkehr oder ganz anderes: Märzfeld oder Marsfeld?', in *Rhythmus und Saisonalität*, eds. P. Dilg – G. Keil – D.-R. Moser (Sigmaringen, 1995), pp. 297*ff.*

46 Gregory of Tours, *Libri historiarum X*, V,28 (ed. Krusch – Levison, pp. 233–4).

47 Gregory of Tours, *Libri historiarum X*, V, 4 (ed. Krusch – Levison, pp. 198–200).

48 See also Murray, 'From Roman to Frankish Gaul', pp. 83–84.

administrative structures of the Eastern Frankish kingdom, but also sheds rare light on the judicial procedures and regulations to be observed when chasing *latrones* and sanctioning their misdemeanor.

Especially telling in this regard are the provisions on how to deal with thieves and robbers caught either by the judge or by the *centena*:[49]

> Likewise it was agreed on the Kalends of March at Cologne, and we have so ordered by our ban (*et ita bannivimus*) that if any judge hears of a criminal robber, the judge is to go to his home and have him bound. Should he be a Frank, he is to be sent to our court; should the thief be of lower status (*debilioris personas*), he is to be hanged on the spot (*in loco pendatur*). If someone refuses to support a *centenarius* or any judge [in pursuing] a malefactor, let the offender by all means be fined 60 *solidi*.

Again, thieves and robbers threatened the public order, prompting the king and his advisers to take remedial action. If the thief was not a Frank, he should be executed immediately, echoing martial law. By contrast, the special procedure prescribed for a Frankish robber, by necessity involving the king's court, points to the higher legal status attributed to the Franks as a military elite in the Frankish law-codes,[50] but again also to Roman models.[51] As the *centenae* apparently did not function properly, the king quite drastically raised the fine for not following the *centenarius* to pursue a criminal to the royal ban of 60

49 *Decretio Childeberti II*, cc. 1 and 2: 'Similiter Kalendas Marcias Colonia convenit, et ita bannivimus, ut unusquisque iudex, criminosum latronem ut audierit, ad casam suam ambulet et ipsum ligare faciat; ita ut si Francus fuerit, ad nostra praesentia dirigatur, et si debilioris personas fuerit, in loco pendatur. Si quis centenario aut cuilibet iudice noluerit ad malefactorem adiuvare, sexaginta solidos omnis modis condempnetur.' (ed. in *Capit. reg. Franc. 1*, no. 7, p. 15; my translation).

50 On the wergild tariffs see most recently Lukas Bothe, 'From Subordination to Integration: Romans in Frankish Law', in *Transformations of Romanness. Early Medieval Regions and Identities*, eds. W. Pohl – C. Gantner – C. Grifoni – M. Pollheimer-Mohaupt (Berlin 2018), pp. 345*ff.*

51 The practice reserved here exclusively for Franks reminds one of the late Roman practice not to kill deserters immediately if they do not resist, but to hear them, and in case of their confession to reserve judgment on their fate to the *magistri militum*, see CTh. VII, 18 (*De desertoribus et occultatoribus eorum*), 11 (403): 'Si qui desertores oberrare in provinciis fuerint conprehensi, eos et conprehendi protinus volumus et ad iudicem deduci, ut auditi, cum de crimine desertionis suae confessi fuerint, carceris custodiae deputentur: de quorum nominibus ad tuam sublimitatem subditis confessionibus referatur, ut instructi magistri militum, quid de his fieri oporteat, pro sui auctoritate constituant.'

solidi – which was a huge sum.[52] The intention was to make clear that not taking part in a search organized by a *centena* was regarded as a severe offence, comparable to not meeting one's obligation to serve in the king's host.

While this clause addressed the role of the *iudex*, who after having been informed about a robber was entitled to immediately inflict capital punishment on the robber (if he was not a Frank), another provision also allowed members of the *centenae* to execute robbers on the spot:[53]

> Concerning thieves and malefactors, we have decreed the following to be observed: If five or seven men (*homines*) of good faith (*bonae fidei*) say under oath (*cum sacramenti interpositione*) – and without personal enmity (*absque inimicitia interposita*) – that someone is a criminal (*criminosum esse dixerint*), let him die without law (*sine lege moriatur*), just as he fell in without law (*quomodo sine lege involavit*). If any judge is convicted of releasing a captured thief, let him lose his life (*vitam suam amittat*), and let by all means order (*disciplina*) be respected among the people.

The Latin term used for public order, *disciplina*, strongly recalls Roman vocabulary used in connection with the behaviour of the army.[54] The justification – who fell in without law should die without law – at first sight seems to echo some idea of *ius talionis*, which is also evident in other provisions of the decree inspired by mosaic legal reasoning.[55] Still, on closer inspection the measure taken here betrays many influences of Roman law.

52 On the royal ban in the Merovingian period see Katrin Bayerle, 'Einsatzfelder des weltlichen Bannes im Frühmittelalter', in *Von den leges barbarorum bis zum ius barbarum des Nationalsozialismus. Festschrift für Hermann Nehlsen zum 70. Geburtstag*, eds. H.-G. Hermann *et al.* (Köln 2008), pp. 13*ff.*

53 *Decretio Childeberti II*, c. 5: 'De furis et malefactoris ita decrevimus observare, ut si quinque aut septem bonae fidei homines absque inimicitia interposita criminosum cum sacramenti interpositione esse dixerint, quomodo sine lege involavit, sine lege moriatur. Si quis iudex comprehensum latronem convictus fuerit relaxasse, vitam suam amittat, et disciplina in populum modis omnibus observetur.' (ed. in *Capit. reg. Franc. 1*, no. 7, p. 16; my translation).

54 See Alfred Richard Neumann, '*Disciplina militaris*', in *Pauly-Wissowa, Realencyclopädie der classischen Altertumswissenschaft*, Suppl. X (1965), pp. 142*ff.*; Vincenzo Giuffrè, '*Militum disciplina* e *ratio militaris*', in *Aufstieg und Niedergang der römischen Welt* (ANRW) II, 13 (1980), pp. 234*ff.*; Murray, 'Pax et disciplina'.

55 On this see Hermann Nehlsen, 'Entstehung des öffentlichen Strafrechts bei den germanischen Stämmen', in *Gerichtslauben-Vorträge. Freiburger Festkolloquium zum 75. Geburtstag von Hans Thieme*, ed. K. Kroeschell (Sigmaringen, 1983), pp. 3*ff.*

Apart from requiring the witnesses' legal integrity under the Roman term of *bona fides*,[56] the somewhat undetermined mention of either five or seven witnesses prompts us to consider Roman civil law. For instance, for making a will, Roman law required the presence of five or seven witnesses,[57] five according to civil law and seven according to praetorian law.[58] A reminiscence of this can also be found in the regulations on donations and testaments in the Burgundian *Liber constitutionum*, where in the *minores causae* even three witnesses were deemed sufficient.[59] It is open to speculation how the established practice of civil law could have entered criminal law, and what precisely the alternative of either five or seven witnesses as delineated in Childebert's decree might have referred to in practice. One could imagine that King Childebert and his legal advisors, among them the former Burgundian referendary Asclipiodotus

---

56 While in Roman law, *bona fides* was most often used as a concept of trust in contract law, it refers here to the integrity and social credit of members of the posse qualified to testify under oath. On the wider implications of *bona fides* in Roman law see Remus Valsan, '*Fides, bona fides,* and *bonus vir.* Relations of Trust and Confidence in Roman Antiquity', in *Journal of Law, Religion and State*, 5 (2017), pp. 48 *ff.*

57 Carl G. Bruns, 'Die sieben Zeugen des römischen Rechts (1877), in Id., *Kleinere Schriften*, vol. 2 (Weimar, 1882), here at pp. 124–127.

58 See CTh. IV,4 (*De testamentis et codicillis*), 1: 'In codicillos, quos testamentum non praecedit, sicut in voluntatibus testamenti septem testium vel quinque interventum deesse non oportet.' See also *Interpretatio* to CTh. IV,4,3: 'Si moriens, cum scribit aut dictat chartulam testamenti, praetermiserit forsitan, ut vocabulum aut poneret aut civilis, id est quod quinque testium, aut praetorii iuris, quod septem testium erit subscriptione firmandum; unde si maiorem vel minorem testium numerum se testator in principio dixerit evocasse, usque ad quinque subscriptorum numerum, etiamsi non relegant testamentum, valere permissum est, ita ut, sicut minor numerus inpedit voluntati, sic, quicquid supra fuerit, non noceat testamentum: quia et lex ipsa constituit, quod superflua iuri inpendire non debeant. Hoc etiam observandum, ut testator signandum testibus offerat testamentum, nec ideo habeatur infirmum, si alicui de subscriptoribus testator aliquid legati nomine derelinquit.' (ed. in *Codex Theodosianus*, I, pp. 170–171).

59 *Lex Burgundionum* 43 (*De donationibus*): '[1.] Licet plura de donationibus lex nostra de superioribus temporibus data praeceperit, tamen quia ex ipso eodemque titulo oriuntur aliqua, de quibus non evidenter legitur constitutum, necesse est, ut praesentis legis adiectio, quae prius fuerant praetermissa, definiantur. Ideoque hoc ordine in populo nostro donationes factae et testamenta valebunt, ut quinque aut septem testes donationi aut testamento, prout possunt, aut signa aut subscriptiones adiciant. [2.] Quod si minor testium numerus interfuisse probabitur, facta donatio aut conditum testamentum nullam habere poterit firmitatem. [3.] In aliis vero, hoc est in minoribus [causis], tres idoneos testes iubemus admitti. [4.] Ceterum de testamentis et donationibus statuta superius forma servabitur.' (ed. de Salis, p. 74). See the analysis by Hans Rüegger, *Einflüsse des römischen Rechtes in der Lex Burgundionum* (Lungern, 1949), pp. 130–133.

who signed the decree,[60] sought to construct a very basic legal procedure that should not be confused with lynch law, and that the number of witnesses may have depended on the size of the group that caught an offender. But the regulation also needs to be interpreted against the backdrop of Frankish law.[61] The prevalence of the oath of purgation in Frankish law did not allow for the possibility of using an oath to convict someone of a crime. It is clear that with regard to Frankish law, it has to be seen as novel regulation, brought about by special circumstances.[62]

The laconic diction that following this procedure 'he who fell in without law' (*sine lege*), should consequently 'die without law' (*sine lege*) clearly betrays an awareness that this was extraordinary procedure, and in no way reconcilable with established legal practice. This was in accordance with the Roman legal handling of robbery.[63] In the Roman imperial period, *latrones* appear as a sort of nightmarish phenomenon in legal and other sources.[64] Often referred to along with deserters,[65] *latrones* was employed as an umbrella term for bandits whose use of violence was illegitimate[66] – a soldier becoming a *desertor* was a robber,[67] and as such regarded as even more lawless than an enemy. As a consequence, using violence to overturn such people was typical and even killing them was considered as legitimate to some extent.[68] A law by the 3rd-century Roman Emperor Gallienus made clear that anyone who killed a *latro* did so in accordance with law.[69] The Roman legal stance against *latrones* has been aptly characterized by Brent Shaw: 'In the pursuit of this obligation the

60 On Asclipiodotus see Detlef Liebs, *Römische Jurisprudenz in Gallien (2. bis 8. Jahrhundert)* (Berlin, 2002), pp. 72–75.

61 Also the exclusion of *inimicitia* could point to Roman law (see, e.g., Dig. 34,9,9 (Ulpian)), as it could also refer to blood revenge in Frankish law.

62 Goebel, *Felony and Misdemeanour*, pp. 74–76.

63 See in particular Luca Loschiavo, 'Autodifesa, vendetta, repressione poliziesca. La lotta al brigantaggio nel passagio dalle province tardo-imperiali ai regni romano-barbarici', in *Il diritto Giustinianeo fra tradizione classica e innovazione*, ed. F. Botta (Torino, 2003), pp. 105*ff.*

64 Brent D. Shaw, 'Bandits in the Roman Empire', in *Past and Present* 105 (1984), pp. 3*ff.*; Id., 'Der Bandit', in *Der Mensch der römischen Antike*, ed. A. Giardina (Frankfurt am Main 1991), pp. 337*ff.*

65 Vincenzo Giuffrè, '*Latrones desertoresque*', in *Labeo* 27, pp. 214–218.

66 Shaw, 'Bandits in the Roman Empire', p. 6.

67 Ramsay MacMullen, 'The Roman Concept Robber-Pretender', in *Revue internationale des droits de l'antiquité* (RIDA), 3. sér. 10 (1963), pp. 221*ff.*

68 See Riess, *Apuleius*, p. 193; Loschiavo, 'Autodifesa', p. 121.

69 Cod. Iust. IX,16 (*Ad legem Corneliam de sicariis*), 3 (a. 265): 'Si, ut adlegas, latrocinantem peremisti, dubium non est eum, qui inferendae caedis voluntate praecesserat, iure caesum videri.'

private individual was authorized to use force, to injure and even to kill such men. And they were also exempted, in doing this, from normal laws on *iniuria* and homicide. Such a carte blanche transfer of the positive use of violent force from the state to private individuals was unusual in Roman law. It went far beyond the normal concession granted by most states to individuals of using reasonable force to protect themselves when attacked. The Roman state clearly regarded this grant of 'public vengeance' (as it was politely phrased) as exceptional, and was manifestly uneasy about this reversal of state power. Still, the practice was justified in the name of the 'common peace'.'[70]

That these ideas may indeed have been channeled through military law to become adopted by the Franks is suggested by late Roman legal practice. The Theodosian Code preserves several imperial laws issued in 403 by the emperor Honorius from the new capital Ravenna, when public order was threatened in Italy and beyond through Alaric's Goths. In this special situation, the emperor granted the population the *ius militiae*:[71]

> We allow by law provincials to overpower deserters. If they dare to resist, we order that everywhere punishment be swift (*velox ubicumque iubemus esse supplicium*). All shall know that, in defence of the common peace (*pro quiete communi*), they have against public robbers and deserters (*adversus latrones publicos desertoresque*) been granted the military right (*militiae ius*) to administer public vengeance (*exercendae publicae ultionis*).

This meant that the provincials, organized in local armed *militiae*, sometimes called *solacium* in Latin,[72] under the *ius militiae* should capture suspected robbers and deserters and have them executed in a summary trial, referred to as *velox supplicium* here, but often included in the more general notion of *severitas*.[73] Of course the dividing line between deserters and robbers could be fluid,

70 Shaw, 'Bandits in the Roman Empire', p. 19.

71 CTh. VII, 18 (*De desertoribus et occultatoribus eorum*), 14 (a. 403): 'Opprimendorum desertorum facultatem provincialibus iure permittimus: qui si resistere ausi fuerint, velox ubicumque iubemus esse supplicium. Cuncti etenim adversus latrones publicos desertoresque militiae ius sibi sciant pro quiete communi exercendae publicae ultionis indultum …' (transl. by Pharr, p. 178, altered).

72 On the uses of the term *solacium* in the Merovingian period see Murray, 'From Roman to Frankish Gaul', pp. 86–88.

73 See Vincenzo Giuffrè, *Iura e arma. Intorno al VII libro del Codice Teodosiano* (Napoli, 1979), pp. 115–116, with important observations on the laws of 403 discussed here. On punishments inflicted upon *latrones* see Shaw, 'Bandits in the Roman Empire', p. 20.

and it is not by chance that these two groups are often referred to within the same laws.[74]

In another law, dating from the same period, these measures were repeated, and only the life of those deserters who did not actively resist these *militiae*, was saved so that the *magister militum* could eventually determine their fate:[75]

> We decree to detain deserters by all means. All provincials, therefore, shall know that they have been granted the right to overpower deserters. In order that deserters may have no hope of punishment being delayed, we order that everywhere execution shall be swift (*velox ubicumque iubemus esse supplicium*). This shall be brought to the notice of the functionaries of the cities (*urbes*), villages (*vici*), and fortresses (*castella*), by which they may know that harbourers (*occultatores*) of such [deserters] shall be subject to the penalty established by the laws of Our divine father.

These laws were enacted for Italy, in a very special situation,[76] but they may give us some idea how already in the late Roman period a mobilization of the provincial population was practised in order to cope with this problem. Moreover, the inclusion of these extraordinary measures into the Theodosian Code points to their continuing relevance into the fifth century.[77] As it seems, the local administration as centered upon *urbes*, *vici* and *castella* provided a basic structure that allowed the recruitment of men from among local communities

---

74 See above fn. 65–67.

75 CTh. VII, 18 (*De desertoribus et occultatoribus eorum*), 13 (a. 403): 'Desertores modis omnibus adtinendos esse censemus. Universi igitur provinciales sciant opprimendorum sibi desertorum facultatem esse permissam: quibus ne sit tarditas poenae solacio, velox ubicumque iubemus esse supplicium. Quod ad notitiam primatium urbium vicorum castellorumque deveniat, quo scire possint etiam occultatores eorum ad subeundam poenam, quae divi genitoris nostri constituta legibus, adtineri.' (transl. by Pharr, p. 178, altered).

76 On the background see Maria Cesa and Hagith Sivan, 'Alarico in Italia: Pollena e Verona', in *Historia* 39 (1990), pp. 361*ff.*; Maria Cesa, 'Römisches Heer und barbarische Föderaten. Bemerkungen zur weströmischen Politik in den Jahren 402–412', in *L'armée romaine et les barbares du IIIe au VIIe siècle*, eds. F. Vallet – M. Kazanski (Saint-Germain-en-Laye, 1993), at pp. 204–206; Ralf Scharf, 'Aufrüstung und Truppenbenennung unter Stilicho. Das Beispiel der *Atecotti*-Truppen', in *Tyche* 10 (1995), at pp. 171–6. For a plausible chronology of events and laws, see John Barrie Hall, 'Pollentia, Verona, and the chronology of Alaric's first invasion of Italy', in *Philologus* 132 (1988), at pp. 250–252.

77 On handling the problem of brigandage and nocturnal theft in the post-Roman kingdoms, see Loschiavo, 'Autodifesa', pp. 123–129.

when security appeared to be threatened by robbers and deserters.[78] It thus can hardly surprise us when, according to a seventh-century Frankish formulary in which a king ordered by ban that the local count should assemble all *pagenses* for swearing the oath of 'manlike-fidelity' (*fidelitas et leudesamio*) to the king in the presence of a royal envoy (*missus*), it was the *civitates, vici* et *castella* where the people should assemble and become the king's *leudes* through their oath.[79] It is on the local administrative level where we can observe in some detail how the new rulers of Gaul governed their territories as integrated into an extended military administration and sought to provide for security.

Late Roman *ius militiae* with its manifold implications, both theoretical and practical, may thus help us to understand the functioning of the provisions taken in the late sixth century by the Frankish king Childebert II together with his nobles and *leudes*. This is not to say that there was simply continuity. Rather, it should remind us that public order was often destabilized in the fifth century, and that taking extraordinary measures to cope with *latrones* was an even older tradition of ancient Rome and its provincial communities. Still, what we can observe both in the Roman and the early medieval contexts is a binary way of handling this problem. In both cases the means and modes employed seem to have operated on two levels: on the one hand, there were methods and measures to persecute theft on a local level that followed established rules of proof which could turn out to be favourable for a culprit, and, on the other hand, there were more *militia*-based chases and procedures presumably directed against robbers and deserters, characterized by a much more rigid method of proof that allowed the persecuting posse to make short work of suspects. Consequently, the first model is typical for a milieu where social control was tighter and where it was easier to arrest a suspect as he was resident within the community or close by. The second model was more appropriate when a suspect was not an inhabitant of the community, but non-local, mobile, ready to use violence, often belonging to a group of like-minded people – and thus treated as an 'outlaw', both socially and with regard to legal procedure. To find this binary model practiced both in the Roman and the Frankish period is more important to my mind than to nuance the extent to which the latter may have been inspired by what in particular German scholars for a long time believed to be 'Germanic law'. Beginning with the Frankish kingdoms, *centenae*, hundreds

78 On the laws of 403 and their exegesis see Giuffrè, *Iura e arma*, pp. 114–124, and Loschiavo, 'Autodifesa', pp. 114–118.

79 *Marculfi Formulae* I, 40 (ed. Zeumer, p. 68).

and 'Zenten',[80] as rooted in Roman military models, would have a significant and long-term impact on how medieval authorities sought to deal with the problem of *latrones* and *mali homines*, later called 'landschädliche Leute' in German.[81] In fact, well-established legal concepts such as the 'handhafte Tat' or the 'Handhaftverfahren',[82] as well as the 'conviction of robbers by seven' (*Übersiebnungsverfahren gegen landschädliche Leute*),[83] in German historiography traditionally developed as an emanation of 'Germanic law' and believed to be ultimately rooted in the practice of blood-feud,[84] deserve to be reconsidered from this vantage point.[85]

6 Conclusion

The establishment of Frankish rule over what had formerly been the Roman provinces of Gaul and Germany went hand in hand with a militarization of these kingdoms' mixed societies and their administrations. While in many places Roman *castra*, fortified cities and villages, became *nuclei* of medieval local life,[86] the *centenae* or *hundred* illustrate this process from a special perspective. A unit superimposed on local districts which appear to have existed in the provinces of Gaul and Germany before the Frankish takeover, the *centena*

80 Meinrad Schaab, 'Die Zent in Franken von der Karolingerzeit bis ins 19. Jahrhundert. Kontinuität und Wandel einer aus dem Frühmittelalter stammenden Organisationsform', in *Histoire comparée de l'administration (IVe–XVIIIe siècles)*, eds. W. Paravicini – K.F. Werner (München, 1980), pp. 345*ff.*

81 Wolfgang Sellert, 'Landschädliche Leute', in *Handwörterbuch zur deutschen Rechtsgeschichte* (*HDR*), 2nd edition, 3 (2016), col. 578*ff.*

82 Wolfgang Schild, 'Handhafte Tat', in *Handwörterbuch zur deutschen Rechtsgeschichte* (*HDR*), 2nd ed., 2 (2012), col. 741*ff.* For some criticism see Thomas Sergej Huck, 'Beobachtungen zur Einleitung des Strafverfahrens in fränkischer Zeit. Am Beispiel von handhafter Tat, Inquisition und Rügeverfahren', in *Hoheitliches Strafen in der Spätantike und im frühen Mittelalter*, ed. J. Weitzel (Köln, 2002), at pp. 194–195.

83 Jürgen Regge, '"Übersiebnen landschädlicher Leute" und "Verfahren auf Leumund" als besondere Prozeßformen gegenüber Fremden?', in *Fremdheit und Reisen im Mittelalter*, eds. I. Erfen – K.-H. Spieß (Stuttgart, 1997), pp. 289*ff.*

84 Franz Beyerle, *Das Entwicklungsproblem im germanischen Rechtsgang*, 1: *Sühne, Rache und Preisgabe in ihrer Beziehung zum Strafprozeß der Volksrechte* (Heidelberg 1915); Hans Planitz, *Handhaft und Blutrache und andere Formen des mittelalterlichen Rechtsganges in anschaulichen Darstellungen* (Leipzig, 1918).

85 See Murray, 'From Roman to Frankish Gaul', pp. 59–64 (on historiography), and Loschiavo, 'Autodifesa'.

86 Michaela Konrad and Christian Witschel, eds., *Römische Legionslager in den Rhein- und Donauprovinzen – Nuclei spätantik-frühmittelalterlichen Lebens?* (München 2011).

illustrates how *militia*-like structures became a central feature of the inner organization of the Frankish kingdoms, in particular in certain rural areas. In establishing *centenae*, the Merovingian kings apparently adapted models used in different parts of the Roman Empire for the maintenance of peace, the upkeeping of security and the persecution of robbers and deserters. In comparison with other kingdoms where *centenarii* as military officers are attested, but not spatial units called *centenae*, this illustrates a peculiar feature of the Frankish kingdoms, in which the dividing line between the military and the civil population appears to have been abandoned in the northern and eastern frontiers regions quite early. As it seems, this paved the way for the emergence of *militia*-like structures imposed on the countryside, which also served as a recruitment base for armies raised by the Franks.

The task of persecuting wrongdoers who were not resident in local communities was an awkward one. If a *centena* succeded in catching a *latro*, they were allowed to practice some sort of summary trial, which seems to have followed models from late Roman military law. The legislation of Childebert II is best understood against the backdrop of a militarized society of former Roman frontier provinces that became reorganized within the Austrasian kingdom of Eastern Francia along the Rhine. Austrasian administration was indeed some sort of extended military organization, as defence against further Germanic and Slavic groups posed a particular challenge here. As in the Bavarian duchy, which was part of the Austrasian kingdom, we observe the adaption of Roman military law in a milieu that was not dominated by Roman law anymore. Large groups of the population, perhaps the majority, were not Romans here, and thus not subject to Roman civil or public law. These groups became integrated into the legal framework of the Austrasian kingdom by the publication of separate law-codes such as *lex Baiuvariorum* or the *lex Ribuaria*, while the oath of fidelity sworn by each adult free man to the Frankish king provided an overarching superstructure which made them acknowledge the king as the legitimate ruler of this kingdom, and obliged them to follow the commands of the extended military administration in the hands of dukes, counts, *centenarii* and others. The oath of fidelity, the royal ban, the *centenae* – they were all part of a political and legal system which allowed for the effective establishment of a new political entity on the remains of Roman civilization. It is this constellation in which the reception and adaption of Roman military law made most ample sense. As it seems, the Merovingian rulers sought to create a synthesis of several layers of Roman and Frankish legal traditions. The processes of transformation that took place in former frontier regions of the Roman Empire were formative in the long-term establishment of the Frankish kingdom. Legislators such as King Childebert II even sought to extend the scope of their legislation, for instance by ordering *centenae* to persecute individuals accused of *raptus*. It is here where we also observe church asylum as a refuge, which allowed the local bishop to be brought into play.[87]

But it also needs to be emphasized that the military superstructure both in the legal and administrative handling of such matters was by no means solely a top-down affair. The 'communitarian' features of the *centena*, visible in the joint liabilities for damage, certainly functioned as an incitement to make these structures work.[88] Still, they can only be explained if we take the underlying horizontal structures of the *centenae* and their rootedness in pre-Frankish conditions seriously.[89]

Bibliography

Primary Sources

Bündner Urkundenbuch, vol. 1 *390–1199*, ed. E. Meyer-Marthaler – F. Perret (Chur 1963)

Capitularia regum Francorum, ed. A. Boretius – V. Krause, 2 voll. [MGH, *Leges* II.1–2] (Hannover 1883/1897)

Codex Theodosianus: *Theodosiani libri XVI cum constitutionibus Sirmondianis et leges Novellae ad Theodosianum pertinentes*, ed. Th. Mommsen – P.M. Meyer, 2 vols. (Berlin, 1905) – Engl. transl. by C. Pharr *The Theodosian Code and Novels and the Sirmondian Constitutions. A Translation with commentary, glossary, and bibliography*, [Princeton, 1952], 10th Print (Union, N.J., 2001)

Decretio Childeberti II (ed. in *Capit. reg. Franc. I*)

Decretio Chlotarii regis (ed. in *Capit. reg. Franc. I*)

Digestum: Iustinianus, *Digesta*, ed. T. Mommsen – P. Krüger, *Corpus Iuris Civilis*, I (Berlin, 1963^{17}) – Engl. Transl. by A. Watson, *The Digest of Justinian*, 4 vols. (Philadelphia, Pennsylvania, 1985)

Flodoardus Remensis, Historia Remensis ecclesiae, ed. M. Stratmann [MGH, *Scriptores* 36] (Hannover 1998)

Gregory of Tours, *Libri historiarum X*, ed. B. Krusch – W. Levison [MGH, *Script. Rer. Meroving.* I.1] (Hannover 1951) – Engl. transl. by L. Thorpe, *Gregory of Tours, The history of Franks* (London 1974)

87 *Decretio Childeberti II*, c. 4 (ed. in *Capit. reg. Franc. I*, no. 7, p. 16).

88 See the important observations on medieval regulations by Landwehr, *Genossenschaftliche Rechtsverfolgung*.

89 I should like to thank Albert Fenton for correcting the English of my article and for bibliographical advice.

Lex Baiuvariorum, ed. E. von Schwind [MGH, *Leges nat. Germ.*, V,2] (Hannover, 1926) – Engl. transl. by Th.J. Rivers, *Laws of the Alamans and Bavarians* (Philadelphia, 1977)

Lex coloniae Iuliae genetivae (ed. S. Riccobono, in Fontes Iuris Romani Antejustiniani, I, *Leges*, no. 21 [Firenze 1968], pp. 177–198

Leges Burgundionum, ed. L.R. de Salis [MGH, *Leges nat. Germ.* II.1] (Hannoverae, 1892)

Marculfi Formulae, ed K. Zeumer, in *Formulae Merowingici et Karolini aevi* [MGH, *Formulae*] (Hannover 1886), pp. 32–112.

Tacitus, *Germania*, ed. A.A. Lund, *The Germania* (Heidelberg 1988)

Secondary Literature

Baltrusch, Ernst, Ignacio Czeguhn, Stefan Esders, Hans Kopp, Cosima Möller, Manfred G. Schmidt and Christian Wendt, 'Infrastructures, Mobility and Water Management. The Iberian Peninsula in the Imperial and Post-Imperial Context', in *eTopoi. Journal for Ancient Studies* 6 (2016), pp. 220–249 [http://edition-topoi.org/publications/details/space-and-knowledge]

Bayerle, Katrin. 'Einsatzfelder des weltlichen Bannes im Frühmittelalter', in *Von den leges barbarorum bis zum ius barbarum des Nationalsozialismus. Festschrift für Hermann Nehlsen zum 70. Geburtstag*, eds. H.-G. Hermann *et al.* (Köln 2008), pp. 13–34

Beyerle, Franz. *Das Entwicklungsproblem im germanischen Rechtsgang*, 1: *Sühne, Rache und Preisgabe in ihrer Beziehung zum Strafprozeß der Volksrechte* (Heidelberg 1915)

Bothe, Lukas. 'From Subordination to Integration: Romans in Frankish Law', in *Transformations of Romanness. Early Medieval Regions and Identities*, eds. W. Pohl – C. Gantner – C. Grifoni – M. Pollheimer-Mohaupt (Berlin, 2018), pp. 345–368

Botta, Fabio and Luca Loschiavo, eds., *Civitas, iura, arma. Organizzazioni militari, istituzioni giuridiche e strutture sociali alle origini dell'Europa* (*sec. III–VIII*) (Lecce 2015)

Bruns, Carl Georg. 'Die sieben Zeugen des römischen Rechts (1877), in Id., *Kleinere Schriften*, vol. 2 (Weimar, 1882), pp. 119–138

Cam, Helen Maud. *Local Government in Francia and England. A Comparison of the Local Administration and Jurisdiction of the Carolingian Empire with that of the West Saxon Kingdom* (London 1912)

Cam, Helen Maud. 'The "Private" Hundred in England before the Norman Conquest', in Ead., *Law-Finders and Law-Makers in Medieval England* (London 1962), pp. 59–70

Cam, Helen Maud. '*Manerium cum hundredo*: The Hundred and the Hundredal Manor' (1932), in Ead., *Liberties and Communities in Medieval England. Collected Studies in Local Administration and Topography* (London 1963), pp. 64–90

Cesa, Maria – Sivan, Hagith. 'Alarico in Italia: Pollena e Verona', in *Historia* 39 (1990), pp. 361–374

Cesa, Maria. 'Römisches Heer und barbarische Föderaten. Bemerkungen zur weströmischen Politik in den Jahren 402–412', in *L'armée romaine et les barbares du IIIe au VIIe siècle*, eds. F. Vallet – M. Kazanski (Saint-Germain-en-Laye, 1993), pp. 203–217

Eckhardt, Wilhelm Alfred. 'Die *Decretio Childeberti* und ihre Überlieferung', in *Zeitschrift der Savigny-Stiftung für Rechtsgeschichte, Germ. Abt.* 84 (1967), pp. 1–71

Esders, Stefan. 'Les implications militaires du serment dans les royaumes barbares (Ve–VIIe siècles)', in *Oralité et lien social au Moyen Âge (Occident, Byzance, Islam): Parole donnée, foi jurée, serment*, eds. M.-F. Auzepy – G. Saint-Guillain (Paris, 2008), pp. 17–24

Esders, Stefan. 'Rechtliche Grundlagen frühmittelalterlicher Staatlichkeit', in *Der Mittelalterliche Staat – Europäische Perspektiven*, eds. W. Pohl – V. Wieser (Vienna, 2009), pp. 423–432

Esders, Stefan. 'Zur Entwicklung der politischen Raumgliederung im Übergang von der Antike zum Mittelalter: Das Beispiel des *pagus*', in *Politische Räume in vormodernen Gesellschaften. Gestaltung – Wahrnehmung – Funktion*, eds. O. Dally – F. Fless – R. Hänsch – F. Pirson – S. Sievers (Berlin, 2013), pp. 195–211

Esders, Stefan. 'Spätantike und frühmittelalterliche Dukate. Überlegungen zum Problem historischer Kontinuität und Diskontinuität', in *Die Anfänge Bayerns. Von Raetien und Noricum zur frühmittelalterlichen Baiovaria*, eds. H. Fehr – I. Heitmeier, 2nd edition (St. Ottilien, 2014), pp. 425–462

Esders, Stefan. 'Late Roman Military Law in the Bavarian Code', in *La forge du droit. Naissance des identités juridiques en Europe (IVe–XIIIe siècles)*, Clio@Themis, 10 (March 2016) [http://www.cliothemis.com/IMG/pdf/3-_Esders-2.pdf]

Esders, Stefan. 'Amt und Bann: Weltliche Funktionsträger (*centenarii, vicarii*) als Teil ländlicher Gesellschaften im Karolingerreich', in *Kleine Welten. Ländliche Gesellschaften im Karolingerreich*, eds. T. Kohl – S. Patzold – B. Zeller (Ostfildern, 2019), pp. 255–307

Esders, Stefan. 'The Church as a Governance Actor in a Period of Post-Imperial Transition: Delegation of Fiscal Rights and Legal Change in tenth-century Churraetia', in *Medieval Worlds* 10 (2019), pp. 17–45 [https://www.medievalworlds.net/medievalworlds_no10_2019?frames=yes]

Esders, Stefan. 'The "Staffelsee inventory". Carolingian manorial economy, mobility of peasants, and 'pockets of functional continuity' in the transition from Antiquity to the Middle Ages', in *Journal of European Economic History* 49 (2020), pp. 206–250 [https://www.jeeh.it/articolo?urn=urn:abi:abi:RIV.JOU:2020;2.207]

Esders, Stefan. 'Man-like discipline and loyalty against the "enemies of God": Some observations on the militarized frontier society of Eastern Francia around 600', in *Early Medieval Militarisation*, eds. E. Bennett – G.M. Berndt – S. Esders – L. Sarti (Manchester, 2021), pp. 181–195

Esders, Stefan – Bassetti, Massimiliano – Haubrichs, Wolfgang. *Verwaltete Treue. Ein oberitalienisches Originalverzeichnis mit den Namen von 174 vereidigten Männern aus der Zeit Lothars I. und Ludwigs II.* (Roma 2024, in press)

Giuffrè, Vincenzo. *Iura e arma. Intorno al VII libro del Codice Teodosiano* (Napoli, 1979)

Giuffrè, Vincenzo. '*Militum disciplina* e *ratio militaris*', in *Aufstieg und Niedergang der römischen Welt* (*ANRW*) II, 13 (1980), pp. 234–277

Giuffrè, Vincenzo. '*Latrones desertoresque*', in *Labeo* 27, pp. 214–218

Glatthaar, Michael. 'Der Edictus Chilperichs I. und die Reichsversammlung von Paris (577)', in *Deutsches Archiv für Erforschung des Mittelalters* 73 (2017), pp. 1–74

Goebel, Julius. *Felony and Misdemeanour. A Study in the History of English Criminal Procedure* (New York, 1937)

Hall, John Barrie. 'Pollentia, Verona, and the chronology of Alaric's first invasion of Italy', in *Philologus* 132 (1988), pp. 245–257

Hennig, Dieter. 'Nyktophylakes, Nyktostrategen und die παραφυλακή της πόλεως', in *Chiron*, 32 (2002), pp. 281–295

Homoth-Kuhs, Clemens. *Phylakes und Phylakon-Steuer im griechisch-römischen Ägypten: Ein Beitrag zur Geschichte des antiken Sicherheitswesens* (München – Leipzig, 2005)

Huck, Thomas Sergej. 'Beobachtungen zur Einleitung des Strafverfahrens in fränkischer Zeit. Am Beispiel von handhafter Tat, Inquisition und Rügeverfahren', in *Hoheitliches Strafen in der Spätantike und im frühen Mittelalter*, ed. J. Weitzel (Köln, 2002), pp. 191–210

Kaiser, Anna Maria. 'Die Fahndung nach Deserteuren im spätantiken Ägypten', in *Actes du 26e Congrès international de papyrologie*, ed. P. Schubert (Génève, 2012), pp. 381–390

Konrad, Michaela and Christian Witschel, eds., *Römische Legionslager in den Rhein- und Donauprovinzen – Nuclei spätantik-frühmittelalterlichen Lebens?* (München 2011)

Landwehr, Götz. 'Genossenschaftliche Rechtsverfolgung im Mittelalter', in *Studien zur Rechtsschutzversicherung in europäischen Ländern und in den Vereinigten Staaten*, ed. H. Möller (Karlsruhe, 1975), pp. 45–78

Liebs, Detlef. *Römische Jurisprudenz in Gallien (2. bis 8. Jahrhundert)* (Berlin 2002)

Loschiavo, Luca. 'Autodifesa, vendetta, repressione poliziesca. La lotta al brigantaggio nel passagio dalle province tardo-imperiali ai regni romano-barbarici', in *Il diritto Giustinianeo fra tradizione classica e innovazione*, ed. F. Botta (Torino, 2003), pp. 105–133

Loyn, Henry R. 'The Hundred in England in the Tenth and Early Eleventh Centuries', in *British Government and Administration. Studies presented to S. B. Chrimes*, eds. H. Hearder – H.R. Loyn (Cardiff, 1974), pp. 1–15

MacMullen, Ramsay. 'The Roman Concept Robber-Pretender', in *Revue internationale des droits de l'antiquité* (RIDA), 3. sér. 10 (1963), pp. 221–225

Manfredini, Arrigo D. '*Municipii* e città nella lotta ai *latrones*', in *Annali dell'Università di Ferrara, Sezione V: Scienze giuridiche*, NS 6 (1992), pp. 23–34

Molyneaux, George. *The Formation of the English Kingdom in the Tenth Century* (Oxford, 2015)

Murray, Alexander Callander. 'The position of the *grafio* in the constitutional history of Merovingian Gaul', in *Speculum* 61 (1986), pp. 787–805

Murray, Alexander Callander. 'From Roman to Frankish Gaul: *Centenarii* and *centenae* in the administration of the Merovingian kingdom', in *Traditio* 44 (1988), pp. 59–100

Murray, Alexander Callandar. '*Pax et disciplina*. Roman public law and the Merovingian state', in *From Roman Provinces to Medieval Kingdoms*, ed. T.F.X. Noble (London, 2006), pp. 376–388

Nehlsen, Hermann. 'Entstehung des öffentlichen Strafrechts bei den germanischen Stämmen', in *Gerichtslauben-Vorträge. Freiburger Festkolloquium zum 75. Geburtstag von Hans Thieme*, ed. K. Kroeschell (Sigmaringen, 1983), pp. 3–16

Neumann, Alfred Richard. '*Disciplina militaris*', in *Pauly-Wissowa, Realencyclopädie der classischen Altertumswissenschaft*, Suppl. X (1965), pp. 142–178

Planitz, Hans. *Handhaft und Blutrache und andere Formen des mittelalterlichen Rechtsganges in anschaulichen Darstellungen* (Leipzig, 1918)

Regge, Jürgen. '"Übersiebnen landschädlicher Leute" und "Verfahren auf Leumund" als besondere Prozeßformen gegenüber Fremden?', in *Fremdheit und Reisen im Mittelalter*, eds. I. Erfen – K.-H. Spieß (Stuttgart, 1997), pp. 289–298

Riess, Werner. *Apuleius und die Räuber. Ein Beitrag zur historischen Kriminalitätsforschung* (Stuttgart, 2001)

Rüegger, Hans. *Einflüsse des römischen Rechtes in der Lex Burgundionum* (Lungern, 1949)

Schaab, Meinrad. 'Die Zent in Franken von der Karolingerzeit bis ins 19. Jahrhundert. Kontinuität und Wandel einer aus dem Frühmittelalter stammenden Organisationsform', in *Histoire comparée de l'administration (IVe–XVIIIe siècles)*, eds. W. Paravicini – K.F. Werner (München, 1980), pp. 345–362

Scharf, Ralf. 'Aufrüstung und Truppenbenennung unter Stilicho. Das Beispiel der *Atecotti*-Truppen', in *Tyche* 10 (1995), pp. 161–178

Schild, Wolfgang. 'Handhafte Tat', in *Handwörterbuch zur deutschen Rechtsgeschichte* (*HDR*), 2nd ed., 2 (2012), col. 741–748

Schmitz, Winfried. 'Die spätantiken und frühmittelalterlichen Grabinschriften in Köln (4.–7. Jahrhundert)', in *Kölner Jahrbuch* 28 (1995) pp. 643–776

Schröder, Richard and Eberhard von Künßberg. *Lehrbuch der deutschen Rechtsgeschichte*, 7th edition (Berlin, 1932)

Sellert, Wolfgang. 'Landschädliche Leute', in *Handwörterbuch zur deutschen Rechtsgeschichte* (*HDR*), 2nd edition, 3 (2016), col. 578–581

Shaw, Brent D. 'Bandits in the Roman Empire', in *Past and Present* 105 (1984), pp. 3–52

Shaw, Brent D. 'Der Bandit', in A. Giardina, ed., *Der Mensch der römischen Antike* (Frankfurt am Main, 1991), pp. 337–381

Springer, Matthias. 'Jährliche Wiederkehr oder ganz anderes: Märzfeld oder Marsfeld?', in *Rhythmus und Saisonalität*, eds. P. Dilg – G. Keil – D.-R. Moser (Sigmaringen, 1995), pp. 297–324

Springer, Matthias. '*Riparii* – Ribuarier – Rheinfranken nebst einigen Bemerkungen zum Geographen von Ravenna', in *Die Franken und die Alemannen bis zur 'Schlacht bei Zülpich'* (*496/97*), ed. D. Geuenich (Berlin – New York, 1998), pp. 200–269

Ubl, Karl. *Köln im Frühmittelalter. Die Entstehung einer heiligen Stadt,* 400–1100 (Köln 2022).

von Schwerin, Claudius. *Die altgermanische Hundertschaft* (Breslau, 1907)

Valsan, Remus. '*Fides, bona fides,* and *bonus vir.* Relations of Trust and Confidence in Roman Antiquity', in *Journal of Law, Religion and State*, 5 (2017), pp. 48–85

Voss, Wulf Eckart. 'Vom römischen Provinzialprozess der Spätantike zum Rechtsgang des frühen Mittelalters', in *Recht im frühmittelalterlichen Gallien. Spätantike Tradition und germanische Wertvorstellungen*, eds. H. Siems – K. Nehlsen-von Stryk – D. Strauch (Köln u. a., 1995), pp. 73–108

Zerjadtke, Michael. *Das Amt dux in Spätantike und frühem Mittelalter. Der Ducatus im Spannungsfeld zwischen römischem Einfluss und eigener Entwicklung* (Berlin and Boston 2018)

CHAPTER 8

Soldiers' Marriages: Before and after the Fall of the Empire

Francesco Castagnino

1 The Family Law of the Roman Soldiers in Late Antiquity[1]

It is well known that in Late Antiquity Roman soldiers had the right to marry during military service. This regime was the result of a reform which radically modified the previous legislation. In fact, until 197 CE a marriage ban prohibited all the *milites* from contracting legal marriages and producing legitimate children to whom they could transmit their citizenship (*status civitatis*) and their assets;[2] only the soldiers belonging to particular units, i.e. the *auxiliares*, the *classiarii* and the *milites* serving in the urban troops,[3] received, at the end of service, the *conubium*, a privilege that allowed them to legitimate their relationships with foreign women under Roman law.[4]

A change in such discipline took place under the emperor Septimius Severus. As Herodianus reports (3.8.5), in 197 he granted to all soldiers permission to marry legally:

> τοῖς τε στρατιώταις ἐπέδωκε χρήματα πλεῖστα ἄλλα τε πολλὰ συνεχώρεσεν ἃ μὴ πρότερον εἶχον. καὶ γαρ τὸ σιτηρέσιον πρῶτος ηὔξησεν αὐτοῖς καὶ δακτυλίοις χρυσοῖς χρήσασθαι ἐπέτρεψε γυναιξί τε συνοικεῖν ...[5]

1 This study is the result of a research funded by the "Fondazione Fratelli Confalonieri" with a post-doc research fellowship. I am grateful to the Foundation for its financial support. I also would like to thank Lorena Atzeri, who has helped me in the revision of the English text.

2 On the soldiers' marriage ban see Sara E. Phang, *The Marriage of Roman Soldiers 13 B.C.–A.D. 235: Law and Family in the Imperial Army* (Leiden-Boston-Köln 2001).

3 Specifically, the *urbaniciani*, the *equites singulares Augusti* and the praetorians.

4 From the early second century the soldiers had to receive an honourable discharge (*honesta missio*) to obtain *ius conubii*. This is attested in the military diplomas, i.e. the documents certifying the privileges conferred to the *milites*. On the military diplomas see Francesco Castagnino, *I diplomata militaria. Una ricognizione giuridica* (Milan 2022).

5 Also in other passages Herodianus uses the verb συνοικέω with the sense of marrying (see Hdn., 1.6.4 and Hdn., 1.8.3). On the reform of Septimius Severus, consult Francesco Castagnino, 'Creare una famiglia. Il ius conubii e i figli dei "soldati" prima e dopo la caduta dell'Impero', in *Ravenna Capitale. L'esercito romano e l'alba dell'Europa. Modelli concettuali*

 | DOI:10.1163/9789004698017_009

> To the soldiers he gave a very large sum of money and many other privileges they did not have before; for he was the first to increase their pay and he also gave them permission to wear gold rings and to marry ...[6]

This new regime was still applied in the Late Roman Empire. However, even in that period soldiers could not easily create a family. Indeed, despite the permission given by Septimius Severus, the ban cohabitation between soldiers and their women had never been lifted,[7] as attested by a constitution of Constans I of 349 (C. 12.35.10 [= CTh. 7.1.3]): according to this law, the *milites* who wanted to reunite with their *familiae* (only including their women, children and slaves bought with the *peculium castrense*) needed a special authorization by the emperor.[8]

In order to mitigate the rigour of military discipline, the later Roman emperors conferred some immunities on the soldiers and their wives. This emerges from a constitution reported in the well-known tables of Brigetio and Durostorum:[9]

> AE 1937, 232; AE 2007, 1124: ... sed et hii qui licet post viginti stipendia adaeque honestam missionem adepti fuerint ab annonario titulo duo capita excusent, id est tam suum quam etiam uxoris suae ... ... etiam si intra viginti stipendia ex ea causa rerum suarum vacationem fuerit consecutus, ad beneficium eiusdem indulgentiae nostrae pertineat, ita ut et suum et uxoris suae caput ...

---

e sperimentazioni sociali tra linguaggi, istituzioni e diritto (*secoli IV–VIII*), ed. G. Bassanelli Sommariva (Santarcangelo di Romagna 2020), pp. 119–120.

6 Engl. transl. by Brian Campbell, 'The Marriage of Soldiers under the Empire', in *Journal of Roman Studies*, 68 (1978), p. 160, revised by me.

7 On this prohibition see Pietro Tassistro, 'Il matrimonio dei soldati romani', in *Studi e Documenti di storia e diritto*, 22 (1901), pp. 23–25; Jacqueline Vendrand-Voyer, *Normes civiques et métier militaire à Rome sous le Principat* (Clermont-Ferrand 1983), pp. 98–118; Werner Eck, 'Septimius Severus und die Soldaten. Das Problem der Soldatenehe und ein neues Auxiliardiplom', in *In omni historia curiosus. Studien zur Geschichte von der Antike bis zur Neuzeit. Festschrift für Helmuth Schneider zum 65. Geburtstag*, eds. B. Onken – D. Rohde (Wiesbaden 2011), p. 76.

8 Another evidence of the existence of the ban can be found in the diplomas delivered to the soldiers of the *classes praetoriae* since 158. According to the new formula of these documents (see, e.g., RMD I 74), the *civitas* was only granted to the children whom the sailor could prove to be born from the woman he had lived with in a permitted cohabitation (*concessa consuetudo*). The need of a permission to cohabit even after 197 confirms that there was still a ban of cohabitation after this date.

9 Werner Eck has recently identified a third copy of this *constitutio*; see Werner Eck, 'Eine dritte Kopie der Tafel von Brigetio aus dem Jahr 311 n.Chr.', in *Quaderni Lupiensi di Storia e Diritto*, 9 (2019), pp. 53–55.

> ... Those men, however, who have acquired honourable discharge but after only twenty years of service, are to be exempted from taxes in kind in respect of two tax units, that is, each man himself and his wife ... ... even if he has secured release from his duties for that reason within a period of less than twenty years of service, is to be eligible for the benefits of this same indulgence of ours, so that he is to be exempted in respect of his own tax unit and that of his wife ...[10]

This was probably not the only measure adopted in favour of the *milites*.[11] Perhaps even the soldiers' exemption from the usual marriage formalities (granted with a constitution of Valentinian III of 426)[12] was conceived with the purpose of ameliorating their position and that of their families.

However, despite these attempts to mitigate military discipline, the soldiers' matrimonial life continued to suffer some hindrances. It must be observed that even after the withdrawal of the marriage ban a *miles* could not marry a woman with whom he did not have *conubium*. This privilege was only granted to him at the end of his term of service, as attested by diplomas.

The Antonine Constitution of 212, which extended Roman citizenship to nearly all free-born residents of the Roman Empire, undoubtedly represented a turning point also in the matter of matrimonial law. Indeed, after the promulgation of this constitution, the *ius conubii* lost much of its significance,[13] and this must have also affected most of the soldiers.

Nevertheless, in spite of the general extension of the *civitas Romana*, some problems still persisted. For example, it was not possible to contract a legal marriage with *Latinae Iunianae* (a *status* formally abolished only by Justinian, many centuries later) without the *ius conubii*. The problem of *conubium* also affected those who were still *peregrini*. It is important to remember that the edict of Caracalla produced its effect only on those subjects who, at the time of its enactment, derived their origin (*origo*) from a community within the Roman Empire (D. 1.5.17, Ulp. 22 *ad ed.*). Therefore, those who had settled in the Roman territory after 212 retained their *status* as foreigners.

---

10 Engl. transl. by Brian Campbell, *The Roman Army 31 BC – AD 337: A Source Book* (London-New York, 1994), pp. 241–244.

11 On the immunities of the soldiers and their wives see also CTh. 7.20.4 (a. 325) and CTh. 7.13.6 (a. 370).

12 C. 5.4.21. According to this *constitutio*, the soldiers' privilege concerned the marriages contracted with the women *ingenuae*.

13 On this point see, among others, Paul Jörs – Wolfgang Kunkel, *Römisches Privatrecht*, 3rd ed. (Berlin 1949), p. 273; Göran Lind, *Common Law Marriage. A Legal Institution for Cohabitation* (Oxford 2008), p. 35 nt. 24.

This is the reason why, even after the *edictum* of Caracalla, the imperial administration perpetuated the practice of granting *conubium* to the veterans of some units, for example to the *classiarii* of the praetorian fleets (until 249/250)[14] to the *equites singulares Augusti* and to the *praetoriani* (respectively until 237/238 and 306).[15]

2 The Marriage Ban between Romans and Barbarians

The members of those barbarian *gentes* that, over the years, were authorized to settle in the Roman Empire were assigned the *status* of *peregrini*. It is known that, on the basis of how they were permitted to settle, barbarian immigrants were grouped in different categories, namely those of *dediticii*, *foederati*, *gentiles* and *laeti gentiles*.[16]

Among these groups, a large number of soldiers was usually recruited in the Roman army. After Constantine, Barbarians enrolled primarily in the *scholae palatinae* (i.e. the Imperial Guard), or in the palatine *auxilia*. However, some of them were also recruited into the Roman legions, i.e. military units normally reserved to the *cives Romani*. From the reign of Valens onwards, especially after a *constitutio* of 375 fixed at 36 solidi the price for every recruit (CTh. 7.13.7), or *tiro*, the *tirones cives Romani* were often replaced with barbarian 'mercenaries'.

All those subjects, and, above all the *foederati*, did not have the right to legally marry Roman women. This lack of *conubium* between Barbarians and *cives Romani* emerges from a well-known constitution issued by Valentinian I between 370 and 373 (CTh. 3.14.1 [= Brev. 3.14.1]).[17] This *constitutio* stated two clear prohibitions, both punished with the *poena capitis*: on the one hand, it forbade the marriage of a *provincialis* (i.e. a Roman citizen of the province) with a barbarian woman; on the other hand, it prohibited any relationship between a *provincialis* and a *gentilis*. About the meaning of this last term,

---

14 See CIL XVI 154.

15 See AE 2015, 1880 and RMD I 78.

16 On the *laeti gentiles* see Valerio Marotta, 'Il problema dei laeti. Fonti e storiografia', in *Civitas, Iura, Arma. Organizzazioni militari, istituzioni giuridiche e strutture sociali alle origini dell'Europa* (*secc. III–VIII*), eds. F. Botta – L. Loschiavo (Lecce 2015), esp. p. 130.

17 On this *constitutio* see, among many others, Mariagrazia Bianchini, 'Ancora in tema di unioni tra barbari e romani', in *Atti dell'Accademia Romanistica Costantiniana. VII Convegno internazionale*, eds. G. Crifò – S. Giglio (Napoli 1988), pp. 225–249; Alessandro Cusmà Piccione, *"Non licet tibi alienigenam accipere". Studio sulla disparitas cultus tra i coniugi nella riflessione cristiana e nella legislazione tardoantica* (Messina 2017), pp. 424–465.

several hypotheses have been advanced. Some scholars believe that the word *gentilis* refers to Barbarians enrolled into Roman military units and headquartered in a certain area of the empire.[18] Others suppose that this term indicates the Barbarians in a wide sense,[19] adducing in support of this hypothesis the provisions of other *constitutiones* (for example, CTh. 12.12.5 [a. 364]), where *gentilis* is used in the sense of *barbarus*. Along with this consideration, it must be observed that the drafters of CTh. 3.14.1 speak of *barbarae* in a general way (this suggests that even the word *gentilis* has a wider connotation). In addition, in the *interpretatio* of the *Breviarium Alaricianum* those affected by the ban are designated as *barbari* and *Romani*.[20]

The regime here described was probably applied also in the following centuries. Indeed, a fragment of the *Institutiones* of Justinian (I. 1.10 pr.) attests that, still in the 6th century, the right to contract *iustae nuptiae* was only held by *cives Romani*:

> Iustas autem nuptias inter se cives Romani contrahunt, qui secundum praecepta legum coeunt, masculi quidem puberes, feminae autem viripotentes …
>
> Roman citizens are joined together in lawful wedlock when they are united according to law, the man having reached years of puberty, and the woman being of a marriageable age …[21]

3 *The Ius Conubii* of Barbarian Soldiers

Once verified the existence of a marriage ban between Romans and Barbarians, we have to investigate whether the Barbarians (as well as the *peregrini* during the Principate) could obtain in some way the privilege of *conubium*.

This topic is strictly connected with the issue of the enlistment in the army, since military service allowed access, even in Late Antiquity, to Roman

18 Rosario Soraci, *Ricerche sui conubia tra Romani e Germani IV–VI s.*, 2nd ed. (Catania 1974), pp. 105–106.

19 Marotta, 'Il problema dei laeti', p. 137 n. 63.

20 Another evidence of the marriage prohibition between Romans and Barbarians is a fragment of Eunapius (fr. 59, ed. Blockley, p. 85) related to the case of the Goth Fravitta. At 392 or 393 he asked the emperor for a permission of marrying a Roman woman. Therefore there could be a marriage ban between Goths and Romans.

21 Engl. transl. by J.B. Moyle, *The Institutes of Justinian*, 5th ed. (1913; repr. Union, New Jersey 2002), p. 12.

citizenship and, consequently, to the right of *conubium* with Roman women. Unfortunately, in the absence of the precious witness borne by the *diplomata militaria*, we have to rely on other evidences. One of them is the use of the *nomen gentilicius* Flavius in the onomastic formula of many officers and soldiers of barbarian origin. Besides the well-known inscriptions of Concordia, this *gentilicius* occurs in several epigraphs and in sources from other areas[22] as well as in some papyri from Egypt.[23] As Mommsen observed,[24] the spread of the name Flavius among the Barbarians would indicate their full integration in the *civitas* after Constantine and his successors.

Other elements denoting the persistence in Late Antiquity of the use of granting the *civitas Romana* to the *milites* emerge from the examination of their privileges. According to some constitutions issued under Constantine (CTh. 7.20.2 [a. 320?][25] [= C. 12.46.1]; 7.20.3 [a. 320];[26] 7.20.4 [a. 325]),[27] soldiers also received fiscal privileges upon discharge.[28] With regard to CTh. 7.20.4, Émilienne Demougeot[29] believes that the exoneration from one personal *caput*, obtained by some categories of veterans here mentioned, would indicate that they had received the *civitas* only for themselves and not also for their children. Indeed, this limitation could have been an echo of the same restrictions imposed in the 2nd century on the concessions to some soldiers. It is known that, after 140, the *auxiliares* no longer received citizenship for their children. Another restriction concerned the veterans of the units of *Mauri equites* and the *Palmyreni sagittarii*, who never obtained the *civitas liberorum*

---

22 See, e.g., CIL V 4369 (Fl. Higgo) and CIL III 7494 (Fl. Stercorico).

23 An example is the case of Flavius Agemundus attested by BGU I 316 (a. 359).

24 Theodor Mommsen, 'Ostgotische Studien', in Id., *Gesammelte Schriften*, 6 (Berlin 1910), pp. 476–477.

25 In contrast to Theodor Mommsen, Otto Seeck, *Regesten der Kaiser und Päpste für die Jahre 311 bis 476 n. Chr.*, (1919; repr. Frankfurt a.M. 1984), pp. 60 and 176 supposes that the constitution dates to 326. A different date (315) is suggested by Pierfrancesco Porena, 'Ancora sulla carriera di Flavius Ablabius, prefetto del pretorio di Costantino', in *Zeitschrift für Papyrologie und Epigraphik*, 190 (2014), pp. 265–266.

26 Seeck, *Regesten*, pp. 82 and 175, excluding the date proposed by Mommsen, believes that the constitution dates back to 325.

27 On the three *constitutiones* see Francesco Castagnino, 'I privilegi fiscali di milites e veterani in età costantiniana', in *Atti dell'Accademia Romanistica Costantiniana, 24. Militia inermis e militia armata. Apparati civili e militari nella tarda antichità*, eds. C. Lorenzi – M. Navarra (Perugia 2021), pp. 188–196.

28 Tax privileges were granted to the veterans even under Valentinian and Valens. See CTh. 7.20.8 (a. 364).

29 Émilienne Demougeot, 'Restrictions à l'expansion du droit de cité dans la seconde moitié du IVe siècle', in *Ktema*, 6 (1981), p. 386.

and the *conubium* with the women they married afterwards, but only the *civitas* for themselves.[30]

The concession of Roman citizenship to the soldiers of the Late Roman army could also be confirmed by a famous inscription from Aquincum, in the *Pannonia Inferior* (CIL III 3576):[31]

> Francus ego cives (sic) Romanus miles in armis / egregia virtute tuli bello …
>
> I, a Frank, a Roman citizen, a soldier in arms, with exemplary courage brought to war …[32]

This inscription commemorates a *miles* of Frank origin. Several scholars interpret it as if the soldier, while remembering his military virtues, would assert to be both a *civis Francus* and a *miles Romanus in armis*.[33] In reality, it has to be assumed that its meaning was different. In particular, as some authors correctly thought,[34] the epigraph of Aquincum celebrates a man serving in the Roman army, who obtained the Roman citizenship during or after the required military service. This privilege would have allowed him to legally marry a Roman woman.

One may wonder whether Late Antique Roman soldiers also obtained the right to marry women of barbarian origin. In this respect, I could not find any evidence attesting that even in this period soldiers received *conubium* with women of foreign *status*. In fact, the only cases we know concern barbarian officers who were allowed to marry Roman women (e.g. Flavius Stilicho). This circumstance, however, depends on the fact that those men were already *cives*, as holders of relevant positions in the Roman military administration.

Why barbarian soldiers were not granted *conubium*? One thing is clear: the Late Antique Roman army was no longer perceived as a vehicle of Romanization for people of different origin.

---

30 On this point see also Marotta in this volume.

31 The dating of the inscription is uncertain. According to Soazick Kerneis, 'Francus civis, miles Romanus: les barbares de l'Empire dans le Code Théodosien', in *Droit, religion et société dans le Code Théodosien*, eds. J.-J. Aubert – P. Blanchard (Geneva 2009), p. 390 no. 54, it dates to the end of the 4th century. Marotta (in this volume) maintains instead that the dating varies between the end of the 2nd and the beginning of the 5th century.

32 Engl. transl. by Kent J. Rigsby, 'Two Danubian Epitaphs', in *Zeitschrift für Papyrologie und Epigraphik*, 126 (1999), p. 176.

33 See Kerneis, 'Francus civis', p. 390.

34 Rigsby, 'Two Danubian Epitaphs', pp. 175–176; Andreas Popescu, *Die Franken und das Römische Militär Identität in Gallien und dem Rheinland in der Spätantike* (Bonn 2017), pp.1, 53. Available at https://bonndoc.ulb.uni-bonn.de/xmlui/handle/20.500.11811/7065.

4 The Provisions on Marriage of Some Romano-Barbarian Kingdoms: A) The Visigoth Kingdom and the Reign of Theodoric

On the issue of possession of *conubium* by the Visigoths and the Ostrogoths I can only formulate some vague conjectures, since, especially in the second case, I could not find sufficient evidence to reconstruct the legal regime in force at that time.

As regards the matrimonial discipline among the Visigoths, it would seem that the marriage ban existing between them and the Romans was expressly confirmed by the constitution of Valentinian I mentioned above and included in the *Breviarium Alaricianum* (*Brev.* 3.14.1). We have seen that the *interpretatio* of the *Breviarium* replaced the words *gentiles* and *provinciales* with 'Barbarians' and 'Romans', making more explicit the content of the provision. However, the marriage prohibition didn't last for the whole reign of the Visigoths, but was later withdrawn, as a *lex antiqua*[35] of the *lex Visigothorum* (a. 654) attests:

> Lex Visig., Ant. 3.1.1: Ut tam Goto romanam romano gotam matrimonio liceat sociari.
>
> Ancient Law: That it is permitted both to a Goth to marry a Roman woman and to a Roman man to marry a Gothic woman.[36]

About the grounds of the marriage proscription, it is possible that the introduction of this ban was inspired not only by ethnic reasons, but also by motives of different nature. As for the first explanation, it must be observed that the Visigothic population represented, in the territories where it settled down, an insignificant minority compared to the Romans. So it is likely that the absence of *conubium* with Romans aimed to defend and consolidate the ethnic identity of that *gens*.[37]

However, it is certain that there were also other reasons behind the prohibition. For example, another possible explanation could be found in the

35 This law was probably issued during Leovigild's reign (569–586); see Karl Zeumer in *Leges Visigothorum* [MGH, *Leges nat. Germ.* 1] (Hannover - Leipzig 1902), p. 121, nt. 1.

36 Engl. trans. by Hagith S. Sivan, 'The Appropriation of Roman Law in Barbarian Hands: "Roman-barbarian" Marriage in Visigothic Gaul and Spain', in *Strategies of Distinction: The Construction of Ethnic Communities, 300–800*, eds. W. Pohl – H. Reimitz (Leiden 1998), p. 200.

37 Wolfgang Liebeschuetz, 'Citizen Status and Law in the Roman Empire and in the Visigothic Kingdom', in *Strategies of Distinction: The Construction of Ethnic Communities, 300–800*, eds. W. Pohl – H. Reimitz (Leiden 1998), pp. 140–141; Sivan, 'The Appropriation of Roman Law', p. 202.

different creed to which Romans and Visigoths belonged;[38] in particular, the Romans followed the symbol of the Niceno–Costantinopolitan faith, while the Visigoths professed the Aryan religion. As it is known, Church canons had constantly banned marital unions between Catholics and non–Catholics. In this respect, the can. 67 of the Council of Agde (a. 506), almost coeval with the enactment of the *lex Alariciana*, is particularly relevant:

> Quoniam non oportet cum omnibus hereticis miscere connubia et vel filios vel filias dare, sed potius accipere, si tamen se profitentur christianos futuros esse catholicos.[39]
>
> It is not proper to mix marriages with any heretics, and to give them sons or daughters, but [it is proper] to accept them, if they promise that they are going to become Catholic Christians.[40]

In that context, the catholic bishops of the Gallic kingdom, in reproducing a provision of the Elvira Council,[41] renewed the marriage prohibition between Catholics and heretics. Since Catholics bishops collaborated in the elaboration of the *Breviarium Alaricianum*,[42] it is extremely likely that the marriage ban was inspired by conciliar deliberations.

Let us now consider the situation in the Ostrogothic kingdom. According to several scholars,[43] in Italy Theodoric pursued a policy of strict separation between Goths and Romans, maintaining in force the traditional law of the *populus-exercitus*, as well as the marriage ban between the Goths *foederati* and

---

38 See Ana M. Jiménez Garnica, 'El origen de la legislación civil visigoda sobre la prohibición de matrimonios entre romanos y godos. Un problema de fundamento religioso', in *Anuario de historia del derecho español*, 55 (1985), pp. 742–746.

39 *Concilium Agathense*, Can. 67 (ed. Mansi, 8, col. 336B). Although the council of Agde took place a few months after the enactment of the *Breviarium*, the legal provisions on marriage provided by this council reveal the tendency of canon law to prohibit mixed religious marriages.

40 Engl. transl. by Ralph W. Mathisen – Hagith S. Sivan, 'Forging a New Identity: The Kingdom of Toulouse and the Frontiers of Visigothic Aquitania (418–507)', in *The Visigoths. Studies in Culture and Society*, ed. A. Ferreiro (Leiden 1999), p. 59.

41 *Concilium Eliberitanum*, Can. 16 (ed. Mansi, 2, col. 8D).

42 The *Commonitorium* to the Breviary of Alaric attests that this *lex* was submitted for approval to a assembly of bishops and *electi provinciales*.

43 Adriano Cavanna, 'Diritto e società nei regni Ostrogoto e Longobardo', now in Id., *Scritti (1968–2002)* (Naples 2007), p. 524; Claudio Azzara – Arianna Bonnini, 'I matrimoni fra barbari e romani nelle leggi di stirpe dell'altomedioevo', in *Matrimoni misti: una via per l'integrazione tra i popoli*, ed. S. Marchesini (Verona 2012), pp. 129–140.

the Romans. Such a policy would have caused in the long run the early disintegration of the Gothic rule on Italy.

However – as Luca Loschiavo[44] observes –, some archaeological finds show that this interpretative model only partially reflects the reality of the Germanic kingdom. Alongside those settlements where the Goths lived in isolation from the Romans, there were other communities in which the two *gentes* lived together.[45] In particular, the integration between the two ethnic groups would be more intense among the higher classes. For example, the excavations at the villa of Settime Desana have brought to light the ring of a man named Stefanius (or Stefanus). This was a rich landowner who got married – so it seems – to a woman of the Gothic aristocracy (Valatrud).[46]

According to glottologists, further elements could emerge from the onomastic data, in particular, from the coexistence of Roman and Gothic names in the same families.[47] Indeed, cases of integration and intermingling between the two cultures are also attested before the reign of Theodoric.[48]

Nevertheless, even though such evidences provides pieces of information about the relationships between the two ethnic groups, they do not permit reconstruction of the real legal regime in force among the Ostrogoths. In fact, they only represent some fragments of daily life, from which it is impossible to deduce the rule of law applied to the unions between Ostrogoths and Romans.

With regard to the subject of this research (i.e. the matrimonial discipline of the soldiers), our sources do not allow the conclusion that even the Ostrogothic

44 Luca Loschiavo, 'Insediamenti barbarici e modelli di coesistenza nell'Italia altomedievale. Il regno degli Ostrogoti', in *Immigrazione e integrazione. Dalla prospettiva globale alle realtà locali*, 1, ed. F. Rimoli (Naples 2014), pp. 331–334.

45 As Luca Loschiavo has stated (see his contribution in this volume), we should see the Ostrogothic kingdom in terms of 'dualism' and not of downright 'separatism'.

46 On the archaeological evidences see Marco Aimone, 'Romani e Ostrogoti fra integrazione e separazione. Il contributo dell'archeologia a un dibattito storiografico', in *Reti Medievali – Rivista*, 13.1 (2012), pp. 31–80. [Published online: Mar 21, 2012. DOI: https://doi.org/10.6092/1593-2214/337].

47 See the cases riported by Nicoletta Francovich Onesti, *I nomi degli Ostrogoti* (Florence 2007), p. 22.

48 A good example comes from a Latin inscription dating back to the mid-4th century found near Nicopolis ad Istrum (*Antike Denkmäler in Bulgarien*, ed. Kalinka, p. 323 no. 412). It was made by a soldier called Marcus Aurelius Maximus for his daughter-in-law Amica. The peculiarity of this inscription is that Amica is mentioned with the gothic term *brutes* instead of the Latin word *nurus*. This also happens in other inscriptions found in areas which were German speaking or bordering on German regions (see CIL III 12377; CIL III 12666). These evidences are a clear example of that process of the mingling of cultures to which the Roman army contributed.

and Visigothic soldiers received (as the *milites* during the Principate) the *conubium* with women of different *status*.

The only evidence I was able to find is a text from the *Variae* of Cassiodorus, which refers to a Gothic soldier (Patsa) who was married to Regina, probably a Roman woman (as her Latin name would suggest):

> Cass., *Variae* 5.33.1: Gravis est Patzenis clementiae nostrae sensibus intimata conquestio. Qui se in expeditione Gallica constituto in eum Brandilam prosiluisse testatur excessum, ut uxorem eius Reginam proprio sociandam duceret esse coniugio et in iniuriam nostrorum temporum adulterium simulata matrimonii fuerit lege commissum.
>
> The intimate complaint of Patza weighs on our sense of forbearance. While Patza was absent and engaged in the Gallic campaign, Brandila is reported to have launched an attack against him, so that Brandila led his wife, Regina, to share his company as his own wife and in an adultery injurious to our reign, he legally contracted the semblance of marriage.[49]

No decisive element seems to emerge from another text of the *Variae*, concerning those *antiqui barbari*, who decided to marry Roman women:

> Cass., *Variae* 5.14.6: Antiqui barbari, qui Romanis mulieribus elegerunt nuptiali foedere sociari ....
>
> Let the former barbarians, who have chosen to be associated with Roman women in the bond of marriage ...[50]

This text refers to German people not belonging to the Gothic race, who settled in *Pannonia* before the reign of Theodoric.[51]

As for the Visigoths, the only evidence available to us is a passage from the Gothic Wars of Procopius related to a general named Theudis. This officer was sent by Theodoric to Spain, where he later became king and married a Roman woman from the region.

49 Engl. transl. by M.S. Bjornlie, *The Variae. The Complete Translation* (Oakland, CA, 2019), pp. 229–230.

50 Engl. transl. by Bjornlie, *The Variae*, p. 215.

51 On this point see Herwig Wolfram, *Die Goten: Von den Anfängen bis zur Mitte des sechsten Jahrhunderts*, 5th ed. (Munich 2009), p. 301.

> Proc., *De bel. got.*, 1.12.50: Μετὰ δὲ Θεῦδις, Γότθος ἀνήρ, ὅνπερ Θευδέριχος τῷ στρατῷ ἄρχοντα ἔπεμψε, γυναῖκα ἐξ Ἱσπανίας γαμετὴν ἐποιήσατο, οὐ γένους μέντοι Οὐισιγότθων, ἀλλ' ἐξ οἰκίας τῶν τινος ἐπιχωρίων εὐδαίμονος …
>
> But afterwards, Theudis, a Goth, whom Theodoric had sent as commander of the army, took to wife a woman from Spain; she was not, however, of the race of the Visigoths, but belonged to the house of one of the wealthy inhabitants of that land …[52]

In this case, an exception from the marriage ban could be made in order to permit a political alliance between the Gothic and the Roman elites.

In the absence of other elements, we could ask whether, in these Romano-Barbaric kingdoms, the provisions on matrimony were influenced by Roman military law in Late Antiquity.

We can notice in the Roman Empire of the 4th and 5th century a strong correlation between recruitment and possession of land. Indeed, the fiscal system conceived the enlistment as a sort of tax that each landowner had to pay to the imperial administration by giving a certain number of recruits.[53] The recruitment also depended on the contingents, given by the communities, of *gentiles* and *laeti gentiles* who had settled in lands belonging to the *res privata*. As Max Weber[54] observed, the settlements of these people reinforced the general tendency to connect personal obligations to provide a service with real rights.

The organizational models employed to govern these communities were similar to those of other *corporati* or *collegiati*[55] designated for carrying out activities of public interest. So, the military administration in charge of their settlement also ruled on fundamental aspects of their family relationships, namely property, marriage, and the choice between revenge or payment of the *wergeld* (both involving an active and passive solidarity of every family group). In particular, they took care of the last of those aspects, but they also did not neglect the first, especially when this was interfering with the other, as it was the case with the hereditary transmission of the land, with the need to provide a regular flow of young soldiers to the imperial army. For this reason the lands of these communities were reserved to male descendants, who were the only ones who could be recruited by the *dilectus*.[56]

---

52 Engl. transl. by H.B. Dewing, *Procopius with an English translation*, 3. *History of the wars, Books V and VI* (London, 1919), p. 131.

53 Yann Le Bohec, *Armi e guerrieri di Roma antica. Da Diocleziano alla caduta dell'impero*, (Ital. transl.) (Rome 2008), p. 81.

54 Max Weber, *Storia agraria romana*, (Ital. transl.) (Milan 1967), pp. 172–173.

55 These were groups of artisans and urban traders organized in corporations called *collegia*.

56 Marotta, 'Il problema dei laeti', pp. 152–156.

With regard to marriage, we can see that, in accordance with a constitution of Libius Severus of 25 September 465,[57] the *laeti* (as well as the *collegiati*) had to get married with women of the same *status*, namely with women of Letic condition or *ingenuae* with a different *origo*.

One may wonder if the Ostrogothic soldiers, especially those not belonging to aristocratic clans, were subject to disciplines similar to those established for the Late Roman *corporati*. Our sources do not allow us to answer this question. Nevertheless, as the administrative apparatus of the Late Roman Italy was not changed by Theodoric, we can believe that the faculty of the Ostrogoths of bequeathing their lands was constantly limited by the obligation, connected with a real right, to provide the continuity of the recruits flow.[58] Therefore, even the inter-ethnic marriages could be tolerated if they did not interfere with this strategic policy for the conservation of the reign.

5 The Provisions on Marriage of Some Romano-Barbarian Kingdoms: B) The Burgundians

The problem of the *ius conubii* with Roman citizens never affected the soldiers of the Burgundian army. Indeed, the legislation of this *gens* (in contrast to the laws of other barbarian populations) did not impose any prohibition of marriages between different races. On the contrary, the rules of Burgundian law show that intermarriages were usually taking place. Some provisions contained in the *Liber constitutionum*, a compilation of Burgundian constitutions,[59] are particularly significant in this regard.

A clause of the Book of Constitutions (XLIV) judged as intolerable a sexual relationship contracted secretly and before the marriage by a Burgundian woman and a man *seu barbarus seu romanus*.[60] From this provision it emerges that the different ethnicity of the lovers was irrelevant for the determination of the sanction, as both the barbarian and the Roman partner had to pay 15 solidi as a penalty. Indeed, such a rule only punished the relationship between

57 Nov. Sev. 2.1.

58 On the lands assigned to the Ostrogohts see Pierfrancesco Porena in this volume (§ 6).

59 On this compilation see Ian Wood, 'The Legislation of Magistri Militum: the laws of Gundobad and Sigismund', in *Clio & Themis. Revue électronique d'histoire du droit*, 10 (2016), pp. 1–16 [Published online: Jun 10, 2021. DOI: https://doi.org/10.35562/cliothemis.1191]; Id., 'Burgundian law-making, 451–534', in *Italian Review of Legal History*, 3 (2017), pp. 2–11 and his contribution in this volume.

60 *Liber Constitutionum* XLIV (ed. de Saliis, p. 74).

a man and a woman who were not united in marriage, while the fact that the man belonged to the same race of the woman or not was of no consequence.

The same consideration also applies for the union contracted by a *mulier burgundia vel romana* without the consent of the parents. The consequence of this action was the same in both cases, or rather, the husband of the woman would exercise the *potestas* on her and her estate.[61]

Another provision to be considered is *Liber Constitutionum* XII, 5. This norm regulated the case of a Roman woman who got married with a Burgundian *sine parentum suorum voluntate*,[62] thus recognizing the possibility for the Burgundians to marry people of different *status*.

The legitimacy of inter-ethnic marriages also emerges from another Burgundian collection, the *Lex Romana Burgundionum* (or *Forma et Expositio legum*). Title XXXVII of this *lex* states, in fact, the conditions for having a legitimate marriage (e.g. the agreement of the parents, the *donatio nuptialis*) without requiring that the betrothed belonged to the same ethnicity.[63]

Hence, even in the absence of a rule explicitly authorising intermarriages, the provisions here illustrated show indirectly that the unions between Burgundians and Romans were not banned by the law.

The decision to adopt this regime was determined by some factors. First, it was consistent with the official policy of the rulers to treat both populations as equally as possible and to establish their equality before the law.[64] Such a political intent is clearly shown by several titles of the *Liber Constitutionum*.[65]

> ... Burgundio et Romanus una conditione teneantur.[66]
>
> Let Burgundians and Romans be held under the same condition ...[67]

---

61 *Liber Constitutionum* C (ed. de Saliis, p. 113).

62 *Liber Constitutionum* XII, 5 (ed. de Saliis, p. 52).

63 *Lex Romana sive forma et expositio legum Romanarum* (*Forma et expositio legum*) XXXVII, 1 (ed. de Saliis, p. 155).

64 On this policy see Odet Perrin, *Les Burgondes: leur histoire, des origines à la fin du premier Royaume* (Neuchâtel 1968), pp. 361–362, 473; Ian Wood, 'Gentes, Kings and Kingdoms: The Emergence of States. The Kingdom of the Gibichungs', in *Regna and Gentes. The Relationship between Late Antique and Early Medieval Peoples and Kingdoms*, eds. H.-W. Goetz – J. Arnut – W. Pohl (Leiden 2003), pp. 263–264; Biagio Saitta, *I Burgundi (413–534)* (Rome 2006), p. 79.

65 David Frye, 'Gundobad, the Leges Burgundionum, and the Struggle for Sovereignty in Burgundy', in *Classica et mediaevalia*, 41 (1990), p. 206, notes that 12 of the first 41 titles proclaim the equality of Burgundians and Romans.

66 *Liber constitutionum* X (ed. de Saliis, p. 50).

67 Engl. transl. by K. Fisher Drew, *The Burgundian Code: Book of Constitutions or Law of Gundobad* (Philadelphia, 1972), p. 30.

> ... Quod tamen inter Burgundiones et Romanos aequali conditione volumus custodiri.[68]
>
> ... Furthermore we wish this to be observed equally among Burgundians and Romans.[69]

As David Boyson[70] observes, this attempt to protect the position and the *status* of the Gallo-Roman subjects was aimed at pursuing a specific goal: in this way, Burgundian kings ensured the provision of skilled administrators and lawyers required to run their kingdom. It is possible that their decision to allow intermarriages was also influenced by the religious context of the kingdom: while in the Burgundian court there were several influential Aryans, Catholics were also quite numerous. Actually, it is likely that the Burgundians soon converted to Catholicism, as it is affirmed by the Greek historian Socrates Scholasticus.[71] Therefore, there is nothing to indicate that religion was a distinguishing mark between Burgundians and Romans, or that it was used to separate the two populations.[72]

It is known that the religious context of the Ostrogothic and the Visigothic kingdoms was completely different, and this (combined with other factors)[73] could have determined the prohibition of marriage between these *gentes* and the Roman people.[74]

In conclusion, in the period between the 4th and 6th century there are no constant features in the legal provisions on marriage of the *ius militare*. Only in Late antiquity we can notice a persistent vitality of the practice of conferring the Roman citizenship upon completion of military service (and, in some cases, also at the moment of recruitment). This privilege would have allowed *milites* of barbarian origin to legally marry Roman women.

68 *Liber constitutionum* XV (ed. de Saliis, p. 54).

69 Engl. transl. by Drew, *The Burgundian Code*, p. 33.

70 David Boyson, 'Romano-Burgundian Society in the Age of Gundobad: Some Legal, Archaeological and Historical Evidence', in *Nottingham Medieval Studies*, 32 (1988), p. 92.

71 *Historia ecclesiastica* 7.30.

72 On the theme of the religion by the Burgundian kingdom see Wood, 'Gentes, Kings and Kingdoms', pp. 263–264; Ian Wood, 'I Burgundi', in *Roma e i barbari. La nascita di un nuovo mondo*, ed. J.-J. Aillagon (Milan 2008), p. 338.

73 See above.

74 As regard the Ostrogothic kingdom, Francesco Giunta, 'Gli Ostrogoti in Italia', in *Magistra barbaritas. I barbari in Italia*, eds. M.G. Arcamone et all. (Milan 1984), p. 84, highlights that the dichotomy between Arians and Catholics represented a serious obstacle for the conservation of the balance pursued by Theodoric.

Bibliography

Primary Sources

AE = *Année épigraphique*

Antike Denkmäler in Bulgarien, ed. E. Kalinka (Vienna, 1906)

BGU I = *Aegyptische Urkunden aus den Königlichen (later Staatlichen) Museen zu Berlin, Griechische Urkunden*, 1 (Berlin, 1895)

Cassiodorus Senatoris, *Variae*, ed. Th. Mommsen [MGH, *Scriptores – Auct. ant.* 12] (Berlin, 1894) *etiam* Flavio Magno Aurelio Cassiodoro Senatore, *Varie*, (dir. A. Giardina), eds. A. Giardina – G.A. Cecconi – I. Tantillo – F. Oppedisano, 6 vols. – Engl. transl. by M.S. Bjornlie, *The Variae. The Complete Translation* (Oakland, CA, 2019)

CIL III = *Corpus Inscriptionum Latinarum*, 3. *Inscriptiones Asiae, provinciarum Europae Graecarum, Illyrici Latinae*, eds. Th. Mommsen, O. Hirschfeld, A. Domaszewski (Berlin, 1873–1902)

CIL V = *Corpus Inscriptionum Latinarum*, 5. *Inscriptiones Galliae Cisalpinae Latinae*, ed. Th. Mommsen (Berlin, 1872–1877)

CIL XVI = *Corpus Inscriptionum Latinarum*, 16. *Diplomata militaria*, post Th. Mommsen ed. H. Nesselhauf (Berlin, 1936–1955)

Codex Justinianus, see Iustinianus *Codex*

Codex Theodosianus: *Theodosiani libri XVI cum constitutionibus Sirmondianis et leges Novellae ad Theodosianum pertinentes*, ed. Th. Mommsen – P.M. Meyer, 2 vols. (Berlin, 1905)

Concilium Agathense, ed. J.D. Mansi, *Sacrorum conciliorum nova et amplissima collectio*, 8 (1762; repr. Graz, 1960)

Concilium Eliberitanum, ed. J.D. Mansi, *Sacrorum conciliorum nova et amplissima collectio*, 2 (1759; repr. Graz, 1960)

Digesta, see Iustinianus

Eunapius, ed. R.C. Blockley, *The Fragmentary Classicising Historians of the Later Roman Empire: Eunapius, Olympiodorus, Priscus and Malchus*, 2 (Liverpool, 1983)

Herodianus, *Ab excessu divi Marci*, ed. K. Stevenhagen (1922; repr. Stuttgart, 1967)

Iustinianus, *Codex*, ed. P. Krüger in T. Mommsen – P. Krüger, *Corpus Iuris Civilis*, 2, 11th ed. (Berlin, 1954)

Iustinianus, *Digesta*, ed. T. Mommsen – P. Krüger, *Corpus Iuris Civilis*, 1, 17th ed. (Berlin, 1963)

Iustinianus, *Institutiones*, ed. P. Krüger, in *Corpus Iuris Civilis*, 1, 17th ed. (Berlin, 1963) – Engl. transl. J.B. Moyle, *The Institutes of Justinian*, 5th ed. (1913; repr. Union, New Jersey, 2002)

Leges Burgundionum, ed. L.R. de Salis [MGH, *Leges nat. Germ.* 2.1] (Hannover, 1892) [available online at www.dmgh.de]

Leges Visigothorum, ed. K. Zeumer [MGH, *Leges nat. Germ.* 1] (Hannover - Leipzig, 1902)

Lex Romana sive forma et expositio legum Romanarum, in *Leges Burgundionum*
Lex Romana Visigothorum, ed. G. Haenel (Leipzig, 1849)
Lex Visigothorum in *Leges Visigothorum*
Liber Constitutionum, in *Leges Burgundionum*; Engl. transl. K. Fisher Drew, *The Burgundian Code: Book of Constitutions or Law of Gundobad* (Philadelphia, 1972)
Novella Severi II, see *Codex Theodosianus*
Procopius Caesariensis, *De bello gothico* in P.C., *De bellis – Tôn Polemôn Logoi*, ed. (and Engl. transl.) H.B. Dewing, *Procopius with an English translation*, 3. *History of the wars, Books V and VI* (London, 1919)
RMD I = *Roman Military Diplomas 1954 to 1977*, ed. M.M. Roxan (London, 1978)
Socrates Scholasticus, *Historia ecclesiastica*, ed. G.C. Hansen, with M. Sirinjan, *Sokrates Kirchengeschichte* (Berlin, 1995)

Secondary Literature

Aimone, Marco, 'Romani e Ostrogoti fra integrazione e separazione. Il contributo dell'archeologia a un dibattito storiografico', in *Reti Medievali – Rivista*, 13.1 (2012), pp. 31–96 [Published online: Mar 21, 2012. DOI: https://doi.org/10.6092/1593-2214/337]
Azzara Claudio and Arianna Bonnini, 'I matrimoni fra barbari e romani nelle leggi di stirpe dell'altomedioevo', in *Matrimoni misti: una via per l'integrazione tra i popoli*, ed. S. Marchesini (Verona, 2012), pp. 139–146.
Bianchini, Mariagrazia, 'Ancora in tema di unioni tra barbari e romani', in *Atti dell'Accademia Romanistica Costantiniana. VII Convegno internazionale*, eds. G. Crifò – S. Giglio (Napoli, 1988), pp. 225–249
Boyson, David, 'Romano-Burgundian Society in the Age of Gundobad: Some Legal, Archaeological and Historical Evidence', in *Nottingham Medieval Studies*, 32 (1988), pp. 91–118
Campbell, Brian, 'The Marriage of Soldiers under the Empire', in *Journal of Roman Studies*, 68 (1978), pp. 153–166
Campbell, Brian, *The Roman Army 31 BC–AD 337: A Source Book* (London-New York, 1994)
Castagnino, Francesco, 'Creare una famiglia. Il ius conubii e i figli dei "soldati" prima e dopo la caduta dell'Impero', in *Ravenna Capitale. L'esercito romano e l'alba dell'Europa. Modelli concettuali e sperimentazioni sociali tra linguaggi, istituzioni e diritto (secoli IV–VIII)*, ed. G. Bassanelli Sommariva (Santarcangelo di Romagna 2020), pp. 115–135
Castagnino, Francesco, 'I privilegi fiscali di milites e veterani in età costantiniana', in *Atti dell'Accademia Romanistica Costantiniana*, 24. *Militia inermis e militia armata. Apparati civili e militari nella tarda antichità*, eds. C. Lorenzi – M. Navarra (Perugia 2021), pp. 183–206
Castagnino, Francesco, *I diplomata militaria. Una ricognizione giuridica* (Milan, 2022)

Cavanna, Adriano, 'Diritto e società nei regni Ostrogoto e Longobardo', now in Id., *Scritti (1968–2002)* (Naples, 2007), pp. 513–574

Cusmà Piccione, Alessandro, *"Non licet tibi alienigenam accipere". Studio sulla disparitas cultus tra i coniugi nella riflessione cristiana e nella legislazione tardoantica* (Messina, 2017)

Demougeot, Émilienne, 'Restrictions à l'expansion du droit de cité dans la seconde moitié du IVe siècle', in *Ktema*, 6 (1981), pp. 381–393

Eck, Werner, 'Septimius Severus und die Soldaten. Das Problem der Soldatenehe und ein neues Auxiliardiplom', in *In omni historia curiosus. Studien zur Geschichte von der Antike bis zur Neuzeit. Festschrift für Helmuth Schneider zum 65. Geburtstag*, eds. B. Onken – D. Rohde (Wiesbaden, 2011), pp. 63–77

Eck, Werner, 'Eine dritte Kopie der Tafel von Brigetio aus dem Jahr 311 n.Chr.", in *Quaderni Lupiensi di Storia e Diritto*, 9 (2019), pp. 53–57

Francovich Onesti, Nicoletta, *I nomi degli Ostrogoti* (Florence, 2007)

Frye, David, 'Gundobad, the Leges Burgundionum, and the Struggle for Sovereignty in Burgundy', in *Classica et mediaevalia*, 41 (1990), pp. 199–212

Garnica, Ana M. Jiménez, 'El origen de la legislación civil visigoda sobre la prohibición de matrimonios entre romanos y godos. Un problema de fundamento religioso', in *Anuario de historia del derecho español*, 55 (1985), pp. 735–748

Giunta, Francesco, 'Gli Ostrogoti in Italia', in *Magistra barbaritas. I barbari in Italia*, eds. M.G. Arcamone et all. (Milan 1984), pp. 53–96

Jörs Paul and Wolfgang Kunkel, *Römisches Privatrecht*, 3rd ed. (Berlin, 1949)

Kerneis, Soazick, 'Francus civis, miles Romanus: les barbares de l'Empire dans le Code Théodosien', in *Droit, religion et société dans le Code Théodosien*, eds. J.-J. Aubert – P. Blanchard (Geneva, 2009), pp. 377–399

Le Bohec, Yann, *Armi e guerrieri di Roma antica. Da Diocleziano alla caduta dell'impero*, (Ital. transl.) (Rome, 2008)

Liebeschuetz, Wolfgang, 'Citizen Status and Law in the Roman Empire and in the Visigothic Kingdom', in *Strategies of Distinction: The Construction of Ethnic Communities, 300–800*, eds. W. Pohl – H. Reimitz (Leiden, 1998), pp. 131–152

Lind, Göran, *Common Law Marriage. A Legal Institution for Cohabitation* (Oxford, 2008)

Loschiavo, Luca, 'Insediamenti barbarici e modelli di coesistenza nell'Italia altomedievale. Il regno degli Ostrogoti', in *Immigrazione e integrazione. Dalla prospettiva globale alle realtà locali*, 1, ed. F. Rimoli (Naples, 2014), pp. 317–348

Marotta, Valerio, 'Il problema dei laeti. Fonti e storiografia', in *Civitas, Iura, Arma. Organizzazioni militari, istituzioni giuridiche e strutture sociali alle origini dell'Europa (secc. III–VIII)*, eds. F. Botta – L. Loschiavo (Lecce, 2015), pp. 117–157

Mathisen Ralph W. and Hagith S. Sivan, 'Forging a New Identity: The Kingdom of Toulouse and the Frontiers of Visigothic Aquitania (418–507)', in *The Visigoths. Studies in Culture and Society*, ed. A. Ferreiro (Leiden, 1999), pp. 1–62

Mommsen, Theodor, 'Ostgotische Studien', in Id., *Gesammelte Schriften*, 6 (Berlin, 1910), pp. 372–482

Perrin, Odet, *Les Burgondes: leur histoire, des origines à la fin du premier Royaume* (Neuchâtel, 1968)

Phang, Sara E., *The Marriage of Roman Soldiers 13 B.C.–A.D. 235: Law and Family in the Imperial Army* (Leiden-Boston-Köln 2001)

Popescu, Andreas, *Die Franken und das Römische Militär Identität in Gallien und dem Rheinland in der Spätantike* (Bonn, 2017). Available at https://bonndoc.ulb.uni-bonn.de/xmlui/handle/20.500.11811/7065

Porena, Pierfrancesco, *L'insediamento degli Ostrogoti in Italia* (Rome, 2012)

Porena, Pierfrancesco, 'Ancora sulla carriera di Flavius Ablabius, prefetto del pretorio di Costantino', in *Zeitschrift für Papyrologie und Epigraphik*, 190 (2014), pp. 262–270

Rigsby, Kent J., 'Two Danubian Epitaphs', in *Zeitschrift für Papyrologie und Epigraphik*, 126 (1999), pp. 175–176

Saitta, Biagio, *I Burgundi (413–534)* (Rome, 2006)

Seeck, Otto, *Regesten der Kaiser und Päpste für die Jahre 311 bis 476 n. Chr.*, (1919; repr. Frankfurt a.M., 1984)

Sivan, Hagith S., 'The Appropriation of Roman Law in Barbarian Hands: "Roman-barbarian" Marriage in Visigothic Gaul and Spain', in *Strategies of Distinction: The Construction of Ethnic Communities, 300–800*, eds. W. Pohl – H. Reimitz (Leiden, 1998), pp. 189–203

Soraci, Rosario, *Ricerche sui conubia tra Romani e Germani IV–VI s.*, 2nd ed. (Catania, 1974)

Tassistro, Pietro, 'Il matrimonio dei soldati romani', in *Studi e Documenti di storia e diritto*, 22 (1901), pp. 3–82

Vendrand-Voyer, Jacqueline, *Normes civiques et métier militaire à Rome sous le Principat* (Clermont-Ferrand, 1983)

Weber, Max, *Storia agraria romana*, (Ital. transl.) (Milan, 1967)

Wolfram, Herwig, *Die Goten: Von den Anfängen bis zur Mitte des sechsten Jahrhunderts*, 5th ed. (Munich, 2009)

Wood, Ian, 'Gentes, Kings and Kingdoms: The Emergence of States. The Kingdom of the Gibichungs', in *Regna and Gentes. The Relationship between Late Antique and Early Medieval Peoples and Kingdoms*, eds. H.-W. Goetz – J. Arnut – W. Pohl (Leiden, 2003), pp. 243–270

Wood, Ian, 'I Burgundi', in *Roma e i barbari. La nascita di un nuovo mondo*, ed. J.-J. Aillagon (Milan, 2008)

Wood, Ian, 'The Legislation of Magistri Militum: the Laws of Gundobad and Sigismund', in *Clio & Themis. Revue électronique d'histoire du droit*, 10 (2016), pp. 1–16 [Published online: Jun 10, 2021. DOI: https://doi.org/10.35562/cliothemis.1191]

Wood, Ian, 'Burgundian Law-making, 451–534', in *Italian Review of Legal History*, 3 (2017), pp. 1–27.

CHAPTER 9

Soldiers' Inheritance: The *Testamentum Militis* and Other Privileges from the Imperial Constitutions to the *Leges Barbarorum*

Iolanda Ruggiero

1 Introduction

In 1927 Nino Tamassia published a small volume entitled '*Testamentum militis* e diritto germanico': a pioneering work in a certain sense – we will soon explain the reasons – for the period in which it was composed.[1] This book starts from a basic premise, suggested by the reading of Tacitus: the total absence, among the Germanic populations, of the very idea of testamentary succession.[2] From here a single thread unravels, characterized by the elements of continuity – in the osmotic process that saw progressively approaching Romanity and *barbaricum*[3] – between the institution of *testamentum militis* (where it would find traces of pre-Roman Italic law) and the establishment of the principle of testamentary succession at the constellation of gentes who populated the territories beyond the Rhine and the Danube.[4] I would like to verify this assumption, as from 1927 onwards no significant contribution on the subject is reported

1 Nino Tamassia, *'Testamentum militis' e diritto germanico* (Venezia, 1927) = Id., *Scritti di storia giuridica*, 3 (Padova, 1969).

2 Tac., *Germ.* 20.3.

3 So, substantially, according to the *communis opinio*. On the progressive fusion of the Roman-Barbarian aristocracies important pages can be read in the collective volume edited by Walter Pohl, *Kingdoms of the Empire. The Integration of Barbarians in Late Antiquity* (Leiden-New York-Köln, 1997), to which can be added the bibliography suggested by Hans-Werner Goets – Jörg Jarnut – Walter Pohl, eds., *Regna and Gentes. The Relationship between Late Antique and Early Medieval Peoples and Kingdoms in the Transformation of the Roman World* (Leiden-Boston, 2003) and, more recently, by Luca Loschiavo, *L'età del passaggio. All'alba del diritto comune europeo (secoli III–VII)*, 2nd. ed. (Torino, 2019) p. 256 on.

4 On the substitution of the label 'German' with that of 'barbarian', cf. Karol Modzelewski, *L'Europa dei barbari. Le culture tribali di fronte alla cultura romano-cristiana*, Ital. transl. (Torino, 2008) (orig. ed., Warsawa, 2004), p. 15 on.; Edward James, *I barbari*, Ital. transl. (Bologna, 2011) (orig. ed. Harlow, 2009), p. 17 on; Stefano Gasparri – Cristina La Rocca, *Tempi barbarici. L'Europa occidentale tra antichità e medioevo (300–900)* (Roma, 2012), p. 17 on.

 | DOI:10.1163/9789004698017_010

(specific exceptions will be indicated in the text). This implies a stringent need for comparison between Roman law scholars and early Middle Ages scholars.

2 The Imperial Privileges

Among the most significant privileges granted to soldiers in the Augustan age were the *testamentum militum* and the *peculium castrense*. The military could therefore draw up a testament (will) completely exempt from the formalities required by the *ius civile* for a will. Augustus also granted the *testamenti factio* to the *filius familias*, limited to what he had acquired *in castris*.[5] The *testamentum militis* took the form of a *privilegium*: the possibility of disposing *quomodo volunt* was attributed to the military by virtue of their very status. Content-wise, this possibility displayed substantial freedoms in comparison to ordinary wills. Let us consider, for example, the disappearance of the principle of incompatibility between testamentary and *ab intestato* delation or the non-application of specific provisions, such as the Falcidia law on legacies and the Augustan legislation on *caelibes* and *orbi*.

The history of this institution is summed up in a famous passage:

> Ulp. 45 *ad ed.* D. 29.1.1pr.[6]: Militibus liberam testamenti factionem primus quidem divus Iulius Caesar concessit: sed ea concessio temporalis erat.

---

5 I. 2.12pr., a text read by Contardo Ferrini, 'Sulle fonti delle Istituzioni di Giustiniano', (1901) now in Id., *Opere*. II. *Studi sulle fonti del diritto romano* (Milano, 1929), p. 366; Antonio Guarino, 'L'oggetto del «castrense peculium»', in *Bullettino dell'Istituto di diritto romano* (*BIDR*), 48 (1941), p. 53 on; Franca La Rosa, *I peculii speciali in diritto romano* (Milano, 1953), pp. 20 on, where further bibliography; Burkhard Lehmann, 'Das Eigenvermögen der römischen Soldaten unter väterlicher Gewalt', in *Aufstieg und Niedergang der römischen Welt* (*ANRW*), II.14 (Berlin-New York, 1982), pp. 266 on, esp. 277; Jacqueline Vendrand-Voyer, 'Origine et développement du «droit militaire» romain', in *Labeo*, 28 (1982), pp. 268 on; Vincenzo Scarano Ussani, *Le forme del privilegio. Beneficia e privilegia tra Cesare e gli Antonini* (Napoli, 1992). Tamassia, *Testamentum militis*, p. 40, linked the Augustan measure to the revocation of the edict of the same emperor recalled by Paul. 3 *sent.* D. 28.2.26.

6 The bibliography on the Ulpian fragment, which is impressive, is summarized by Valerio Marotta, *Mandata principum* (Torino, 1991), p. 20 fn. 56 and Scarano Ussani, *Le forme del privilegio*, p. 83 on fn. 1, to which can be added Ines De Falco, 'I giuristi e il testamentum militis. L'orientamento di Iavolenus Priscus', in *Studia et documenta historiae et Iuris* (*SDHI*), 80 (2014), pp. 419 on and Andrea Lovato, 'P.Fay. 10 (Ulpianus 45 ad edictum) + P.Berol. inv. P. 11533', in D. Mantovani – S. Ammirati, eds., *Giurisprudenza romana nei papiri. Tracce per una ricerca* (Pavia, 2018). Already Sibylle von Bolla, 'Zum römischen Militärtestament', in *Studi in onore di Vincenzo Arangio-Ruiz nel XLV anno del suo insegnamento*, I (Napoli, 1953), pp. 273 on attributed the Flavian measure to the increasing provincialization of the army: to the

> postea vero primus divus Titus dedit: post hoc Domitianus: postea divus Nerva plenissimam indulgentiam in milites contulit: eamque Traianus secutus est et exinde mandatis inseri coepit caput tale. Caput ex mandatis: 'Cum in notitiam meam prolatum sit subinde testamenta a commilitonibus relicta proferri, quae possint in controversiam deduci, si ad diligentiam legum revocentur et observantiam: secutus animi mei integritudinem erga optimos fidelissimosque commilitones simplicitati eorum consulendum existimavi, ut quoquomodo testati fuissent, rata esset eorum voluntas. faciant igitur testamenta quo modo volent, faciant quo modo poterint sufficiatque ad bonorum suorum divisionem faciendam nuda voluntas testatoris'.
>
> The deified Julius Caesar was in fact the first to concede unrestricted *testamenti factio* to soldiers; but that concession was temporary. However, later the deified Titus first gave [it]; after this Domitian; thereafter, the deified Nerva conferred the fullest indulgence on soldiers; and Trajan followed this, and thenceforth such a chapter came to be inserted in [imperial] mandates. A chapter from the mandates: 'As it has been submitted to my notice that wills left by our fellow soldiers, which could be open to dispute if regard were had to the diligent observance of the laws, are repeatedly submitted; following the openess of my heart toward those excellent and most faithful fellow soldiers, I thought that provision should be made for their inexperience [in legal matters], so that whatever the way in which they made their wills, their wishes should be confirmed. Therefore, let them make their wills in any way they wish, let them make them in any way they can, and let the bare wishes of the testator suffice to settle the distribution of their property.

In this essay I will focus on the famous *caput mandatorum* of the Trajan era.[7] Informed of how, sometimes soldiers left wills which – if subjected to a diligent

---

emergence, therefore, of numerous problems, including legal (many of the wills drawn up by soldiers were invalid because they lacked the requisites normally required). According to Scarano Ussani, *Le forme del privilegio*, pp. 87 on, in the perspective of Ulpian the action of Nerva occupied an important place in reconstructing the history of the institution: the scholar identifies in the Nerva's principate an essential moment in the process of consolidation of the privilege. It is therefore not far-fetched to believe that military will and granting to soldiers, subject to the *patria potestas*, of the faculty to dispose by will of the *peculius castrensis* constituted two moments of a single plan. A fitting example of how the political element – in our case the power relations between the emperor and the army – concretely affected the law, giving rise to two important juridical figures.

7 *P. Fayoum* 10. On the papyrus, most recently, Lovato, 'P.Fay. 10', with bibliography. Cf. also Scarano Ussani, *Le forme del privilegio*, pp. 92 on.

scrutiny of the laws – could be the subject of controversy, the emperor decided to help the *simplicitas* of the 'excellent and most faithful comrades': in whatever manner they had tested, their will was ratified. Therefore, soldiers made their wills as they wished. The *nuda voluntas testatoris* would have been sufficient to carry out the division of their assets.[8] The *optimus princeps* thus gave a definitive status to two legal figures establishing a privileged situation for the soldiers vis-à-vis the civil society.[9]

Later on, Trajan's concession required some clarification. I refer to the famous rescript of the emperor to Statilius Severus, reported by Florent 10 *inst.* D. 29.1.24.[10] Trajan wrote to the then *legatus* of *Thracia*[11] as follows. First of all, the existence of a will *quoquo modo factum* wills had to be established. Therefore, if the soldier whose case was being discussed, having asked some comrades to attest to his will, indicated his heir and the slaves to whom he would grant freedom, this sufficed for him to make his will. If, on the contrary – as it could often happen when speaking – the miles said to someone: 'I establish you as heir' or 'I leave you my goods', such statements could not be construed as a proper will. No one – and here the rescript ends – cared more than the very ones to whom the privilege had been granted that such an example should not be admitted. This avoided an unwelcome possibility: the case of a witness claiming to have heard a soldier saying, prior to his death, that he wanted to leave the *bona cui visum est, et per hoc iudicia vera subvertuntur.*

The measure was probably intended to clarify the previous provision, which had undoubtedly created numerous interpretative problems. Trajan established the limits and criteria to be applied to the concession. The interpretative

8 Scarano Ussani, *Le forme del privilegio*, p. 93 fn. 24, on the footsteps of Tamassia, *Testamentum militis*, p. 456 on, observes how the phrase *faciant ... poterint* had to become an authentic 'motive' in the language of jurists and chancelleries: this would be deduced from a series of significant textual coincidences (Gai. 2.114: 'militibus ... quomodo velint vel quomodo possint, permittitur testamentum facere' ; Paul. 11 *resp.* D. 29.1.40pr.: 'respondi militibus, quoquo modo velint et quo modo possunt, testamentum facere concessum esse' ; Ulp. 45 *ad ed.* D. 29.1.44: 'rescripta principum ostendunt omnes ... quomodo velint et quomodo possint, testari ...'; Ulp. 45 *ad ed.* D. 37.13.1pr.: 'omnes ... quomodo velint et quomodo possint, testabuntur' ; Tit. Ulp. 23.10: 'nam principalibus constitutionibus permissum est illis, quomodocumque vellent, quomodocumque possent, testari'; C. 6.21.15pr. Imp. Costantinvs A. ad popvlvm: 'milites ... quomodo possint ac velint testentur' [a. 334]. Lovato, 'P.Fay. 10', p. 129, adds C. 6.21.3pr. Imp. Antoninvs A. Vindiciano: 'quomodo velint et quomodo possint' [a. 213].

9 A separation, according to Tamassia, *Testamentum militis*, p. 453 on, prior to the empire: the rule, in short, formally recognized what was already practiced in *castra.*

10 Cf. I. 2.11.2. For the readings of D. 29.1.24 refer to Scarano Ussani, *Le forme del privilegio*, p. 97 on, fn. 31. Cf. also De Falco, 'I giuristi e il testamentum militis', pp. 420 on.

11 For all, Giuseppe Camodeca, 'Il giurista L. Neratius Priscus cos. Suff. 97. Nuovi dati su carriera e famiglia', in *Studia et documenta historiae et iuris* (SDHI), 73 (2007), p. 303 and fn. 40.

rigor reveals the concern of the princeps in framing the new legal figure, mitigating its subversive potential and anchoring it to a criterion of respect for the actual will of the deceased.[12]

3 How Soldiers Made Their Wills

The sources mention the use of a common will by the *milites*, despite the practical difficulties and their privilege of freedom of disposition. It is rare the case of a soldier who died intestate; on the contrary, substitutions of heirs were very frequent.[13] Often, a comrade-in-arms was designated as such.[14] The heir could also be established in a form reminiscent of the *fedecommissum*: a figure suitable, for its flexibility, to the needs of military life.

In Iav. 11 *ep*. D. 36.1.48 we find a case quite revealing of the castrensian customs.[15] *Seius Saturninus archigubernus* of the *classis Britannica*[16] established the trierarch *Valerius Maximus* as trustee, with the obligation to return the inheritance to his son *Oceanus*[17] when he turned sixteen. *Iavolenus* skillfully extracts from the figure of the trustee that of the *tutor*. And precisely because of this function (which was not new) of *fedecommissum*, *Maximus* will lose his inheritance: *Oceanus* died earlier and *Maximus*, taking his task too seriously, contested the goods to the legitimate heir of the child. From *fedecommissum* to deposit the step is short. Some imperial constitutions of the third century

12 Scarano Ussani, *Le forme del privilegio*, p. 99 hypothesized an influence of Neratius on the rescript, a jurist 'con un orizzonte teorico dominato dall'idea del *ius finitum* e dal timore di un totale sovvertimento delle certezze'. On these themes Vincenzo Scarano Ussani, *Valori e storia nella cultura giuridica fra Nerva e Adriano. Studi su Nerazio e Celso* (Napoli, 1979), pp. 5 on, 58 on; Id., *Empiria e dogmi. La scuola proculiana fra Nerva e Adriano* (Torino, 1989), pp. 68 on. Opposite direction Sergio Nappi, *Ius finitum* (Bari, 2005).

13 A good repertory of sources in Tamassia, *Testamentum militis*, pp. 465 on.

14 The frequency of substitutions was probably due to the need of the testator to ensure that the death of the first person instituted heir did not invalidate the entire will.

15 On the passage, for all, Bernd Eckardt, *Iavoleni Epistulae* (Berlin, 1978), p. 21 on. Most recently Harries, 'Saturninus the Helmsman, Pliny and Friends: Legal and Literary Letter Collections', in A König – C. Whitton, eds., *Roman Literature under Nerva, Trajan and Hadrian. Literary Interactions,* AD *96–138* (Cambridge, 2018), pp, 262 on.

16 On the origins of the Roman military fleet cf. Giovanni Forni, 'Esperienze militari nel mondo romano', in *Nuove questioni di storia antica* (Milano, 1972) and Michel Reddé, *Mare nostrum. Les infrastructures, le dispositif et l'histoire de la marine militaire sous l'empire romain* (Roma, 1986), pp. 3 on.

17 It is striking the name given by the Roman navarch to his son: in *CIL* III 1301b another *Oceanus* is mentioned, while CIL X 2872 preserves traces of names of rivers attributed to soldiers: Rhine, Danube, Euphrates.

addressed to soldiers deal with this case.[18] When conflicts were imminent, the *miles* provided security of *res castrenses* by depositing them with a fellow soldier.

There was no lack of other *instrumenta voluntatis*, of a more distinctly military nature. In the first place, the *adoptio in fratrem* carried out through a will. Diocletian frowned on it,[19] but at the same time he attested a use of it that was certainly not limited to the *peregrini*. Another significant passage is C. 2.3.19, a *constitutio* addressed to the soldier *Vittorianus* in the year 290.[20] Conjunctive (two people making their will in a single document) and reciprocal wills (a joint will in which two people established each other as heir), although a predominantly Eastern practice, were not unknown in the West. The case concerns the mutual donation *mortis causa* of two brothers, but the principle also applies to strangers. However, the exceptional character of the concession is explicitly stated: the emperor approves of the military custom. The privilege therefore consists in recognizing legal effects to an act proper to the castrensian customs. It is easy to draw a comparison between the mutual donation *mortis causa* among soldiers and the donation between spouses without offspring, proven by the same etymology of *consortes*. Spouses and comrades share the same fate. This is clear from a well-known Constantinian constitution (preserved in C. 6.21.15) of 334.[21] Already its first part presupposes an order among those called to the inheritance, people for whom the soldier is supposed to care: wife, children, friends and fellow soldiers. The second part of the constitution establishes that soldiers can make a will as they wish and as circumstances allow. They can write on any material, and in any way: with their own blood on the shield or on the sheath of the sword or in the dust. The Constantinian image of the will written in blood – wrongly considered rhetorical by Tamassia[22] – would hide the imaginary creation of a legal relationship based not on the will but on the external form it assumes.[23] In other words,

18 Cf. C. 4.34.1 [a. 234]; C. 4.34.2 [a. 238] and C. 4.34.3 [a. 239].

19 C. 6.24.7 [a. 285].

20 C. 2.3.19 [a. 290], a text at the time discussed by Pietro De Francisci, 'Quelques observations sur l'histoire du testament conjonctif réciproque', in *Revue historique du droit français et étranger* (RHDFE), 3 (1924), pp. 57 on: pages in which the connection of the passage under examination to the figure of the conjunctive will is excluded.

21 Tamassia, *Testamentum militis*, pp. 471 on.

22 Tamassia, *Testamentum militis*, pp. 472.

23 In this way, Jacob Grimm, *Deutsche Rechtsalterthümer*, vol. 1 (Leipzig, 1899), pp. 153 on, an impressive work, composed between 1828 and 1854, in which the author tried to reconstruct 'le 'istituzioni' dei popoli germanici' (thus Loschiavo, *L'età del passaggio*, p. 123 fn. 7).

in such a will one could see that symbolism typical of the most ancient juridical institutions (whose legacy is indeed attested by the *milites*) in contrast to the *testamentum iure civili*. In the wards of Constantine's army, the goods of a soldier were probably transmitted to his faithful comrade in the fatal hour through *traditio*, with the symbolic delivery of a weapon or any other material mark available: in the hands of the person designated to the succession, that symbol was a clear token of a right.[24] At the same time, the Constantinian *constitutio* aimed rather at validating the last will of the soldier in the imminence of death, expressed in the only possible way in those tragic circumstances.

The problem of testamentary alterations, therefore, ought not to be overlooked. Often in the history of Roman wills one finds concerns regarding the fate about the testator's will and the ways to ensure its fulfillment, something that also characterizes life in *castra*. In its broadest meaning, the *testamentum militis* is the expression of a given society, not an artificial construction bestowed because of one's status. In this regard the case found in C. 3.44.5, a rescript of 224 by Alexander Severus to *Cassius* the soldier. In all probability, it is a *testamentum militis* whose execution was grounded by *Cassius* in the *commilitium*.

In this regard, the core of Tamassia's 'continuist' thesis finds its expression in a clear statement: the story of the military will is connected to the broader story of 'popular' law (without, however, resorting to so-called 'vulgarization'),[25] which finds its most faithful representation in the army. Proximity with the military structures of the empire magnified the penetration of its usages among the Barbarian tribes at its service, thereby providing the *humus* for the transplant of such usages in the law of those barbarian populations – which, in the original indeterminacy of its legal arrangements, was rather weak and, therefore, unable to oppose much resistance to Roman influence.

24 Consider the importance of the *anulus signatorius* – the imprint makes it the most characteristic and significant object of the person to whom it belongs – in the history of legal symbolism: to whomever it is solemnly transferred, it passes all rights of the previous owner. Cf. Béatrice Fraenkel, *La signature. Genèse d'un signe* (Paris, 1992) and Claude Jeay, *Signature et pouvoir au Moyen Age* (Paris, 2015).

25 Tamassia, *Testamentum militis*, pp. 477 on. Tamassia inserted himself with originality in the intense debate opened by Heinrich Brunner and Ludwig Mitteis at the close of the nineteenth century on 'vulgar law' and 'popular law (the latter was a term used by the historians of German law to designate the customs of the Germanic peoples as written down from the fifth century onwards); in this regard, it is always useful reading Francesco Calasso, *Medio Evo del diritto*. 1. *Le fonti* (Milano, 1954), pp. 57 on. Significantly, to clarify the position of Tamassia, Calasso [pp. 61 on] takes into account precisely the work on *testamentum militis* here discussed.

4 Barbarians in the Imperial Army

The entry of Franks and Alamanni into the imperial army is of particular relevance: what did the barbarian take with him and what did he keep when he went back home? Because of their very geographical conditions, the territories occupied by the barbarian tribes forced those who lived there to subsistence warfare. Weapons dominate the whole legal symbolism, from the status of an army man and the declaration of war with the roar of arms, to the concession (always done with arms) of freedom to the slave or to the grant of full legal capacity to the *filius familias*.

Those peoples were initially concerned with the preservation of family groups, so as to provide for the continuity of their military power. Therefore, among them, the main significance of inheritance was not economic but rather political. The rise of a central and stable public power changed this logic, as it led to the distribution by lot among those who had contributed to put together the spoils of war first, and then to the control of territories. Tacitus informs us of the absence of testamentary succession among the Germans, who knew only of legitimate succession: *heredes tamen successoresque sui cuique liberi, et nullum testamentum.*[26]

Nino Tamassia found traces of 'Roman crusts' in the *adoptio per arma* and in title 46 of the *Pactus Legis Salicae* (*de adfathamire*), to which we will return shortly. In the *adoptio*, the Salic Law knows the rite of the embrace (*fathum*). While evidently done without weapons, its validity required that it was carried out in the *mallus* (the judicial assembly held periodically in each village) duly convened. Although there is no mention of the *traditio* of weapons, we are nevertheless in the presence of armed people. It is almost a commonplace: the imperial army is nearly unanimously considered a 'vehicle' of Romanity and civilization. Enlistment brought in recruits from peripheral or barbarian populations. Those new recruits fought side by side with the Romans; they learned their language, observed their hierarchy and discipline. They would have been fascinated by the complexity of the structure of the Roman army and have hoped to emulate the lifestyle of its generals. The well-known inscription from *Aquincum* (*Pannonia inferior*, CIL III, 3576)[27] – *Francus ego cives* (*sic!*) *Romanus miles in armis / egregia virtute tuli bello mea dextera sem*[*p*]*er*, regardless of its

26 *Supra*, fn. 2.

27 Paolo Cugusi, *Per un nuovo corpus dei Carmina Latina Epigraphica. Materiali e discussioni* (Roma, 2007), pp. 108, 172.

actual reconstruction,[28] captures well the fluid sense of the reality in which the *miles* lived.

This situation was greatly influenced by the role played by the spread of Christianity among the ranks of the army and, consequently, its effect on succession law. Suffices to think of the insistence – both in the Gospels and in patristic literature – on the character of the new *militia*: *miles* is the Christian who fights for the faith.[29] The edict of Constantine and the following provisions inaugurated the dispositions *pro anima* in favour of the Church.[30] This way, the testament acquired a new simplification of forms which, in the case of *milites*, was added to the more ancient ones. All together, these dispositions constituted a further inroad of the principle of testamentary succession in the barbaric society. *Commilitium*, military customs, changes in religion[31] (with the sole but significant exception of the Langobardic resistance) can therefore be considered causes of this process of transformation.

---

28 Usually, it is interpreted as if the first-person speaker claimed, in recalling his military virtues, to be a *civis Francus* and *a miles Romanus in armis*: in this sense Soazick Kerneis, 'Francus civis, miles Romanus: les barbares de l'Empire dans le Code Théodosien', in J.-J. Aubert – P. Blanchard, eds., *Droit, religion et sociétè dans le Code Théodosien* (Genève, 2009). For a different reading (Kent J. Rigsby, 'Two Danubian Epitaphs', in *Zeitschrift für Papyrologie und Epigraphik* [*ZPE*], 126 [1999], p. 175 on; Andreas Popescu, *Die Franken und das römische Militär Identität in Gallien und dem Rheinland in der Spätantike* (Bonn, 2017), pp. 1 on, 53, 256). According to Marotta (in this volume at p. ++) 'This inscription honours someone deceased who served in the army, obtaining *civitas* either during the prescribed service or when he was discharged'.

29 On the very complicated subject of *militia Christi*, a classic reading is the work by Adolf von Harnack *Militia Christi, Die christliche Religion und der Soldatenstand in den ersten drei Jahrhunderten* (Tübingen, 1905). With his basic thesis – the definitive conversion to Christianity took place precisely through the army – the German scholar mapped out an itinerary that has yet to be traveled. Cf. John Helgeland, 'Christians and the Roman Army from Marcus Aurelius to Constantine', in *Aufstieg und Niedergang der römischen Welt* (*ANRW*), II 23.1 (Berlin-New York, 1979), 724 on and Sergio Tanzarella, *I cristiani e il servizio militare nella Chiesa antica*, in *Asprenas* 31 (1984),75 on.

30 For the early Middle Ages mention should be made at least of the contributions by Josiane Barbier, 'Testaments et pratique testamentaire dans le royaume franc (VIe–VIIIe siècles)', pp. 37 on; Stefano Gasparri, 'I testamenti nell'Italia settentrionale fra VII e IX secolo', pp. 97 on; Eliana Magnani, 'Don aux églises et don d'églises dans le sud-est de la Gaule: du testamen d'Abbon (739) aux chartes du début du XIe siècle', pp. 379 on and Wendy Davies, 'Buying with Masses: "donation" pro remedio animae in Tenth-Century Galicia and Castile-León', in F. Bougard – C. La Rocca – R. Le Jan, eds., *Sauver son âme et se perpétuer. Transmission du patrimoine et mémoire au haut moyen âge* (Roma, 2005), pp. 401 on.

31 See Ian Wood, *The Trasformation of the Roman West* (Kalamazoo and Bredford, 2017), pp. 37–40.

Already with Diocletian, military power was separated from civil power: this widened the legal jurisdiction on the *milites*. For Tamassia, out of the 'cedimenti del diritto romano' the first epiphanies of the barbarian institutions came forth. But the Roman *ius singulare*, in turn, overcame the aversion of these peoples to testamentary succession.[32] During the empire, the 'vittoria del testamento' was not limited to the barbarians in military service. The most ancient inscriptions identify the first barbarian 'testators' in the so-called *corporis custodes*, the Germanic cavalry corps constituted by Augustus and then reconstituted by Tiberius.[33] Arguing from the particularities of the name and other circumstances, the inscriptions concerning *custodes* under the principate of Tiberius are probably to be referred to slaves; those of the period between Claudius and Nero seem instead to suggest free people.[34] These last inscriptions represent the oldest example of wills drawn up by barbarian soldiers, the first instance in the history of testamentary succession among Germanic populations, whereas the last one is found in the Theoderic Edict (E.T. 32).[35] The epigraphs provide a series of data, such as the name of the *corporis custos*; that of the head of the *decuria*; the ethnicity; the years of age and service; the name of the heir who placed the title and his affiliation to the *collegium Germanorum*.[36] It is remarkable that the heirs mentioned there are unequivocally *ex testamento*: there was no blood tie with the deceased. In the later military inscriptions, instead, relatives heirs, especially brothers, appear more frequently.[37] This shift likely depends on the way in which the soldiers were recruited: in the later inscriptions a soldier would move from the single family in the corps to groups, thereby maintaining the initial unity; the *corpores custodes*, on the contrary, probably came from various *gentes* or were confused among the prisoners.

The Roman example therefore also placed the *commilitium* at the basis of succession for the barbarians, and this was further strengthened by the collegial bond. Those tribes ended up in practice identifying themselves with the cohorts. The hereditas left by the *corporis custodes* was composed only of the *peculium castrense* of those people, increased by imperial donations and by

---

32 Tamassia. *Testamentum militis*, pp. 511 on.

33 Cf. De Ruggiero, 'Corporis custodes', in *Dizionario epigrafico di antichità romane*, II.ii (Spoleto, 1910), pp. 1237 on.

34 De Ruggiero, 'Corporis custodes', p. 1238.

35 On this chapter, se *infra* Loschiavo in this volume at p. 428.

36 *CIL* VI, 8802, 8803, 8804, 8806, 8807, 8808, 8809 and *ILS* 1724–1730.

37 *ILS* 2029, 2056, 2065, 2167, 2197, 2225, 2257, 2306, 2341, 2344, 2365, 2897, 2898 etc. and *CIL* X, 246.

the sums forfeited by the those moving around the imperial palace. Not even death could sever the corporate bond.[38]

5 Ancient *Leges Barbarorum*

Continuing our long-distance dialogue with Tamassia, we need to think about the elements that favored the establishment of the principle of testamentary succession in the Roman-Barbarian *leges*. After the fall of the empire, *militia*, romanity and *barbaricum* constituted a fruitful trinomial with political, military and legal consequences.[39] The situation of fluidity in the political and institutional structures described above led to the strengthening of the barbarian commanders' power on their *gentes*, the so-called 'signori della guerra'.[40] Those *signori* imposed themselves as necessary interlocutors of Rome, thus giving life over time to that plurality of Roman-Barbarian kingdoms whose formation represents the political outcome of the crisis in those years.[41] In Tamassia's perspective, Germanic life slowly penetrated into the forms of Roman institutions and was marked by them. The military function undertaken by those populations 'si connette(va) necessariamente e legalmente all'estensione ad esse di quei privilegi militari, che vanno (andavano), dal diritto di alloggio, alle

38 This is evidenced by the examination of a number of inscription titles: a selection in Tamassia, *Testamentum militis*, p. 515.

39 The complex of military units comprising the *scholae palatinae* and the *comitatenses* was strengthened by the populations – whether or not barbarian – established on the borders of the empire or inland, along strategic lines. *Laeti, Gentiles, Sarmati* made up these 'militarised' peoples, who were granted land for the defence of the *limes* or for the burden of recruitment and service imposed on them (cf., in this regard, Valerio Marotta, 'Il problema dei laeti. Fonti e storiografia', in F. Botta – L. Loschiavo, eds., *Civitas, Iura, Arma. Organizzazioni militari, istituzioni giuridiche e strutture sociali all'origine dell'Europa* (*secc. III–VIII*) [Lecce, 2015], pp. 117 on). The raids of Alaric, Attila and Genseric upset the Roman military order. The ephemeral resurgence of the city militias, framed by *numeri* and *bandi*, was of little use: the history of this anxious resistance is summed up in the figures of the 'generalissimi' Stilicho and Aetius. The banks were now definitively broken. Remains of the army survived sporadically even after 476, but did not retain any strategic importance.

40 According to an effective description by Loschiavo, *L'età del passaggio*, pp. 170 on.

41 In this way Pierfrancesco Porena, 'La posizione dell'elemento militare nell'Impero romano e i "regni romano-barbarici"', in Botta – Loschiavo, eds., pp. 221 on. On aspects common to the genesis of the so-called Romano-Barbarian kingdoms, for all, Santo Mazzarino, *L'impero romano*, vol. 2 (Roma-Bari, 1973 = 2005), pp. 812 on.

concessioni della *sors barbarica*, alla giurisdizione castrense; e infine alla *testamenti factio*'.[42]

In my opinion, the merit of Nino Tamassia's 'continuist' approach consists precisely in recognizing the extreme risks involved in attributing the spread of testamentary succession among those tribes to single causes, such as the emulation of Rome, *commilitium*, and, last but not least, the religious impulse to care for the soul. On the way to *favor testamenti*, the particular nature of *bona castrensia* played a crucial role. In this regard, a text of *Vegetius*[43] stands out. In order to save the soldier from easy prodigality, it was established that half of the donation received from the soldiers should be deposited *apud signa*, where it should be kept not to be wasted in unnecessary expenses or in sheer vices. This deposit of money was considered to be for the benefit of the troops themselves; since their maintenance was at public expense, the military salary was increased by half of each donation. Moreover, the soldier who was aware that his earnings were deposited (and, consequently, administered) by the *signiferi* was less likely to desert; on the contrary, that deposit worked as an incentive to his attachment to the insignia, spurring him to fight bravely. As for the distribution of the sums, for each cohort were placed ten saddlebags – that is, ten bags in which the sum was kept. An eleventh sack, to the filling of which the whole legion contributed, was then destined to cover the expenses for the burial of comrades fallen in service. The *signiferi* kept this sum in a basket. They were selected not only for their honesty, but also for their level of education: they had to be able to keep the deposit, accounting for each one.

In Iuven. *Sat.* 16.50 on[44] we find confirmation both of the concession of the *testamenti factio* to the *filius familias*, and of the particular nature of the *bona castrensia*. Those who returned (typically not empty-handed) from military service were reluctant to share their combat earnings with their families. There was therefore a conflict of interest between those who remained at home and those who preferred to enlist.[45] The status of the *pater familias miles* also had a significant impact on the path towards testamentary dispositions. The sudden enrichment and the profound changes in the living conditions of the barbarian

42 Tamassia, *Testamentum militis*, p. 519 on.

43 Veg. *Epit. rei militaris* 2.20.

44 'Solis praeterea testandi militibus ius vivo patre datur. Nam quae sunt parta labore militiae, placuit non esse in corpore census, omne tenet cuius regimen pater. Ergo Coranum signorum comitem castrorumque aera merentem quamvis iam tremulus captat pater; hunc favor aequus provehit et pulchro reddit sua dona labori. Ipsius certe ducis hoc referre videtur ut qui fortis erit, sit felicissimus idem, ut laeti phaleris omnes et torquibus, omnes ...'.

45 Other examples of 'approximation' to testamentary provisions found in *leges barbarorum* in Tamassia, *Testamentum militis*, p. 522 on.

populations in contact with the Roman environment posed an obstacle to intestate succession, as inheritance now acquired a mainly economic content.

Even so, Tamassia finds a tenacious resilience of the Germanic institutions in the succession of Athalaric to Theoderic, as recalled by his minister Cassiodorus.[46] On his deathbed, Theoderic gathered around him the great men of the kingdom, representatives of the *exercitus Gothorum*, and put his nephew ut *dominus* on the throne, transferring to him the king's powers without interruption. Those present expressed their approval and took an oath of fidelity to Athalaric.[47] The consent of the *populus-exercitus* therefore healed the irregularity of the election: even if for a few moments, to all intents and purposes, the Goths had two kings.[48]

On the other hand, the influence of barbaric elements increased moral disorder in the army. In many cases it was impossible to apply the rules of intestate succession because of the lack of a marriage tie between those whose inheritance was involved. This was a context in which wills were more suited to provide for the needs of soldier' family.[49] Often – it is good to remember – the soldiers joined *focariae*, slaves, prisoners of war, sometimes already married. The condition of the presumed children of the *miles* remained always suspended, given his faculty of *praeterire* and *exheredare* . With the will, children were acknowledged or disavowed: in the institution of the heir, or in its opposite (disinheritance), the testator's judgment on his possible paternity was implicit. In the chaos of the soldier' family, wills constituted a powerful weapon in the hands of the father against those children whom the environment and the penetration of Germanic law made reckless or rebellious.[50]

6 Title 46 of the *Pactus Legis Salicae*

Within the context outlined above, I will focus on a particular title of the *Pactus Legis Salicae*. First of all, I will look for clues to confirm the hypothesis that

46 Tamassia, *Testamentum militis*, p. 533 on.

47 Cass., *Variae* 8.2–7.

48 Cassiodorus' account bears a striking analogy to Gregory of Tours' account of Gontranus' adoption of Childebert, an act accomplished through the *traditio* of the auction. Like Theoderic, Gontranus had no male heirs: Greg. Tur. *Lib. hist.* 5.17; 7.33. For an excursus of the provisions *on testamentum militis* in the *leges Romana Wisigothorum* and *Romana Burgundionum*, see Tamassia, *Testamentum militis*, p. 534 on.

49 On the composition of soldiers' families: C. 12.35.10.

50 For the dissemination of the will in the *Codex Euricianus* cf. Tamassia, *Testamentum militis*, p. 541 on.

the structure of the provisions on testamentary matters in the *lex Salica* came from Roman law.

With the generic name of Franks – as we know – the Romans indicated a plurality of Rhineland people, often in competition with each other: not yet a unitary people, much less a compact and defined one. The singular denomination of the *Salii* (after the name of tribe that imposed its rule on the others) seems quite ancient. It could derive from their residence or, as suggested more recently, from the link with the veneration of the god Odin: thus, *Sal* would be the cord worn around the neck as a sign of devotion.[51] Having prevailed militarily over them, Julian then allowed some groups of that tribe to settle as *laeti* in the Roman region which today lies astride Holland and Belgium, so as to repopulate it, make it again productive and defend it. For several generations the Salian Franks lived there, respecting the alliance with Rome: the *Notitia Dignitatum* recalls the military cooperation and therefore the training with Roman weapons and the adoption of Roman customs.[52] They never led an open war against the imperial government, but rather distinguished themselves among other tribes and gained the trust of the emperors (the figure of the 'generalissimo' Arbogast is well known). A century later, between 450 and 480, the Salian Franks were led by Chlodio, Merovech and Childerich in turn: it is with them that the Merovingian dynasty began. In 481, Childerich was succeeded by his son Clovis. Remigius, bishop of Reims, behaved with the young sovereign in a way not dissimilar to what Ambrose had done previously with Gratian and Theodosius. He drew up a precise plan of government for Clovis, a plan marked by an alliance with the Catholic clergy. During the thirty years in which he remained in power, Clovis was able to unify the Franks and subjugate the entire Roman Gaul, thus inaugurating the kingdom of the Franks, the *Francia*.[53] This way, the Gallo-Roman institutions were born.

Today, the most credited hypothesis on the origin of the *Pactus Legis Salicae*[54] – on the basis of its content and the environment in which it was

---

51 On the origin of the name of *Salii* see, for all, Jean-Pierre Poly, 'La corde au cou. Les Francs, la France et la loi salique', in *Genèse dell'Etat moderne en Méditerranée* (Roma – Paris, 1993), pp. 287 on.

52 *Or.* 5.10.29; *Occ.* 7.62.67 (Salii seniores, Iuniores Gallicani).

53 Gregory of Tour compared Clovis' conversion to that of Constantine. This was, after all, an important transition: at the end of the fifth century most of the Romanesque inhabitants of the West were professing the Catholic faith, while the other minorities of barbarian origin remained faithful to Arianism: cf. Loschiavo, *L'età del passaggio*, p. 177.

54 For all, Jean-Pierre Poly, 'Sous les chênes de Salaheim. La loi salique, l'armée romaine et le bilan de la barbarie', in Botta – Loschiavo, eds., *Civitas, Iura, Arma*, 193 on; Id., 'Leges barbarorum. La création des lois des nations', in S. Kerneis, ed., *Une historie juridique de*

drafted – dates it earlier than the reign of Clovis and ascribes a hybrid character to it – both ethnic and military.[55] In particular, according to Jean Pierre Poly, the oldest nucleus of clauses would have been devised by people who might have fulfilled at the same time both the role of Roman army officers and that of leaders of barbarian military bands, in order to provide for the settlement of the Salian tribe. The *Pactus*, therefore, would have promoted peaceful coexistence in this community, preventing the continuous outbreak of feuds and ensuring the cultivation of fields and the regular supply of army recruits. The same ungrammatical denomination of *pactus legis* could arise from an original pacificatory function of the principles contained in it.[56]

According to Nino Tamassia, Roman elements would have penetrated into the PLS both directly and indirectly. Directly, with the acceptance of typically Roman institutes (such as *fides facta*, *verbum regis*, the rules on the second marriage of the widow, manumission *ante regem*, the penalties for the union of free with servants). Indirectly, by imitation of the Eurician codex, the oldest Visigothic codex.[57] But it is in the title 46 PLS, on 'affatomia',[58] that Tamassia

---

l'Occident. Le droit et la coutume (*III*^E^–*IX*^E^ *siècle*) (Paris, 2018), pp. 216 on. The reference edition of the *PLS*, is edited by Eckardt for the *MGH*.

55 Poly, 'La corde au cou'; Soazick Kerneis, 'Rome et les barbares. Aux origines de la personnalité des lois', in Botta – Loschiavo, eds., *Civitas, Iura, Arma*, p. 114; Jean-Pierre Poly, 'Freedom, warrior's bond, legal book. The "Lex Salica" between Barbarian custom and Roman Law', in *Clio@Themis* 10 (2016).

56 This interpretation is substantially accepted and enriched with new arguments by Loschiavo, *L'età del Passaggio*, p. 178 on, who, in particular, points out the use of a simple and approximate Latin and the constant recourse, in the drafting of the rules, to a compositional scheme characterised by the introductory formula *si quis* followed by the case considered and the corresponding economic compensation provided by the sanction. A scheme that serves a twofold purpose: easy memorization and easier application by the military judge.

57 Tamassia, *Testamentum militis*, p. 557 on.

58 *PLS* 46: 1. 'Hoc convenit observare, ut thunginus aut centenarius mallum indicant et scutum in ipso mallo habere debent et <postea> tres homines tres causas <in mallo ipso> demandare debent. Et postea requirant hominem, qui ei non pertineat, et sic festucam in lesum <su(um)> iactet. Et ipse, in cuius lesum festucam iactavit, dicat verbum, de fortuna sua quantum ei voluerit dare, aut si totam aut mediam furtunam <suam>, cui voluerit dare. 2. Postea ipse, in cuius lesum festucam iactavit <ipse>, in casa ipsius manere debet et hospites tres <vel amplius> suscipere debet, et de facultate sua, quantum ei datur, in potestate sua habere debet. Et postea ipse, cui iustum creditum est, ista omnia cum testibus collectis agere debet. 3. <Et> sic postea aut ante regem aut in mallo legitimo illi, cui fortunam suam deputavit, reddere debet et accipiat festucam <et> in mallo ipso ante duodecim menses ipsi, quos heredes deputavit, in lesum iactet, nec minus nec maius <et> nisi quantum ei creditum est. 4. Et si contra hoc aliquis aliquid dicere voluerit, debent tres testes iurati dicere, quod ibi<dem> fuissent in mallo, quem thunginus aut centenarius

saw the clearest evidence of such an influence. First of all, in the title: *adfathamire, adfatomire* would derive from *fathom, fathum,* that is *amplexus.* It is also necessary, however, to take into account a very complicated manuscript tradition.[59] Grimm had already described the formalities of adoption fulfilled by the embrace,[60] through which the adopted child would be assimilated to the natural child. First, the child was 'created'; only then was he armed. *Adfathamire* would include *traditio, donatio post obitum,* reciprocal dispositions of last will, *adoptio in hereditatem* and – this is the heart of Tamassia's thesis – also *fideicommissum*: a 'nome germanico (che) copre merce romana'. Hence the need for a fruitful comparison between Roman law historians and medieval law scholars (another 'pioneering' merit to be acknowledged to the author of *Il testamentum militis*).

The rite of the *adfathamire* took place in the *mallus* – the legal assembly held periodically in each village – presided over by the *thunginus*[61] or *centenarius*.[62] In front of the army (or part of it) assembled together to settle disputes,

indixerunt, et quomodo vidissent hominem illum, qui fortunam suam dare voluerit <et> in lesum <ille> iam quem elegit festucam iactare. <Et> nominare debent denominatim illum, qui fortunam suam in lesum iactat; et sic illum, in cuius lesum iactatur, et illum, quem heredem appellat, publice similiter nominent. 5. Et alteri tres testes iurati dicere debent, quod in casa illius, qui fortunam suam donavit, ille, in cuius lesum festuca iactata est, ibidem mansisset et hospites tres vel amplius ibidem collegisset et pavisset et hospites illi tres vel amplius <ei> de susceptione gratias egisset et in beodo pultes manducassent et illi testes <tres> collegissent. 6. <Et> ista omnia illi alii tres testes iurati dicere debent et <de> hoc, quod in mallo ante regem vel <in> legitimo mallo publico <et> ille, qui accepit in lesum fortunam ipsa(m), aut ante regem aut in mallo publico legitimo, hoc est in mallobergo anttheoda aut thungino, fortunam illam, quos heredes appellavit, publice coram omnibus festucam in lesum illius iactasset, hoc est ut novem <testimonia vel> testes ista omnia debent adfirmare.'

59 Cf. ed. Eckhardt on p. 176 on.

60 Grimm, *Deutsche Rechtsalterthümer*, pp. 638 on.

61 The *thunginus* was dentified in a subject linked to the religious world, of which he knew and perpetuated the rituals. He expressed, to all effects, the values of the old tribal society.

62 Next to the *thunginus*, the *PLS* mentions the *centenarius*. Unlike the former, the *centenarius* comes from the Roman military world, an army non-commissioned officer, the late-antique heir of the old centurion: cf. Veg. *Epit. rei militaris*, 2.13. The *centenarius* commanded a *centena*, over which he exercised territorial control and police duties. However, by delegation of his superiors, he could also carry out the task of the judge, especially in the most remote places, both in the case of the involvement of a soldier and, more generally, in judgments of modest value. It was therefore often the case that private individuals themselves had recourse to this military judge: in the eyes of those living near a garrison, the centenarian's 'justice' had to appear effective in terms of speed and economy. In essence, the *centenarius* best represented the new order imposed by the sovereigns of the Merovingian dynasty. The *PLS* still bears traces of this moment of transition; by contrast, the sons of Clovis seem to know only the centenary. On the centenarians, their derivation

they performed what would be called today acts of voluntary jurisdiction. At the beginning of the 1900s, German historiography recognized in title 46 PLS the 'germ' of the medieval testamentary executor.[63] There are three subjects involved in it: the person who wanted to provide for his estate, the trusted person to whom the estate was transferred, and finally the heir. As such, we can identify many moments. The first: in the *mallus*, duly convened and constituted, the first person came forth, advancing towards another *libero, qui ei non pertineat, et sic festucam in lesum <su(um)> iactet. Et ipse, in cuius lesum festucam iactavit dicat verbum, de fortuna sua quantum ei voluerit dare, aut si totam aut mediam furtunam <suam>, cui voluerit dare*. The ritual of throwing the *festuca in lesum* (in the breast) of the one who received it was the material expression of the *traditio* of the thing.

The second moment: the trustee had to stay in the house of the person who threw the *festuca* in his breast, he had to receive three guests, in the presence of as many witnesses, who could swear to have seen him together with the three guests, sitting at the table eating the *puls*.

Third moment: in the *mallus*, the trustee had to make it clear that *ante regem … fortunam illam, quos heredes appellavit, publice coram omnibus festucam in lesum illius iactasset, hoc est ut novem <testimonia vel> testes ista omnia debent adfirmare*. All this had to be carried out within twelve months. The *traditio* to the heir, moreover, had to be *nec minus nec maius* <et> *nisi quantum ei creditum est*.

Some have seen in the affatomy the *adoptio per arma*: the former, i.e. the institution based on the embrace, could constitute a step in the adoption process, and it would have as its logical and necessary sequel the delivery of weapons.[64] According to Nino Tamassia, however, with this name the Salic law indicated another institution, which was characterized by a different origin and structure.[65] The Salic law divides in many real scenes the choice of the trustee, the conferment of the mandate and its execution – all acts that would properly belong to the *fideicommissum*. For this law, in *contrast* to the Roman one, legal relationships must be created materially. Therefore, it is possible to

---

from the old centurions, and their function as judges, see Alexander Callander Murray, 'From Roman to Frankish Gaul: 'Centenarii' and 'Centenae' in the Administration of the Merovingian Kingdom', in *Traditio* 44 (1988), pp. 59 on and now Esders in this volume.

63 Robert Cailleimer, *Origines et développement de l'exécution testamentaire* (Lyon, 1901), p. 29.

64 Other scholars have traced the affatomy in the *mancipatio familiae*: thus Schmidt-Recla, *Mancipatio familiae und Affatomie*. In the bibliography mentioned, however, the significant absence of Nino Tamassia's work should be noted.

65 Tamassia, *Testamentum militis*, pp. 563 on.

see in the affatomy the main lines of the *fideicommissum* (an institution so 'malleable' that could be shaped after the needs of the *castra*), expressed and implemented with special forms, but Roman in its substance.

7 A Long-Distance Dialogue

Let us attempt, at this point, to answer the question asked in the introduction of this essay. Is there any evidence that the testamentary provisions contained in the *Pactus Legis Salicae* borrowed their structure from Roman law?

In my opinion, if we look at it thinking of military law as *ius singulare*, then the answer is positive. The influence of Roman law on the PLS can be clearly seen only in the rule on the inheritance in the absence of male offspring, forbidding daughters to succeed in the *terra Salica*: PLS 59.[66] This is the land received in concession from the emperor at the time of the settlement of these populations. This singular exclusion – it is well known that among the Franks matrilinearity was favored[67] – follows an ancient Roman military principle, according to which the possession of the lands assigned to the *limitanei*[68] was conditional upon the exercise of the *militia*. In other words, these plots of land could only be transmitted to male descendants, since each of them would have had to provide one or more recruits. The *Historia Augusta* points to this practice in two third-century emperors *vitae*, although there is no explicit reference to settlements by groups of barbarian origin. Alexander Severus assigned territories to *milites* on the condition that those territories remained their 'property' only if their heirs would also pursue a military career, and that they could never come into private possession. Moreover, the emperor supplied the soldiers with cattle and slaves, to cultivate what they had received and to avoid that the lack of arms or the owners' advanced age would cause

66 (*De alodis*) 'De terra vero Salica nulla in muliere [portio aut] hereditas est, sed ad virilem sexum, qui fratres fuerint, tota terra pertineat.' On the exclusion of daughters from the inheritance of terra Salica cf. Thomas Anderson, 'Roman military Colonies in Gaul, Salian Ethnogenesis and the forgotten meaning of "Pactus Legis Salicae" 59.5', in *Early Medieval Europe*, 4 (1995), pp. 129 on and Jean-Pierre Poly, 'Terra salica. De la société franque à la société féodale: continuité et discontinuité', in J. Pérez – S. Aguadé Nieto, eds., *Les origines de la féodalité. Hommage à Claudio Sánchez Albornoz* (Madrid, 2000), pp. 189 on.

67 See Valerio Marotta, 'Prefazione. Esercito, cittadinanza e integrazione tra principato ed età tardo antica', in D.A.N. Costa, ed., *Civitas et conubium. Integrazione degli stranieri e politica militare nell'Impero romano* (Milano, 2019) pp. 23 on.

68 Cf. Benjamin Isaac, 'The Meaning of the Terms Limes and Limitanei', in *Journal of Roman Studies* (*JRS*), 78 (1988), pp. 125 on.

the abandonment of the lands bordering with those of the barbarians.[69] In the biographer's vision, the allocation of tax lands to the *limitanei* would therefore have had a twofold purpose: to solve the problem of their livelihood and, at the same time, ensure a regular flow of recruits to the wards. A passage from the *Vita Probi* contains a similar measure: Probus gave to the veterans as their private possessions all those places that were difficult to access, on condition that their sons at the age of eighteen would be enlisted in the army.[70]

I wonder, then – despite the undoubted merits of the work of Nino Tamassia – how much of his 'continuist' approach (as we would call it nowadays) is to be ascribed to the ideological setting dominant at the beginning of the nineteenth century, namely the desire to recognize in the barbarian gentes as many peoples-nations, with a distinct individuality and a specific character, identifying in them the ancestors of many of the current European populations. The most paradigmatic case of this ideology, needless to say, concerns the Germanic tribes in their different articulations.[71]

I also wonder if, by looking at this subject from the different perspective of ethnogenesis[72] – according to which the great barbarian tribes would have

69 Hist. Aug., *Sev. Alex.* 58.4–5: '4. Sola, quae de hostibus capta sunt, limitaneis ducibus et militibus donavit, ita ut eorum essent, si heredes eorum militarent, nec umquam ad privatos pertinerent, dicens attentius eos militaturos, si etiam sua rura defenderent. 5. Addidit Sne his et animalia et servos, ut possent colere, quod acceperant, ne per inopiam hominum vel per senectutem possidentium desererentur rura vicina barbariae, quod turnpissimum ille ducebat.'

70 Hist. Aug., *Prob.* 16.6: 'Veteranis omnia illa, quae anguste adeuntur, loca privata donavit addens, ut eorum filii ab anno octavo decimo, mares dumtaxat, ad militiam mitterentur, ne latrocinare umquam discerent.' Marotta, *Prefazione*, p. 27 and fn. 55, 56, has noted how this passage finds a valuable term of comparison in the devices of many constitutiones collected in C. 11.60 (*De fundis limitotrophis et terris et paludibus et pascuis limitaneis vel castellorum*).

71 See Loschiavo, *L'età del passaggio*, pp. 122 on.

72 Starting with the studies of Ernesto Sestan, *Stato e nazione nell'Alto Medioevo. Ricerche sulle origini nazionali in Francia, Italia, Germania* (Napoli, 1952) and Reinhard Wenskus, *Stammesbildung und Verfassung. Das Werden der frühmittelalterlichen gentes* (Köln-Graz, 1961), the inconsistency of historical reconstructions based solely on a biological concept of ethnicity was proved, demonstrating the complexity and variability of the mechanisms that accompany the formation of peoples. On this historiographical turning point, cf. Stefano Gasparri, *Prima delle nazioni. Popoli, etnie e regni fra Antichità e Medioevo* (Roma, 1997), pp. 11 on, 63 on; Walter Pohl, *Le origini etniche dell'Europa. Barbari e Romani tra antichità e medioevo* (Roma, 2000); Patrick Geary, *Il mito delle nazioni. Le origini medievali dell'Europa*, It. transl. (Roma, 2009) (orig. ed., Princeton N.J., 2002); Goets – Jarnut – Pohl, *Regna and Gentes*; James, *I barbari*, pp. 149 on. More recent interventions attest the liveliness of the debate on the ethnic identity of the barbarians: see Gasparri – La Rocca, *Tempi barbarici*, pp. 83 on; Rosalba Arcuri, 'Etnogenesi, «entelechia barbarica» e attuali

been formed as a result of particular events, where a decisive role was played by the cultural factor, by the commonality of interests and, above all, by the political and military skills of some leaders in exploiting contingent situations, in short, by their ethnic and cultural pluralism – the answer to this question continues to be positive, and how much of Nino Tamassia's reconstruction was influenced by the ideology just set out (in this regard, however, Tamassia himself excluded the existence of the presumed unitary 'Germanic law').[73] As Walter Pohl reminds us, in many cases the barbarian formations that lived permanently in the empire did not have a consolidated history behind them. Consequently, today it is clear to all that 'nessuno dei popoli europei moderni può affermare di essere erede diretto di una *gens* altomedioevale'.[74]

Bibliography

Primary Sources

Cassiodorus Senatoris, *Variae*, ed. Th. Mommsen [MGH, *Scriptores – Auct. ant.* XII] (Berlin, 1894) *etiam* Flavio Magno Aurelio Cassiodoro Senatore, *Varie*, (dir. A. Giardina), eds. A. Giardina – G.A. Cecconi – I. Tantillo – F. Oppedisano, **6 vols.** – Engl. transl. by M.S. Bjornlie, *The Variae. The Complete Translation.* (Oakland, CA, 2019)

CIL = *Corpus Inscriptionum Latinarum*, consilio et auctoritate Academiae litterarum Borussicae editum, Berlin, 1863-

Edictum Theoderici regis, ed. F. Bluhme [MGH *Leges* V] (Hannoverae, 1875–1889); Engl. transl. S. Lafferty, *Law and Society in the Age of Theoderic the Great. A Study of the* Edictum Theoderici (Cambridge, 2013), pp. 243–294

Gaius, *Institutiones*, ed. (and Engl. transl.) W.M. Gordon – O.F. Robinson, *The Institutes of Gaius*, London 1988

Gregorius Turonensis, *Libri Historiarum X*, ed. B. Krusch – W. Levison [MGH, *Script. Rer. Meroving.* I.1] (Hannover, 1951) *etiam* [*Historiarum libri decem*, ed. Rudolf Buchner, *Ausgewählte Quellen zur Geschichte des Mittelalters*, 2 (Darmstadt, 1974)] – Engl. transl. by L. Thorpe, *Gregory of Tour, The history of Franks* (London, 1974)

Historia Augusta, ed. D. Magie, *The Scriptores Historiae Augustae*, with an english translation by D. Magie (London, 1921)

orientamenti storiografici sulla Völkerwanderungszeit', in *Koinonia*, 37 (2013), pp. 107 on; Stefano Gasparri, 'I barbari, l'impero, l'esercito e il caso dei Longobardi', in Botta –. Loschiavo, eds., *Civitas Iura, Arma*, pp. 91 on.

73 Tamassia, *Testamentum militis*, p. 545.

74 Pohl, *Le origini etniche*, p. 57. For an excellent overview of the recent historiographical debate between 'catastophists' and 'continuists' see the *Introduction* of Chris Wickham, *Framing the early Middle Ages. Europe and the Mediterranen, 400–800*, (Oxfdord, 2005).

ILS = *Inscriptiones Latinae Selectae*, ed. H. Dessau. Weidmann, Berlin 1892–1916

Iustinianus, *Codex*, ed. P. Krüger in T. Mommsen – P. Krüger, *Corpus Iuris Civilis*, II (Berlin, 1954[11]) – Engl. transl. B.W. Frier, *The Codex of Justinian. A New Annotated Translation*, I–III, (Cambridge, 2016)

Iustinianus, *Digesta*, ed. T. Mommsen – P. Krüger, *Corpus Iuris Civilis*, I (Berlin, 1963[17]) - From: *The Digest of Justinian*, English translation edited by Alan Watson, 4 vols. (Philadelphia, Pennsylvania, 1985)

Iustinianus, *Institutiones*, ed. P. Krüger, in *Corpus Iuris Civilis*, I (Berlin, 1963[17]) – Engl. transl. S. Scott, *The civil law*, vol. 2 (Cincinnati, 1932)

Iuvenalis, Saturae, ed. W.V. Clausen, Oxford Classical Text (Clarendon Press, 1992)

Notitia Dignitatum: accedunt Notitia urbis Constantinopolitanae et laterculi prouinciarum, ed. O. Seeck (Berlin 1876 = Frankfurt, 1983); ed., C. Neira Faleiro, *La Notitia Dignitatum. Nueva edicion critica y comentario historico* (Madrid, 2005); ed. in progress, I. Maier, *The compilation 'notitia dignitatum', Cnd* (//www.notitiadignitatum.org). Units: Hoffmann, 1969–1970

Pactus Legis Salicae, ed. K.A. Eckhardt [MGH, *Leg. Nat. Germ.*, 4.1] (Hannoverae, 1962)

Pauli Sententiae, ed. M. Bianchi Fossati Vanzetti, *Pauli Sententiae. Testo e interpretatio* (Padova 1995)

Tacitus (Publius Cornelius), *Germania*, ed. J. Perret (Paris, 1949 = 1983).

Tituli ex corpore Ulpiani, ed. J. Baviera [Fontes Iuris Romani Antejustiani, III. *Auctores*] (Florentiae, 1968), pp. 259–301

Vegetius, *Epitoma rei militaris*, ed. A. Önnerfors (Stuttgart, 1995)

Secondary Literature

Anderson, Thomas. 'Roman military Colonies in Gaul, Salian Ethnogenesis and the forgotten meaning of "Pactus Legis Salicae" 59.5', in *Early Medieval Europe*, 4 (1995), pp. 129–144

Arcuri, Rosalba. 'Etnogenesi, «entelechia barbarica» e attuali orientamenti storiografici sulla Völkerwanderungszeit', in *Koinonia*, 37 (2013), 107–142

Barbier, Josiane. 'Testaments et pratique testamentaire dans le royaume franc (VIe–VIIIe siècles)', in F. Bougard – C. La Rocca – R. Le Jan, eds., *Sauver son âme et se perpétuer. Transmission du patrimoine et mémoire au haut moyen âge* (Roma, 2005), pp. 7–79

Caillemer, Robert. *Origines et développement de l'exécution testamentaire* (Lyon, 1901)

Calasso, Francesco. *Medio Evo del diritto. 1. Le fonti* (Milano, 1954)

Camodeca, Giuseppe. 'Il giurista L. Neratius Priscus cos. Suff. 97. Nuovi dati su carriera e famiglia', in *Studia et documenta historiae et iuris* (*SDHI*), 73 (2007), pp. 291–311

Cugusi, Paolo. *Per un nuovo corpus dei Carmina Latina Epigraphica. Materiali e discussioni* (Roma, 2007)

Davies, Wendy. 'Buying with Masses: "donation" pro remedio animae in Tenth-Century Galicia and Castile-León', in F. Bougard – C. La Rocca – R. Le Jan, eds., *Sauver son âme et se perpétuer. Transmission du patrimoine et mémoire au haut moyen âge* (Roma, 2005), pp. 401–416

De Falco, Ines. 'I giuristi e il testamentum militis. L'orientamento di Iavolenus Priscus', in *Studia et documenta historiae et Iuris* (*SDHI*), 80 (2014), pp. 419–446

De Francisci, Pietro. 'Quelques observations sur l'histoire du testament conjonctif réciproque', in *Revue historique du droit français et étranger* (*RHDFE*), 3 (1924), pp. 57–82

De Ruggiero, Ettore. 'Corporis custodes', in *Dizionario epigrafico di antichità romane,* II.II (Spoleto, 1910), pp. 1237–1240

Eckardt, Bernd. *Iavoleni Epistulae* (Berlin, 1978)

Ferrini, Contardo. 'Sulle fonti delle Istituzioni di Giustiniano', (1901) now in Id., *Opere.* II. *Studi sulle fonti del diritto romano* (Milano, 1929) pp. 307–419

Forni, Giovanni. 'Esperienze militari nel mondo romano', in *Nuove questioni di storia antica* (Milano, 1972), pp. 815–863

Fraenkel, Béatrice. *La signature. Genèse d'un signe* (Paris, 1992)

Gasparri, Stefano – La Rocca, Cristina. *Tempi barbarici. L'Europa occidentale tra antichità e medioevo (300–900)* (Roma, 2012)

Gasparri, Stefano. *Prima delle nazioni. Popoli, etnie e regni fra Antichità e Medioevo* (Roma, 1997)

Gasparri, Stefano. 'I testamenti nell'Italia settentrionale fra VII e IX secolo', in F. Bougard – C. La Rocca – R. Le Jan, eds., *Sauver son âme et se perpétuer. Transmission du patrimoine et mémoire au haut moyen âge* (Roma, 2005), pp. 97–114

Gasparri, Stefano. 'I barbari, l'impero, l'esercito e il caso dei Longobardi', in F. Botta – L. Loschiavo, eds., *Civitas Iura, Arma. Organizzazioni militari, istituzioni giuridiche e strutture sociali all'origine dell'Europa (secc. III–VIII)* (Lecce, 2015), pp. 91–102

Geary, Patrick. *Il mito delle nazioni. Le origini medievali dell'Europa*, It. transl. (Roma, 2009) (orig. ed., Princeton N.J., 2002)

Goets, Hans-Werner – Jarnut, Jörg – Pohl, Walter, eds., *Regna and Gentes. The Relationship between Late Antique and Early Medieval Peoples and Kingdoms in the Transformation of the Roman World* (Leiden-Boston, 2003)

Grimm, Jacob. *Deutsche Rechtsalterthümer*, vol. 1 (Leipzig, 1899)

Guarino, Antonio. 'L'oggetto del «castrense peculium»', in *Bullettino dell'Istituto di diritto romano* (*BIDR*), 48 (1941), pp. 41–73

Harries, Jill. 'Saturninus the Helmsman, Pliny and Friends: Legal and Literary Letter Collections', in A König – C. Whitton, eds., *Roman Literature under Nerva, Trajan and Hadrian. Literary Interactions, AD 96–138* (Cambridge, 2018), pp. 260–279

Helgeland, John. 'Christians and the Roman Army from Marcus Aurelius to Constantine', in *Aufstieg und Niedergang der römischen Welt* (*ANRW*), II 23.1 (Berlin-New York, 1979), pp. 724–834

Isaac, Benjamin. 'The Meaning of the Terms Limes and Limitanei', in *Journal of Roman Studies* (*JRS*), 78 (1988), pp. 125–147

James, Edward. *I barbari*, Ital. transl. (Bologna, 2011) (orig. ed. Harlow, 2009)

Jeay, Claude. *Signature et pouvoir au Moyen Age* (Paris, 2015)

Kerneis, Soazick. 'Francus civis, miles Romanus: les barbares de l'Empire dans le Code Théodosien', in J.-J. Aubert – P. Blanchard, eds., *Droit, religion et sociétè dans le Code Théodosien* (Genève, 2009), pp. 377–399

Kerneis, Soazick. 'Rome et les barbares. Aux origines de la personnalité des lois', in F. Botta – L. Loschiavo, eds., *Civitas, Iura, Arma. Organizzazioni militari, istituzioni giuridiche e strutture sociali all'origine dell'Europa (secc. III–VIII)* (Lecce, 2015), pp. 103–116

La Rosa, Franca *I peculii speciali in diritto romano* (Milano, 1953)

Lehmann, Burkhard. 'Das Eigenvermögen der römischen Soldaten unter väterlicher Gewalt', in *Aufstieg und Niedergang der römischen Welt* (ANRW), II.14 (Berlin-New York, 1982), pp. 183–284

Loschiavo, Luca. *L'età del passaggio. All'alba del diritto comune europeo (secoli III–VII)*, 2nd. ed. (Torino, 2019)

Lovato, Andrea. 'P.Fay. 10 (Ulpianus 45 ad edictum) + P.Berol. inv. P. 11533', in D. Mantovani, S. Ammirati, eds., *Giurisprudenza romana nei papiri. Tracce per una ricerca* (Pavia, 2018), pp. 131–136

Magnani, Eliana. 'Don aux églises et don d'églises dans le sud-est de la Gaule: du testamen d'Abbon (739) aux chartes du début du XIe siècle', in F. Bougard – C. La Rocca – R. Le Jan, eds., *Sauver son âme et se perpétuer. Transmission du patrimoine et mémoire au haut moyen âge* (Roma, 2005), pp. 379–400

Marotta, Valerio. *Mandata principum* (Torino, 1991)

Marotta, Valerio. 'Il problema dei laeti. Fonti e storiografia', in F. Botta – L. Loschiavo, eds., *Civitas, Iura, Arma. Organizzazioni militari, istituzioni giuridiche e strutture sociali all'origine dell'Europa (secc. III–VIII)* (Lecce, 2015), pp. 117–157

Marotta, Valerio. 'Prefazione. Esercito, cittadinanza e integrazione tra principato ed età tardo antica', in D.A.N. Costa, ed., *Civitas et conubium. Integrazione degli stranieri e politica militare nell'Impero romano* (Milano, 2019), pp. 11–30

Mazzarino, Santo. *L'impero romano*, vol. 2 (Roma-Bari, 1973 = 2005)

Modzelewski, Karol. *L'Europa dei barbari. Le culture tribali di fronte alla cultura romano-cristiana*, Ital. transl. (Torino, 2008) (orig. ed., Warsawa, 2004)

Murray, Alexander Callander. 'From Roman to Frankish Gaul: 'Centenarii' and 'Centenae' in the Administration of the Merovingian Kingdom', in *Traditio* 44 (1988), pp. 59–100

Nappi, Sergio. *Ius finitum* (Bari, 2005)

Pohl Walter, ed., *Kingdoms of the Empire. The Integration of Barbarians in Late Antiquity* (Leiden-New York-Köln, 1997)

Pohl Walter. *Le origini etniche dell'Europa. Barbari e Romani tra antichità e medioevo* (Roma, 2000)

Poly, Jean-Pierre. 'La corde au cou. Les Francs, la France et la loi salique', in *Genèse dell'Etat moderne en Méditerranée* (Roma – Paris, 1993), pp. 287–320

Poly, Jean-Pierre. 'Terra salica. De la société franque à la société féodale: continuité et discontinuité', in J. Pérez – S. Aguadé Nieto, eds., *Les origines de la féodalité. Hommage à Claudio Sánchez Albornoz* (Madrid, 2000), pp. 183–197

Poly, Jean-Pierre. 'Sous les chênes de Salaheim. La loi salique, l'armée romaine et le bilan de la barbarie', in F. Botta – L. Loschiavo, eds., *Civitas, Iura, Arma. Organizzazioni militari, istituzioni giuridiche e strutture sociali all'origine dell'Europa (secc. III–VIII)* (Lecce, 2015), pp. 193–219

Poly, Jean-Pierre. 'Freedom, warrior's bond, legal book. The "Lex Salica" between Barbarian custom and Roman Law', in *Clio@Themis* 10 (2016): http://journals.openedition.org/cliothemis/1149

Poly, Jean-Pierre. 'Leges barbarorum. La création des lois des nations', in S. Kerneis, ed., *Une historie juridique de l'Occident. Le droit et la coutume (IIIE–IXE siècle)* (Paris, 2018), pp. 187–246

Popescu, Andreas. *Die Franken und das römische Militär Identität in Gallien und dem Rheinland in der Spätantike* (Bonn, 2017) http://hss.ulb.uni-bonn.de/2017/4760/4760.pdf

Porena, Pierfrancesco. 'La posizione dell'elemento militare nell'Impero romano e i 'regni romano-barbarici'', in F. Botta – L. Loschiavo, eds., *Civitas, Iura, Arma. Organizzazioni militari, istituzioni giuridiche e strutture sociali all'origine dell'Europa (secc. III–VIII)* (Lecce, 2015), pp. 221–257

Reddé, Michel. *Mare nostrum. Les infrastructures, le dispositif et l'histoire de la marine militaire sous l'empire romain* (Roma, 1986)

Rigsby, Kent J. 'Two Danubian Epitaphs', in *Zeitschrift für Papyrologie und Epigraphik* (ZPE), 126 (1999), pp. 175–176

Scarano Ussani, Vincenzo. *Valori e storia nella cultura giuridica fra Nerva e Adriano. Studi su Nerazio e Celso* (Napoli, 1979)

Scarano Ussani, Vincenzo. *Empiria e dogmi. La scuola proculiana fra Nerva e Adriano* (Torino, 1989)

Scarano Ussani, Vincenzo. *Le forme del privilegio. Beneficia e privilegia tra Cesare e gli Antonini* (Napoli, 1992)

Schmidt-Recla, Adrian. 'Mancipatio familiae und Affatomie. Überlegungen zu Parallelentwicklungen im römischen und fränkischen Recht und zu Rezeptionsbedingungen im Frühmittelalter', in G. Dilcher – E.-M. Distler, eds., *Leges – Gentes – Regna. Zu Rolle von germanischen Rechtsgewohnheiten und lateinischer Schrifttradition bei der Ausbildung der frühmittelalterlichen Rechtskultur* (Berlin, 2006), pp. 461–486

Sestan, Ernesto. *Stato e nazione nell'Alto Medioevo. Ricerche sulle origini nazionali in Francia, Italia, Germania* (Napoli, 1952)

Tamassia, Nino. *Testamentum militis e diritto germanico*, (Venezia 1927 = in Id., *Scritti di storia giuridica*, 3, Padova 1969, 423–581

Tanzarella, Sergio. *I cristiani e il servizio militare nella Chiesa antica*, in *Asprenas* 31 (1984), pp. 75–88

Vendrand-Voyer, Jacqueline. 'Origine et développement du «droit militaire» romain', in *Labeo*, 28 (1982), pp. 259–277

von Bolla, Sibylle. 'Zum römischen Militärtestament', in *Studi in onore di Vincenzo Arangio-Ruiz nel* XLV *anno del suo insegnamento*, I (Napoli, 1953), pp. 273–278

von Harnack, Adolf. *Militia Christi, Die christliche Religion und der Soldatenstand in den ersten drei Jahrhunderten* (Tübingen, 1905)

Wenskus, Reinhard. *Stammesbildung und Verfassung. Das Werden der frühmittelalterlichen gentes* (Köln-Graz, 1961)

Wickham, Chris. *Framing the early middle ages. Europe and the Mediterranen, 400–800*, (Oxfdord, 2005)

Wood, Ian. *The Trasformation of the Roman West* (Kalamazoo and Bredford, 2017)

PART 3

Symbols, Rituals and Identity Models

CHAPTER 10

The *Cingulum Militiae* in the Early Middle Ages: Between Status and Function

Andrea A. Verardi

1 Introduction

Modern historiography emphasises the central role of the army in the new cultural shape of the Latin West between the periods of Late Antiquity and the Early Middle Ages. If on the one hand the army was a vehicle of Roman culture, on the other hand, due to its internal evolution and its increasing openness to the recruitment of non-Romans from the fourth century, it was permeable to alien customs and traditions.[1] The Roman army underwent a reciprocal acculturation that, starting with the armies stationed along the fluid Roman *limes*, gradually extended to urban societies. This acculturation nourished the transformation of the Roman society and characterised these centuries.

An indicator of these dynamics is the persistence and re-semantization of traditional Roman forms of self-representation. Indeed, since late Roman society was strongly hierarchically structured, the Romans used a set of external symbols to emphasise the status and functions of people. These symbols functioned as a formal language, represented by specific clothes and accessories regulated by the public authority that distinguished men on the basis of their profession or social role.[2] The most famous examples of this are the use of the white toga for members of the senate and the chlamys and onion-head *fibula* worn by the military and public officials.[3] The state directly produced

1 Philipp von Rummel, *Habitus barbarus: Kleidung und Repräsentation spätantiker Eliten im 4. und 5. Jahrhundert* (Berlin - New York 2007).

2 Javier Arce, 'La ostentació del rango en la vida y en la muerte', in *Archeologia e società tra tardo antico e alto medioevo*, eds. G.P. Brogiolo – A. Chavarría Arnau (Mantova 2007), pp. 265–278.

3 Elisa Possenti, 'Abbigliamento e rango in Italia settentrionale tra v e vi secolo', in, *Archeologia e società tra tardo antico e alto medioevo*, eds. G.P. Brogiolo – A. Chavarría Arnau (Mantova 2007), pp. 279–298.

© KONINKLIJKE BRILL BV, LEIDEN, 2024 | DOI:10.1163/9789004698017_011 and distributed some of these distinctive accessories, such as *fibulae*, because they were linked to the exercise of a public function.[4]

Among these symbolic attributes, the so-called *cingulum militiae* or *balteus* stands out: the leather belt with metal inserts was a distinctive sign of soldiers and, from the fourth century onwards, of all those performing public service (*militia*) for the Roman State.[5]

Indeed, throughout the Middle Ages, documentary, narrative and legal sources often referred to the *balteus*. It also became central in the tenth century to the liturgy of knightly adornment.[6]

This continuity has led historians to question its actual scope, especially for the early Middle Ages. Karl Ferdinand Werner saw in the use of the *cingulum* the expression of a direct derivation of the medieval nobility from the Roman one.[7] More recent historiography has acknowledged the role of the *cingulum* as a marker of status for elites but has also stressed the difficulty of understanding its real value.[8]

For this reason, my paper aims to investigate whether an evident continuity in lexical use is matched by a similar conceptual and functional identity; that is, whether the link between the object (the belt) and what it represented (the office held by the wearer) remains the same despite the institutional and social changes of the early Middle Ages. The starting point of my analysis is the late Roman context. Then, I consider some of the main early medieval institutional governments (Gothic, Frankish, Lombard and Carolingian kingdoms). The

---

4 On *fibulae* see Dominic Janes, 'The golden clasp of the Late Roman State', in *Early Medieval Europe*, 5 (1996), pp. 127–153 and Simon Esmonde, *The Roman west. AD 200–500: An Archaeological Study* (Cambridge 2013), pp. 55–60, 79–90.

5 On the use of the *cingulum* in the Middle Ages see Johanna M. Winter, 'Cingulum Militiae', in *Revue d'histoire du droit*, 44 (1976), pp. 1–92; Jean Flori, 'Les origines de l'adoubement chevaleresque: étude des remises d'armes et du vocabulaire qui les exprime dans les sources historiques latines jusqu'au début du XIIIe siècle', in *Traditio*, 35 (1979), pp. 209–272; Karl Ferdinand Werner, *Naissance de la noblesse* (Paris 1998), pp. 187–225; Karl Leyser, 'Early Medieval Canon Law and the Beginnings of Knighthood', in *Institutionen, Kultur und Gesellschaft im Mittelalter. Festschrift für Josef Fleckenstein zu seinem 65., eds. Geburtstag* L. Fenske – W. Rösener – T. Zotz, (Sigmaringen 1984), pp. 549–566; Sabine Krüger, 'Character militaris und character indelebilis', in ibid., pp. 567–580. For a summary of the archaeological evidence see Markus Sommer, *Gürtel und Gürtelbeschläge des 4. und 5. jahrhunderts im römischen Reich* (Bonn 1984), pp. 83–118.

6 Flori, 'Les origines'.

7 Werner, *Naissance*, pp. 187–225. For a criticism of Werner, see Igor Mineo, 'Di alcuni usi della nobiltà medievale', in *Storica*, 7 (2001), pp. 9–58.

8 See Chris Wickham, 'The changing composition of early élites', in, *Théorie et pratiques des élites au Haut Moyen Âge. Conception, perception et réalisation sociale*, eds. F. Bougard – H.-W. Goetz – R. Le Jan (Turnhout 2011), pp. 5–18 esp. p. 10.

chronological span of this paper is between the fourth and ninth centuries. This is for two reasons: first, the fourth century coincided with the extension of the use of this symbol in its application to civil officials; second, the ninth century coincided with a new regulation of the use of the *cingulum* by Carolingian sovereigns.

2 Between Otherness and Assimilation: The *Cingulum's* Shift from Sign of Service to Sign of Status (Fourth-Fifth Centuries)

It is impossible to date precisely the introduction and adoption of the *cingulum* in the Roman army. Originally, as attested in sources between the 1st century BC and the 1st century AD, it seems to have been used for practical reasons. The belt held weapons or tools, or even a pouch to hold one's salary.[9] However, it soon took on a more distinctive function. The belt denoted the person of the soldier, his position with his own society and his legal status.[10]

The shape of the army in the imperial age influenced this development. It consisted of small corps, directly subject to the emperor and physically distant from civil life. Indeed, in the imperial and late imperial period, the army was a kind of society separate from the urban one, from which it differed in culture, virtues and norms.

It was in this milieu that the *cingulum* became a symbol: it represented, *tout court*, the military component of the society and, by metonymy, the very concept of service.

The iconography of the funerary *stelae* of *milites* in the empire between the second and third centuries and its uniformity shows that this society produced well-structured forms of self-representation. In this iconography of the ideal soldier, the distinctive attributes are the *cingulum*, the onion-head *fibula*, the shield and the lance.[11]

---

9 Sommer, *Gürtel*, pp. 83–118.

10 Pierfrancesco Porena, 'La posizione dell'elemento militare nell'Impero romano e i 'regni romano-brbarici'', in *Civitas, Iura, Arma. Organizzazioni militari, istituzioni giuridiche e strutture sociali alle origini dell'Europa* (*secc. III–VIII*), eds. F. Botta – L. Loschiavo (Lecce 2015), pp. 221–258 e Sylvain Janniard, 'L'esercito del tardo impero romano. Dalla tetrarchia a Giustiniano', in *Storia d'Europa e del Mediterraneo 7,3: L'Ecumene romana. L'impero tardoantico*, ed. A. Barbero (Roma 2009), pp. 495–523.

11 See e.g. Giovanni Lettich, *Itinerari epigrafici aquileiesi. Guida alle iscrizioni esposte nel Museo archeologico nazionale di Aquileia* (Trieste 2003) e Sofie Waebens, 'The Representation of Roman Soldiers on Third-Century AD Funerary Monuments from Nikopolis (Egypt)', in *Revue internationale d'histoire militaire ancienne*, 1 (2014), pp. 61–75.

Belts, which were of course already known and used in the Roman world, became a fundamental part of the *habitus militaris* mainly through the mediation of the *foederati* contingents of barbarian origin. The emperor had progressively regulated and standardised the use and style of these belts, thanks also to the centralised production by state-owned workshops.[12]

The emperor wore the *cingulum* as the supreme head of the army. So did the military commanders and legionnaires, as they perceived the bond between the emperor and commanders as reciprocal. The *ordines* were marked by the shape and the materials of the *cingulum*; belts were made of gold, silver and bronze, although the custom of embellishing them with precious stones was the privilege of the emperor alone.

By accepting this badge, the soldiers, forced to live for a jubilee (20 or 25 years) on the borders of Roman *civilitas*, made an unbreakable bond of trust with the Augustus. As a reward for their loyalty and service, he guaranteed them food, provisions and a regular salary from the imperial treasury.[13]

As a symbol of bond and service, the *cingulum militiae* also became part of the religious language. The virtues of self-denial and service that it represented were well suited to late antique monastic Christianity. Indeed, the monks of this period represented themselves as the new and true *militia*, the one at the service of Christ. The monks represented this service by the *cingulum*, attributing to the belt the role of service and the bond of chastity.[14]

The use of the *cingulum* changed during the fourth century due to the reforms made by Diocletian and Constantine.

The division between civil and military careers had excluded members of the senatorial aristocracy from positions in the army. The reorganisation of the army established specialised troops, the *comitatensi*, led by the *magistri* and *comites*, who were better paid and stationed in the empire. Men who joined these new divisions as senior military officers were permitted to join the senate at the end of their service.[15]

---

12 Dominic Janes, 'The golden clasp of the Late Roman State', in *Early Medieval Europe*, 5 (1996), pp. 127–153.

13 Porena, 'La posizione', pp. 221–258.

14 On the use of a military terminology in Christianity see Adolf Von Harnack, *Militia Christi. Die Chrisliche Religion und der Soldatenstand in den ersten drei Jahrhunderten* (Tübingen 1905); Friedrich Prinz, 'Primi stadi della militia Christi altomedioevale nella tarda antichità e nel sistema ecclesiastico imperiale del periodo carolingio e degli Ottoni', in *Militia Christi e Crociata nei secoli XI–XIII*. Atti della XI Sett. Intern. di studio (Mendola, 28 agosto-1 settembre 1989) (Milano 1992), pp. 49–63.

15 Janniard, *L'esercito*, pp. 495–523. On the quartering of the Comitatense troops cf. Whitby, Michael. 'Army and Society in the Late Roman World: Context for Decline?', in *A companion to the Roman Army*, ed. P. Erkdamp (Hoboken - NJ 2010) pp. 515–531.

Moreover, there was a greater presence of barbarian contingents both in the structured bodies of the Roman army, such as the *laeti* and *dediticii*, and, starting with Honorius (395–423), the *foederati*.[16]

The increasing non-Roman presence in the army had a deep influence on military society, changing its characteristics. As Von Rummel pointed out, the barbarian *habitus*, once thought to be feature of barbarian populations, seems instead to indicate military status.[17]

Permeable to internal and external influences, the military society poured its acculturation potential into the empire. The permanent war that affected the West between the end of the fourth and the first two decades of the fifth century had changed Roman society. The inhabitants of the cities invaded were influenced by the military habits after coming into direct contact with the army.

Three laws of the Emperors Gratian, Valentinian and Theodosius illustrate this change. These laws forbade senators from wearing military footwear and trousers in Rome and Constantinople, a sign of the spread of these elements outside their traditional context.[18]

Theodosius' law of 382 explicitly mentions the *cingulum*. The emperors Gratianus, Valentinianus and Theodosius forbade the use of *cingulum* by those who did not hold public office.[19] The rule, addressed to the senators of Constantinople, emphasises the introduction of military customs into city life, perhaps for cultural and fashionable reasons. It also indicates that the senators might perceive it convenient or normal to do military dress at certain public events (such as a meeting of the Senate or a public judgment). It is possible that the senators realized the potential of the *habitus miltiae* in urban society and decided to use it to demonstrate their status at a time of transition, a milieu in which the old *insignia* might no longer be sufficient.

If on the one hand, the emperors Gratianus, Valentinianus and Theodosius tended to repress these actions, calling the senators to their *mores*, they also extended the use of the *cingulum* to public officials. This initiative was part of a process of militarisation of the Roman state aimed at increasing the

16 Luca Loschiavo, *L'età del passaggio. All'alba del diritto comune europeo (secoli III–VII)* (Torino 2019²), pp. 135–138.

17 Von Rummel, *Habitus barbarus*. *Contra* Liebeschütz, Wolfgang. 'Habitus barbarus: did barbarians look different from Romans?', in *Expropriation et confiscations dans les royaumes barbares. Une approche régionale*, eds. P. Porena – Y. Rivière (Roma 2012), pp. 13–28, which emphasises the importance of written sources between the fourth and sixth centuries.

18 On connection between ostentation of rank and dress on the basis of Roman law see Arce, 'La ostentación', pp. 257–264.

19 CTh. 14.10.1 e CTh. 14.10.2.

obligations of public officials and their submission to the emperor. They also subjected civil officials to legislation modelled on the military one.

The emperors Theodosius and Valentinianus themselves gives us a precise indication of the relationship between *dignitas* and *cingulum* in a law of 440.[20] They indicate the rank attributed to high dignitaries of the empire and enumerates five positions of dignity of the *viris illustres*.[21] The highest rank belonged to those who had actually held offices and received the *cingulum* proper to indicate the rank of *illustris*. Then came those who were titular office-holders and who, while at court (*in comitatu*), received the *cingulum*. In the third rank were those who, although holding the same rank as the second, were not at court but received a *cingulum* from the emperor. The last two ranks belonged to those who had held honorary offices either in the presence of the court or away from it, and received it as a certificate of their office, a *codicillus* of the rank of *illustris*. In the cases mentioned, the granting of the *cingulum* depended exclusively on the possession of an effective (not honorary) public office, granted by the emperor.

It is not clear whether soldiers and officials had to return the *cingulum* at the end of their service or if they were permitted to keep it. A letter dating from 401 prescribing the leave of eleven soldiers from an *ala* established at Psoftis in Egypt required the prefect to ensure the return of the soldiers' equipment, which may have included the *cingulum*.[22] Some hagiographic texts written at the end of the fourth and fifth centuries however contradict this practice. In these writings, the *cingulum* remained in the possession but not in use of the soldier.[23]

Obviously, those who occupied leading positions or were closer to the emperor could obtain further privileges. A law of Emperor Anastasius, dating between 492 and 497, granted the promotion to *viris illustribus* to those who had served as chamberlains to the imperial couple. These men could continue wearing the *cingulum* at the end of their service if associated with the senatorial *ordo* when they went to inspect their lands or for other matters. This example emphasises that the chamberlains conferred particular value to the

20 C. 12.8.2.

21 According to *the notitia dignitatum*, dating from the early fifth century, the *viris illustres* held the following offices: *Praefectus praetorio, Praefectus urbi, Magister militum, Praepositus sacri cubiculi, Magister officiorum, Quaestor sacri palatii, Comes sacrarum largitionum, Comes rerum privatarum.*

22 It is the Papyrus Vindob. Inv. L. 8 +125 edited by John Rea, 'A Cavalryman's Career A.D. 384 (?)-401', in *Zeitschrift für Papyrologie und Epigraphik* ZPE), 56 (1984), pp. 79–88.

23 See David Woods, 'The Ownership and Disposal of Military Equipment in the Late Roman Army', in *Journal of Roman Military Equipment Studies*, 4 (1993), pp. 55–65.

cingulum as a means of communicating their status as high dignitaries close to the emperor, even after the end of their office.

A homily by Maximus of Turin (who was bishop from 380 to 423) gives a concise picture of the use of the *cingulum* in relationships between the emperor, his officials and the aristocracy. The prelate describes how the *militiae*, civil and military, prepared for the celebrations of the emperor's *dies natalis*. The text offers a vivid description of the public use of the *cingulum* as a symbol of status.[24]

The bishop depicts the urban elites in the context of the hierarchical structure of Rome. In the upper ranks are the *duces* and *principes*. Then come those who hold positions in the administration (*militantes*). Elaborate silk robes and belts embellished with shining gold adorned all these persons for the solemn festivities of the Emperor's birthday. The prescribed uniforms faithfully reflect the structure of the late Roman aristocracy. There are those who held public offices, performing both civil and military service, and the provincial aristocracy of the city *curia*. Interestingly, the use of the *cingulum* is not only the prerogative of the two *militiae*, but also of the aristocracy, which expressed its status in a similar way to the other classes.

The main recipient of the ostentation of these luxury items by the elites was the emperor. The urban elite aimed in this way to impress the emperor, with the aim of receiving more gifts from him. Emperors traditionally gave gifts to their loyalists at such events.[25]

In general, the homily emphasises two key points about the use of the *cingulum*. First, all the actors involved used a belt as a symbol of status. Here, the

24 Maximus Taurinensis, sermon 60 edited in PL 57, col. 0221, 'Laetitia quanta sit, quantusque concursus, cum imperatoris mundi istius natalis celebrandus est, bene nostis, quemadmodum duces eius et principes omnes, etiam militantes accurati sericis vestibus, accincti operosis cingulis, auro fulgente pretiosis, ambiant solito nitidius in conspectu regis incedere. Credunt enim maius esse imperatoris gaudium, si viderit maiorem suae apparitionis ornatum, tantoque illum laetum futurum quanto ipsi fuerint in eius solemnitate devoti: ut quia imperator tanquam homo corda non conspicit, affectum eorum erga se probet vel habitum contuendo: ita fit, ut splendidius se accuret, quisquis regem fidelius diligit. Deinde quia in die natalis sui sciunt eum largum futurum, ac donaturum plurima vel ministris suis, vel his, qui in domo eius abiecti putantur, et viles, tanta prius thesauros eius replere divitiarum varietate festinant; ut in quantum praerogativa voluerit, in tantum praerogatio copiosa non desit, et ante voluntas donandi deficiat, quam substantia largiendi. Haec autem ideo sollicite faciunt, quia maiorem sibi remunerationem pro hac sollicitudine sperant futuram.', engl. transl. B. Ramsey, *The Sermons of St. Maximus of Turin* (New York 1989), pp. 144–147.

25 Roland Delmaire, *Largesses sacrées et res privata. L'aerarium impérial et son administration du* IVe *au* VIe *siècle* (Rome 1989).

cingulum symbolises more a concept of 'ruling classes' than of service. Second, these groups of people use the *cingulum* to consolidate their positions, also economically, through their relationship with the emperor. The *cingulum* is therefore a tool to attract imperial attention and favour, and to increase the *Königsnähe*.

These strategies centred on the *cingulum* and similar objects, however, they still required imperial regulation. Emperor Leo, who ruled between 457 and 474, forbade the use of pearls, emeralds, and sapphires to embellish bridles, saddles, and belts, reserving their use exclusively to himself.[26] The rule also forbade the imitation of accessories or clothes that belonged only to the emperor, even if a gift for the emperor motivated their creation. The reason for this prohibition, apart from the obvious desire to avoid dangerous imitations of the *habitus imperialis* (socially destabilising and demythologising his authority) was that the emperor had no need of gifts (he was the only one who could exercise *largitas*). Moreover, the production of this special type of luxury object was the sole responsibility of the court artisans.

Let us now see how much of these dynamics remain in some of the new Romano-Barbarian societies of the Latin West.

3 The Cingulum in Ostrogothic Italy, from Theodoric to the Byzantines

The Amal Theodoric, who had spent some ten years (from eight to eighteen) as a hostage at the imperial court (Leo I) before returning to Pannonia, arrived in Italy at the head of an army of *foederati* warriors in 484. This was after he had received the title of *patricius* from Emperor Zeno in 476 and then those of *magister militum presentalis* and *consul* in 478.[27] Theodoric was a warlord of barbarian origin, fully integrated into the Roman imperial structure of power.

The Valameriac Goths had elected Theodoric king in 474 and, after a series of displacements in the Danubian regions, he came to Italy and made an agreement with the emperor Zeno. Zeno offered Theodoric the rule of the region if he defeated Odoacer.[28]

After his victory over the Herul in 493, Theodoric took over the government of the Italic territories, seeking the stabilisation and legitimisation of his new

26 C. 9.11.12.

27 Jonathan J. Arnold, *Theoderic and the Roman Imperial Restoration* (Cambridge 2014).

28 Anonymus Valesianus, *Pars posterior*, § 11.49 (ed. Mommsen, p. 316).

conquest. Having asked in vain from the emperor to recognise his regency,[29] he assumed the title of *rex* from the Gothic troops he led, and thus combined the Roman military title of *magister militum* with the title of king.[30]

Conquering Italy, Theodoric ruled a population of Goths and Romans formally subordinate to the Emperor of Constantinople, who recognised his royal authority only in 498. He based his reign on a formal dualism: military roles were exclusive to the Goths, while civil offices were reserved for the Roman elite. This was a central distinction, as the military connotation imposed itself as the identity of the Goths in Italy.[31] For this reason, it is not surprising that the Goths used the *cingulum militiae* in continuity with the Roman world.

During the reign of Theodoric, a letter dated between 507 and 511 (Varie III, 23) mentioned the *cingulum* for the first time. The letter dealt with the designation of Colosseus as governor of *Pannonia Sirmiensis*, designating him *inlustris cinguli dignitate*.[32] Further fame, however, depended on the quality of the service provided and on the success of the campaign entrusted to him. Indeed, only this would allow Colosseus to obtain the royal favour.

The *cingulum* was also central in the attribution of the honour of the patriciate.[33] In Varie (6, 2), the administration of justice motivated the granting of the *cingulum*. In this case, however, Theodoric gave the *cingulum* to the patrician in perpetuity. Indeed, unlike the other offices, the *patricius* did not have to remove his belt at the end of the service.[34] Even under Theodoric, the patriciate was the highest honour – higher than the prefecture and all the other *dignitates*, second only to the consulate, which, unlike the patriciate, was of limited duration.[35]

Another passage in the Varie (9, 18) speaks of the relationship between the exercise of justice on behalf of the sovereign and the exhibition of the

29 *Ibid.*, § 12.57 (p. 322).

30 Stefano Gasparri, 'I barbari, l'impero, l'esercito e il caso dei Longobardi', in *Civitas, Iura, Arma. Organizzazioni militari, istituzioni giuridiche e strutture sociali alle origini dell'europa* (*secc. III–VIII*), eds. F. Botta – L. Loschiavo (Lecce 2015), pp. 91–102.

31 Patrick Amory, *People and identity in Ostrogothic Italy* (*489–554*), (Cambridge 1997). *Contra* Halsall, Guy. 'The Ostrogothic Military', in *A Companion to Ostrogothic Italy*, eds. J. Arnold – S. Bjornlie – K. Sessa (Leiden 2016), pp. 173–199.

32 See Cassiodorus, *Variae* (ed. Giardina *et all.*, II: Libri III–V, pp. 243–245).

33 See Wilhelm Enßlin, 'Aus Theoderichs Kanzlei', in *Würzburger Jahrbücher für die Altertumswissenschaft*, 2/1 (1947), pp. 75–85.

34 See Cassiodorus, *Variae*, and engl. transl. by Bjornlie, at p. 247.

35 *Ibid.*

cingulum. In Athalaric's edict of 533, the leave of a negligent *iudex* came with the loss of the *cingulum* and the *dignitas* attached to it.[36]

In all attestations in the *Variae*, the *cingulum militiae* represents the public function. The *cingulum* represents a dignitas or office delegated by the king with a clear reference to the regulation of previous imperial legislation. The honorary office connected to the granting of the *cingulum* thus seems to be assigned according to the rules of Roman law, with the novelty that it is granted by Theodoric in the fulfilment of his government functions.

The lack of information on Gothic burials with weapons makes it difficult to establish a material correspondence with the conceptual continuity of the use of the *cingulum*.[37] In this period, we know of only two male tombs with weapons attributable to the Goths: tomb 25 in Dravlje, Slovenia, dated between the fifth and sixth centuries, and tomb 11 in Globasnitz, Carinthia, dated to the late sixth century. Of the two, the Austrian find is an exceptional case. It contains a *cingulum* preserved both in its metal part, consisting of a plate and cloisonné trimmings, and of its support partially made of leather. In the grave there is also an onion-head *fibula* that contained a woollen *chlamys*.[38] Therefore, from a material point of view, there seems to be a continuity in the use of distinctive markers of the Roman military/administrative system in the Theodoric era.[39]

---

36 The rule is a citation of Valentinian II's provision punishing the iudex's bad exercise of justice. Gothic law, however, made it more severe. In addition to a fine, the Gothic sovereign provided for the dismissal of the judge. In the same context, in the Roman area, we also find a reference to the *cingulum*, both in the so-called apocrypha of the Simmachian period and in the first three versions of the *Liber Pontificalis*, to identify public judges (*iudices cincti*), see Andrea Antonio Verardi, *La memoria legittimante. Il liber pontificalis e la chiesa di Roma nel secolo VI* (Roma 2016), pp. 122–123.

37 See Volker Bierbrauer, 'Goti dell'Oriente e dell'Occidente europeo tra la fine del V e la prima metà del VI secolo alla luce delle fonti archeologiche', in eds., *Archeologia e storia delle migrazioni. Europa, Italia, Mediterraneo fra tarda età romana e alto medioevo*, eds. C. Ebanista – M. Rotili (Cimitile 2011), pp. 9–30; Marco Aimone, 'Romani e Ostrogoti fra integrazione e separazione. Il contributo dell'archeologia a un dibattito storiografico', in *Reti Medievali*, 13/1 (2012) pp. 1–66; Joan Pinar Gil, 'Chlamyds e cingulum nel tardo V secolo. Tre rinvenimenti dall'Emilia Romagna', in *Oreficeria in Emilia Romagna. Archeologia e storia tra età romana e medioevo*, eds. A.L. Morelli – I. Baldini Lippolis (Bologna 2010), pp. 229–256; Caterina Giostra, 'L'età di Teodorico. I reperti goti di Tortona', in *Onde nulla si perda. La collezione archeologica di Cesare Di Negro-Carpani*, eds. A. Crosetto – M. Venturino Gambari (Alessandria 2007), pp. 287–326.

38 Franz Glaser, *Ostgotisches Militär in Kärnten. Erste Funde aus der Zeit des Königs Theoderich des Großen (493–536)* (Klagenfurt 2003).

39 Glaser, Franz, 'Elementi di corredo di un ufficiale ostrogoto', in *Guerrieri, principi ed eroi fra il Danubio e il Po dalla preistoria all'alto medioevo*, eds. F. Marzatico – P. Gleirscher (Trento 2004), p. 739.

It is also significant that the *cingulum* in question has similar characteristics to some belt buckles found both in the Danubian area and in Italy. It belongs to a formal Roman *koiné*. The Austrian belt is similar to the one described by John Lydus in his Περὶ ἀρχῶν τῆς Ῥωμαίων πολιτείας.[40] Lydus describes the *cingulum* of the praetorian prefect as a very simple gold belt with a rectangular plate buckle made of folded foil mounted on a strip of red-dyed leather.

In considering this phenomenon, however, it should be noted that relations between the Gothic and Byzantine worlds were continuous, and the emperors used objects such as belts throughout the early Middle Ages to define the hierarchy of their officials and to establish and consolidate diplomatic relations.[41]

The question arises whether the outcome of the Greek-Gothic War and the consequent extension of the *pragmatica sanctio* in 554 over the Italic province further strengthened the traditional formalism of the *cingulum*, the use of which seems to have remained on the whole unaltered during the Gothic reign. At a first glance, looking to Gregory the Great's *exeplum legis,* it seems to be possible to answer affirmatively.[42] From an iconographic point of view, the Ravenna mosaics of S. Vitale show us a series of court dignitaries characterised by the traditional symbols of *dignitas*, such as the *chlamys*, the onion-head *fibula* and a red *cingulum* under their robes. The re-establishment of Byzantine authority in Italy, after the war had profoundly changed the local social structures, supported the emergence of a new aristocracy of soldiers and officials. This new military elite may have contributed to the emphasis placed on traditional symbols such as the *cingulum* and the *fibula* as elements of social-political legitimacy first and second status.[43]

40 Ioannes Lydus, *De magistratibus*, II § 13 ed. and transl. Jean le Lydien, *Des magistratures de l'État romain*, I–II, texte établi, trad. Michel Dubuisson – Jacques Schamp (Paris 2006).

41 Notorious is the case of the Gepids or the Avars, cf. Walter Pohl, *The Avars: A Steppe Empire in Central Europe* (567–822) (Ithaca NY 2018). Concerning the use of belts and *cingulum* in Constantinople see Stefan Albrecht, 'Warum tragen wir einen Gürtel? Der Gürtel der byzantiner – symbolik und function', in *Byzanz – das Römerreich im Mittelalter, 1: Welt der Ideen, Welt der Dinge*, eds. F. Daim – J. Drauschke (Mainz 2010), pp. 79–95.

42 Gregory uses the term *cingulum* in Ep. XIII. 50 (*Gregorii I papae registrum epistularum*, ed. Ewald – Hartmann, II, p. 416): an exemplum legis, addressed to the defensor of Spain John, in which Justinian's Nov. 123 is quoted literally. Gregory then uses the lexeme on two other occasions: in Ep. X.16, (*ivi*, p. 251) to indicate the civil functions assumed by Innocent as prefect of Africa, and in *Dialogorum libri*, Lib. IV,31 (ed. de Vogüé, T. III, (Paris 1980), pp.104–105) to describe Theodoric's loss of his royal *dignitas*.

43 Schoolman, Edward M., 'Nobility, aristocracy and status in early medieval Ravenna', in *Ravenna: Its Role in Early Medieval Change and Exchange*, in eds. J. Herrin – J. Nelson (London, 2016), pp. 211–238.

4 The Use of the *Cingulum* by the Franks

It is possible to compare by analogy the Italic context with the early phases of the Frankish kingdom, established in the northern territories of former Roman Gaul between the fifth and the sixth centuries.

In the southern territories, south of the Loire, were the stable kingdoms of the Visigoths and Burgundians where the elite of Roman origin continued to exercise a central role in society.

The senatorial aristocracy of this region linked their status to their nobility of birth. This enabled them to preserve their position, supported by the continuity of their function in the cities and their land holdings, even after the decline of imperial authority.[44] On a regional basis, however, the members of the Gallo-Roman elite continued to maintain their social status.

In the northern regions the situation was very different. Greater political uncertainty had favoured a wider militarisation of society. The fight for political and social pre-eminence had a number of military leaders of both Gallic Roman and barbarian origin as its main protagonists.[45] Both groups claimed their authority by submission to the empire as *foederati*, although imperial control was very difficult. The structure of these groups, independent of their ethnicity, conformed to Roman military law and represents the regulatory basis for the militarised society of northern Gaul in the fifth century.[46]

Here, the Gallic Roman aristocracy had only two ways to maintain their status. They could take on a militaristic attitude, using their land holdings as a power base and maintaining their own army. Alternatively, they could join the

44 Chris Wickham, *Framing the early Middle Ages: Europe and the Mediterranean (400–800)* (Oxford 2006), pp. 168 e sg.

45 For just over a century, the region was ruled by people who were part of the Roman military system. This was the case with the Roman *Aegidius*, who acted as *magister militum* on behalf of the Emperor Majoran and managed to control the entire region north of the Loire. When he died in 461, he was succeeded by his son Siagrio. They formally had control of a number of foederate military leaders, including the founder of the Merovingian dynasty, Childeric.

46 See Jean-Pierre Poly, 'Sous lex chênes de Salaheim. La loi salique, l'armée romaine et le bilan de la barbarie', in *Civitas, Iura, Arma. Organizzazioni militari, istituzioni giuridiche e strutture sociali alle origini dell'europa (secc. III–VIII)*, eds. F. Botta – L. Loschiavo (Lecce 2015), pp. 193–219 and Soazick Kerneis, 'Garants et compensations: Romanité ou barbarie dans la tres ancienne loi des Bretons d'Armorique', in *Traditio iuris. Permanence et/ou discontinuité du droit romain durant le haut Moyen Âge*. ed. A. Dubreucq (Lyon 2005), pp. 77–92.

clergy, taking control of episcopal sees and abbeys, which gave them the control of major urban centres where the bishops had taken over the government.[47]

The rise of the Frankish Salians and the Merovingian dynasty brought with them a new state structure. The light structure of the Roman-Barbaric state required far fewer men and focused mainly on officials capable of leading armies and administering justice. In the early generations the assembling of armies, which were multi-ethnic and no longer based on a standing militia, were ruled by the king alone, but later these functions were delegated. *Duces* and patricians (in the more Romanized area), acting on a central or subregional level, and *comites*, whose range of action was predominantly urban, exercised these functions instead. Their legitimacy, as well as their chances of coming into possession of new wealth, depended on their relationship with the king. This was also true at the local level, where the counts, appointed by the king, often came from pre-existing local aristocracies that confirmed their social positions through their inclusion in the structure of the kingdom.[48] Bishops also played a central role in this new structure. They often came from the Gallo-Roman elite, and the kings often supported their election.

The establishment of the Merovingians in the seventh century throughout the former Roman Gaul led to a strong militarisation of the society's customs and habits.

For these reasons, it is not surprising to find an apparent continuity in the use of and references to the *cingulum militae*. From the point of view of material culture, the Frankish case has provided us with an exceptional example of a complete set from the tomb of Childericus I, found in a burial site discovered in the 17th century in Tournai, Belgium.[49]

This is a burial complete with military equipment, including a ring with the image and title of a king, a shield, an onion-head *fibula* and, indeed, sections of a multiple belt (*cingulum*) of high quality and decorated in *cloisonné*.

The burial equipment, dated before the 482, allows us to make some important observations. First, the social aspect of burial with grave goods; Childeric represented himself with military equipment, which made him look similar to his superior and main opponent, the *magister militum* and *rex romanorum* Syagrius. This confirms the hypothesis that the military function supports the

47 Laury Sarti, 'The Military and Its Role in Merovingian Society', in *The Oxford Handbook of the merovingian world*, eds. B. Effros – I. Moreira (Oxford 2020), pp. 255–277.

48 Yitzhak Hen, 'The Merovingian Polity: A Network of Courts and Courtiers', in *The Oxford Handbook of the merovingian world*, eds. B. Effros – I. Moreira (Oxford 2020), pp. 217–237.

49 Michel Kazanski – Patrick Périn, 'Le mobilier funéraire de la tombe de Childéric 1er ; état de la question et perspectives', in *Revue archéologique de Picardie*, 3–4 (1988), pp. 13–38.

use of these objects as status symbols by an individual and his family. Indeed, Childeric's burial and the objects interred with him underline the shifting nature of the ethnic identity of the buried. Besides the military symbols of the late Roman military *koiné*, the grave also features the burial of horses, in accordance with a Germanic/Nomadic practice.

The funeral ceremony also served to emphasise the rank of the deceased and the status of his family. The purpose of this ritual was not only to define the parental group of the king, but also to facilitate the succession to the royal office by Clovis. It also emphasised that the military office of Roman origin held by the deceased was in the full possession of his family group. Part of the purpose of the rite was to avoid a renegotiation of the positions held by Childeric – both the royal one in the Franco military group and the Roman military one – with his direct superior in the army ranks. Although weak, contacts between the local authorities and the empire were still alive. Indeed, after his victory over Syagrius, Clovis received the insignia of the consulate (*codicilli*) directly from Constantinople.[50]

Belts, whole and in parts, are nevertheless generally present in Merovingian burials. This suggests their continuing value as a symbol of status as well as function.[51]

Returning to the written sources, the concept of *cingulum militiae* is widely attested, both in the southern and northern territories.

Venantius Fortunatus mentions the *cingulum* in both of his *carmina*, dedicated respectively to Lupus, Duke of Champagne, and Count and domesticus Conda, and the life of St. Martin and Germanus.[52] In the *carmina*, although the language is characterized by a strongly Romanizing rhetoric, the references are interesting. In both cases the *dignitas* and *honus* of the functions performed on behalf of the king motivates the use of the *cingulum*. Moreover, in Conda's case, his assumption of the dignity of count did not justify the granting of the *cingulum*; instead, it was a concession made by the king (Teudobert) to emphasise the value of the service carried out.[53]

Ulfus, minister of Queen Chrodosind, Count Nicasius, of the *castrum avalonis* on the border of the Dauphiné and the *vir illustris* Attila, *domesticus regalis*

50 Gregorius Turonensis, *Libri historiarum* X, II, cap. 38 (ed. Krush – Levison, p. 88).

51 Ralph J.Patrello, 'Belt Buckles and Burials in Soutwestern Gaul', in, *The Oxford Handbook of the merovingian world*, eds. B. Effros – I. Moreira (Oxford 2020), pp. 916–936.

52 Venantius Fortunatus, *Carmina*, Lib. VII, 7 e 16 dedicated respectively to the Duke Lupus and the *domesticus* Conda (ed. Leo, pp. 159–161 and 170–172).

53 'For Theudebert granted you the prize of being a count and added a belt in recognition of your services. He saw that excellent characters deserve better and was soon willing to raise the rank you had earned' (transl. by George, pp. 65–66).

aulae, were all given a *cingulum* (*balteum*).[54] All these people laid down their insignia in exchange for the intervention of Bishop Germanus. In some cases, he used them to free prisoners or to feed the poor.

It therefore appears that the king granted the *cingulum* to the high officials of the kingdom, dukes and counts, with both military and jurisdictional functions. The belt is also a symbol of prestige, which in the cases mentioned depends exclusively on *Königsnähe*.

The stable relationship between the *cingulum* and *dignitas* also recurs in the canons of the Council of Clermont of 535 and in that of Clichy of 626–627.[55] In this case, the *cingulum militiae* indicates people who hold public office in a list that also includes ranks and authorities.

The works of Gregory of Tours provide a more articulated picture of the use of the *cingulum* as a badge of function and status.

In the *Historiarium Libri* x, lib. vii c. 38, Gregory tells of the dispute between King Guntram and his brother Gundovald. The latter wears a golden *balteus* and a sword lent to him by the Duke of Avignon, Mummolus, as a sign of friendship.[56] Mummolus was a duke who had escaped from the royal court to take refuge in Avignon and offered Gundovald hospitality. Gundovald did not carry out any function on behalf of the duke whose guest he was. However, the example narrated by Gregory of Tours could be an example of the Germanic practice of *adoptio per arma*. As it is well known, this practice took place through the donation of a sword and cingulum.[57]

By offering these objects to Gundovald, Mummolus legitimised himself through his ability to offer hospitality and status to a member of the royal family. His belt and sword are symbols that the Duke wants to recover before delivering Gundovald into the hands of his enemies. By taking back these objects, Mummulus regains his goods of high economic value and publicly signals the severing of ties with Gundovald.

Two other parts of the *Historia* highlighted the giving of the *cingulum* and precious weapons as a means to establish or strengthen political relations. In the first case, the sons of the former palace master Waddo, who had been found

54 Venantius Fortunatus, *Vita sancti Germani*, respectively § 21 (ed. p. 16 – transl. George, pp. 384–385), § 31 (ed. p. 18 – transl. George, p. 391 and § 60 (ed. p. 23 – transl. George, pp. 408–409).

55 References to the cingulum are Council of Clermont (535) c.13 and Council of Clichy (626–627) c. 18 (edited both in *Les Canons*, 1, p. 216 and 2, p. 538). In the first case the reference sanctions the abandonment of civil offices before receiving ecclesiastical dignitas.

56 Gregorius Turonensis, *Libri hist.* x, Lib. vii, 38 (ed. Krush – Levison, pp. 359–362).

57 Emmanuelle Santinelli, 'Continuité ou rupture? L'adoption dans le droit mérovingien', in *Médiévales*, 35 (1998), pp. 9–18.

guilty of a number of acts of banditry, went to the king to anticipate the actions of Count Macco, who intended to stop their incursions.[58] They attempted to establish a direct relationship with the king through the donation of a *balteum* and a sword made of gold and precious stones.

The second reference concerns the cubicularius Eberulf, who locked up in the basilica of St. Martin in Tour, formed a bond with a 'cheap' man from the city through the donation of a piece of his golden *balteum.*[59]

Archaeologically, the Frankish region demonstrates the practice of breaking and distributing pieces of a *cingulum.* Archaeologists and historians have generally interpreted belt breakage as a ritual to emphasise the relationship between the deceased and his heirs at the time of burial with the aim of encouraging the intergenerational transmission of identifying elements of status. In Gregory's work, however, it was the creation of a network of social relations of dependence/friendship/collaboration that motivated this practice *inter vivos.*[60]

The success of the *cingulum* as a status symbol unrelated to the exercise of a public function emphasises its economic value. In addition to its power as a symbol, the belt was a luxury good capable of underlining the rank of its owner. At the same time, because it was obviously personal and recognisable, it could be an instrument of the social strategies of its owner, in life or after death.

The gift of belts or parts of belts to living persons favoured the consolidation of the possessor's social network and the public manifestation of his power.

However, the *cingulum* was also an object of pious donations to ecclesiastical bodies. This practice is widely attested in the Frankish world. The abovementioned cases, reported by Venantius Fortunatus, refer to public officials donating in *ex voto* for grace received (in reality, the bishop imposed these donations on them in exchange for his intervention). Still, Gregory of Tour reports the donation of 'a belt that had been most carefully crafted out of pure gold and its fittings' on the altar of the Church of St. Nazarius near Nantes.[61] The *ex voto* was from a pious man who offered the gift so that the martyr would assist him in his affairs. The function of the gift was to ensure the sustenance of the poor and of those who performed the service at the martyr's grave.

A further noteworthy aspect is that the belt became the target of Britto, the most influential of the retainers of count Waroch of Brittany, a high-ranking

58 Gregorius Turonensis, *Libri hist. x*, Lib. x, 21 (ed. Krush – Levison, pp. 513–514).

59 Ibid., Lib. VII, 28 (p. 346).

60 Santinelli, 'Continuité ou rupture?'

61 Greogorius Turonensis, *Miracula*, c. 60 (ed. Krush, p. 79 – Engl. transl. by van Dam, pp. 58–59).

figure in the service of a count who in this case acted completely autonomously. He was keen on the possibility of emphasising his status autonomously, i.e., without being loyal to a superior.

The Queens Radegund and Balthild used the *cingulum* in the same way. The former, wife of Clotarius I, placed her *cingulum auri ponderaturm* on the altar and had it broken to distribute to the poor.[62]

The second had a royal belt, probably belonging to her husband, King Clovis II. She had probably worn it during her regency for her infant son. She then donated the belt and distributed it to the poor when she entered a monastery.[63]

In these cases, the belt was part of the movable property of the Merovingian aristocracy as a sign of status. Indeed, it guaranteed the persistence and replicability of a family's rank regardless of the exercise of a function delegated by some authority.[64]

The use of the *cingulum* as a pious donation is also significant. While in the case of men it seems to be an *ex voto* donation, without necessarily implying a change in the donor's status, in the case of women, the dismissal of the *cingulum* is associated with the change of status of the Queens, who embraced a monastic life.[65] Through this act, those who donated *the cingulum* guaranteed the family of origin the emphasis of their status, the chosen institution the means of support for the poor, and themselves the salvation of their souls.

5 The *Cingulum* in Lombard Italy: A Discontinuity?

The use of *cingulum* was significantly different in Lombard Italy. This use can partly be attributed to a geographical differentiation and a continuous evolution over the three centuries.

Upon their arrival in Italy in 569, the Lombards were a group composed of several bands of armed men led by a king-warrior, Alboin.[66] They had crossed the Julian Alps from Pannonia. In this region, they had come in contact with the Byzantine Empire, which they had served as *foederati*. They had also

62 Venantius Fortunatus, *Vita sanctae Radegundis*, c. 29 (ed. Krusch, pp. 41–42).

63 *Vita Sanctae Balthildis*, c. 8 (ed. Krusch, pp. 492).

64 Cristina La Rocca, 'Segni di distinzione. Dai corredi funerari alle donazioni 'post obitum'', in *L'Italia centro-settentrionale in età longobarda*, ed. Lidia Paroli, (Firenze 1997), pp. 31–54.

65 Traditionally, since the 3rd century, the removal of the *cingulum*, and of the functions associated with it, was a necessary condition for entry into religious, ecclesiastical or monastic life.

66 See Stefano Gasparri, 'Culture barbariche, modelli ecclesiastici, tradizione romana nell'Italia longobarda e franca', in *Reti Medievali Rivista*, 6/2 (2005), pp. 1–56.

encountered other groups (Germanic, nomadic and provincial from Noricum and Pannonia), which the main Lombard core incorporated.[67] At this stage the Lombards consisted of the ruler and his family and aristocratic groups organised by lineage. At the top of the Lombards was the king, immediately below him were the dukes, who were also military leaders. The Lombards were different from the basic military cell, which was often an armed or simply military kinship group.[68]

Unlike the Goths and Franks, the Lombards came to Italy without any official agreement with the Roman authority. The conquest of the Italic territories was characterised as a violent military occupation.

The army led by Alboin had conquered some cities in Veneto and Pavia. After the first conquests, Byzantine resistance and lack of internal cohesion slowed their expansion. Lombard factions assassinated both Alboin (572) and his successor, Clefi (574), and for about ten years following, there were no kings. In this time, the dukes independently led the different groups. Those active in northern Italy led their militias into Piedmont and Tuscany, while at the same time other bands had independently federated with the empire and had settled in central and southern Italy, founding the Duchy of Spoleto and possibly that of Benevento.[69]

At the time of their arrival in Italy, the lineage of the Lombards essentially coincided with that of the armed freemen (*arimanni/exercitales*). It was from these men that, after the reconfiguration of the kingdom from 584 onwards, came the direct representatives of royal power. The duces, military leaders inferior only to the king, and the gastalds, former royal officials with modest administrative functions, were promoted to the highest rank in the public hierarchy.[70]

---

67 See Pohl, *The Empire and the Lombards*; Stefano Gasparri, *Prima delle Nazioni. Popoli, etnie e regni fra antichità e medioevo* (Roma, 1997).

68 On the *farae* see Stefano Gasparri, in this volume.

69 On the first phase of the two independent Duchies see Stefano Gasparri, *I duchi longobardi* (Roma, 1978) and Claudio Azzara, 'Spoleto e Benevento', in *I Longobardi dei ducati di Spoleto e Benevento: atti del XVI Congresso Internazionale di Studi sull'Alto Medioevo* (Spoleto, 2004), pp. 105–123. About their later developments see Simone Collavini, 'Duchi e società locali nei ducati di Spoleto e Benevento nel secolo VIII', in *I Longobardi dei ducati di Spoleto e Benevento. Atti del XVI Congresso Internazionale di Studi sull'Alto Medioevo* (Spoleto 2003), pp. 125–166 and Giulia Zornetta, *Italia meridionale longobarda* (Roma 2020).

70 Gasparri, *I duchi longobardi*. On the Longobard military organisation in comparison with the Roman tradition, with reference also to the contemporary Frankish examples, see Gasparri, 'I barbari, l'impero'.

Below them were the counts, in charge of governing the cities, and some minor officials, the *centenarii* or decans and the *sculdasci*, whose functions are difficult to define.

Lombards' *militiae* had settled in Italy and violently expropriated large sections of land.[71] They eliminated much of the Roman aristocracy and marginalised the rest from positions of power. In the towns, which were the core of the Lombards' social and political activity after their settlement, they did not eliminate the Roman population. They also spared the small and medium-sized rural landowners. Within a century and a half, the two populations assimilated. From an initial distinction based on the practice of arms, they moved on to one based on economic factors. The possession of land now supported the social rise of the kingdom's elites. The acquisition of social hegemony, initially linked to participation in the war and proximity to the king or duke, from the eighth century onwards depended on wealth and holding public office.[72]

The socio-political evolution of Lombard society influenced how the elites displayed their status. This is particularly evident from the archaeological data. The grave equipment of the Lombard men's tombs from the first phase of occupation shows a marked military character.[73] This was already visible in the Pannonian phase, but the practice of burials with knightly connotations increased after the Lombards' arrival in Italy. In burials from this phase, one finds a belt with trimmings, a sword and shield as well as spurs. The weapons, and their ostentation at the time of burial, marked the social status of the bearer. These traits, which became characteristic of the Lombard elites, persisted even after the weakening of the social role of military functions – in the case of Collegno, in Piedmont, a burial with military equipment of a person clearly unfit for arms is significant.

---

71 See Walter Goffart, *Barbarians and Romans* (*A.D. 418–584*) (Princeton 1981), pp. 176–189; Paolo Delogu, 'Longobardi e romani: altre congetture', in *Il regno dei longobardi in Italia. Archeologia, società e istituzioni*, ed. S. Gasparri, (Spoleto 2004), pp. 111–120; Walter Pohl, 'Per hospites divisi. Wirtschaftliche Grundlagen der langobardischen Ansiedlung in Italien', in *Römische Historische Mitteilungen*, 43 (2001), pp. 197–200 ; Id., *'I lognobardi e la terra', in Expropriations et confiscations dans les royaumes barbares : une approche régionale*, eds. P. Porena – Y. Rivière (Rome 2012), pp. 279–294; Stefano Gasparri, 'Le basi economiche del potere pubblico in età longobarda', in *Taxation and Rent. Fiscal problems from Late Antiquity to Early Middle Ages*, eds. P.C. Díaz – I. Martín Viso (Bari 2011), pp. 71–85.

72 Paolo Cammarosano, *Nobili e re. L'Italia politica dell'alto medioevo* (Roma 2009), chapters 3–4.

73 Caterina Giostra, 'Luoghi e segni della morte in età longobarda: tradizione e transizione nelle pratiche dell'aristocrazia', in *Archeologia e società tra tardo antico e alto medioevo*, eds. G.P. Brogiolo – A. Chavarría Arnau (Mantova 2007), pp. 311–339.

Within these burials, belts are a common item: in gold, silver, bronze or iron, and made with different techniques.

Since the end of the sixth century, and increasingly so until the first thirty years of the seventh, Lombard tombs contained multiple gold belts and some five-piece belts together with other accessories for suspending weapons. The Lombards considered belts essential to define the prestige of their owner and his role in society.[74] Later, belts made of less precious metal joined and then replaced the gold and silver ones.[75]

Although Lombard burials included belts modelled on late antique military ones as symbols of status, it is not possible to find references in the sources to these belts identifying them as *cingula militiae*.

Indeed, on a lexical level, documents and historiographic works do not mention the *cingulum militiae*. Paul the Deacon, for example, uses the term in his *Historia romana*, probably deriving it from his sources, but never uses it in the *historia langobardorum*.

Only one testament dating from the first half of the eighth century (745) refers explicitly to a belt, referred to by the Germanic term *ringa*. A *vir magnificus* by the name of Rottpert from Agrate (Monza), evidently of high rank, donated his movable and immovable property *pro anima*, including a gold belt.[76] The donation followed similar dynamics as in the Frankish zone, but in this circumstance, the object could remain in the family inheritance. Indeed, the male son of Rottpert could redeem the belt, paying *solidos centum*. The family would donate the proceeds to the poor. This process preserved the pious intent of the donor and did not undermine the possibility of the son inheriting the symbols of his father's rank. All this underlines the symbolic value of the belt as symbol of personal and family status.

For the Lombards, therefore, the belt was an essential element in the strategies of status representation.[77] It is likely that in the early phase of settlement

74 Karl Joseph Leyser, 'Early medieval warfare', in Id., *Communications and power in the Middle Ages. 1: The Carolingian and Ottonian centuries* (London, 1994), pp. 29–50 and Barbiera, Irene. 'Remembering the Warriors: Weapon Burials and Tombstones between Antiquity and the Early Middle Ages in Northern Italy', in *Post-Roman Transitions Christian and Barbarian Identities in the Early Medieval West,* eds. W. Pohl – G. Heydemann, eds. (Turnhout 2013), pp. 407–437.

75 Caterina Giostra, 'Tre nuovi anelli-sigillo aurei longobardi', in *I signori degli anelli. Aggiornamento sugli anelli sigillo aurei longobardi in memoria di Otto von Hessen e Wilhelm Kurze*, ed. M.S. Lusuardi Siena (Milano 2004), pp. 89–96.

76 *Codice Diplomatico Longobardo*, I, nr. 82, (ed. Schiaparelli, pp. 238–244). Cristina La Rocca studied this document in 'Segni di distinzione'.

77 Caterina Giostra, 'Rompere e distribuire sulle tombe longobarde: le cinture come veicolo di conservazione della memoria e di Trasmissione dello status', in, *Antropologia e*

these belts may have had some relationship with the *cingulum militiae* in use by the Roman armies. However, the strong familial connotation of the Lombard military groups, as well as the substantial autonomy of its military leaders, may have disconnected the object from a specific office. The absence of a centralised governmental structure in early medieval Italy, the marginalisation of Roman elites and, last but not least, the traditional continuity of ancient war ritual practices – with systems of status and rank affirmation based on different symbols than those of Roman origin – may also have contributed to this.[78] The Lombards are all free men who can bear arms and, therefore, wear belts.

This shared military function and relations with worlds that used the *cingulum* in a more codified way (notably its Byzantine neighbours) may have favoured the continuity of the belt's use and its typological differentiation. Large landowners, for example, could purchase military equipment inspired or exemplified by those with greater political and social relevance in neighbouring contexts.[79] Above all, they could do so autonomously and without royal and/or ducal control. There seems to be no direct relationship between the office held and the type of belt worn. This accentuates the dimension of belts primarily as indicators of status, both personal and familial. Finally, it is noteworthy that although a family group or lineage may transmit a belt, each individual tends to have his own. The belt, at the time of the owner's death, through the practice of fracturing and distribution of pieces, may become part of the belt of one or more members related to the deceased.[80] This custom contributed to the sharing of the status of the dead person with a restricted group of individuals linked to them, but at the same time left them wide margins of autonomy in structuring their personal form of social self-representation. The use is very similar to that seen in the Frankish context, so it is possible that the sharing of pieces of the belt took place even while the owner was alive, to seal or underline a network of social relations.

---

archeologia a confronto: archeologia e antropologia della morte. 2. Corpi, relazioni e azioni: il paesaggio del rito, ed. V. Nizzo (Roma 2018), pp. 225–231.

78 The Lombard king held a lance at the time of his election and did not hold a sword. The ritual changed after the Carolingians conquered the kingdom. See Stefano Gasparri, 'La regalità longobarda. Dall'età delle migrazioni alla conquista carolingia', in *Alto medioevo mediterraneo*, ed. S. Gasparri (Firenze 2005), pp. 207–232, esp. p. 220.

79 In the tombs of Castel Trosino we find, for example, belts similar to those made by the craftsmen of Rome and found in the excavations of the Crypta Balbi. See Ricci, Marco. 'La produzione di merci di lusso e di prestigio a Roma da Giustiniano a Carlomagno', in *Roma. Dall'antichità al medioevo. Archeologia e Storia nel Museo Nazionale Romano Crypta Balbi*, eds. M.S. Arena *et all.* (Firenze 2001), pp. 79–87.

80 Giostra, 'Rompere e distribuire'.

6 Between Public Function and Christian Kingship: The *Cingulum Militiae* in the Carolingian Milieu

The social structure of Merovingian Gaul, based on the delegation of jurisdictional functions by the king to the local nobility, also remained in place under the Arnolfingian/Carolingian dynasty. The same modes of transmission of status and its ostentation by the elites also remained in place. The warlike values that were at the basis of the emergence of the elites in the Merovingian kingdom remained unchanged in the Carolingian era. The elites and the kings themselves continued to show their positions according to traditional semantics: that is, through the gift and possession of weapons (sword, spear, shield) and their accessories. Weapons reflected the social hierarchy of the kingdom and were instrumental in developing bonds of loyalty and dependence.

In this mileu, the *cingulum* was again at the centre of representations of status, attributions of functions and bonds of *fidelitas*. The eighth century coincided with an increase in attestations of the *cingulum* in medieval sources prior to the year 1000. The biographies of the emperors, Charlemagne and Louis the Pious, as well as councils, capitularies and annals are full of references to the *cingulum*.[81]

Carolingians look to the belt, traditionally, as an expression of social preeminence. Compared to the past, however, the religious spirit that characterised Carolingian rule placed the *cingulum* and its granting in a new ritual system that underlined the contemporary social order. The sovereign, anointed by God, received the royal ministry to bring about the Christian kingdom/empire. As the person at the top of the social hierarchy, the king ordered the members of *militia saecularis,* including the Frankish nobility, through arms and the *cingulum*.[82] Louis the Pious, for example, offered gifts to all his palace servants, each according to his rank, to the *nobilioribus aut balteos aut fascilones preciossisimaque vestimenta a latissimo imperio perlata distribui iubet*, while to the *inferiores* he bestowed *saga Fresconica omnimodi coloris.*[83] For this reason, the Carolingian rulers, strengthened by imperial self-representation and the link

81 Eginardus informs us that Charlemagne always wore a richly adorned *balteum* see *Ehinardi Vita Karoli*, c. 23 (ed. Older-Egger, p. 28). He had invested Louis, while still a child, with a *cingulum* as King of Aquitaine. The investiture did not include the child taking up arms. In fact, they repeated the ceremony when Louis was a adolescent cfr. Astronomus, *Vita Hludowici imperatoris*, c. 4 (ed. Tremp, pp. 294, 296). Louis, too, never separated from his *cingulum* in adulthood cfr. Notkerus, *Gesta Karoli Magni*, II,22 (ed. Rau, p. 426).

82 See Regine Le Jan, *Femmes, pouvoir et société dans le haut Moyen Age* (Paris 2001), pp. 171–189.

83 Notker, *Gesta Karoli*, II, 21 (Ed. Rau, p. 426).

with Rome, had exclusive use of the sword and exercised the power associated with it.

It was precisely the Christian dimension, which ambiguously characterised the Carolingian institutions, that put the emperor himself in check. In the past, only the imperial or royal authority that could dismiss officers by depriving them of the *cingulum*. In the Carolingian world, however, the episcopal authority could also judge and punish the negligent, even the emperor, in a strange mixture of public and private, political and moral evaluation.

In 833, the bishops of the kingdom accused Louis of creating scandal in the church and undermining the stability of the state. They put the emperor on trial and declared him unfit to exercise *militia saecularis*. After a public purgation, he deposited the *militiae cingulum super altare*. Only after he had atoned for his sins by putting on penitent clothes could he return to his former position.[84]

This practice echoed the practice of public penance in the late antique and early medieval Roman liturgy, which involved abandoning one's public functions before entering the *ordo poenitentium*, symbolically renouncing the *cingulum*.[85]

7 Conclusions

Born as a distinctive symbol of the militia, an institution characterised by a well-defined juridical and social physiognomy, the *cingulum* became by extension the insignia of public service and the term to indicate it (in a progressive passage from object, to symbol, to concept). It is with that sense that it recurs both in the civil sphere and in Christian patristics, where it is used to indicate the otherness of the ecclesiastical career with respect to the civil one. In this same religious context, the *cingulum* was adopted (both as an object and as a concept) to also indicate the total self-denial and service of the monk and then of the clergy.

84 Mayke De Jong, 'Power and humility in Carolingian society: the public penance of Louis the Pious', in *Early Medieval Europe*, 1 (1992), pp. 29–52.

85 See Michel Sot, 'Concordances et discordances entre culture des élites laïques et culture des élites cléricales à l'époque carolingienne: Jonas d'Orléans et Dhuoda', in, *La culture du haut moyen âge une question d'élites?*, eds. F. Bougard – R. Le Jan – R. McKitterick (Turnhout 2009), pp. 341–363.

Imperial choices, the extension of the militia to public officials as well, and a progressive militarisation of Roman society influenced the development of the *cingulum*.

The duality between the evolution of the object and the use of the concept characterised the entire course of the early Middle Ages in the Gothic as well as in the Frankish contexts. In both cases, despite a variable adherence to the original function, the *cingulum* accentuated its meaning as a status symbol. While the attestations in the sources of the Gothic area connect the *cingulum* to traditional functions and uses, the examples from the Frankish area show an evolving situation. For the Franks, the *cingulum* was one of the main ways to represent status and social bonds.

The examples from the Lombard area, on the other hand, show a clear conceptual discontinuity. None of the Lombard sources uses the term *cingulum*/ *cingulum militiae* to define the functions held by the king or officers or the belt. This may have been due to the lack of members of the ancient Roman elites in the administration of the kingdom or the duchies, as well as to the particular internal structure of the Lombard military groups in the first Italian decades. Despite this, the Lombards, like the Franks, used the belt as a symbol of their status and as a marker of social ties, but did so outside a codified system of Roman-military inspiration.

A turning point for the *cingulum* is the Carolingian era within a broader reorganisation of honours by the Pipinids. Here, the *cingulum* regained its value as a formal and institutional sign of a power delegated and managed centrally by the ruler. It connoted the *militia saecularis*, belonging to the sovereign and the nobility: both at the service of the heavenly God and for the construction of the Christian empire.

Bibliography

Primary Sources

Anonymus Valesianus, *Pars posterior*, ed. Th. Mommsen [MGH Auct. Ant., 9], (Berolini, 1892), pp. 306–328.

Astronomus, *Vita Hludowici imperatoris*, ed. E. Tremp, [MGH Scrip. Rer. germ. N.S., 64], (Hannover 1995), pp. 280–555.

Cassiodorus, *Varie*, a cura di A. Giardina - G. Alberto Cecconi – I. Tantillo, Vol. II: Libri III–V, Roma 2014, english transl in Cassiodorus, The Variae, trans. M. Shane Bjornlie, (Oakland CA, 2019).

Codice Diplomatico Longobardo, ed. Luigi Schiaparelli, I, (Roma, 1929).

Einhardi Vita Karoli Magni, ed. O. Holder-Egger [MGH Script. Rer. germ., 25], (Hannoverae-Lipsiae, 1911), pp. 1–41.

Gregorii I papae registrum epistularum, ed. P. Ewald- L. Hartmann, [MGH Epp., 1–2, I–II], (Berolini, 1891–1899), english tanslation *The Letters of Gregory the Great*, trad. John R. C. Martyn (Turnhout, 2004).

Grégoire le Grand, *Dialogues*, ed. A, de Vogüé, tr. P. Antin [Sources chrétiennes, 251, 260, 265], (Paris, 1978–1980).

Gregorii episcopi turonensis Libri historiarum X, ed. B. Krush –W. Levison [MGH SS. Rer. Mer., 1.1], (Hannoverae, 1951) engl. transl. Gregory of Tour, *The history of Franks*, trans. Lewis Thorpe, (London, 1974).

Gregorii episcopi turonensis Miracula et opera minora, ed. B. Krush, MGH SS. Rer. Mer., 1.2, (Hannoverae, 1885), pp. 34–111 engl. Transl. Gregory of Tours, *Glory of the Martyrs*, trans. Raymond Van Dam, (Liverpool, 2004)

Jean le Lydien, *Des magistratures de l'État romain*, I–II, texte établi, trad. M. Dubuisson – J. Schamp (Paris 2006).

Les Canons des Conciles mérovingiens (*VIe–VIIe siècles*), Texte latin de l'édition C. De Clercq, Introduction, traduction et notes par J. Gaudemet- B. Basdevant, I–II, (Paris, 1989).

Notker the Stammerer, *Gesta Karoli Magni*, ed. R. Rau, [*Quellen zur Karolingischen Reichsgeschichte*, III, (Ausgewählte Quellen zur deutschen Geschichte des Mittelalters. Freiherr-vom-Stein-Gedächtnisausgabe, 7), (Darmstadt, 1960), pp. 322–426, engl. Transl. in Einhard and Notker the Stammerer, *Two lives of Charlemagne*, transl. by David Ganz, (Harmondsworth, 2008).

Rea, John. 'A Cavalryman's Career A.D. 384 (?)-401', *Zeitschrift für Papyrologie und Epigraphik*, 56 (1984), pp. 79–88.

The Codex of Justinian. A New Annotated Translation with Parallel Latin and Greek Text, ed. Bruce W. Frier, I–III, (Cambridge, 2016).

Codex Theodosianus: *Theodosiani libri XVI cum constitutionibus Sirmondianis et leges Novellae ad Theodosianum pertinentes*, ed. Th. Mommsen – P.M. Meyer, 2 vols. (Berlin, 1905) – Engl. transl. by C. Pharr *The Theodosian Code and Novels and the Sirmondian Constitutions. A Translation with commentary, glossary, and bibliography*, [Princeton, 1952], 10th Print (Union, N.J., 2001).

Maximus Taurinensis, *Homiliae*, ed. V. Amedevo, [PL 57] (Paris 1862), engl. trans.The Sermons of St. Maximus of Turin, trasl. Boniface Ramsey, (New York 1989).

Venanti Honori Clementiani Fortunati presbyteri Italici Opera poetica, ed. F. Leo[MGH, Auct. Ant., 4.1] (Berolini, 1881) engl. Transl. Venantius Fortunatus, *Personal and Political Poems*, trans. J. George (Norwich, 1995).

Venantius Fortunatus, *Vita Germani episcopi parisiaci*, ed. B. Krusch [MGH SS. Rer. Mer. 7/1], (Hannoverae-Lipsiae, 1920) pp. 372–418.

Venantius Fortunatus, *Vita sanctae Radegundis*, ed. B. Krusch [MGH Auct. Antiq 4/2], (Berolini, 1885), pp. 38–49.
Vita Sanctae Balthildis, ed. B. Krusch [MGH Scr. Rer. Mer. 2], (Hannoverae, 1888), pp. 475–508.

Secondary Literature

Aimone, Marco. 'Romani e Ostrogoti fra integrazione e separazione. Il contributo dell'archeologia a un dibattito storiografico', in *Reti Medievali*, 13/1 (2012) pp. 1–66
Albrecht, Stefan, 'Warum tragen wir einen Gürtel? Der Gürtel der byzantiner –symbolik und function', in F. Daim – J. Drauschke, eds., *Byzanz – das Römerreich im Mittelalter, 1: Welt der Ideen, Welt der Dinge* (Mainz, 2010), pp. 79–95
Amory, Patrick. *People and identity in Ostrogothic Italy (489–554)*, (Cambridge, 1997)
Arce, Javier. 'La ostentació del rango en la vida y en la muerte', in G.P. Brogiolo – A. Chavarría Arnau, eds., *Archeologia e società tra tardo antico e alto medioevo* (Mantova, 2007), pp. 265–278
Arnold, Jonathan J. *Theoderic and the Roman Imperial Restoration* (Cambridge, 2014)
Azzara, Claudio. 'Spoleto e Benevento', in *I Longobardi dei ducati di Spoleto e Benevento: atti del XVI Congresso Internazionale di Studi sull'Alto Medioevo* (Spoleto, 2004), pp. 105–123
Barbiera, Irene. 'Remembering the Warriors: Weapon Burials and Tombstones between Antiquity and the Early Middle Ages in Northern Italy', in W. Pohl – G. Heydemann, eds., *Post-Roman Transitions Christian and Barbarian Identities in the Early Medieval West* (Turnhout, 2013), pp. 407–437
Bierbrauer, Volker. 'Goti dell'Oriente e dell'Occidente europeo tra la fine del V e la prima metà del VI secolo alla luce delle fonti archeologiche', in C. Ebanista – M. Rotili, eds., *Archeologia e storia delle migrazioni. Europa, Italia, Mediterraneo fra tarda età romana e alto medioevo* (Cimitile, 2011), pp. 9–30
Cammarosano, Paolo. *Nobili e re. L'Italia politica dell'alto medioevo* (Roma, 2009)
Colalvini, Simone. 'Duchi e società locali nei ducati di Spoleto e Benevento nel secolo VIII', in *I Longobardi dei ducati di Spoleto e Benevento. Atti del XVI Congresso Internazionale di Studi sull'Alto Medioevo* (Spoleto; 2003), pp. 125–166
De Jong, Mayke. 'Power and humility in Carolingian society: the public penance of Louis the Pious', in *Early Medieval Europe*, 1992 (1), pp. 29–52
Delmaire, Roland. *Largesses sacrées et res privata. L'aerarium impérial et son administration du IVe au VIe siècle* (Rome, 1989)
Delogu, Paolo. 'Longobardi e romani: altre congetture', in S. Gasparri, ed., *Il regno dei longobardi in Italia. Archeologia, società e istituzioni* (Spoleto, 2004), pp. 111–120
Enßlin, Wilhelm. 'Aus Theoderichs Kanzlei', in *Würzburger Jahrbücher für die Altertumswissenschaft*, 2/1 (1947), pp. 75–85

Esmonde, Simon. *The Roman west.* AD *200–500: An Archaeological Study* (Cambridge, 2013)

Flori, Jean. 'Les origines de l'adoubement chevaleresque: étude des remises d'armes et du vocabulaire qui les exprime dans les sources historiques latines jusqu'au début du XIIIe siècle', in *Traditio*, 35 (1979), pp. 209–272

Gasparri, Stefano. *I duchi longobardi* (Roma, 1978)

Gasparri, Stefano. *Prima delle Nazioni. Popoli, etnie e regni fra antichità e medioevo* (Roma, 1997)

Gasparri, Stefano. 'La regalità longobarda. Dall'età delle migrazioni alla conquista carolingia', in S. Gasparri, ed., *Alto medioevo mediterraneo* (Firenze, 2005), pp. 207–232

Gasparri, Stefano. 'Culture barbariche, modelli ecclesiastici, tradizione romana nell'Italia longobarda e franca', in *Reti Medievali Rivista*, 6/2 (2005), pp. 1–56

Gasparri, Stefano. 'Le basi economiche del potere pubblico in età longobarda', in P.C. Díaz – I. Martín Viso, *Taxation and Rent. Fiscal problems from Late Antiquity to Early Middle Ages* (Bari, 2011), pp. 71–85

Gasparri, Stefano. 'I barbari, l'impero, l'esercito e il caso dei Longobardi', in F. Botta – L. Loschiavo, eds., *Civitas, Iura, Arma. Organizzazioni militari, istituzioni giuridiche e strutture sociali alle origini dell'europa (secc. III–VIII)* (Lecce, 2015), pp. 91–102

Gil, Joan Pinar. 'Chlamyds e cingulum nel tardo V secolo. Tre rinvenimenti dall'Emilia Romagna', in A.L. Morelli – I. Baldini Lippolis, eds., *Oreficeria in Emilia Romagna. Archeologia e storia tra età romana e medioevo* (Bologna, 2010), pp. 229–256

Giostra, Caterina. 'L'età di Teodorico. I reperti goti di Tortona', in A. Crosetto – M. Venturino Gambari, eds., *Onde nulla si perda. La collezione archeologica di Cesare Di Negro-Carpani* (Alessandria, 2007), pp. 287–326

Giostra, Caterina. 'Luoghi e segni della morte in età longobarda: tradizione e transizione nelle pratiche dell'aristocrazia', in G.P. Brogiolo – A. Chavarría Arnau, *Archeologia e società tra tardo antico e alto medioevo* (Mantova, 2007), pp. 311–339

Giostra, Caterina. 'Tre nuovi anelli-sigillo aurei longobardi', in M.S. Lusuardi Siena, ed., *I signori degli anelli. Aggiornamento sugli anelli sigillo aurei longobardi in memoria di Otto von Hessen e Wilhelm Kurze* (Milano, 2004), pp. 89–96

Giostra, Caterina. 'Rompere e distribuire sulle tombe longobarde: le cinture come veicolo di conservazione della memoria e di Trasmissione dello status', in V. Nizzo, ed., *Antropologia e archeologia a confronto: archeologia e antropologia della morte. 2. Corpi, relazioni e azioni: il paesaggio del rito* (Roma, 2018), pp. 225–231

Glaser, Franz, 'Elementi di corredo di un ufficiale ostrogoto', in F. Marzatico – P. Gleirscher, eds., *Guerrieri, principi ed eroi fra il Danubio e il Po dalla preistoria all'alto medioevo* (Trento, 2004), p. 739

Glaser, Franz. *Ostgotisches Militär in Kärnten. Erste Funde aus der Zeit des Königs Theoderich des Großen (493–536)* (Klagenfurt, 2003)

Goffart, Walter, *Barbarians and Romans* (*A.D. 418–584*) (Princeton, 1981)

Halsall, Guy. 'The Ostrogothic Military', in J. Arnold – S. Bjornlie – K. Sessa, eds., *A Companion to Ostrogothic Italy* (Leiden, 2016), pp. 173–199

Hen, Yitzhak. 'The Merovingian Polity: A Network of Courts and Courtiers', in B. Effros – I. Moreira, eds., *The Oxford Handbook of the merovingian world* (Oxford, 2020), pp. 217–237

Janes, Dominic. 'The golden clasp of the Late Roman State', in *Early Medieval Europe*, 5 (1996), pp. 127–153

Janniard, Sylvain. 'L'esercito del tardo impero romano. Dalla tetrarchia a Giustiniano', in A. Barbero, ed., *Storia d'Europa e del Mediterraneo 7,3: L'Ecumene romana. L'impero tardoantico* (Roma, 2009), pp. 495–523

Kazanski, Michel – Périn, Patrick. 'Le mobilier funéraire de la tombe de Childéric 1er ; état de la question et perspectives', in *Revue archéologique de Picardie*, 3–4 (1988), pp. 13–38

Kerneis, Soazick. 'Garants et compensations: Romanité ou barbarie dans la tres ancienne loi des Bretons d'Armorique', in A. Dubreucq, ed., *Traditio iuris. Permanence et/ ou discontinuité du droit romain durant le haut Moyen Âge* (Lyon, 2005), pp. 77–92

Krüger, Sabine. 'Character militaris und character indelebilis', in F. Lutz – W. Rösener – T. Zotz, eds., *Institutionen, Kultur und Gesellschaft im Mittelalter. Festschrift für Josef Fleckenstein zu seinem 65. Geburtstag* (Sigmaringen, 1984), pp. 567–580

La Rocca, Cristina. 'Segni di distinzione. Dai corredi funerari alle donazioni 'post obitum'', in Lidia Paroli, ed., *L'Italia centro-settentrionale in età longobarda* (Firenze, 1997), pp. 31–54

Le Jan, Regine. *Femmes, pouvoir et société dans le haut Moyen Age* (Paris, 2001)

Lettich, Giovanni. *Itinerari epigrafici aquileiesi. Guida alle iscrizioni esposte nel Museo archeologico nazionale di Aquileia* (Trieste, 2003)

Leyser, Karl. 'Early Medieval Canon Law and the Beginnings of Knighthood', in L. Fenske – W. Rösener – T. Zotz, eds., *Institutionen, Kultur und Gesellschaft im Mittelalter. Festschrift für Josef Fleckenstein zu seinem 65. Geburtstag* (Sigmaringen, 1984), pp. 549–566

Leyser, Karl Joseph. 'Early medieval warfare', in Id., *Communications and power in the middle ages. 1: The Carolingian and Ottonian centuries* (London, 1994), pp. 29–50

Liebeschütz, Wolfgang. 'Habitus barbarus: did barbarians look different from Romans?', in P. Porena – Y. Rivière, eds., *Expropriation et confiscations dans les royaumes barbares. Une approche régionale* (Roma, 2012), pp. 13–28

Loschiavo, Luca. *L'età del passaggio. All'alba del diritto comune europeo (secoli III–VII)* (Torino, 2019^2)

Mineo, Igor. 'Di alcuni usi della nobiltà medievale', in *Storica*, 7 (2001), pp. 9–58

Patrello, Ralph J. 'Belt Buckles and Burials in Soutwestern Gaul', in B. Effros – I. Moreira, eds., *The Oxford Handbook of the merovingian world* (Oxford, 2020), pp. 916–936

Pohl, Walter. 'The Empire and the Lombards: treaties and negotiations in the sixth century', in W. Pohl, ed., *Kingdoms of the Empire. The integration of Barbarians in Late antiquity* (Leiden-New York-Köln 1997), pp. 75–133

Pohl, Walter. 'Per hospites divisi. Wirtschaftliche Grundlagen der langobardischen Ansiedlung in Italien', in *Römische Historische Mitteilungen*, 43 (2001), pp. 197–200

Pohl, Walter. 'I lognobardi e la terra', in P. Porena – Y. Rivière, *Expropriations et confiscations dans les royaumes barbares : une approche régionale* (Rome, 2012), pp. 279–294

Pohl, Walter, *The Avars: A Steppe Empire in Central Europe (567–822)* (Ithaca, NY 2018)

Poly, Jean-Pierre. 'Sous lex chênes de Salaheim. La loi salique, l'armée romaine et le bilan de la barbarie', in F. Botta – L. Loschiavo, eds., *Civitas, Iura, Arma. Organizzazioni militari, istituzioni giuridiche e strutture sociali alle origini dell'europa (secc. III–VIII)* (Lecce, 2015), pp. 193–219

Porena, Pierfrancesco. 'La posizione dell'elemento militare nell'Impero romano e i 'regni romano-brbarici'', in F. Botta – L. Loschiavo, eds., *Civitas, Iura, Arma. Organizzazioni militari, istituzioni giuridiche e strutture sociali alle origini dell'Europa (secc. III–VIII)* (Lecce, 2015), pp. 221–258

Possenti. Elisa. 'Abbigliamento e rango in Italia settentrionale tra V e VI secolo', in G.P. Brogiolo – A. Chavarría Arnau, eds., *Archeologia e società tra tardo antico e alto medioevo* (Mantova, 2007), pp. 279–298

Prinz, Friedrich. 'Primi stadi della militia Christi altomedioevale nella tarda antichità e nel sistema ecclesiastico imperiale del periodo carolingio e degli Ottoni', in *Militia Christi e Crociata nei secoli XI–XIII*. Atti della XI Sett. Intern. di studio (Mendola, 28 agosto-1 settembre 1989), (Milano, 1992), pp. 49–63

Rea, John. 'A Cavalryman's Career A.D. 384 (?)-401', in *Zeitschrift für Papyrologie und Epigraphik* ZPE), 56 (1984), pp. 79–88

Ricci, Marco. 'La produzione di merci di lusso e di prestigio a Roma da Giustiniano a Carlomagno', in M.S. Arena *et all.*, eds., *Roma. Dall'antichità al medioevo. Archeologia e Storia nel Museo Nazionale Romano Crypta Balbi* (Firenze, 2001), pp. 79–87

Santinelli, Emmanuelle. 'Continuité ou rupture? L'adoption dans le droit mérovingien', in *Médiévales*, 35 (1998), pp. 9–18

Sarti, Laury. 'The Military and Its Role in Merovingian Society', in B. Effros – I. Moreira, eds., *The Oxford Handbook of the merovingian world* (Oxford, 2020), pp. 255–277

Schoolman, Edward M., 'Nobility, aristocracy and status in early medieval Ravenna', in J. Herrin – J. Nelson, eds., *Ravenna: Its Role in Early Medieval Change and Exchange* (London, 2016), pp. 211–238

Sommer, Markus. *Gürtel und Gürtelbeschläge des 4. und 5. jahrhunderts im römischen Reich* (Bonn 1984)

Sot, Michel. 'Concordances et discordances entre culture des élites laïques et culture des élites cléricales à l'époque carolingienne : Jonas d›Orléans et Dhuoda', in

F. Bougard – R. Le Jan – R. McKitterick, eds., *La culture du haut moyen âge une question d'élites?* (Turnhout, 2009), pp. 341–363

Verardi, Andrea Antonio. *La memoria legittimante. Il liber pontificalis e la chiesa di Roma nel secolo VI* (Roma, 2016)

Von Harnack, Adolf. *Militia Christi. Die Chrisliche Religion und der Soldatenstand in den ersten drei Jahrhunderten* (Tübingen, 1905)

von Rummel, Philipp. *Habitus barbarus: Kleidung und Repräsentation spätantiker Eliten im 4. und 5. Jahrhundert* (Berlin - New York, 2007)

Waebens, Sofie. 'The Representation of Roman Soldiers on Third-Century AD Funerary Monuments from Nikopolis (Egypt)', in *Revue internationale d'histoire militaire ancienne*, 1 (2014), pp. 61–75

Werner, Karl Ferdinand. *Naissance de la noblesse* (Paris, 1998)

Wickham, Chris. *Framing the early Middle Ages: Europe and the Mediterranean (400–800)* (Oxford, 2006)

Wickham, Chris. 'The changing composition of early élites', in F. Bougard – H.-W. Goetz – R. Le Jan, eds., *Théorie et pratiques des élites au Haut Moyen Âge. Conception, perception et réalisation sociale* (Turnhout, 2011), pp. 5–18

Whitby, Michael. 'Army and Society in the Late Roman World: Context for Decline?', in P. Erkdamp, ed., *A companion to the Roman Army* (Hoboken - NJ, 2010) pp. 515–531

Winter, Johanna M. 'Cingulum Militiae', in *Revue d'histoire du droit*, 44 (1976), pp. 1–92

Woods, David. 'The Ownership and Disposal of Military Equipment in the Late Roman Army', in *Journal of Roman Military Equipment Studies*, 4 (1993), pp. 55–65

Zornetta, Giulia. *Italia meridionale longobarda* (Roma, 2020)

CHAPTER 11

Answering the Call to Arms: *Lex Visigothorum* 9.2

Esperanza Osaba

1 Introduction

This article centres on military affairs (*in rebus bellicis*) in Visigothic times.[1] In particular we will deal with the period of the Visigothic Kings of Hispania, who ruled throughout the sixth and seventh centuries before they were overthrown by Muslim forces in early eighth century, starting in 711.

I will approach this subject by examining the laws contained in the most important legal text of the period, the *Lex Visigothorum* (LV), first promulgated in 654 during the reign of Recceswinth ([649]–653–672).[2] I will focus on the laws comprised in Title 2 *De his, qui ad bellum non vadunt aut de bello refugiunt* (Book IX *De fugitivis et refugientibus*).

This second title consists of nine laws: seven *antiquae* (LV 9.2, 1–6) and the other two from kings Wamba (LV 9.2.8) and Erwig (LV 9.2.9), although in addition to the title specifically devoted to the *res militaris*, we also have Egica's law LV 5.7.19 (687–702), which supplements Erwig's earlier law and is, in fact, the

1 This study is part of the research line developed by the 'Grupo de Investigación de Derecho Romano' of the University of the Basque Country (UPV/EHU), Reference GIU19/083, 2020–2023. For a broad overview of the different traditional historiographical positions on the Visigothic army and its components, see Sabrina Soledad Orlowski, 'Principales tendencias historiográficas sobre la organización militar visigoda del reino de Toledo', in *Medievalista* [on line], 19 (2016), pp. 1–25; Dionisio Pérez Sánchez, *El ejército en la sociedad visigoda* (Salamanca, 1989), pp. 93–103; Amancio Isla Frez, *Ejército, sociedad y política en la Península Ibérica entre los siglos VII y XI* (Madrid, 2010), pp. 9–24.

2 Hereinafter LV. I will use the edition by Zeumer, *Leges Visigothorum*. On Visigothic legislation, see Yolanda García López, *Estudios críticos y literarios de la de la "Lex Wisigothorum"* (Alcalá de Henares, 1997); Carlos Petit, 'Derecho visigodo del siglo VII. Un ensayo de síntesis e interpretación', in *Derecho, Cultura y Sociedad en la Antigüedad Tardía*, ed. E. Osaba (Bilbao, 2013), pp. 205–218; Pérez-Prendes Muñoz-Arraco, José Manuel Pérez-Prendes Muñoz-Arraco, 'Historia de la legislación visigótica', in *San Isidoro Doctor Hispaniae* (Sevilla, 2002), pp. 51–67; Karl Zeumer, *Historia de la legislación visigoda*, Span.transl. Carlos Clavería (Barcelona, 1944). Michael J. Kelly, *Isidore of Seville and the Liber Iudiciorum. The Struggle for the Past in the Visigothic Kingdom* (Boston-Leiden, 2021); Céline Martin, 'Le Liber Iudiciorum et ses différentes versions', in *Mélanges de la casa de Velázquez*, 41 (2) (2011), pp. 17–34.

 | DOI:10.1163/9789004698017_012

last military law in the LV.[3] In the following pages I will analyse five of these *antiquae*, LV 9.2.1; LV 9.2.2; LV 9.2.3; LV 9.2.4; LV 9.2.5 and LV 9.2.7.[4] However, I have not dwelt on LV 9.2.6, which deals with provisioning of the army and the punishment for fraud in its distribution.

2 The *Leges Antiquae* in LV 9.2

2.1 *LV 9.2.1 Ant*

The law starts by setting out the penalty for any *thiufadus*[5] who accepted a bribe to permit a member of his unit (*thiufa*) to return home. He was to be compelled to pay the *comes civitatis* of the territory in which he was stationed an amount equivalent to ninefold the bribe received. If, on the other hand, the *thiufadus* had received no payment (*nullam mercedem*), but had allowed the soldier to return home or failed to compel him to answer the call to arms without a legitimat pretext (i.e. illness) he was ordered to pay 20 *solidi*. For the commanders of smaller units, *quincentenarius*, *centenarius* or *decanus*, the sanction was 15, 10 and 5 *solidi* respectively, which was to be distributed among the *centena* to which they were assigned.[6]

---

3 Elsewhere in the LV, military issues are also addressed in: LV 2.5.13, Chind.; LV 4.2.15, *ant.* [= CE, 323]; LV 4.2.16 Recc.; LV 4.5; LV 8.1.7 *ant.*; LV 8.1.8 Recc.; LV 8.1.9 *ant.*, and LV 10.2.6, Chind. See Claudio Sánchez Albornoz, 'El ejército visigodo: su protofeudalización', in *Investigaciones y documentos sobre las instituciones hispanas* (Santiago, 1970), pp. 53 ff.; Edorta Córcoles Olaitz, 'La crisis militar visigoda de finales del siglo VII a través de la legislación', in *El ejército, la paz y la guerra: Jornadas de la Sociedad de Historia del Derecho*, eds. J. L. De Los Mozos, I. Szászdi León-Borja (Valladolid, 2009), pp.165–178.

4 A preliminary study of these laws was published in Esperanza Osaba, '*Lex Visigothorum* 9,2: *De his qui ad bellum non vadunt aut de bello refugiunt*', in *Civitas, Iura, Arma. Organizzazioni militari, istituzioni giuridiche e strutture sociali alle origini dell'Europa* (*secc. III–VIII*), eds. F. Botta – L. Loschiavo (Lecce, 2015), pp. 161–194.

5 Probably the equivalent to the Roman *millenarius*, the figure of the *thiufadus*, has long attracted interest: Edward Arthur Thompson, *Los godos en España*, 5th ed. (Madrid, 2007), pp. 174 *ff.*; Luis A. García Moreno, 'Hispania Visigoda (Siglos V a VII)', in *Historia Militar de España*, dir. H. O'Odonnell, v. 2, coord. M. Á. Ladero Quesada (Madrid, 2010), pp. 41 *ff.*; Paul D. King, *Derecho y sociedad en el reino visigodo* (Madrid, 1981), pp. 93, 101 and, with regard to their jurisdictional role, 100–104; Sánchez Albornoz, 'El ejército visigodo', 23 ff.; Céline Martin, *La géographie du pouvoir dans l'Espagne visigothique* (Lille, 2003), p. 151 *ff.*

6 An example of the sale of the *commeatus* can be found in the Code of Justinian of 534 addressed to Belisarius, CJ.1.27.2.9: See Salvatore Puliatti, *Ricerche sulla legislazione "regionale" di Augusto. Lo statuto civile e l'ordinamento militare della prefettura africana* (Milano, 1980), pp. 111–113.

The purpose of this first law is therefore to punish any corrupt commander issuing unjustified leave and thus enabling a member of the *thiufa* to shirk his military duty. If in addition he received a bribe (*merces*), he was subject to the payment of nine times the amount received. However, even in cases where he had received no bribe or payment, he was also punished, although the fine was considerably smaller at 20 *solidi*. It is only in this category of offence that the law mentions other ranks of officer, in descending order, with their respective penalties (all considerable).

2.2 LV *9.2.2 Ant*

The second law, LV 9.2.2, deals with the appropriation by the enlistment officers named *servi dominici – id est compulsores exercitus* –[7] of the property of those called to arms: *quando Gotos*[8] *in hostem exire conpellunt*. The term *hostis* is employed here to refer to the Visigothic army's military expeditions.[9] In later laws in the LV, although only from Chindaswinth onwards, it is also used to refer to the enemy.[10]

The elevenfold penalty, provided for *servi dominici* who commit robbery while serving the draft – a punishment rarely applied in the LV[11] – is to be returned to the victim. It also establishes that venality amongst the *compulsores exercitus* responsible for delivering the call to arms is punishable by 50 lashes.

2.3 LV *9.2.3 Ant*

The next law in this title that deserves special attention is *antiqua* LV 9.2.3, whose heading refers to both (*prepositi*) *exercitus* and (*relicto bello*) *bellum*. In this law, attributed to Leovigild,[12] we come to the crux of the matter. Although

7 On the *servi dominicis* or slaves of the king, see King, *Derecho y sociedad*, p. 97; García Moreno, *Hispania visigoda*, pp. 49, 58; Javier Arce, *Esperando a los árabes: Los Visigodos en Hispania* (507–711) (Madrid 2011), p. 124 *ff.*; Thompson, *Los godos*, pp. 174 ff.; Felix Dahn, *Die Könige der Germanen*, v. 6, 2nd ed. (Leipzig, 1885, p. 211; Pérez Sánchez, *El ejército*, pp. 105 *ff.*

8 Dahn, *Die Könige*, pp. 211, 219; Pérez Sánchez, *El ejército*, pp. 105 *ff.* García Moreno, 'Hispania visigoda', p. 59.

9 Dahn, *Die Könige*, pp. 216–217.

10 Osaba, '*Lex Visigothorum* 9,2', pp. 170–171.

11 In addition to the two laws in this title, LV 9,2,2 and LV 9,2,5, the elevenfold penalty is also applied in the cases of theft provided for in LV 6.4.2 *ant.*; LV 8.1.10 *ant.* and LV 8.1.6 *ant.*

12 Álvaro D'Ors, *El Código de Eurico* (Roma-Madrid, 1960), pp. 52 and 84; Rafael Ureña Y Smenjaud, *La legislación gótico-hispana* (*Leges Antiquiores - Liber Iudiciorum*). *Estudio crítico*, ed. C. Petit (Pamplona, 2003), pp. 246 *ff.*, 268 and 281; Dahn, *Die Könige*, pp. 207–222 and Sánchez Albornoz, 'El ejército visigodo', pp. 8 *ff.*, believe the *antiquae* in Title

the heading refers to cases in which a *praepositus exercitus* abandons the war, and returns home, or permits others to do so, the remainder of the law focuses exclusively on the *centenarius*.

The law is structured in different parts. The first deals with the case of *centenarius* who desert his unit *in hostem* and flee to his home, an offence punishable by death. Noteworthy is the use of the verb *refugere* to designate the action of desertion itself. Another relevant feature, in my opinion, is the similarity between the description of the punishment, *capitali supplicio subiacebit*, and the wording used in the law of the emperors Gratian, Valentinian, Theodosius of 380: *capitali supplicio subiugetur.*[13] A final aspect of interest is the term *domus*, signifying the place the deserter returns to after leaving the campaign or which he refuses to leave in answer to the call to arms. According to Isla Frez, this suggests a reference to property-owning officers or those related to them.[14]

The penalty for a *centenarius* who deserted and returned home was death. However, if he sought sanctuary from the church or the bishop (*si ad altaria sancta vel ad episcopum forte confugerit*), this sentence could be commuted to a fine of 300 *solidi* – to be paid to the *comes civitatis* – and demotion, after which he could only perform the functions of a *decanus*. The wording of the law is that of the right of asylum, with the verb *confugere* used for the act of seeking sanctuary in a holy place. The right of sanctuary is regulated in LV 9.3, which appears immediately after this title and is headed *De his, qui ad ecclesiam confugiunt.*[15]

In any case, given the severity of the punishment, I am inclined to think that this provision refers to desertion from a military expedition. As I have suggested elsewhere, the origins of this law may lie, *inter alia*, in the constitution of Constantine of 323, which would not have included the possibility of seeking sanctuary.[16]

The law emphasises that the deserters in this case are attempting to return home, although it does not go into details, nor does it address the destabilising

LV 9.2 are in effect Leovigildian laws. However, according to García Moreno, 'Hispania visigoda', p. 58, they belong to Euric.

13 Under the Title *De desertoribus et occultatoribus eorum* in the Code of Justinian, CJ. 12.45, 1pr. (= CTh. 7.18.5).

14 Isla Frez, *Ejército*, p. 18 n.29.

15 I analyze these laws in Esperanza Osaba, 'Responsabilité et droit d'asile dans l'Hispania visigothique', in *Responsabilité et Antiquité* I, *Méditerranées* 34/35 (2003), pp. 77–105.

16 CTh. 7.12.1 (= CJ. 12.42.1), see Esperanza Osaba, 'Influenza delle leggi costantiniane nella Lex Visigothorum', in *Diritto @ Storia. Quaderni di scienze giuridiche e tradizione romana*, 2 (2003).

factor posed – for the Roman legislator at least – by thieving deserters.[17] One indication of the gravity with which the offence is viewed in LV is that the *comes* is charged with notifying the king himself of the desertion.

In the second part of the law the *centenarius* is held responsible, not for his own desertion, but that of the men in his unit, the *centena*. Two possible situations are considered: first, when, unbeknownst to the *prepositus hostis* or *thiufadus* (*prepositis exercitui*, in the terms used in the heading of the previous law), the *centenarius* allows any (*ab aliquo*) of his subordinates to return home (*ad domum suam*); and second, when he fails to compel him to join the army (*in hostem*), either in exchange for a bribe or in answer to an entreaty. If a bribe had been received, he was required to pay ninefold the amount received to the *comes civitatis* of the corresponding territory, who would distribute it to the *centena* to which he was attached. If no bribe had been received (*nullam mercedem acceperit*), this commander was obliged to pay a sum of ten *solidi*, analogous therefore to that provided for in the preceding law. The law also clearly reflects the command chain: in such cases the *centenarius* would have issued the permit without the knowledge or consent of his superior, the *thiufadus*.

Interestingly, in this law the monarch again speaks in the first person (*sic cum nostra ordinatione partiantur ... et sicut superius diximus*), and expresses a particular interest in being informed of the case by the *comes civitatis*, to ensure that his dictate is being properly applied, etc. This would seem to be in keeping with the keen legislative style of Leovigild.[18] In all other aspects, the framework of sanctions is similar to that of the first law, LV 9.2.1, which is also repeated in subsequent ones: nine times the bribe is to be paid to the *comes* and distributed to the *centena*; where no bribe has been taken, there is a scale of fines, depending on the rank.

One very significant aspect of this law is that it explicitly refers to the *centenarius* own desertion, for which it imposes the death penalty, commutable if he seeks sanctuary. On the other hand, desertion by a *thiufadus* or any other officer is not addressed by the law, and we therefore cannot tell whether it

17 However, damage caused by those called to a *publica expeditio* and the corresponding punishment are dealt with in LV 8.1.9; LV 8.1.8 Recc. and LV 8.1.7 *ant.* See Pérez Sánchez, *El ejército*, p. 132. For the relationship between thieves, defectors and deserters, see Vincenzo Giuffrè, *Letture e richerche sulla "res militaris"*, 2 (Napoli, 1996), pp. 517–526.

18 On the relationship between the monarch and the law, see Enrique Álvarez Cora, '*Qualis erit lex*: La naturaleza jurídica de la ley visigótica', in *Anuario de Historia del Derecho Español* AHDE, 66 (1996), pp. 11–117; Patrick Wormland, '*Lex scripta and Verbum Regis*. Legislation and Germanic Kingship from Euric to Cnut', in *Early Medieval Kinhship*, eds. H.-W. Goetz – J. Jarnut – W. Pohl (Leeds, 1977), pp. 105–138.

applied equally to them. I would also note the lack of any mention of the term *miles* in addressing the case of a *centenarius* permitting the desertion of someone (*ab aliquo*) from his hundred.

Two aspects of this law are significant: the interest expressed in notice being given to the king, and the required fine distribution among the *centena*.

2.4 *LV 9.2.4 Ant*

This law is in complete consonance with the preceding ones and its heading is almost identical to that of LV 9.2.3 *ant*. The first provision turns its attention to another rank, considering the case of a *decanus* who leaves his command and abandons his unit (*de hoste ad domum refugerit* ...) or, despite being in good health, refuses to join the expedition. As in the previous law, we find the term *refugere* used. For this offence, he must pay ten *solidi* to the *comes civitatis* or, if a bribe has changed hands, five *solidi* – i.e. the same amount provided for in LV 9.2.1 for cases in which no bribe was received. The money was to be distributed among the members of the hundred, by the express order of the king, who once again leaves his imprint upon the text.

It seems surprising that in the case of the *decanus*, the sanction was smaller precisely in the event that a bribe had been paid. One may perhaps assume this to have been an error or bad drafting, given that the habitual fine for bribery was ninefold the amount received (LV 9.2.1; LV 9.2.3, and also LV 9.2.5) and that in LV 9.2.1 the *decanus* was to pay five *solidi*, precisely when there was no *merces* involved.

The second provision of the law establishes sanctions for any presumably unranked soldier – *si aliquis* (note again the absence of the word *miles*) – attached or *numeratus* to the *thiufa*, who without the permission of either his *thiufadus* or the *quingentenarius*, *centenarius* or *decanus* flees to his house or refuses to leave it to enlist (*de hoste ad domum suam refugerint aut de domo sua in hostem proficisci noluerit*). Thus it again addresses the two circumstances envisaged in previous laws. Anyone thus deserting was to receive one hundred lashes publicly (*in conventu mercantium publice*) and a fine of ten *solidi*. As well as punishing the deserter, the law also seeks presumably to achieve the maximum visibility for the dissuasive and exemplary punishment, which in many cases would result in the death of the offender.

Whereas the previous laws addressed the responsibility of commanders in order of rank, this second provision prescribes punishment for any unranked soldiers who evaded enlistment or returned home without the permission of the respective *thiufadus*, *quingentenarius*, *centenarius* or *decanus*. Once again, then, it includes the two cases of desertion discussed previously.

2.5 LV 9.2.5 *Ant*

This complex and sometimes nebulous text also contains two separate provisions. The first deals with *compulsores exercitus* (*servi dominici*): any such officer found to have accepted a bribe to release another from enlistment was to pay the *comes civitatis* ninefold the amount received. If no bribe had changed hands, the penalty was five *solidi*, the same as that provided for *decanus* in similar circumstances in LV 9.2.1.

The second provision of this text, which is in keeping with the previous one, with which it is closely related, requires the *thiufadus* to investigate his *centenarii*, and the *centenarius* to investigate his *decani*, to determine whether any have deserted the army or avoided enlisting, again, either through bribery or entreaty (*per precem aut per redemtionem*). In addition – we are told in somewhat unclear terminology[19] – he must convey the results of his inquiries to the *preposito comitis*.

It is then stipulated that the *thiufadus* should address to the *comes civitatis* by writing (again specifying *in cuius est territorio constitutus*) who in turn should deliver to the *thiufadi*, the *centenarii*, the *decani* or the *servis dominicis* the *vindicta* prescribed by law for those who desert or fail to heed the call to service, either through bribery or entreaty.

However, in the event that after receiving the *vindicta*, he concealed or failed to declare it, he must pay ninefold the amount received. And if, either through bribery or entreaty, he should fail to demand the penalty, he must pay twice the amount out of his own property to the officers charged with distributing it among the *thiufa*. And finally, if the *comes* should divert the *vindicta* for his own benefit or failed in his duty to notify to the king or its appropriate distribution to the *thiufa*, he must pay elevenfold the amount by way of reparation.[20]

As we can see, this intricate law begins with the same scheme of sanctions, applied here to the *servi dominici*, and goes on to address the investigation of deserters. It concludes with a very different section, implying direct control by the king over his *comes civitatis*. The law gives details on the procedure to be followed in the event of non-compliance and how it should be punished.

---

19 As noted by Zeumer, *Leges Visigothorum*, the phrase *thiufadus preposito comitis* appears to be a mistake. According to Zeumer, p. 369, it should read: *preposito hostis vel comiti exercitus*. See also Eugen Wohlhaupter, *Gesetze der Westgoten* (Weimar, 1936), p. 268.

20 Manuel Torres López, *Lecciones de Historia del Derecho Español*, 2nd Ed. (Barcelona, 1934), p. 324, considers that the ninefold and elevenfold amount are the penalties for the *comes* and the twofold amount to be the sanction for the officers who have deceived him.

Clearly the *comes civitatis* —the maximum authority and the king's representative in the cities— was not immune to entreaties and other pressures in this respect.

2.6 LV 9.2.7 *Ant*

Although this last of the *antiquae* laws does not directly deal with punishments for desertion as such or failure to answer the call to arms, it is the first to address the case of those who, in a desperate situation, cross enemy lines and rescue slaves and other property. The term *inimicus* appears three times in the text, whereas the word *hostis* – used in the meaning of 'enemies' rather than of 'army' as in the other *antiquae* in this title – is only employed in the heading, which we may presume to have been added at the time of Recceswinth's promulgation.[21]

It is noteworthy, in my opinion, the treatment given in the law to such people, who are rewarded for their profits (*mancipia vel quodeumque genus pecunie aut aliarum rerum*) captured from the enemy. As Pérez Sánchez argues, this provision might be related to the magnates' interest in retrieving slaves and other goods they had taken with them to war.[22] Of the total amount recovered the retriever was only required to restore two thirds to their owner, *pro intuito misericordia*, keeping the remainder as a reward *pro laboris sui.* The law also stipulates that if anyone recovers a slave from the enemy (*ad inimicis sollicitaverit*) through *patrocinium aut consilium,* upon returning him to his master, he should be entitled to a reward of one tenth of his value.

3 Wamba LV 9.2.8

Wamba acceded to the throne in 672 and scarcely a year later, in 673, he had already issued his wide-ranging military law, upon returning from an expedition to put down a revolt led by the *dux* Paulus in the province of Narbonne.[23]

---

21 LV 9.2.7 (...) *qui mancipia vel quaslibet res de manu hostis excusserit.* The expression *de manu hostis* is coincident with Biblia Vulgata, Iob 6.23, *Liberate me de manu hostis*, also in Gregory the Great, *Moralia in Iob*, 7.34.1.

22 Pérez Sánchez, *El ejército*, p. 116.

23 Isabel Velázquez Soriano, 'Wamba y Paulo: dos personalidades enfrentadas y una rebelión', in *Espacio, Tiempo y Forma*, Serie 2, *Hispania Antigua*, 2 (1989), pp. 213–222; José Ángel Castillo Lozano, 'Los tiranos de la Historia Wambae Regis: Ilderico y Paulo, el *rex perditionis*', in *Categorías de poder en el reino visigodo de Toledo: los tiranos en las obras de Juan de Biclaro, Isidoro de Sevilla y Julián de Toledo, Antigüedad y Cristianismo,* 2016–2017 (Murcia, 2019), pp. 173–182.

A detailed account of these events, which culminated with the defeat of the *tyrannus* Paulus, is given by Julian of Toledo (644–690), a key figure of the Visigothic era and author of the *Historia Wambae regis*,[24] which provides first-hand information on the period.

We do not know who ordered this and Wamba's other laws to be incorporated into the LV, although it was most likely on the initiative of Erwig when the new LV was promulgated in 681. There is no doubt that the law was widely known, and Erwig himself even raised the need for it to be reviewed in his address to the XII Council of Toledo in 681. Yolanda García considers the possibility that Wamba's laws may have been sent out to the different parts of the kingdom as separate orders, precisely because in the LV they stand out for maintaining the original structure intact.[25]

3.1 *Preamble*

Right from the preamble of this important law LV 9.2.8,[26] we see how the art of rhetoric enhances the literary vigour of this singular text. One outstanding example is the metaphor of the trumpet of the law which, as outlined, should spur on this *fraternitas* rosed up in arms in defense of the king, the people and the country. Julian of Toledo, who probably inspired this law, draws attention to the effect of the *tuba* in order to encourage fighting spirit in the battlefield.[27] Interestingly, Augustine of Hippo, who greatly influenced Julian of Toledo, resorted to a similar metaphor already three centuries before when

24 Julian, metropolitan of Toledo probably from 680, attended the councils of XII Toledo (681), XIII Toledo (683), XIV Toledo (684) and XV Toledo (688) and presided over the last three. Gregorio García Herrero, 'El reino visigodo en la concepción de Julián de Toledo', in *Antigüedad y Cristianismo*, 12 (Murcia, 1995), pp. 385–420; Suzanne Teillet, *Des goths à la nation gothique. Les origins de l'idée de nation en Occident du* V^e^ *au* VIIe *siècle*, 2nd ed. (Paris, 2011), pp. 585 *ff.*; Roger Collins, 'Julian de Toledo and the Education of Kings in Late Seventh-Century Spain', in *Law, Culture and Regionalism in Early Medieval Spain*, (Hampshire, 1992), pp., 1–22; Jocelyn Nigel Hillgarth, 'Introduction', in *Sancti Iuliani Toletanae sedis episcopi opera*. Pars I, CCSL (Turnhout, 1976), pp. VIII–XXI; Joaquín Martínez Pizarro, *The Story of Wamba. Julian of Toledo's Historia Wambae Regis* (Washington, 2005).

25 García López, *Estudios*, pp. 11–13. In addition to LV 9.2.8, dating from 673, examined here, two other laws from Wamba are preserved in the LV: LV 4.5.6 and LV 4.5.7, both from 675.

26 An extended version of this study in Esperanza Osaba, '*Quid debeat observari.* En torno a LV 9,2,8 Wamba', in *L'esercito romano e l'alba dell'Europa. Modelli concettuali e sperimentazioni sociali tra linguaggi, istituzioni e diritto (secoli* IV–VIII), ed. G. Bassanelli Sommariva, Ravenna Capitale (Santarcangelo di Romagna, 2020), pp. 149–184.

27 Julian of Toledo, *Historia Wambae regis*, 13, and also Isidore, *Eymologiae*, 18.4.2–4; 3.21.3.

describing the coenobitic ideal: the psalter is like the tuba calling the brethren to community life.[28]

Notably, the preamble places the law at centre stage and expresses the desire to bring an end to the old and unresolved problem of desertion, which has led Wamba to legislate once more. From the outset it also draws a clear distinction between lawsuits intended to settle the *negotia populorum* and matters of war (*in rebus bellicis*), which leads us to think that he is explicitly expressing a desire to legislate on military issues. The text also links the function and objective of this law with the fraternal union between men at arms – *fraternitas, fraterno, fratrum* –, repeated three times in the preamble, even though *fraternitas* or *fraternus* are quite unusual terms.[29] Finally, it is worth noting the stress on military action in the interest of the *publica utilitas*. Brethren who shirk their duty to their country cause either the desertion of many or the death of the most courageous and act against the common good of the kingdom and its people. I believe that this reference to the *utilitas publica pro gentis et patrie,* doubly underlined (*qui prestare debuit publicis utilitatibus, pro gentis et patrie utilitatibus*) is decisive. Right from the preface, Wamba is insisting on the public duty of defending the *patria* and the *gens*, which, alongside safeguarding the king, was one of the elements of the oath of allegiance sworn to the monarch from c. 75 of the IV Council of Toledo (633), albeit here it is transferred to a military context. I will return to this issue below. The preamble is followed by the detailed dispositive provisions of this innovative law.

3.2 *First Provision*

This first provision of LV 9.2.8 refers to calls to arms made to repel hostile attacks by enemies from other kingdoms. The call expressly extends to the bishop and other members of the clergy, the *dux* or *comes*, the *thiufadus* or *vicarius*, and the *gardingus,* and any other persons entrusted with the defence of the territory subject to attack, either in the vicinity of the conflict or within a radius of up to a hundred miles. Upon being alerted by any of these officers, or by any means whatsoever, they were to answer the call without delay, seconded and supported by all their forces (*cum omni virtute sua*).

The specified individuals should come as soon as required upon first notice or *admonitio* from their duke or count, and likewise if summoned by the *thiufadus* or his *vicarius* or by any person with authority to do so, as was primarily

28 Augustine, *In Psalmun 132, Enarratio. Sermo ad plebem.* See García López, *Estudios*, 11–13; Hillgarth, *Las fuentes de San Julián*, pp. 113–115.

29 Botta, notes in this volume (p. ++) the use of the term in relations amongst comrades in arms.

the case of the *compulsores exercitus* (*servi dominici*), as we see in the *antiquae* LV 9.2.2 and LV 9.2.5. The law also requires them to respond to the appeal upon having notice of the attack, regardless of how they had been made aware of it (*vel quocumque modo ad suam cognitionem pervenerit*), and thus, we should presume, no official call was in such a case preceptive.[30]

This reference to the *admonitio* is of special interest. Apart from Wamba's military law, the term is only found in the later military law of Erwig, LV 9.2.9, and in relation to a judicial summons especially in LV 2.1.19 of Chindaswinth.[31] The verb *admonere*, on the other hand, is less infrequent, and is generally also very closely in relation with to summonses.[32]

For all these reasons, and as we shall see below, I believe there is an important link between this law and Chindaswinth's LV 2.1.19,[33] with this specific law on judicial summons influencing both the law of Wamba and the later law of Erwig, with the precise meaning of a subpoena or demand to fulfil a service of public order,[34] This connection has already been noted by Yolanda García,[35] who links the enactment of the law to attempts by Julian of Toledo and Erwig to centralise control. The form of judicial summons developed by Chindaswinth in LV 2.1.19, and also in LV 2.2.4 and LV 2.2.7 was used not only in the military laws of Wamba and Erwig, but also in the councils' canons, with Canon 8 of XIII Council of Toledo (683), in the reign of Erwig, being a good example. We have notice of this *admonitio* in council documents, albeit not in the LV, including no less than the III Council of Toledo in 589 (*gloriosissimi principi admonitione, regiam admonitionem*).[36] Furthermore, this summons

30 Thompson, *Los godos*, p. 309.

31 LV 2.1.19, Chind.; LV 8.1.7, *ant.* and LV 6.1.3, Wittiza.

32 *Admonere*, in different verb forms, in LV 9.2.8 Wamba, LV 9.2.9 Erw. and LV 2.1.19 Chind., but also in LV 2.1.28 *ant.*; LV 2.2.2 Chind.; LV 2.2.7 Chind.; LV 2.2.8 *ant.*; LV 2.4.2 *ant.*; LV 12.1.3 Erwig. It also features with other meanings in: LV 2.1.9 Recc.; LV 6.4.3 Chind. LV 6.5.15 Chind. LV 7.2.14 ant; LV 8.3.3 *ant.* LV 8.4.23 *ant.*; LV 8.5.1 *ant.*; LV 12.2.18 Egica.

33 On the judicial summons, see Carlos Petit, *Iustitia Gothica. Historia Social y Teología del Proceso en la Lex Visigothorum* (Huelva, 2001), pp. 203*ff.*

34 One of the meanings of the term is the fulfilment of a duty of public order, see Jan Frederik Niermeyer – Co van de Kieft, '*Admonitio*' and '*Admonere*', in *Mediae Latinitatis Lexicon Minus*, 2nd ed. Jan W. J. Burgers, II (Leiden, 2002), p. 28. A text from the *Histoire général de Languedoc* is particularly illuminating: *Omnes burguenses totius villae Bitterrensis, sui scilicet et vicecomitis, in expeditionem eum sequantur ad admonitionem episcopi per se vel per suum nuntium factam.*

35 García López, *Estudios*, pp. 374–378. This form of call had a basic structure, comprising: 1. The *admonitio*, or notice given by the authority; 2. The offence, that is, the breach itself; 3. The pretexts adduced to escape the call.

36 III Toledo (589:) c. 18, also in IV Toledo (633) c. 32 and c. 75; VII Toledo (646) c. 6; Toledo. XIII (683) c. 8; XV Toledo (688) tomus regius of Egica; XVI Toledo (693) c. 7.

or *admonitio* is pronounced by the monarch, a factor that Manuel Torres sees as being a manifestation of the king's power in his capacity as supreme commander of the army.[37]

As we shall see below, Wamba's law shares other features with that of Chindaswinth on judicial summons, namely the call addressed to the church authorities and the requirement upon any person within a 100 miles radius to answer the call. With all the necessary caveats, I believe that in general this law essentially matches the earlier one.

Wamba emphatically – and in the first person – turns the spotlight on those who do not answer the summons promptly, with all forces available to them (*omne virtute sua*) to repel the enemy and in support of their people and the country (*ad defensionem gentis vel patrie nostre*).[38] This *patria* and *gens*, together with the king, represent the unity which is broken by the desistance of the *fratres* in the face of external attack, and furthermore are the institutions they have sworn to defend.[39] Anyone acting in this way commits the criminal offence of failing to answer the royal summons.

And speaking of such reprehensible breaches, the solemn words of King Wamba take the form of a rousing harangue to his troops. Three times he

37 Torres López, 'El estado visigodo', p. 376.

38 These resources must be taken to include the members of their retinues, as noted in Pérez Sánchez, *El ejército*, p.156.

39 Alexander Pierre Bronisch, 'Precisiones sobre algunas informaciones históricas en la 'Crónica de Alfonso III'', in *Edad Media. Rev. Hist.*, 12 (2011), pp. 31 *ff.*, consider it to be very significant that the term *rex, gens vel patria* is always used in laws and council canons when the monarchs or council fathers are referring to high treason. See also the origin and analysis of this very significant expression in Teillet, *Des goths*, pp. 524 *ff.*; Martin, *Géographie*, pp. 361 *ff.*, who further argues that in addition to the individual meaning of each word, there is a political and institutional significance and value in Visigothic society of combining the three elements, the human *gens*, the territorial *patria* and the institutional *rex*; also Céline Martin, 'La notion de gens dans la péninsule Ibérique des VIe -VIIe siècles: quelques interprétations', in *Identité et Ethnicité, Concepts, débats historiographiques, exemples (IIIe–XIIe siécle)*, V. Gazeau – P. Bauduin – Y. Modéran (Caen, 2008), pp. 75–89; Walter Pohl – Philipp Dörler, 'Isidore and the gens gothorum', in *Antiquité tardive*, 23 (2015), pp. 133–141; Aquilino Iglesia Ferreirós, *Historia de la traición. La traición regia en León y Castilla* (Santiago de Compostela, 1971). A recent development can be found in Stefan Esders, '*Regem iura faciunt*. Der westgotische Treueid im Kräftefeld personaler und Trasnpersonaler Konzepte der Legitimität politischer Herrschaft', in *Die Macht des Herrschers. Personale und Transpersonale Aspekte*, eds. A. Mechthild – E. Brüggen – K. Klaus, (Göttingen, 2019), pp. 78 *ff.*; Id., Esders, Stefan. 'Grenzüberschreitende Mobilität als Problem politischer Loyalität. Das Hochverratsgesetz des westgotischen Königs Chindasvinth (642/643) und das 7. Konzil von Toledo (646)', in *Wasser-Wege-Wissen auf der iberische Halbinsel. Eine interdisziplinäre Annähderung im Verlauf der Geschichte,* eds. I. Czeguhn – C. Möller – Y. Quesada Morillas – J.A. Pérez Juan (Baden-Baden, 2018), pp. 92 *ff.*

reiterates the expressions which stresses the same idea: *ad defensionem gentis vel patrie nostre prestus* (…) *non fuerit*; *pro vindicatione patrie non existat*; *ad prestitum vel vindicationem gentis et patrie exire vel intendere contra inimicos nostre gentis* (…) *distulerit*.

It then goes on to describe the different causes and circumstances underlying the contravention of the law, some of which have already been mentioned in the preamble: the sole desire to go elsewhere or the use of excuses or pretexts, manifestations of will or character, such as laziness, fear, malice, irresolution or half-heartedness. It also establishes the inevitable consequence of abandoning *fratres* to their fate (*aut qui prestare debuit publicis utilitatibus fratrum destitutus adiutorio retrahatur*) which is one of the *leitmotivs* of this law.

If, due to a sudden attack from without, the inhabitants or the provinces of the kingdom (*in populos vel provincias regni nostri*) should suffer injure or captivity, those responsible through disloyal desistance shall be subject to a series of punishments, which the law goes on to list.

Bishops (*sacerdotes*), priests, deacons and clerics, who did not have the means to pay for the damages incurred with their own property, were sentenced to exile. All other clergy would undergo the same sentence as the laity. Nobles (*nobilis*) and those of middle or lower rank – *mediocriter vilioque persona* – as well as the clergy, would forfeit their right to testify and be reduced to slavery, to be disposed of as the king saw fit.

In pecuniary terms, in light of the gravity of their crimes, they would have to satisfy any damages caused by the enemy to the country (*damna rerum terre nostre ab inimicis*) and those suffering harm. In this case, however, the amount was not to be paid to the hundred or other military unit, as in the *antiquae*,[40] as it would undoubtedly have been difficult to quantify and distribute. Rather, the property confiscated from the wrongdoers was – largely – to be given to the king, who would use it to reward his faithful subjects, as we know from XII Toledo (681) and XIII Toledo (683), which in the reign of Erwig discussed the issue of restitution.

Finally, bringing this section on external attacks to a close, Wamba adds a short conclusion, with a Chindaswinthian ring: any man who, through malice or cowardice, fails to repel the aggressors or to show virility[41] in battle against the adversary should justly suffer (*recte se doleat*) the loss of their noble dignity and the disposal of their property. This vehement statement that the guilty

40 For example, in *antiquae* LV 9.2.1 and LV 9.2.3. See Osaba, '*Lex Visigothorum* 9,2', pp. 170 *ff.*

41 References to virility (in contrast to womanliness) are a constant in such military contexts. See, for example, Julian of Toledo, *Historia Wambae regis*, 9, in the context of Wamba's harangue, the indignity of being branded as *effeminatos et molles*.

must suffer for their actions seems to have been taken from Chindaswinth: this unusual mention to the punishment to be meted out (*et amarissimam vitam ducere se perenniter doleat*) is only found here and in LV 2.1.8.[42]

3.3 *Second Provision*

In the second and important provision of LV 9.2.8, Wamba – again speaking in the first person – turns to the defence of the people, the country (*patria*), his reign and that of his successors (*nostrumque regnum vel etiam successorum nostrorum*) against internal attack – i.e. rebellions and plotting –, a constant feature of Visigothic history. The context in which this law was promulgated and the link to Paulus' revolt is clear. Three decades earlier, Chindaswinth, in LV 2.1.8, had deplored the fact that the military forces expended more effort in quelling internal rebellions than external ones, with the ensuing suffering for the country. This emphasis in the king's own kingdom and that of his successors reflects a traditional desire for permanence and the future upholding of the law, which is also common in Chindaswinth.

This whole paragraph, I believe, owes much to LV 2.1.8 of Chindaswinth, which Wamba appears to have taken as his point of departure. As well as the passages that explicitly mirror the earlier law, we can also see its influence in the wording and, of course, in the meaning (*scandalum ... contra gentem vel patriam nostrumque regnum vel etiam successorum nostrorum moverit aut movere voluerit*).[43] Significantly, the term *scandalum*, implying an internal revolt, is also employed by Chindaswinth in an analogous context in LV 2.1.8.[44]

Also relevant is the mention of rebellion as a cause *contra gentem vel patriam nostrumque regnum*, which also follows Chindaswinth, who appears throughout the LV to have been who finally poured into a legal mould this crime of treason,[45] with these characteristics.

Following a description of the nature of the internal attack, the law goes on to regulate the call to arms to supress the rebellion, using the same framework as that established for external attacks. Those same social groups are again

---

42 Gerhard Köbler, '*dolere*', in *Wörterverzeichnis zu den Leges Visigothorum* (Giessen-Lahn, 1981). This term only appears in LV 2.1.8 Chind. and in LV 9.2.8.

43 LV 2.1.8: *perrexit sive perrexerit aut* (...) *voluit vel quandoque voluerit* (...) *extitit vel extiterit* (...) *intendere vel intendisse* (...).

44 LV 2.1.8, Chind.: (...) *sive ab anno regni nostri primo vel deinceps quispiam infra fines patrie Gotorum quamcumque conturbationem aut scandalum in contrarietatem regni nostri facere voluerit* (...).

45 Iglesia Ferreirós, *Historia*, pp. 43 *ff.*; Abilio Barbero – Marcelo Vigil, *La formación del feudalismo en la Península Ibérica*, 4th ed. (Barcelona, 1986), pp. 126 *ff.*; Esders, 'Regem iura faciunt', p. 97 n.88; Teillet, *Des goths*, pp. 624–625.

requested to join up: priests, clergy, dukes, counts, *thiufadi*, *vicarii* and any other persons, in the same order as above. In general, anyone in neighbouring territories up to a radius of one hundred miles must also join in the quelling of the *scandalum*.

They were exhorted to respond without delay and I believe that we must presume that there would have been an official summons, conveyed perhaps by the *compulsores exercitus*. Likewise, as in the previous provision of the same LV 9.2.8, anyone who becomes aware of the attack by any other means is also required to take action, even if they have received no official request to that effect (*admonitus fuerit, vel ad suam cognitionem quoquo modo pervenerit*). Both circumstances, the receipt of an *admonitio* or the simple notice of the aggression, require the person to respond to the attack and defend (*ad vindicationem*) the king, the people, the country and the king's other subjects (*fidelium*).[46]

All those who evaded the call to arms – be they bishops, clergy or even members of the *officium palatinum*, of whatever rank or title, together with the *inferiores* (*aut fortasse ex officio palatino, in quocumque sit ordine constitutus vel quelibet persona fuerit dignitatis, aut fortasse inferior*) – were considered to be implicated and would therefore be subject to the *scelus* of breach of faith (*infidelitas*). The punishment was banishment, following confiscation of their property, which the king was to use at his discretion.

The mention to *scelus infidelitatis* is very significant. The offence pertains to the area of treason, as suffered by Wamba in Paulus' revolt, which, as pointed out above, had been suppressed very shortly before this law was enacted (and may have formed the motivation for it). The punishment is not directly imposed on the conspirators themselves, but on those who evade fighting by failing to answer the appeal to arms. Naturally, there could be some overlap, with some avoiding the call either because they were themselves the instigators of the revolt or because they supported it, obviously leading their dependents to offend with them.

Betrayal of the king, the *gens* and the *patria* hangs over all of this law.[47] The crime of *infidelitas*, precisely in this dispositive part, is a clear expression of the breaking of an oath. There is no clear evidence of a military oath at this

46 On the *fideles*, see King, *Derecho*, p. 78; Dahn, *Die Könige*, pp. 141–142. Martin, *Géographie*, p. 368.

47 Floyd Seyward Lear, *Treason in Roman and Germanic Law* (Austin, 1965), pp. 147–149; Isla Frez, *Ejército*, pp. 51–54; Renan Frighetto, '*Incauto et inevitabili conditionum sacramento*: juramento de fidelidad y limitación del poder regio en la Hispania visigoda en el reinado de Egica (688)', *Intus-Legere Historia*, 1, ½, (2007), pp. 70–71.

time, but we do know that subjects swore fealty to the monarch, and we have an example of the possible wording of the oath in a later law, LV 2.1.7 of Egica. It seems reasonable in any case to assume that this oath of allegiance extended to providing military support where necessary to defend the king, the people and the country,[48] and this law of Wamba attempts to clarify this issue. In the event of an internal revolt, anyone refusing to answer the call to arms was in breach of the *fides* of their oath to the monarch, for which reason this offence was deemed as *scelus infidelitatis*; by extension, failure to act in the event of external attacks or aggressions would be categorised in the same way. As Barbero and Vigil have pointed out, the measures established in this law show that military obligations were a component of the general duty of allegiance.[49] Yolanda García[50] believes that it was Julian of Toledo who restored and transposed this terminology, reusing King Reccared's argument against Arianism at the III Council of Toledo (589) and applying it at a time when it was felt important to emphasize the sacral nature of kingship.

Transgressors were condemned to exile and the loss of their property, to be disposed of at the monarch's discretion. As we have seen, exile was also the punishment imposed on the clergy – bishops, priests and deacons. However, there is no mention of the forfeiture of their right to testify or their enslavement, which are central aspects of the dispositive part on the call to arms in response to external attacks. As already discussed, while in the *antiquae* the confiscated properties were given to the military units (generally to the *centenae*),[51] here they are in all cases assigned to the king.

As with exile, this was a specific punishment applied throughout Visigothic history in cases of treason, and was later incorporated by Erwig into LV 2.1.8 in his new promulgation of the LV in 681.[52]

---

48 Carlos Petit, '*Rex iudex*. El momento judicial del rey de Toledo, 'Procesos, inquisiciones, pruebas', in *Homenaje a Mario Sbriccoli*, eds. E. Conte – M. Madero (Buenos Aires, 2009), p. 48; Pérez Sánchez, *El ejército*, p. 137.

49 Barbero – Vigil, *La formación*, pp. 142–143.

50 García López, *Estudios*, pp. 331–339.

51 For example, in *antiquae* LV 9.2.1; LV 9.2.3 and LV 9.2.4.

52 Margarita Vallejo Girvés, 'Los exilios de católicos y arrianos bajo Leovigildo y Recaredo', in *Mundo medieval. Hispania sacra*, 55 (2003), pp. 35–47. See also Céline Martin, 'Ervig and Capital Penalties: The Way of Exile', in *Framing Power in Visigothic Society. Discourses, Devices and Artifacts*, eds. E. Dell'Elicine – C. Martin (Amsterdam, 2020), pp. 133 *ff.*; Petit, '*Rex iudex*', pp. 42 *ff.*; King, *Derecho*, pp. 110–111; Iglesia Ferreirós, *Historia*, pp. 43 *ff.*; Augusto Priego de Lis, 'La pena de exilio en la legislación Hispanogoda', in *Espacio y tiempo en la percepción de la Antigüedad Tardía*, eds. E. Conde Guerri – R. González Fernández – A. Egea Vivancos, *Antigüedad y Cristianismo. Monografías históricas sobre la Antigüedad Tardía*, 23 (Murcia, 2006), pp 515–529; Pablo C. Díaz, 'Confiscation in the Visigothic Reign of

3.4 *Exemptions and Pretexts*

The last provision of this law deals with admissible exemptions and pretexts for failure to attend and provide indemnity for those failing to answer the call. Subjects who were unable to join due to infirmity or illness must nonetheless send all their resources and forces (*omnem tamen suam virtutem*), which would be allocated to the troop of the bishops and the clergy and brothers, serving sincerely and faithfully the cause of the king, the people and the country. As well as the repeated reference to the *utilitas,* it is also noteworthy that the forces of those who were unable to enlist in person were to join the church retinue, while those failing to comply were liable to the same penalties as the actual transgressors.

The wording of the law of Chindaswinth, LV 2.1.19, on judicial summons was similar: illness (*egritudo*), as proved by competent witnesses, was also ground for exemption.

3.5 *Closing*

Wamba concludes by returning to the stance set in the preamble of the law, and as well as in the *exordium* he had stated his intention to legislate in order to rectify what he describes an inveterate bad habit (*huius male usitate consuetudinis mores*), in the final lines he expresses once more his confidence that this vicious custom would finally be rooted out by the severe sanction of the law.

Having developed this idea, he concludes with a second one, which is repeated throughout the rest of the text: together with the rigour of the law, it is concord and unanimous agreement that will bring serenity to the people and the defence of the country (… *quietem plebium et patrie defensionem adquirat*).

4 The Time of Erwig (680–687)

In October 680, King Wamba fell ill and was believed to be about to die. He received canonical penance, and was tonsured and dressed in penitential garb, as it was customary in such cases.[53] He agreed to abdicate in favour of Erwig, calling on Julian of Toledo to anoint the new king swiftly, in order to legitimize the appointment. Most historians believe that the whole episode was the result of a carefully designed conspiracy to disqualify him from office, headed

Toledo', in *Expropiations et Confiscations dans les royaumes barbares. Une approche regionale*, eds. P. Porena – Y. Rivière (Roma, 2012), pp. 93–112.

53 Roger, Collins, *Visigothic Spain 409–711*, 2nd ed. (Oxford, 2004), p.98.

by Erwig with the backing of key sectors of the nobility and the church. However, other scholars dispute this interpretation.[54]

The XII Council of Toledo was held in 681, scarcely three months after Erwig's accession to the throne. In his *tomus regius*, the new king announced his intention to introduce a series of firm measures, *inter alia* amendments to Wamba's military law. Julian of Toledo, attended the council as the metropolitan of Toledo – a position he had taken up in 680. He was to retain close links to Erwig throughout the rest of his reign. One year later, in 682, Erwig promulgated his extensive military law, LV 9.2.9, which I will now examine.

5 LV 9.2.9 Erwig

From the outset, Erwig's law distances itself from that of his predecessor, Wamba.[55] The heading contains terms not found in Wamba's military law, such as *exercitus* and *expedition*, while deserters are referred to with the word *refugere*, thus establishing a link with the *antiquae* of the same title. The law is also even more specifically military in nature. Whereas Wamba's law focused on the *scandalum* (*Quid debeat observari, si scandalum infra fines Spanie exsurrexerit*), Erwig's law lays down punishment for those who sought to evade the call to arms, either by refusing to join the army or by deserting from the war front. It also stipulates how many of his slaves each landlord was required to bring to the expedition.[56]

---

54 Thompson, *Los godos*, pp. 271–273; Collins, *Visigothic Spain*, pp. 98 *ff.*; Claude, Dietrich. *Adel. Kirche and Königtum im Westgotenreich* (Sigmaringen, 1971); José Orlandis, *Historia del reino visigodo español*, 2nd ed. (Madrid, 2006), pp. 175–176; Suzanne Teillet, 'La déposition de Wamba, un coup d'état au VIIe siècle', in *De Tertullien aux Mozarabes*, II: *Antiquité tardive et christianisme ancien (VIe–IXe siècles), Mélanges offerts à Jacques Fontaine à l'occasion de son 70e anniversaire, par ses élèves, amis et collègues*, eds. L. Holtz – J.-C. Fredouille – M.-H. Jullien (Paris, 1992), pp. 99–113; Pérez Sánchez, *El ejército*, p. 159. Others, however, deny the alleged conspiracy against Wamba: Francis X. Murphy, 'Julian of Toledo and the fall of the Visigothic Kingdom', in *Speculum* 27, 1 (1952), pp. 1–27; Bronisch, 'Precisiones'; Martínez Pizarro, *The Story of Wamba*, pp. 67 *ff.*

55 A previous study of this law, in Osaba, 'El Ejército de Ervigio'.

56 Martin, 'Le Liber', 22 ff., conjectures that when Erwig announced at the XII Toledo his intention to amend the law of Wamba, he was not referring to LV 9.2.8, but to another law on desertion, headed *De progressione exercitus*, which is mentioned in c. 7 of the XII Toledo, and that it was in fact this law that Erwig replaced with his new LV 9.2.9. He suggests that it was complementary in nature: whereas Wamba's law LV 9.2.8 sought to regulate the call to arms for situations of emergency arising from an unforeseen external or internal attack, Erwig's LV 9.2.9 concerned regular enlistment to the army.

The preamble begins with a powerful rhetorical question. An eloquent and effective contrast is drawn between these *desertores*[57] and their clear antithesis, the *amatores* – those who risk their lives for love of country – and, finally, the *vindicatores*, or defenders of the *patria*. Interestingly, Erwig's law refers exclusively to *vindicatores* of the *patria*, whereas in Wamba *vindicatio* extends to *regnum*, *patria* and *gentes*, and even to *fideles regis*.

The expression *amatores patrie* (used here for the only time in the entire LV) is closely related to the notion of love of the country, which we find – in reference to the call of the *tirones* – in two constitutions of Arcadius and Honorius dating from 406, compiled in the Codex Theodosianus, CTh. 7.13.16 and CTh. 7.13.17.[58] This same syntagma is also used by Julian of Toledo in his *Prognosticum futuri saeculi*.[59]

There follows a *narratio*, which is largely explanatory and sets out the purpose of the law. It shares certain points in common with Wamba's law, although Erwig's fingerprint is not as visible, and unlike his predecessor, he does not speak in the first person. The *narratio* laments, recriminates and marks out those who avoid or delay going to war – *profectio bellica*[60] – and what is worse, having been duly notified, fail in their duty to join the troops and stay home, even knowing that their participation is urgently required. The law also focused on those who do answer the appeal, but fail to do as ordered, joining the ranks without the arms and equipment required for battle. Other show more interest in their own lands,[61] and leave their slaves to work the fields, bringing less than the twentieth part of them to the fight, thus leaving themselves underprotected in battle. This is the first mention of the proportion of slaves each master was to bring – as we shall see, it will be not the last. The *narratio* refers to the same idea: such masters are giving preference to their estates and their belongings over their own safekeeping (*fruge quam corporis sospitate*). This provides us with an illustrative picture of the way the retinues functioned.

---

57 The term desertores is not found anywhere else in the LV: Osaba, '*Lex Visigothorum* 9,2'.

58 Amator only appears once in the LV, when it is used – again by Erwig – in the context of an *exordium*, as part of a rhetorical interrogation in LV 12.3.5. Giuffrè, *Letture*, pp. 444–445; Patrick Le Roux, 'L'amour patrie dans les cités sous l'Empire romain', in *Le toge et les armes. Rome entre Méditerranées et Oceán*, c. 31 (Rennes, 2011), pp. 565–581; José Luis Cañizar Palacios, 'Populus y provinciales: la población bajoimperial en la inscriptio de las constituciones del Codex Theodosianus', in *Veleia*, 27 (2010), pp. 285–303.

59 For example, in Julian of Toledo, *Pronosticorum futuri saeculi libri tres* 1.11.

60 The term *bellica profectio* is entirely absent from Wamba's law, although it is used by Julian of Toledo in the preface to his work *Pronosticorum futuri saeculi libri tres*.

61 In Roman precedent, a constitution of Leo from 458, CJ 12, 35 (36), 15, addresses the issue.

5.1 *First Provision*

5.1.1 General Call to Arms

In the first provision of the law, a general *constitutio*[62] for all (*cunctis populis regni nostri*), Erwig orders (*precipimus*) that whensoever the king (styled here as *princeps*) shall issue an order for anyone to join the army, upon a certain day or date, or when his *duces* or *comites*, sent out *in publica utilitae*, do so upon his order,[63] it shall be mandatory to come in answer to the call.

The position of the summoning monarch (and *princeps*) at the head of his troops becomes more visible here than in LV 9.2.8. There follows a list of persons required to attend without delay, stating that anyone receiving the order (*admonitio*) shall be required to present themselves. The requirement to join the *bellica profectio* also extends to anyone who becomes aware of the call by word of mouth or by any other means.

As we have seen, the term *admonitio* is also found in Wamba's law LV 9.2.8,[64] which most probably influenced Erwig's military law, since in both cases it is used with the precise meaning of a summons to fulfil a service of public order.[65] The term is repeated here, and given the legal nature of the *admonitio*, it is striking that in establishing the grounds for non-fulfilment and the corresponding sanction, the law refers to circumstances so vague and very hard to prove as mere indications or hearsaying or knowledge of it by any means.[66]

Once the subject had received the call – officially, or by any other means – he was required to present himself without delay in the place where the army was to fight (*in loco exercitus bellaturus*). The law then explicitly states that none must remain at home – a common fear – nor shall they proffer in their defence any delay or excuse that might prevent their answering the call. Once the place and time (*locis adque temporibus*) have been made known, either on the direct order of the king himself (*iussio principalis*) or on the order (*admonitio*) of the *dux*, *comes*, *thiufadus*, *vicarius* or any other authorised official (*seu cuiuslibet curam agentis tetigerit*), they must all proceed without delay to the place and at the time established.

---

62 It is interesting to note the term *constitutio* used as a synonym of *lex*. See Álvarez Cora, '*Qualis erit lex*', pp. 77–81, for its detailed examination of the *lex* in the LV.

63 Most of these posts have their origins in the Roman bureaucracy of the Later Empire; Díaz, 'El esquema provincial'.

64 For a lengthier discussion of this issue, Osaba, '*Quid debeat observari*', pp. 164 *ff.*

65 See the full development of the proposal in García López, *Estudios*, pp. 374–378.

66 Thompson, *Los godos*, p. 309.

This last section lists the individuals and posts authorised to issue the *admonitio*, which are the same as those given in Wamba's law.[67] Finally, the law stipulates that any other person invested with authority may also legitimately issue the call. We may presume that this task could be carried out by the *compulsores exercitus*, the bearers of the summons. However it may also refer, as the law states, to *cuiuslibet curam agentis tetigerit*, i.e. persons specifically authorised for this purpose, as we also see in LV 2.1.19 Chind.

One very significant difference with Wamba's law is that Erwig's call to arms no longer extends to any members of the clergy.[68] The inclusion of clergy in the appeal to arms under Wamba's military law has been cited, *inter alia*, as one of the reasons for the church's displeasure with the king, and even for its possible complicity in his overthrow.

We shall now examine the circumstance of non-compliance, deemed to occur when someone received the order or *admonitio*, or became aware that it had been issued by any of the other means or indications established in the law, but nonetheless failed to go immediately or neglected to present himself at the established place and time. As we shall see, this was the offence for which punishment was meted out.

5.1.2 Penalties

In establishing the penalties for non-compliance, this law – like Wamba's – draws a distinction between the following groups:

5.1.2.1 *Maioris Loci Personae*

The first group are the *maiores loci*, the palatine nobility, comprising only the military and the civil authorities, with no mention of the clergy. The list begins

67 Wamba's law LV 4.5.6 – a text with no military character, which establishes the punishment for the appropriation of Church goods by bishops – stipulates that *duces*, *comites*, *thiufadi*, *vicarii* and, likewise any person having legal cognisance of the matter shall be authorised to act as the accuser. On this point, see Pablo C. Díaz, 'El esquema provincial en el contexto administrativo de la monarquía visigoda de Toledo', in *Mélanges de la Casa de Velázquez*, 49–2 (2019), pp. 77–108; Martin, *Géographie*, pp. 143*ff.*; García Moreno, 'Hispania Visigoda', pp. 41–78; Thompson, *Los godos*, pp. 297*ff.*; King, *Derecho*, pp. 71–104; Sánchez Albornoz, 'El ejército visigodo', pp. 23*ff.*

68 However, Manuel Torres López, *Historia de España Menéndez Pidal*, v. 3 (Madrid, 1994), pp. 123–131, believes they would have been included anyway in the general call, *cunctis populis regni nostri*. At the same time, certain manuscript copies of the LV (designated as V 15. 16.17 and 20), compiled by Zeumer and highlighted in the edition of this law, suggest that the bishops played a leading role in examining those who gave pretexts for not answering the call to arms, Zeumer, *LV*, p. 376. This version of Erwig's law is the one contained in the Fuero Juzgo 9.2.8.

with the duke (*dux*), followed by the count (*comes*) and the *gardingus.*[69] In this case, failure to answer the summons was punishable by exile and deprivation of property, which was to be placed at the disposal of the *sublimitas principalis* – i.e. the monarch.[70]

Throughout Visigothic history, the punishment of exile was imposed in cases of treason. Erwig also applied it as a sanction for treason in the revised text of Chindaswinth's LV 2.1.8, contained in his new promulgation of the LV of 681. Specifically, Erwig's LV 2.1.8 amended the penalties established by Chindaswinth for this crime – death, commutable to blinding and confiscation of property –, replacing them with decalvation and a hundred lashes, exile and servitude, in addition to forfeiture of property. From documented cases, it can be seen that a sentence of exile might involve either banishment outside the borders of the kingdom or confinement to a specific place within it or to a monastery.[71] However, given the political nature of the crime of treason, on occasions the penalties were either not enforced or applied in an arbitrary manner not provided for in law.

5.1.2.2 *Inferiores Sane Vilioresque Personae*

These *inferiores* and *viliores* comprised the *thiufadi*, the *compulsores exercitus* and any others ordered by the former to answer the call to arms. The inclusion of the *thiufadi* in this second tier of this short, but very general, list has long attracted attention, given that they can be seen to play a similar role in the *antiquae* laws in this title.[72] The fact that they were added to this group, rather than that of the *maiores*, has been linked to the progressive feudalization of the kingdom of Toledo, whereby they became agents of the king or the nobles.[73] Next, we find the *compulsores exercitus*, responsible for the *admonitio*, as well as the group of summoned individuals or subjects identified only by the expression *hi, qui conpelluntur* – i.e. those subject to conscription. This is closer and more in line with the situation presented in the *antiquae*, that is, *thiufadi*, bearers and receivers of the order. Or perhaps we might speak of recruiters and recruits, given that the picture we get from this list of *inferiores*

---

69 For an overview of the system of government and its organizational structure, see Pérez-Prendes Muñoz-Arraco, José Manuel Pérez-Prendes Muñoz-Arraco, *Historia del Derecho Español* I, 9th ed. (Madrid, 2004), pp. 550–561, and particularly the gardingus, pp. 540–550.

70 Manuel Díaz y Díaz, 'Títulaciones regias en la monarquía visigoda', in *Revista Portuguesa de História*, 16 (1976), pp. 133–141.

71 See bibliography in fn. 52.

72 See, for example, *antiquae*, LV 9.2.1; LV 9.2.3; LV 9.2.4 or LV 9.2.5.

73 Pérez Sánchez, *El ejército*, p. 163.

and *viliores* is more akin to the bases of the regular king's army rather than private retinues – which, as we shall see, are also included in this law.[74]

Let us now look at the punishment imposed on those who did not join the army (*in exercitum venire*), failed to march, or simply neglected to report at the established place and time, or who gave fraudulent pretexts to excuse themselves. This variation on the offence is of particular interest, consisting as it does of avoiding joining the expedition, but in this case by fleeing (*effugere*), with the terminological implication of desertion. The last case, that of giving fraudulent pretexts for avoiding combat, is also addressed in Wamba's law.

Anyone who disregarded the king's call under any of the above circumstances was to be sentenced to flogging (two hundred lashes), decalvation and forfeiture of a pound of gold[75] (a very significant amount) which the monarch could use at his own discretion. Should the *transgressor* be unable to raise the payment, he was to be reduced to perpetual servitude, and he and his property would be disposed of as the king saw fit.[76]

Again, a link can be drawn between these penalties and Erwig's amendment to Chindaswinth's LV 2.1.8 on treason, which – with the exception of death penalty and blinding – lists down the same punishments: exile, decalvation, scourging (100 lashes rather than 200) and perpetual servitude. There is also no mention of any pecuniary payment. In this law the punishments, which are the same as those for treason (apart from the fine), have clearly been ramped up.

5.1.3 Exemptions and Excuses

The first provision of the law concludes with a list of admissible exemptions whereby the established penalties might be circumvented and a reference to justifying excuses or pretexts. Thus, with the express authorisation of the monarch, minors (*minoris etatis*), elderly (*vetustas senectutis*) and infirms were deemed to be excused from the call and thus exempt from the serious consequences of not answering it. As in the law of Wamba, anyone claiming exemption by reason of illness had to prove their case before a competent witness

---

74 Guy Halsall, *Warfare and Society in the Barbarian West, 450–900* (London, 2003), p. 60, cites a text in Isidore, *Rule for Monks*, 4.2, suggesting that a roll of soldiers existed.

75 Felix Retamero, 'La moneda del Regnum gothorum (CA. 575–714). Una revisión del registro numismático', in *Entre el impuesto y la renta. Problemas de la fiscalidad tardoantigua y altomedieval*, eds. P.C. Díaz – I. Martín Viso (Bari, 2011), pp.189–212. For a detailed analysis of this type of sanction in Rome, see Gloria Maragno, *'Punire e sorvegliare'. Sanzioni in oro imperatori burocrazia* (Napoli, 2020).

76 I will not dwell here on the interesting variant found in manuscripts V 15; 16; 17 and 20. See Zeumer, *LV*, pp. 375–376; García López, *Estudios*.

(*per legitimum testem*). However, the dispensation applied only to the infirm person and not to the troops he was required to provide. The law stresses that he must promptly offer up his available forces and property (*omnem tamen virtutem rei sue*) to the army, who were to join the troops of his *dux* or *comes* in public service (*in publicis utilitatibus*).

This implies in a way the existence of a well-organised force in which each group would know in advance under which commanders it was to march and fight *in exercitu*. As we have seen, under Wamba's law, they were required to join the ecclesiastic retinue. In some manuscripts of the LV (V 15, 16.17 and 20), the task of inspecting and substantiating the veracity of the claims made by those claiming sickness to avoid serving appears to have lain with the bishops.[77] This is also the prevailing version in the text of the Fuero Juzgo, FJ 9.2.8.

Under the law of Chindaswinth LV 2.1.19, which Wamba partly takes as his template, a person could be exempted from answering judicial summons by reason of age or illness (*etatis aut egritudinis instantia*). The same was presumably true in the case of Erwig, although perhaps in this case he takes his model from Wamba's law, which included illness, but not age, as grounds for exemption or dispensation.

5.2 *Second Provision*

Having concluded the general exposition, Erwig goes on to legislate on specific cases related to the troops and resources of those enlisting, in a *decretum specialis*.

This new, specific section has two clear objectives: firstly, to establish the percentage of dependents with which each person must answer the call; and secondly, to clarify in what circumstances and with what weapons they were to present themselves.

Anyone joining the army, be they a *dux*, *comes* or *gardingus*, Goth or Roman, freeman or manumitted or *servi fiscales*, must bring with them to the *bellica expeditio* one tenth of their slaves,[78] their dependents, who must also be suitably armed. This mention to 'Goths and Romans' is quite unusual in the LV[79]

---

77 Zeumer, LV 9.2.9 (V 15. 16.17 and 20), pp. 375–376.

78 In consonance with the variant found in manuscripts V 15 and 16, instead of a tenth, they were required to contribute half of all serfs aged between twenty and fifty years of age. This is the version included in FJ 9.2.8.

79 According to Köbler, *Wörterverzeichnis*, the only other references to Goths and Romans are in: LV 3.1.1 *ant.*; LV 10.1.8 *ant.*; LV 10.1.9 *ant.*; LV 10.1.16 *ant.*; LV 10, 3, 5, *ant.* = CE 276. There is an exclusive mention to Romans in the laws of Recceswinth, LV 2.1.10, and Chindaswinth, LV 3.1.5, as well as two laws of Sisebut, LV 12.2.13 and LV 12.2.14, and other to Goths alone in LV 2.1.8 Chind.; LV 3.1.5 Chind./Erwig; LV 7.4.2 *ant.*; LV 9.2.2 *ant.*

and has been remarked on. Carlos Petit, for example, considers it to be purely rhetorical, a pleonasm intended to amplify the message that absolutely no one is free from the duty to answer the call, and argues that it has no specific social or institutional significance.[80] It is striking that while Wamba appeals constantly in his law to the *regna*, *patria* and *gentes* without ever using the genitive *gothorum*, Erwig alludes to a distinction between Goths and Romans that was already somewhat anachronistic by this time.[81] It is also noteworthy that the long list of those summoned under this concrete provision, comprises *servi fiscales*, but not *liberti fiscales*, who later, in the reign of Egica (687–702), were to be addressed in law LV 5.7.19.

However, this special provision does not end here: its second purpose, as we have said, is to specify what weaponry each subject was required to bring to the confrontation. A detailed list is given of the different kinds of arms that were to be carried and paraded either before the king or the *dux* or *comes* – literally *duci vel comiti suo* (again, this use of the possessive would appear to reflect the different retinues they were to join).[82] Different weapons are listed for different social groups: the first – the most experienced and highest ranking – were to bear cuirasses, i.e. *zaba* and *loricis*, which was costly equipment. The second, more numerous, group were to be armed with a shield (*scutum*), sword (*spatha*), a gladius or short sword (*scrama*),[83] as well as spears and arrows (*lanceae*, *sagittae*). Finally, the least prepared and larger group should bear slings (*fundae*) and any other (unspecified) weapons they may have received for the occasion from their masters and lords.[84]

5.2.1 Penalties

The law goes on to establish the punishment to be imposed on those who, despite answering the call to arms and reporting for duty, failed to bring the requisite ten percent of their serfs (which would be constituted both by slaves

80 Petit, *Iustitia Gothica*, p. 381.

81 For a recent contribution to this complex area, see Loschiavo, in this volume.

82 Isla Frez, *Ejército*, p. 86.

83 The word *scrama* is possibly of Frankish origin. See Haubrichs, in this volume, at pp. 238* and 245*. On the practices developed by the Roman army in Hispania, see María José Bravo Bosch, 'Exercitus hispanicus', in *Atti dell'Accademia Romanistica Costantiniana*, XXIV (2021), pp. 153–170.

84 Isla Frez, *Ejército*, 79 ff.; Ángela Parra Romo, 'El ejército visigodo, en campaña. Wamba y la sedición de la Narbonense', in *Studia Storica Historia Antigua*, 36 (2018), pp. 221–251pp. 240 ff.; Halsall, *Warfare*, p. 164; Pat Southern – Karen R. Dixon, *El ejército romano del Bajo Imperio*, trans. Y. Balbás (Madrid, 2018), pp. 133–199; Yann Le Bohec, *El ejército romano*, trans. I. Hierro (Barcelona, 2004), pp. 164 ff.; Haubrichs, in this volume.

and freemen). An estimate was be made of the full number and whatever portion of the required tenth part that was lacking, i.e. the defrauded amount, was to be surrendered to the king to dispose of at his pleasure. Once again, this measure seems to imply the existence of an administrative and fiscal framework for controlling property: it has been apparently possible, without great difficulty, to determine whether there had been strict compliance and identify those who reported without the required ten percent (*extra hanc decimam partem*). As Isla Frez notes,[85] this ability to keep tabs on the troops of each subject suggests that appropriate records were being kept, and we must therefore presume that updated rolls were taken of the serfs and subjects of each *dominus*.[86] Accounts of similar records for Merovingian areas are reported, for example, by Gregory of Tours.[87]

For fiscal purposes, presumably, the punishment consists of forfeiture of the missing serfs, who were handed over to the monarch to be disposed of as his convenience.[88] In this case, the punishment does not entail a pecuniary or afflictive penalty, but the appropriation of dependents, who were to be reallocated to more faithful subjects. It is an indication of the importance of these retinues that they were later to become the subject of a specific law under Egica, LV 5.7.19.

5.3 *Third Provision*

There follows a short section, dealing exclusive with the special case of palatine officers[89] engaged in an armed expedition, who failed to perform diligently

85 Isla Frez, *Ejército*, p. 86.

86 This whole area of property censuses and registration came under the remit of the *comes thesaurorum*, the successor of the *comes sacrarum largitionum*. See Pablo C. Díaz, 'El reino visigodo en vísperas del 711: sistema político y financiero', in *711. Arqueología e historia entre dos mundos*, I (Madrid, 2011), p. 38; Roland Delmaire, 'La caisse des largesses sacrées et l'armée au bas empire', in *Armées et fiscalité dans le monde Antique* (Paris, 1977), pp. 311–329.

87 There is a reference to the existence of tax-lists for the Kingdom of the Franks in Gregory of Tours, *Historia francorum*, 9.30. See Halsall, *Warfare*, p. 250 n.112.

88 Díaz, 'El esquema provincial', pp. 77–108; Pérez-Prendes Muñoz-Arraco, *Historia del Derecho*, pp. 542 ff.; Id., 'Resonancias jurídicas en las pizarras visigodas', in *En la pizarra. Los últimos hispanorromanos de la Meseta*, eds. I. Velázquez Soriano, M. Santonja Gómez (Burgos, 2005), pp. 127–141; Abilio Barbero – Marcelo Vigil, 'Algunos aspectos de la feudalización del reino visigoda en relación con su organización financiera y militar', in *Sobre los orígenes sociales de la Reconquista*, 2th ed. (Barcelona, 1988), pp. 105–137; María Rosa Valverde Castro, 'La ideología fiscal en el reino visigodo de Toledo', in *Entre el impuesto y la renta. Problemas de la fiscalidad tardoantigua y altomedieval*, eds. P.C. Díaz – I. Martín Viso (Bari, 2011), pp. 163–187.

89 On the composition of the *wardia*, see Díaz, 'El reino visigodo', pp. 30 ff.; Pérez-Prendes Muñoz-Arraco, *Historia del derecho*, pp. 550 ff.; Remedios Morán Martín, 'La monarquía

the tasks assigned to them in the service and custody of the king (*in principali servitio*) or the royal guard (*wardia*), with their *fratres*. This reference to their 'brethren' appears to have been borrowed from Wamba, who used it in his military law to allude to comrades-in-arms. Equally interesting is the term *wardia*, of Germanic origin,[90] which designates the special guard entrusted with protecting the king, and drawn from the Visigothic elites.[91] Palatine officers found guilty of such negligence were subject to the same penalties as those who failed to answer the summons, delayed joining, etc.

5.4 *Fourth Provision: Concerning the* Exercitales

This section gives us a clear idea of the real situation of the time, focusing on the *exercitales*, a group of soldiers made up of free men, named after its Roman predecessor.[92] Among the Lombards these individuals were also known as *arimanni*.[93] The *exercitales* were included in this provision because on occasions,[94]

---

visigoda y Administración. Rupturas y permanencias respecto al Imperio romano', in *Historia de la Administración en España: Mutaciones, sentidos y Rupturas*, eds. R. Morán Martín – J. García Martín, I (Madrid, 2018), pp. 137–140; Luis A. García Moreno, 'El estado protofeudal visigodo: precedente y modelo para la Europa carolingia', in *L'Europe héritier de l'Espagne wisigothique*, eds. J. Fontaine – C. Pellistrandi (Madrid, 1992) pp. 40–43.

90 Cf. Haubrichs and Falluomini, in this volume, respectively at pp. 228–229* and 8*20–22*. The term *warda* is attested – although later – in the *Lex Chamavorum* and in the *Capitulare Bonnoniense* of 811. Niermeyer – Kieft, '*warda*', pp. 1469–1470. Very few words of Germanic origin are preserved in the Visigothic laws, a feature that has been attributed to the integrated nature of Visigothic society. See Dietrich Claude, 'Remarks about relations between visigoths and hispano-romans in the seventh century', in *Strategies of distinction. The Construction of Ethnic Communities, 300–800*, eds. W. Pohl – H. Reimitz (Leiden-Boston-Köln, 1998), pp. 117–130; Gisella Ripoll, 'The arrival of the Visigoths in Hispania: Population problems and the process of acculturation', *ibidem*, pp. 153–179; Wolf Liebeschuetz, 'Citizen status and Law in the Roman Empire and the Visigothic Kingdom', *ibidem*, pp. 131–152.

91 Isla Frez, *Ejército*, pp. 62 *ff.*

92 According to Stefan Esders, 'Spätrömisches Militärrecht in der Lex Baiuvariorum', in *Civitas, Arma, Iura: Arma: Organizzazioni militari, istituzioni giuridiche e strutture sociali alle origini dell'Europa* (*secc. III–VIII*), eds. F. Botta – L. Loschiavo (Lecce, 2015), esp. pp. 54–55, from the end of the fifth century, groups of Romans – the *Romani exercitales* or *tributarii* – continued to exercise military functions, even though the Roman administration as such had pulled out of the area. Although this practise is only documented in the area around present-day Salzburg, it is likely to have existed in other areas as well.

93 Claudio Azzara – Stefano Gasparri, *Le leggi dei Longobardi. Storia, memoria e diritto di un popolo germánico* (Roma, 2005); Stefano Gasparri, 'I barbari, l'impero, l'esercito e il caso dei Longobardi', in *Civitas, Arma, Iura*, p. 100. They also appear as modest landowners who participate in the Lombard army as *exercitales*; Esders, 'Spätrömisches Militärrecht', p. 54; Michael Zerjadtke, *Das Amt "Dux" in Spätantike und frühen Mittelalter: der "ducatus" im Spannungsfeld zwischen römischem Einfluss und eigener Entwicklung* (Berlin/Boston, 2019) pp. 194–197, 204, 208.

94 Sánchez Albornoz, 'El ejército visigodo', 53 *ff.*

although they did take part in the battle, they failed to join the forces of their allocated *dux* or *comes* or of their landlord, and either delayed joining up or entered the retinues of other patrons. This was a breach of their obligation of *wardia* towards their lord and a failure to fulfil their duty of public service. This reflects the stamp of the networks of dependence and the importance of the retinues, and the image of strength and finery it sought to project, as clearly evidenced in Egica's LV 5.7.19.

Any *exercitales*, who, having agreed to enlist, delayed by protecting parties other than their *dux*, *comes* or lord, were to receive the same punishment as that stipulated for *inferior sane vilioresque persone*, namely two hundred lashes, decalvation and a fine of one pound of gold, to be disposed of at the king's discretion. Should they be unable to pay this hefty fine, they were to be condemned to perpetual servitude, and they and their property were to be forfeited to the king. The reason for such a severe punishment here is clearly that this constituted breach of faith to their patron, and although the offender has indeed joined the *expeditio* (and is not stated to have been in breach of his duty), he is punished as if he had failed to do so, evidencing the importance placed on the bonds of dependence.

5.5 *Fifth Provision: Corrupt Practices*

The last link in the legislative chain is preceded by a short preliminary section, indicating that the law is nearing its end. It is the closest thing we find to an epilogue or colophon in this law, and it also serves as a small opening or preamble for the final issue to be addressed. Having settled and determined the basic issues, Erwig says, it only remains to curtail the avarice of those acting in matters related to the king's service (*utilitatis nostre*). Whereas earlier sections of the law made mention of the private *utilitas* or *utilitas propria* of those called to arms, the final part sanctions certain corrupt practices which, in Erwig's words, threaten the *utilitas nostra*.[95]

This provision addresses the issue of bribery and other corrupt practices, already covered in the *antiquae* of this title, thus to some extent completing the circle. The text now stresses that *duces*, *comites*, *thiufadi* and those entrusted with issuing the call to arms were under no circumstances to excuse any of the king's subjects from their military duty, in exchange for a bribe or out of ill will. We are then given some indications —albeit complex to interpret— on the

95 For the concept of *publica utilitas* from the Later Roman Empire. See Maria Luisa Navarra, '*Utilitas publica – utilitas singulorum* tra IV e V sec. d. C. Alcune osservazioni', in *Studia et Documenta Historiae et Iuris* (SDHI), 63 (1997), pp. 269–291. Jean Gaudemet, 'Utilitas publica', in Études de droit romain (Camerino, 1979), pp. 163–197.

usual nature of such fraudulent practices, which appear to have taken place either in the *admonitiones* (the summons to the army, *de progressione exercitus*) or the *inductiones armorum*, which may have been the requisite procedure for parading arms. However, manuscript E1 in Zeumer's edition refers to the *indictio armorum*, which in Dahn's interpretation involves a fiscal charge.[96]

Naturally, the *dux*, the *comes*, the *thiufadus* and all other authorised officers, especially the *compulsores exercitus*, were best placed to indulge in bribery and corruption. As mentioned, this situation had already been addressed in the *antiquae* in this title: LV 9.2.1: *Si thiufadus ab aliquo de thiufa sua fuerit beneficio corruptus*; LV 9.2.3: which also targeted military leaders *ab aliquo per beneficio persuasus aut rogitus*; and LV 9.2.5: *Si conpulsores exercitus beneficio accepto aliquem* (...). We know that historically widespread corruption of this kind was identified and punished in Roman military law and, as we can see, continued to plague the Visigothic army until the end, in spite the efforts to suppress such practice from the kingdom's earliest laws.

The punishment for such corrupt behaviour varied depending on whether the offenders were *primates palatii* or *minores*. *Primates palatii* were required to return fourfold the amount received to the person from whom they had received it and forfeit a pound of gold to the monarch as penalty for their unlawful enrichment.[97] *Minores loci personae*, on the other hand, who had neither the office nor the dignity of free men, were to be delivered up to the king, who would rule on their fate and that of their possessions at his own discretion.

The issue of military commanders fraudulently charging for unjustified leave was also common in Rome. Evidence of this practice during the late period can be found in Justinian's constitution of 534, CJ 1.27.2, on the civil and military organization of Africa, which deals directly with such offences.[98] This practice was therefore a continuance from an earlier age, albeit adjusted to changing times.[99]

---

96 Dahn, *Die Könige*, 209. See also Isla Frez, *Ejército*, pp. 57 *ff.*

97 In her recent study, 'Punire e sorvegliare', Gloria Maragno examines the role of fines in gold in the period between Constantine and Justinian, with emphasis on cases of corruption related to the *militia*.

98 The extensive constitution is compiled in CJ 1.27.2 and addressed to Belisarius. See Arnold Hugh Martin Jones, *Le declin du monde Antique 284–619*, I (Paris, 1970), pp. 214–216; Maragno, 'Punire e sorvegliare', pp. 282 *ff.*

99 For this common thread running from the post-classical Roman period, fundamentally by way of imperial constitutions demonstrating the nobles' power over their dependents, see Dionisio Pérez Sánchez, 'Defensa y territorio en la Sociedad península Hispana durante la Antigüedad Tardía (SS. V–VII)', in *Studia Historica, Hª Antigua*, 16 (1998), pp. 281–300.

6 Summary: Connections between the Antiquae, LV 9.2.8 Wamba and LV 9.2.9 Erwig

a) Firstly, the information on army ranks provided in the *antiquae* analysed here (LV 9.2.1; 3; 4; 5 and 7) includes concrete references to the *thiufadus*, who led the *thiufa*, under whose command were the *quingentenarius*, *centenarius*, and *decanus*. Of these decimal units, only the *thiufadus* will be present in later laws. The *servi dominici*, or *compulsores exercitus*, are also mentioned. This detailed image of a defined, graded military hierarchy, with penalties varying according to rank, is striking. It probably took its inspiration from the Roman army, but we must assume that listed groups were most likely more theoretical than real. The presence of the monarch (probably Leovigild) is also noteworthy, always closely related to the *comes civitatis*. The *duces* do not figure in these laws.

The military ranks as defined in the *antiquae* are not present yet in the laws of Wamba and Erwig. In their place, as we see in Wamba's LV 9.2.8, ecclesiastics appear alongside *duces*, *comites*, *thiufadi*, *vicarii*, and *gardingi*. With the exception of the ecclesiastics, all of these ranks are also found in Erwig's LV 9.2.9.

b) In both the *antiquae* and the laws of Wamba and Erwig, the notion of *domus* plays a prominent and constant role. It is a refuge, the place that deserters and those obtaining fraudulent permits did not want to leave or to which they sought to return. The long preambles to Wamba and Erwig's laws give detailed information on their reasons for wishing to stay behind: they would rather prefer to work in their fields, in defence of their interests, than face the enemy on the battlefield.

c) As pointed out above, in the *antiquae*, the penalties raised from pecuniary sanctions were distributed on the monarch's orders, generally amongst the hundred (LV 9.2.1; LV 9.2.3, and LV 9.2.4), although *thiufa* also appears in LV 9.2.5. This system of sanctions was also a way of helping to provision and support military units always in need of financing.

Under Wamba and Erwig, the fate of both the offenders, their property and their slaves, as well as any fines they had to pay, lay in the hands of the monarch, who could dispose of them at will. The penalties are similar to those for treason.

d) A comparison of the laws of Wamba LV 9.2.8 and Erwig LV 9.2.9 reveals some notable differences. I believe Wamba's LV 9.2.8 owes much to the laws of Chindaswinth, specifically LV 2.1.8 and LV 2.1.19. Wamba's military law is a rather peculiar one for the constant stress on common integration: it resembles a rousing speech calling for unity and concord between *regnum*, *patria* and *gens*, and opposing treason. Its constant appeal to fraternity among combatants, the relative lack of military terminology and the fact that it extends to both external and internal attacks, are distinctive features of the law. For all of these reasons, indeed, from the wording of the law, one might infer that Wamba was merely extending in more explicit terms, the law on political treason promulgated three decades before by Chindaswinth. Although it is included in the military title, at no point does it employ the terms *exercitus* or *expeditio.*

Erwig's law, on the other hand, is more in line with the *antiquae,* despite some notable differences. It is a purely military law, intended to make an effective reorganisation of the administration. Terms alluding to the army and the call to arms occur in almost every paragraph. However, Erwig's law does not mention the triad *regnum, gens, patria.*

The overriding main purpose of all of these laws was to get to grips with a long-standing and deeply entrenched military problem, namely the repeated failures to answer the call to arms and the recourse to bribery and corruption for this purpose.

Bibliography

Primary Sources

Codex Euricianus , ed. D'ORS, Álvaro, *El Código de Eurico,* in *Cuaderno del Instituto Jurídico Español,* 12 (Roma-Madrid 1960 = Madrid 2014)

Codex Justinianus, v. Iustinianus Codex

Codex Theodosianus: *Theodosiani libri XVI cum constitutionibus Sirmondianis et leges Novellae ad Theodosianum pertinentes,* ed. Th. Mommsen – P.M. Meyer, 2 vols. (Berlin, 1905)

Concilios Visigóticos e Hispano-romanos, ed. J. Vives (Barcelona-Madrid, 1963)

Fuero Juzgo, in *Códices Antiguos de España,* ed. M. Martínez Alcubilla (Madrid 1885)

Gregorius Magnus, Moralia sive Expositio in Iob (*libri I–X*), in *Corpus Christianorum. Series latina,* ed. M. Adriaen, 143 (Turnhout,1979)

Gregorius Turonensis, Libri Historiarum X, ed. B. Krusch – W. Levison [MGH, *Script. Rer. Meroving.* I.1] (Hannover, 1951)

Isidori Hispalensis Episcopi, Etymologiarvm sive originvm libri XX, ed. M. Lindsay, Scriptorum Classicorum Bibliotheca Oxoniensis (Oxford, 1957 [=1911])

Iustinianus, *Codex,* ed. P. Krüger in T. Mommsen – P. Krüger, *Corpus Iuris Civilis,* II (Berlin, 1954[11])

Leges Visigothorum, ed. K. Zeumer [MGH, *Leges nat. Germ.* I] (Hannoverae et Lipsiae, 1902)

San Leandro, San Fructuoso, San Isidoro: Santos Padres Españoles II. *Reglas monásticas de la España visigoda. Los tres libros de las "Sentencias",* eds. J. Campos Ruíz – I. Roca Meliá, Biblioteca de Autores Cristianos, 321 (Madrid, 1971)

Sancti Aurelii Augustine, Enarrationes in Psalmos CI-CL (eds. E. Dekkers, J. Fraipont), in *Corpus Christianorum. Series Latina*, 40 (Turnhout, 1956)

Sancti Iuliani Toletanae sedis episcopi Opera, Pars I. *Prognosticon futuri saeculi libri tres. Apologeticum de tribus capitulis. De comprobatione sextae aetatis. Historia Wambae regis. Epistula ad Modoenum*, eds. N. Hillgarth, B. Bischoff, W. Levison, *Corpus Christianorum. Series Latina*, 115 (Turnhout, 1976)

Secondary Literature

Álvarez Cora, Enrique, '*Qualis erit lex*: La naturaleza jurídica de la ley visigótica', in *Anuario de Historia del Derecho Español AHDE*, 66 (1996), pp. 11–117

Arce, Javier, *Esperando a los árabes: Los Visigodos en Hispania* (507–711) (Madrid, 2011)

Azzara, Claudio – Gasparri, Stefano, *Le leggi dei Longobardi. Storia, memoria e diritto di un popolo germánico* (Roma, 2005)

Barbero, Abilio – Vigil, Marcelo, *La formación del feudalismo en la Península Ibérica*, 4th ed. (Barcelona, 1986)

Barbero, Abilio – Vigil, Marcelo, 'Algunos aspectos de la feudalización del reino visigoda en relación con su organización financiera y militar', in *Sobre los orígenes sociales de la Reconquista*, 2th ed. (Barcelona, 1988), pp. 105–137

Botta, Fabio – Loschiavo, Luca (eds.). *Civitas, Arma, Iura. Organizzazioni militari, istituzioni giuridiche e strutture sociali alle origini dell'Europa* (*secc. III–VIII*), (Lecce, 2015)

Botta, Fabio. "Group' criminal liability and solidarity. Military formations and armed bands betweem Roman law and the early Middle Ages', in this volume*

Bravo Bosch, María José, 'Exercitus hispanicus', in *Atti dell'Accademia Romanistica Costantiniana*, XXIV (2021), pp. 153–170

Bronisch, Alexander Pierre, 'El concepto de España. La historiografía visigoda y asturiana', in *Norba. Revista de Historia*, 19 (2006), pp. 9–42

Bronisch, Alexander Pierre, 'Precisiones sobre algunas informaciones históricas en la 'Crónica de Alfonso III'', in *Edad Media. Rev. Hist.*, 12 (2011), pp. 35–66

Cañizar Palacios, José Luis, 'Populus y provinciales: la población bajoimperial en la inscriptio de las constituciones del Codex Theodosianus', in *Veleia* 27 (2010), pp. 285–303

Castillo Lozano, José Ángel, 'Los tiranos de la Historia Wambae Regis: Ilderico y Paulo, el rex perditionis', in *Categorías de poder en el reino visigodo de Toledo: los tiranos en las obras de Juan de Biclaro, Isidoro de Sevilla y Julián de Toledo, Antigüedad y Cristianismo*, 2016–2017 (Murcia, 2019), pp. 173–182

Claude, Dietrich, *Adel. Kirche and Königtum im Westgotenreich* (Sigmaringen, 1971)

Claude, Dietrich, 'Remarks about relations between visigoths and hispano-romans in the seventh century', in *Strategies of distinction*, Pohl – Reimitz, pp. 117–130

Collins, Roger, 'Julian de Toledo and the Education of Kings in Late Seventh-Century Spain', in *Law, Culture and Regionalism in Early Medieval Spain* (Hampshire, 1992), pp. 1–22

Collins, Roger, *Visigothic Spain 409–711*, 2nd ed. (Oxford, 2004)

Córcoles Olaitz, Edorta, 'La crisis militar visigoda de finales del siglo VII a través de la legislación', in *El ejército, la paz y la guerra: Jornadas de la Sociedad de Historia del Derecho*, eds. José Luis De Los Mozos, I. Szászdi León-Borja (Valladolid, 2009), pp.165–178

D'Ors, Álvaro, *El Código de Eurico* (Roma-Madrid, 1960)

Dahn, Felix, *Die Könige der Germanen*, v. 6, 2nd ed. (Leipzig, 1885)

Delmaire, Roland, 'La caisse des largesses sacrées et l'armée au bas empire', in *Armées et fiscalité dans le monde Antique* (Paris, 1977), pp. 311–329

Díaz y Díaz, Manuel, 'Títulaciones regias en la monarquía visigoda', in *Revista Portuguesa de História*, 16 (1976), pp. 133–141

Díaz, Pablo C., 'Confiscation in the Visigothic Reign of Toledo', in *Expropiations et Confiscations dans les royaumes barbares. Une approche regionale*, eds. P. Porena – Y. Rivière (Roma, 2012), pp. 93–112

Díaz, Pablo C., 'El reino visigodo en vísperas del 711: sistema político y financiero', in *711. Arqueología e historia entre dos mundos*, I (Madrid, 2011), pp. 31–40

Díaz, Pablo C, 'El esquema provincial en el contexto administrativo de la monarquía visigoda de Toledo', in *Mélanges de la Casa de Velázquez*, 49–2 (2019), pp. 77–108

Esders, Stefan, 'Spätrömisches Militärrecht in der Lex Baiuvariorum', in Botta – Loschiavo, *Civitas, Arma, Iura*, pp 43–78

Esders, Stefan, 'Grenzüberschreitende Mobilität als Problem politischer Loyalität. Das Hochverratsgesetz des westgotischen Königs Chindasvinth (642/643) und das 7. Konzil von Toledo (646)', in *Wasser-Wege-Wissen auf der iberische Halbinsel. Eine interdisziplinäre Annähderung im Verlauf der Geschichte*, eds. I. Czeguhn – C. Möller – Y. Quesada Morillas – J.A. Pérez Juan (Baden-Baden, 2018), pp. 89–120

Esders, Stefan, '*Regem iura faciunt*. Der westgotische Treueid im Kräftefeld personaler und Trasnpersonaler Konzepte der Legitimität politischer Herrschaft', in *Die Macht des Herrschers. Personale und Transpersonale Aspekte*, eds. A. Mechthild – E. Brüggen – K. Klaus (Göttingen, 2019), pp. 69–154

Frighetto, Renan, 'Incauto et inevitabili conditionum sacramento: juramento de fidelidad y limitación del poder regio en la Hispania visigoda en el reinado de Egica (688)', in *Intus-Legere Historia*, 1, ½, (2007), pp. 67–79

García Herrero, Gregorio, 'El reino visigodo en la concepción de Julián de Toledo', in *Antigüedad y Cristianismo*, 12 (Murcia, 1995), pp. 385–420

García López, Yolanda, *Estudios críticos y literarios de la de la "Lex Wisigothorum"* (Alcalá de Henares, 1997)

García Moreno, Luis A., 'El estado protofeudal visigodo: precedente y modelo para la Europa carolingia', in *L'Europe héritier de l'Espagne wisigothique*, eds. J. Fontaine – C. Pellistrandi (Madrid, 1992) pp., 17–43

García Moreno, Luis A., 'Hispania Visigoda (Siglos V a VII)', in *Historia Militar de España*, dir. H. O'Odonnell, v. 2, coord. M. Á. Ladero Quesada (Madrid, 2010), pp. 41–78

Gasparri, Stefano, 'I barbari, l'impero, l'esercito e il caso dei Longobardi', in Botta – Loschiavo, *Civitas, Arma, Iura*, pp. 91–102

Gaudemet, Jean, 'Utilitas publica', in Études de droit romain (Camerino, 1979), pp. 163–197

Giuffrè, Vincenzo, *Letture e richerche sulla "res militaris"*, 2 (Napoli, 1996)

Halsall, Guy. *Warfare and Society in the Barbarian West, 450–900* (London, 2003)

Haubrichs, Wolfgang, "Warriors' Names and Military Language of the Westgermanic Peoples (Franks and Langobards)', in this volume*

Hillgarth, Jocelyn Nigel, 'Las fuentes de San Julián de Toledo', in *Anales Toledanos*, 3 (1971), pp. 97–118

Hillgarth, Jocelyn Nigel, 'Introduction', in *Sancti Iuliani Toletanae sedis episcopi opera*, Pars I, CCSL (Turnhout, 1976), pp. VIII–XXI

Iglesia Ferreirós, Aquilino, *Historia de la traición. La traición regia en León y Castilla* (Santiago de Compostela, 1971)

Isla Frez, Amancio, *Ejército, sociedad y política en la Península Ibérica entre los siglos VII y XI* (Madrid, 2010)

Jones, Arnold Hugh Martin, *Le declin du monde Antique 284–619*, I (Paris, 1970)

Kelly, Michael J., *Isidore of Seville and the Liber Iudiciorum. The Struggle for the Past in the Visigothic Kingdom* (Boston-Leiden, 2021)

King, Paul D., *Derecho y sociedad en el reino visigodo* (Madrid, 1981)

Köbler, Gerhard, Wörterverzeichnis zu den Leges Visigothorum (Giessen-Lahn, 1981)

Le Bohec, Yann, *El ejército romano*, trans. I. Hierro (Barcelona, 2004)

Le Roux, Patrick. 'L'amour patrie dans les cités sous l'Empire romain', in *Le toge et les armes. Rome entre Méditerranées et Oceán*, c. 31 (Rennes, 2011), pp. 565–581

Lear, Floyd Seyward, *Treason in Roman and Germanic Law* (Austin, 1965)

Liebeschuetz, Wolf, 'Citizen status and Law in the Roman Empire and the Visigothic Kingdom', in *Strategies of distinction*, Pohl – Reimitz, pp. 131–152

Loschiavo, Luca, 'Personality of Law or ius speciale militum? Around the origins of the *leges barbarorum*', in this volume

Maragno, Gloria, *'Punire e sorvegliare'. Sanzioni in oro imperatori burocrazia* (Napoli, 2020)

Martin, Celine, *La géographie du pouvoir dans l'Espagne visigothique* (Lille, 2003)

Martin, Céline, 'La notion de gens dans la péninsule Ibérique des VIe–VIIe siècles: quelques interprétations', in *Identité et Ethnicité, Concepts, débats historiographiques, exemples* (*III*[e]–*XII*[e] *siécle*), eds. V. Gazeau – P. Bauduin – Y. Modéran (Caen, 2008), pp. 75–89

Martin, Céline, 'Le Liber Iudiciorum et ses différentes versions', in *Mélanges de la casa de Velázquez*, 41 (2) (2011), pp. 17–34

Martin, Céline, 'Ervig and Capital Penalties: The Way of Exile', in *Framing Power in Visigothic Society. Discourses, Devices and Artifacts*, eds. E. Dell'Elicine – C. Martin (Amsterdam, 2020), pp. 133–157

Martínez Pizarro, Joaquín, *The Story of Wamba. Julian of Toledo's Historia Wambae Regis* (Washington, 2005)

Morán Martín, Remedios, 'La monarquía visigoda y Administración. Rupturas y permanencias respecto al Imperio romano', in *Historia de la Administración en España: Mutaciones, sentidos y Rupturas*, eds. R. Morán Martín – J. García Martín, I (Madrid, 2018), pp. 137–140

Murphy, Francis X, 'Julian of Toledo and the fall of the Visigothic Kingdom', in *Speculum* 27, 1 (1952), pp. 1–27

Navarra, Maria Luisa, '*Utilitas publica – utilitas singulorum tra* IV e V sec. d. C. Alcune osservazioni', in *Studia et Documenta Historiae et Iuris* (SDHI), 63 (1997), pp. 269–291

Niermeyer, Jan Frederik –Kieft, Co van de, *Mediae Latinitatis Lexicon Minus,* 2nd ed. Jan W. J., Burgers (Leiden, 2002)

Orlandis, José, *Historia del reino visigodo español,* 2nd ed. (Madrid, 2006)

Orlowski, Sabrina Soledad, 'Principales tendencias historiográficas sobre la organización militar visigoda del reino de Toledo', in *Medievalista* [on line], 19 (2016), pp. 1–25

Osaba, Esperanza, 'Influenza delle leggi costantiniane nella Lex Visigothorum', *Diritto @ Storia. Quaderni di scienze giuridiche e tradizione romana* (revista electrónica), 2 (2003)

Osaba, Esperanza, 'Responsabilité et droit d'asile dans l'Hispania visigothique', in *Responsabilité et Antiquité* I, *Méditerranées*, 34/35 (2003) pp. 77–105

Osaba, Esperanza, '*Lex Visigothorum* 9,2: *De his qui ad bellum non vadunt aut de bello refugiunt*', in Botta – Loschiavo, *Civitas, Iura, Arma*, pp. 161–194

Osaba, Esperanza, '*Quid debeat observari.* En torno a LV 9,2,8 Wamba', in *L'esercito romano e l'alba dell'Europa. Modelli concettuali e sperimentazioni sociali tra linguaggi, istituzioni e diritto* (*secoli* IV–VIII), Gisella Bassanelli Sommariva ed, Ravenna Capitale (Santarcangelo di Romagna, 2020), pp. 149–184

Osaba, Esperanza, 'El Ejército de Ervigio a través de LV 9,2,9', in *Seminarios Complutenses de Derecho Romano*, XXXIV (2021), pp. 321–361

Parra Romo, A., 'El ejército visigodo en campaña. Wamba y la sedición de la Narbonense', in *Studia Storica Historia Antigua*, 36 (2018), pp. 221–251

Pérez Sánchez, Dionisio, *El ejército en la sociedad visigoda* (Salamanca, 1989)

Pérez Sánchez, Dionisio, 'Defensa y territorio en la Sociedad península Hispana durante la Antigüedad Tardía (ss. v–vii)', in *Studia Historica, Hª Antigua*, 16 (1998), pp. 281–300

Pérez-Prendes Muñoz-Arraco, José Manuel, 'Historia de la legislación visigótica', in *San Isidoro Doctor Hispaniae* (Sevilla, 2002), pp. 51–67

Pérez-Prendes Muñoz-Arraco, José Manuel, *Historia del Derecho Español* i, 9th ed. (Madrid, 2004)

Pérez-Prendes Muñoz-Arraco, José Manuel, 'Resonancias jurídicas en las pizarras visigodas', in *En la pizarra. Los últimos hispanorromanos de la Meseta*, eds. I. Velázquez Soriano - M. Santonja Gómez (Burgos, 2005), pp. 127–141

Petit, Carlos, *Iustitia Gothica. Historia Social y Teología del Proceso en la Lex Visigothorum* (Huelva, 2001)

Petit, Carlos, '*Rex iudex*. El momento judicial del rey de Toledo, Procesos, inquisiciones, pruebas', in *Homenaje a Mario Sbriccoli*, eds. E. Conte – M. Madero (Buenos Aires, 2009), pp. 39–71

Petit, Carlos, 'Derecho visigodo del siglo vii. Un ensayo de síntesis e interpretación', in *Derecho, Cultura y Sociedad en la Antigüedad Tardía*, E. Osaba (Bilbao, 2013), pp. 205–218

Pohl, Walter – Dörler, Philipp, 'Isidore and the gens gothorum', in *Antiquité tardive*, 23 (2015), pp. 133–141

Walter, Pohl – Reimitz, Helmut eds., *Strategies of distinction. The Construction of Ethnic Communities, 300–800* (Leiden-Boston-Köln, 1998)

Priego de Lis, Augusto, 'La pena de exilio en la legislación Hispanogoda', in *Espacio y tiempo en la percepción de la Antigüedad Tardía*, eds. E. Conde Guerri – R. González Fernández – A. Egea Vivancos, *Antigüedad y Cristianismo. Monografías históricas sobre la Antigüedad Tardía*, 23 (Murcia, 2006), pp 515–529

Puliatti, Salvatore, *Ricerche sulla legislazione "regionale" di Augusto. Lo statuto civile e l'ordinamento militare della prefettura africana* (Milano 1980)

Retamero, Felix, 'La moneda del Regnum gothorum (ca. 575–714). Una revisión del registro numismático', in *Entre el impuesto y la renta. Problemas de la fiscalidad tardoantigua y altomedieval*, eds. P.C. Díaz – I. Martín Viso (Bari, 2011), pp.189–212

Ripoll, Gisella, 'The arrival of the Visigoths in Hispania: Population problems and the process of acculturation', in *Strategies of distinction*, Pohl – Reimitz, pp. 153–179

Sánchez Albornoz, Claudio, 'El ejército visigodo: su protofeudalización', in *Investigaciones y documentos sobre las instituciones hispanas* (Santiago, 1970), pp. 5–56

Southern, Pat – Dixon, Karen R., *El ejército romano del Bajo Imperio*, trans. Y. Balbás (Madrid, 2018)

Teillet, Suzanne, 'La déposition de Wamba, un coup d'état au viie siècle', in *De Tertullien aux Mozarabes*, ii: *Antiquité tardive et christianisme ancien (vie–ixe siècles), Mélanges offerts à Jacques Fontaine à l'occasion de son 70e anniversaire, par ses élèves, amis et collègues*, eds. L. Holtz – J.-C. Fredouille – M.-H. Jullien (Paris, 1992), pp. 99–113

Teillet, Suzanne, *Des goths à la nation gothique. Les origins de l'idée de nation en Occident du* V^e^ *au* VIIe *siècle*, 2nd ed. (Paris, 2011)

Thompson, Edward Arthur, *Los godos en España*, 5th ed. (Madrid, 2007)

Torres López, Manuel, 'El estado visigodo', in *Anuario de Historia del Derecho Español* AHDE, 3 (1926), pp. 307–475

Torres López, Manuel, *Lecciones de Historia del Derecho Español*, 2nd Ed. (Barcelona, 1934)

Torres López, Manuel, *Historia de España Menéndez Pidal*, v. 3 (Madrid, 1994), pp. 123–131

Ureña Y Smenjaud, Rafael, *La legislación gótico-hispana* (*Leges Antiquiores - Liber Iudiciorum*). *Estudio crítico*, ed. Carlos Petit (Pamplona, 2003)

Vallejo Girvés, Margarita, 'Los exilios de católicos y arrianos bajo Leovigildo y Recaredo', in *Mundo medieval. Hispania sacra*, 55 (2003), pp. 35–47

Valverde Castro, María Rosa, 'La ideología fiscal en el reino visigodo de Toledo', in *Entre el impuesto y la renta. Problemas de la fiscalidad tardoantigua y altomedieval*, eds. P.C. Díaz – I. Martín Viso (Bari, 2011), pp. 163–187

Velázquez Soriano, Isabel, 'Wamba y Paulo: dos personalidades enfrentadas y una rebelión', *Espacio, Tiempo y Forma*, Serie 2, *Hispania Antigua*, 2 (1989), pp. 213–222

Wohlhaupter, Eugen, *Gesetze der Westgoten* (Weimar, 1936)

Wormland, Patrick, 'Lex scripta and Verbum Regis. Legislation and Germanic Kingship from Euric to Cnut', in *Early Medieval Kinhship*, eds. H.-W. Goetz – J. Jarnut – W. Pohl (Leeds, 1977), pp. 105–138

Zerjadtke, Michael, *Das Amt "Dux" in Spätantike und frühen Mittelalter: der "ducatus" im Spannungsfeld zwischen römischem Einfluss und eigener Entwicklung* (Berlin/ Boston, 2019)

Zeumer, Karl, *Historia de la legislación visigoda*, trans. Carlos Clavería (Barcelona, 1944)

CHAPTER 12

'Traditionskern', 'Gefolgschaft': More Questions Than Answers

Francesco Borri

1 Two Words to Begin With[1]

I was kindly invited to Rome in the late spring of 2019 to talk about *Traditionskern*, a concept discussed by the German historian Reinhard Wenskus in different sections of his *Stammesbildung und Verfassung. Das Werden der frühmittelalterlichen Gentes* (1961).[2] *Traditionskern* describes a group that arose within the Germanic peoples of the Migration Age. It was an aristocratic inner circle close to the royal lines which preserved the tribal self-consciousness of the *gens*. I thought at the time that *Traditionskern* should have been discussed together with the warrior retinues of late Antiquity and the early Middle Ages called *Gefolgschaft* in German historiography, another of Wenskus' central ideas. I also knew that the *Traditionskern* was a largely evanescent social group and I presumed that the *Gefolgschaft* shared many of the aspects which Wenskus attributed to the *Traditionskern*. Both were social groupings inside larger entities (armies or *gentes*) close to the powerholders and, I believed, they were the preferred bearers of traditions. This train of thought, however, led me to underestimate not only the level of abstraction behind both the *Gefolgschaft* and the *Traditionskern*, but also the enormous influence of late nineteenth and early twentieth centuries historiographies, which heavily encumbered them as research tools. Nevertheless, I have here maintained the structure of my Rome paper, while changing the title in praise of uncertainty.

This article outlines Reinhard Wenskus' interpretations of *Traditionskern* and *Gefolgschaft* together with my understanding of the use we can still make of them. While focusing largely on historiography, I have also made references

1 I am particularly grateful to Katharina von Winckler and Stefano Gasparri for going through my paper in the different stages of its composition and helping me clear my thoughts. Ça *va sans dire* the mistakes are all mine.

2 Reinhard Wenskus, *Stammesbildung und Verfassung. Verfassung. Das Werden der Frühmittelalterlichen Gentes* (Köln – Graz, 1961). On Wenskus: Heinrich Beck, 'Wenskus, Reinhard', in *Reallexikon der Germanischen Altertumskunde*, 33 (2006), pp. 454–455.

 | DOI:10.1163/9789004698017_013

to late antique and medieval narratives. The present article was difficult to write, walking the fine line between the ancient, the modern and the non-existent Germans, as well as between practice, discourse and fantasy.

2 Reinhard Wenskus and His Work

Stammesbildung und Verfassung is a history of Germanic tribes (*Stämme*) from the late Iron Age up to the foundation of the Barbarian Kingdoms in the aftermath of Rome's fall. Up to the middle of the twentieth century, it was believed that the Barbarian people were communities of descent whose habits and identity stretched back to a time immemorial. Because of this approach, it was supposed that the costumes and institutions narrated primarily by Tacitus, along with other Roman authors, echoed in much later periods. The Germans were, thereafter, *Randvölker von Hochkulturen*, marginal peoples of High Cultures (or a *Randkultur* as Karl Hauck wrote in the same years as Wenskus'), eventually becoming dominant at the heart of the Roman world.[3] Wenskus painstakingly moved between old scholarship and innovation. In his idea these tribes were mercurial, constantly being formed and rapidly disbanded, joined by new warriors or crushed by war. The traditions and norms that held them together were, on the other hand, very ancient.

Migrating groups shared a tribal mindset which Wenskus called *Gentilismus*, which was 'die besonderes völkerwanderungszeitlich germanische Form des ethnische Bewußtseins', the peculiar Germanic form of ethnic awareness during the migration age. In his view, this described a consciousness antithetical to the Roman *Reichsbewußtsein*, and which would, eventually, substitute it.[4] Driven by their *Gentilismus*, the Germanic tribes grew progressively larger and institutionally more complex to the extent that they ultimately built large political structures on the terrain of the former Roman Empire. It has been also proposed that *Gentilismus* was a *Begriff*, a concept which conveniently recast the burdened idea of *Volkstum*.[5] It is true that the notion was not an idea of Wenskus', stemming from early twentieth century debate on the ancient Germans and their identity instead. We know that Wolfgang Fritze, the renowned

---

3 Wenskus, *Stammesbildung*, p. 7; Karl Hauck, 'Von einer spätantiken Randkultur zum karolingischen Europa', in *Frühmittelalterliche Studien*, 1 (1967), pp. 3–93.

4 Wenskus, *Stammesbildung*, p. 2; quoted in Walter Pohl, 'Gentilismus', in *Reallexikon der Germanischen Altertumskunde*, 10 (1998) pp. 90–101, at p. 91.

5 Andrew Gillet, 'Ethnogenesis: A Contested Model of Early Medieval Europe', in *History Compass*, 4 (2006), pp. 241–260, at p. 244. On the concept of *Volkstum*: Emmerich, Wolfgang, *Zur Kritik der Volkstumsideologie* (Frankfurt am Main, 1971).

historian of the *Elbenslawen*, like Wenskus a Marburg graduate, used the concept of *Gentilismus* in his 1952 dissertation, as Wenskus himself acknowledged; similar doctrines spread along a larger circle of German historians.[6] Nevertheless, *Gentilismus* offered an alternative understanding of the medieval *gentes*, where the will of the individual played a decisive role in their ethnic allegiances. In fact, *Gentilismus* was actively expressed in terms of the pride of belonging and will to be part of the *gens* rather than a passive reflection of it.[7] While working on the history of the *Stämme*, Wenskus added further significant concepts, however, he seldom introduced them clearly. Some stemmed from previous ethnographical studies, where the ancient Germans were compared to the *Naturvölkern*, and others were drawn from older historical works. The term *Traditionskern* apparently owns something of both.

Due to its great length, 656 pages, and the fact that it is a *Dissertation* discussing *Stammesbildung und Verfassung* is an ungrateful task. Compared to non-German monographs it is awfully long and, further, has never been translated. Moreover, the issue of its linguistic complexity, making it of difficult comprehension even for scholars rather proficient in German, has been raised.[8] Native speakers may also find comprehending the work a challenge because of some excesses in detail.[9] In fact, the book's obscurity was perhaps the reason for its initial ambivalent reception. Renowned scholars reviewed the volume, but opinions were apparently mixed. It was evaluated by Michael Wallace-Hadrill, Michael Kobler and Rafael von Uslar. Rolf Hachmann and František Graus gave Wenskus' work more lengthy consideration including a discussion of it in their monographs on Scandinavian and Merovingian societies, respectively.[10] Von Uslar and Hachmann highly praised the volume; Graus,

6 The work was published only forty years later: Wolfgang H. Fritze, *Untersuchungen zur frühslawischen und frühfränkischen Geschichte bis ins 7. Jahrhundert, in Europäische Hochschulschriften* (Frankfurt am Main, 1994), p. 129; quoted at Wenskus, *Stammesbildung*, p. 2. Moreover: Pohl, 'Gentilismus', p. 91, who also recall Alfred Dove, Heinz Löwe and Helmut Baumann among Wenskus' predecessors.

7 Wenskus; *Stammesbildung*, p. 2; Pohl, 'Gentilismus', p. 96.

8 Peter Heather, 'Ethnicity, Group Identity, and Social Status in the Migration Period', in, *Franks, Northmen, and Slavs: Identity and State Formation in Early Medieval Europe*, ed. I. Garipzanov – P. Geary – U. Urbańczyk (Turnhout, 2008), pp. 17–49, at p. 28: 'Wenskus' great work, which is not that easy to read, even for those fluent in German'.

9 Rafael v. Uslar, 'Stämme und Fundgruppen: Bemerkungen zu Stammesbildung und Verfassung von R. Wenskus', in *Germania*, 43 (1965), pp. 138–147, at p. 139.

10 John Michael Wallace-Hadrill, Rev. of R. Wenskus, *Stammesbildung und Verfassung. Das Werden der Frühmittelalterlichen Gentes*, in *English Historical Review*, 79 (1964), pp. 137–139; Michael Kobler, Rev. of R. Wenskus, *Stammesbildung und Verfassung. Das Werden der Frühmittelalterlichen Gentes*, in *Deutsches Archiv für Erforschung des Mittelalters*, 19

in an otherwise sympathetic review, criticized the text which he saw as being haunted by nineteenth and early twentieth centuries historiography; while Wallace-Hadrill dryly remarked that the book 'is twice as long as it need be and half as clear as it might have been'.[11]

As Ian Wood recently remarked, Wenskus' book experienced renewed interest at a much later date thanks to Herwig Wolfram's studies, particularly the *Geschichte der Goten* in the English translation of its second edition.[12] It was through Wolfram's and later Walter Pohl's perspectives that *Stammesbildung und Verfassung* was placed among the founding literature of the so-called Vienna School of Medieval History, although its centrality has diminished through the decades.[13] It is because of its closeness to the Viennese medievalists that Wenskus' work was afterwards somehow overexposed, particularly between the end of the 1990s and the beginning of the 2000s. At this time, it became fundamental in the controversy, involving many European and American historians, over Barbarian identities, the role of early medieval narratives for the study of the oldest history of the *gentes*, and the so-called 'ethnogenesis theory'.[14] In 2000, an entire section of the International Medieval Congress in Kalamazoo was dedicated to the Vienna School and was eventually published by Brepols with the title *On Barbarian Identity*. This remains a very influential book if, occasionally, a misrepresentation of the subject. Alexander Callander Murray's contribution to the volume entirely focused on Reinhard Wenskus.[15]

(1963), pp. 253–254; Rafael v. Uslar, Rev. of R. Wenskus, *Stammesbildung und Verfassung. Das Werden der Frühmittelalterlichen Gentes*, in *Nachrichten aus Niedersachsens Urgeschichte*, 34 (1965), pp. 143–144; Rolf Hachmann, Rev. of R. Wenskus, *Stammesbildung und Verfassung. Das Werden der Frühmittelalterlichen Gentes*, in *Historische Zeitschrift*, 198 (1964), pp. 663–674, Id., *Die Goten und Skandinavien* (Berlin, 1970); František Graus, Rev. of R. Wenskus, *Stammesbildung und Verfassung. Das Werden der Frühmittelalterlichen Gentes*, in *Historia*, 7 (1963), pp. 185–191, Id., *Volk, Herrscher und Heiliger im Reich der Merowinger. Studien zur Hagiographie der Merowingerzeit* (Praha, 1965), pp. 314–336.

11 Wallace-Hadrill, Rev. of R. Wenskus, p. 138.

12 Ian Wood, *The Modern Origins of the Early Middle Ages* (Oxford, 2013), p. 301; Herwig Wolfram, *History of Goths* (Berkley – Los Angeles – London, 2000), but see now: Id., *Die Goten. Von den Anfängen bis zur Mitte des sechsten Jahrhunderts. Entwurf einer historischen Ethnographie* (München, 2009[5]).

13 Walter Pohl, 'Von der Ethnogenese zur Identitätsforschung', in *Neue Wege der Frühmittelalterforschung*, eds. M. Diesenberger – W. Pohl – B. Zeller (Wien, 2018), pp. 9–34.

14 Wolf Liebeschuetz, 'Debate about the Ethnogenesis of the Germanic Tribes', in *From Rome to Constantinople: Studies in Honour of Averil Cameron*, eds. H. Amirav – B. ter Haar Romeny (Leuven, 2007), pp. 341–356.

15 *On Barbarian Identity: Critical Approaches to Ethnicity in the Early Middle Ages*, ed. A. Gillet (Turnhout, 2002); Alexander Callander Murray, 'Reinhard Wenskus on "Ethnogenesis", Ethnicity, and the Origin of the Franks', in *On Barbarian Identity*, pp. 339–368.

There, Wenskus was accused of relying on older, ideologically loaded literature, while following a research path which, under a reassuring surface, was rooted in positivist historiography and the Romantic as well as in troubling *Rassenkunde* and National Socialist ideology.

Since then, Reinhard Wenksus and his scholarship have attracted further contributions. Prominent are the writings of his colleagues and pupils such as Heinrich Beck, Herwig Wolfram and Walter Pohl. Significant papers have also been written by scholars stemming from diverse academic traditions and working in distant countries, such as Francesco Filotico, who penned an attentive yet unfortunately rather neglected article, and Robert Kaspersky.[16] The major concern was, and still is, Wenskus' degree of innovation, rather than his dependency on a previous, ideologically burdened scholarship.

Stammesbildung und Verfassung is indeed open to different interpretations which, perhaps, originate from the entangled historiography of those years. In fact, reviewers such as Michael Wallace-Hadrill or Rolf Hachmann have offered rather different readings of the same work. As Stefano Gasparri stated, it was perhaps beyond Wenskus' intentions to write a revolutionary book, anchored as he was to traditional literature and methodologies.[17] And, as Walter Pohl remarked, *Stammesbildung und Verfassung* was written 'noch stark auf dem Boden der traditionellen „Germanischen Altertumskunde"'; still strongly anchored to traditional 'German Antiquities'. Indeed, the book appears to be in tension between a contemporary understanding of ethnic processes and the old myths of descent of the irrational, and more sombre thoughts belonging to the nineteenth century. It is due to this twofold nature that the monograph occupies a peculiar position in the historian's mind and, perhaps, is more quoted than read, retaining a halo of mystery having become a receptacle of diverging interpretations.

If Wenskus' *Stammesbildung und Verfassung* did not represent the watershed it was thought to be until a few years ago, it was nevertheless successful in bringing both older and newer ideas to a broader audience. For this

16 Francesco Filotico, 'Reinhard Wenskus e il concetto di etnicità nell'età delle invasioni barbariche', in *Nuova rivista storica*, 95 (2011), pp. 786–826; Robert Kaspersky, 'Ethnicity, Ethnogenesis, and the Vandals: Some Remarks on a Theory of Emergence of the Barbarian gens', in *Acta Poloniae Historica*, 112 (2015), pp. 201–242.

17 Stefano Gasparri, 'Tardoantico e alto Medioevo: metodologie di ricerca e modelli interpretativi', in *Il Medioevo (secoli V–XV), VIII. Popoli, poteri, dinamiche* (Roma, 2006), pp. 27–61, at p. 32: 'forse al di là delle intenzioni stesse del suo autore che era molto legato a una letteratura di tipo tradizionale'.

reason, it remains a milestone.[18] I will now move to discuss *Traditionskern* and *Gefolgschaft*.

3 A Kernel of Tradition

Traditionskern meant 'kernel of tradition' where the kernel refers to an aristocratic group close to the royal house forming a cohesive element of a late antique and early medieval *gens*. In Wenskus' reading, the tradition embodied a whole set of customs and knowledge ideally stemming from the past and shaping the identity of the *gens*; 'die Abstammungsmythen und Erzählungen vom gemeinsamen Schicksal sowie Lebensnormen enthielt'.[19]

Investigating the actual nature and role of the *Traditionskern* highlights the same issues we find when discussing Wenkus' work. Andreas Schwarz memorably wrote that 'the main problem [...] is that this damned book of his [Wenskus'] is too hard to read and that nowhere does he give an explicit theory of *Traditionskern*'.[20] With this dispassionate remark, Schwarz emphasizes the absence of any clear definition. As a result, the essence and function of the *Traditionskern* must be deduced from among its many pages. In fact, a clear description of the concept was only made many decades later by Herwig Wolfram in a crucial contribution dedicated to the research terminology on ethnic identity.[21]

In Wenskus' view, *Stämme* were held together by faith in a common descent, *Abstammungsglaube*, and the belief in biological proximity, which were parts of a shared *Tradition*. Paramount to this tradition was the name of the *gens*, which survived throughout the centuries and over enormous distances

18 As said Hachmann, Rev. of R. Wenskus, p. 673: 'Sein Buch ist ein Meilenstein in der Geschichte der Erforschung der frühen Germanen'. Also, Heather, 'Ethnicity', p. 26: 'The single most influential work here was Reinhard Wenskus' masterpiece of 1961, *Stammesbildung und Verfassung*'.

19 Pohl, 'Von der Ethnogenese', p. 10.

20 Andreas Schwarz, 'Discussion to Heather', in *The Visigoths from the Migration Period to the Seventh Century: An Ethnographic Perspective*, ed. P. Heather (Woodbridge, 2003), pp. 73–92, at p. 75. Also: Pohl, 'Von der Ethnogenese', pp. 10–11: 'Das hier skizzierte Modell lässt sich aus den über 600 Seiten von „Stammesbildung und Verfassung" ableiten, wird aber nirgends konzis zusammengefasst'.

21 Herwig Wolfram, 'Terminologisches', in *Nomen et fraternitas. Festschrift für Dieter Geuenich zum 65. Geburtstag*, eds. U. Ludwig – T. Schilp (Berlin – New York, 2008), pp. 787–802, at pp. 793–797; Id., *Gotische Studien. Volk und Herrschaft im frühen Mittelalter* (München, 2005), pp. 231–33, pp. 231–233. See, moreover: Giuseppe Albertoni, 'Intervista a Herwig Wolfram', in *Reti Medievali. Rivista*, 9 (2008), pp. 1–28, at pp. 9–10.

accompanying its rituals, mysteries, and sacred history.[22] The tradition was mostly related to *Herkunftssagen*, which the author saw as ancient, almost *enthistorisiert*, or 'out of history'.[23] However, not every member of the *Stamm* was equally committed to his tribal identity, tradition, and mission. The *Traditionskern* was a smaller group within the *Stamm*; the repository of the *gens*' identity, cementing its *Zusammengehörigkeitsbewußtsein* that is, a feeling of belonging together. Wenskus thought that it was formed by women and men close to the royal lines, such as the Amal and Balt dynasties, the Merovingians and the Agilolfings as well as the numerous clans listed in the *Origo gentis Langobardorum* or the laws of king Rothari (636–654).[24] Wenskus posited that these groups were limited in numbers; working in the years when Gustav Kossinna's teaching was still paramount, he wrote on two occasions that the *Traditionskern* were too small to be archeologically visible.[25]

It was around these small nuclei that, in Wenskus' reading, the Barbarian *gentes* gathered, drawing elements also stemming from different linguistic groups such as the Celts, the Slavs, the Balts or the 'Scythians'.[26] They were the aristocratic cores that through *Anschlußbewegungen*, a movement of affiliation, attracted the vast confederations in the last days of the Roman Empire.[27] In his words, 'kleine Splitter, die sich schneller der ethnischen Tradition des Kernverbandes anschlossen', that is, some of the warriors gathering around the *Traditionskern* became part of it.[28] Although Wenskus' was open to the fluidity of the barbarian *gentes*' ethnic composition, the Germanic element remained prominent. In his view, the *Traditionskern* remained essentially a German

22 Wenskus, *Stammesbildung*, pp. 14–17, 54–82.

23 Pohl, 'Von der Ethnogenese', p. 14; Graus, Rev. of R. Wenskus, p. 188; Wolfram, *Gotische Studien*, p. 215.

24 Balt and Amal dynasties: Wolfram, *Die Goten*, pp. 41–46; Hans Kuhn – Reinhard Wenskus, 'Amaler', in *Reallexikon der Germanischen Altertumskunde*, 1 (1973), pp. 246–248; and Hans Kuhn – Reinhard Wenskus, 'Balthen', in *Reallexikon der Germanischen Altertumskunde*, 2 (1973), pp. 13–14; on Merovingians and Agilolfings: Ian Wood, 'Deconstructing the Merovingian Family', in *The Construction of Communities in the Early Middle Ages: Texts, Resources and Artefacts*, eds. R. Corradini – M. Diesenberger – H. Reimitz (Leiden –Boston – Köln, 2003), pp. 149–171; Jörg Jarnut. *Agilofingerstudien. Untersuchungen zur Geschichte einer adligen Familie im 6. und 7. Jahrhundert* (Stuttgart, 1986); on the Lombard lineages see, newly, Eduardo Fabbro, *Warfare and the Making of Early Medieval Italy (568–652)* (Abingdon, 2020).

25 Wenskus, *Stammesbildung*, p. 466: 'Zur archäologischen Unerkennbarkeit trägt weiterhin bei, daß wir – wie schon oben bemerkt – vielfach nur mit kleinen Traditionskernen rechnen können'. See also, Hachmann, *Die Goten*, pp. 8–9.

26 Wenskus, *Stammesbildung*, pp. 76–77.

27 Wenskus, *Stammesbildung*, pp. 66–87.

28 Wenskus, *Stammesbildung*, p. 441.

phenomenon at its core, although, returning on the subject later in life, Wenskus questioned the legitimacy of the label 'Germans', or *Germanenbegriff*.[29]

Herwig Wolfram, the man who made Wenskus famous, loved to repeat that a concept similar to *Traditionskern* was already in Hector Munro Chadwick's writings.[30] Chadwick is best known for his *Heroic Age*, a fascinating and influential work, born out of the discoveries made in the nineteenth century of the similarities between distant mythologies.[31] In a further book of his, *The Nationalities of Europe and the Growth of National Ideologies*, Chadwick suggested that there was a royal family in every 'state' of the Heroic Age, which formed 'its nucleus or backbone'; before the written word, this *stirps regia* kept 'together with genealogy, a traditional account of its origin and early history'. According to Chadwick, it was a feature of the 'northern peoples', listing Picts, Dál Riata, but also Rus, Croats and Serbs along the 'Teutons'.[32]

Wenskus quoted Chadwick's *The Nationalities of Europe* in his *Stammesbildung und Verfassung* once, in relation to the Celtic origin of the name *Germani*.[33] He made no refence to the shared view of his *Traditionskern* and Chadwick's nucleus. It would thus appear that Wenskus coined the concept independently from the English scholar. And, in fact, *Traditionskern* may have been rooted deep in the darkest layers of German scholarship rather than in Chadwick's work. Heinrich Beck suggested that Wenskus was inspired by the anthropologist Willhelm Mühlmann, an idea also firmly upheld by Francesco Filotico.[34] More recently, Robert Kasperski highlighted the striking similarities between Wenskus' work and the one of Walter Schlesinger, one of the major exponents of the *Neue Verfassungs Geschichte*. In a skilfully selected passage of

---

29 Reinhard Wenskus, 'Über die Möglichkeit eines allgemeinen interdisziplinären Germanenbegriffs', in *Germanenprobleme in heutiger Sicht*, ed. H. Beck (Berlin – New York, 1986), pp. 1–21.

30 Wolfram, 'Terminologisches', p. 794; Albertoni, 'Intervista', p. 10. Moreover: Walter Pohl, 'Ethnicity, Theory, and Tradition: A Response', in *On Barbarian Identity*, pp. 221–239, at p. 11 note 8; Murray, 'Reinhard Wenskus', p. 53.

31 Hector Munro Chadwick, *The Heroic Age* (Cambridge, 1912); Stefan Arviddson, *Aryan Idols: Indo-European Mythology as Ideology and Science* (Chicago, 2006), p. 331.

32 Hector Munro Chadwick, *The Nationalities of Europe and the Growth of National Ideologies* (Cambridge, 1945), p. 94.

33 Wenskus, *Stammesbildung*, p. 218, fn. 527: 'sieht in „Germani" wie schon einzelne Forscher vor ihm einen keltischen Namen'.

34 Beck, 'Wenskus', p. 457; Filotico, 'Reinhard Wenskus', pp. 821–824; Pohl, 'Von der Ethnogenese', p. 12; Albertoni, 'Intervista', p. 10. On Mühlmann: Andre Gingrich, 'Ruptures, Schools, and Nontraditions: Reassessing the History of Sociocultural Anthropology in Germany', in *One Discipline, Four Ways: British, German, French and American Anthropology*, eds. F. Barth – A. Gingrich – R. Parkin – S. Silverman (Chicago, 2005), pp. 61–153, at pp. 118–120. I would like to thank Francesco Filotico for discussing the subject with me.

Schlesinger's most quoted contributions, which 'one could readily identify [...] as coming from Wenskus', Kaspenski highlights the usage of *Kern*, to signify a centre of 'crystallization' for warriors (*Kristallisationskern*). A *gentiler Kern* of the *Stamm* struggled to cement tribal identity as well as preserving the people's name.[35]

Notwithstanding its origin, the concept of *Traditionskern* assumed a different meaning in the new context where '[the] model made possible the overthrow of the deeply entrenched paradigm that ethnicity was essentially biological and immutable, and permitted the development of a more dynamic understanding of ethnic processes in Late Antiquity and the early Middle Ages', to quote Walter Pohl.[36] I would add that if it is important to scrutinize the climate and the origin of different historical thoughts, accusations of contiguity with National Socialist ideology could be often misleading, shaping mystifications and reducing the whole of a scholar's production to racist ideology. It seems a sort of *Blues Brothers* syndrome, where the modern academic impersonates Jake or Elwood Blues (the actors John Belushi and Dan Aykroyd) in their mission for God against all present National Socialists, who, in their case, are the Illinois Nazis.[37]

What remains is that *Traditionskern* is an obscure concept. Alexander Callander Murray correctly wrote how 'the *Traditionskern* idea will always elude demonstration and refutation [...], it leaves no trace. It is unattested in sources: the evidence for it is always indirect'.[38] We may add that:

1. It is presented as an original barbarian institution, developed with no contact to the Empire and its culture.[39]
2. It is very elitist a concept because of its almost exclusive linkage to aristocratic and royal authorities.[40] In fact, we know that Barbarian identities survived the fall of royal lines as in the case of the Amals; or even the demise of kingship itself, as occurred with the Lombards in

35 Walter Schlesinger, 'Über germanisches Heerkönigtum', in *Das Königtum. Seine geistigen und rechtlichen Grundlagen*, ed. T. Mayer (Lindau – Konstanz, 1956), pp. 105–141, at p. 123: 'Ein gentiles Moment hatte sich also erhalten, und man gewinnt vor allem den Eindruck, daß vielfach ein gentiler Kern vorhanden war, der eine stammhafte Tradition wahrte, ein Kristallisationskern gleichsam, an den sich die anderen Bestandteile anfügten, und zwar bei den verschiedenen Stämmen verschieden fest'.

36 Pohl, 'Ethnicity, Theory, and Tradition', p. 221.

37 *The Blues Brothers*, dir. by John Landis (Universal Pictures. 1980).

38 Murray, 'Reinhard Wenskus', p. 68.

39 Pohl, 'Von der Ethnogenese', pp. 14–15.

40 Pohl, 'Von der Ethnogenese', p. 15; Albertoni, 'Intervista', p. 9; Heather, 'Ethnicity', p. 30.

the years following the death of king Cleph of 574.[41] Elsewhere, identities and traditions seem to have also existed with a great multiplicity of sovereigns, as did amid the continental Saxons and the Frisians, or many of the Scandinavian peoples of the high Middle Ages.[42] Further ethnic groups, while beyond the Germanic world, apparently also gathered in the *absence* of royal lines, such as the Leutici did since the tenth century.[43] How traditions were passed from one generation to the next has never been explained; we could think of songs and sagas, but Wenskus does not make this clear.

3. Transmission seems to be male dominated. It is, however, well known that women shared an important role in the commemoration of the dead, and their deeds and roles in society: the cores around which identity unfolded.[44]
4. Finally, and most problematic, is that a *Traditionskern* implies strong continuities of some sort.[45] As Heather himself remarked, the different attestations of ethnic names (the ethnonym *Gothi* is continuously recorded from the first to the seventh century A.D.) were often explained as chains of unbroken accounts passed from one generation to the next.[46] Murray suggested that Wenskus' reading depended on 'the inventive interpretation of late sources, and a profound faith in the undetectable,

---

41 Fall of the Amal dynasty: Massimiliano Vitiello, *Theodahad: A Platonic King at the Collapse of Ostrogothic Italy* (Toronto – Buffalo – London, 2014), pp. 163–168. Clefi: Francesco Borri, 'Re Clefi (572–74)', in *I Longobardi a Venezia. Studi per Stefano Gasparri*, eds. I. Barbiera – F. Borri – A. Pazienza (Turnhout, 2020), pp. 39–50.

42 Mathias Becher, '*Non enim habent regem idem Antiqui Saxones*... Verfassung und Ethnogenese in Sachsen während des 8. Jahrhunderts', in *Sachsen und Franken in Westfalen. Zur Komplexität der ethnischen Deutung und Abgrenzung zweier frühmittelalterlicher Stämme*, eds H.-J. Häßler – J. Jarnut – U. Lund Hansen (= *Studien zur Sachsenforschung*, 12 [1999]), pp. 1–31.

43 Christian Lübke, 'Christianity and Paganism as Elements of Gentile Identities to the East of the Elbe and Saale Rivers', in *Franks, Northmen and Slavs*, pp. 189–203.

44 Literature here is legion. See: Cristina La Rocca, 'Donare, distribuire, spezzare. Pratiche di conservazione della memoria e dello status in Italia fra VIII e IX secolo', in *Sepolture tra IV e VIII secolo*, ed. G. P. Brogiolo (Mantova, 1998), pp. 77–87; Patrick Geary, *Women at the Beginning: Origin Myths from the Amazons to the Virgin Mary* (Princeton, 2006). I am grateful to Annamaria Pazienza for discussing the topic with me.

45 Heather, 'Ethnicity', p. 26: 'By re-reading the old sources without nationalist preconceptions, he was able to show just how much evidence there was for huge discontinuities in the history of Germanic-speaking groups in the Roman and early medieval periods'.

46 Ivi, p. 27: 'For, amidst all the clear discontinuities, evidence suggesting some continuity remained. The term "Goth" turns up in literary evidence from the first to the seventh centuries'.

> subterranean passage of ancient traditions'.[47] This observation is meticulous because Wenskus' interpretation depends on an enduring oral tradition, able to pass stories from one generation to the next; a model which has been seriously challenged in recent decades. We do know that names could have recurred without the existence of a *Traditionskern*. In fact, the Elbe Lombards of the Velleius Paterculus, who wrote at the beginning of the first century A.D., were hardly the same as those of the sixth century Danube recorded by Procopius. How names wandered is a matter of opinion, but we can imagine barbarian groups at the imperial fringes, competing to claim and monopolize prestigious ethnonyms, labels of power strongly embedded in the politics of the years. Even in later centuries names such as Saxons, Bulgarians or Croats are attested in different European regions at the same time: explaining these recurrences due to the movement of people only is unsatisfactory and sometimes trivial.[48]

As such, *Traditionskerne* probably never existed. Notwithstanding, the concept remains *bon à penser*, 'good to think with', to quote the famous dictum of Claude Lévi-Strauss.[49] It represents an *Idealtypus*, a historical abstraction suited to explaining our evidence. Barbarian traditions did exist and some members of the *gens*, perhaps aristocrats, must have transmitted it. The *Traditionskern* absolves this function. However, in defying both tradition and the kernel we face another set of issues with things becoming fuzzy in that:

1. We can no longer see the Barbarian *Tradition* in the far-fetched stories which survived, the so-called *origines gentium* of the early Middle Ages. Suggestions such as Patrick Amory's, mocking an imagined oral tradition of the Goths singing their songs around a bonfire, helped us to draw a line at the excesses of previous historiography.[50] Rather, we know that these histories were mostly built out of the Roman discourse on alterity. We could, nonetheless, reach for what have been called *vor-ethnographische*

47 Murray, 'Reinhard Wenskus', p. 62.

48 On the names, see: Walter Pohl, 'Die Namen der Barbaren: Fremdbezeichnung und Identität in Spätantike und Frühmittelalter', in *Zentrum und Peripherie. Gesellschaftliche Phänomene in der Frühgeschichte*, eds. H. Friesinger – A. Stuppner (Wien, 2004), pp. 95–104.

49 Marjorie Garber, *Loaded Words* (New York, 2012), pp. 94–103.

50 Patrick Amory, *People and Identity in Ostrogothic Italy 489–554* (Cambridge, 1997), p. 295: 'Need we think of Goths going round and telling tales or singing songs of Goths in the past, at dinners or around campfires, and regrouping them according to their present-day geographical locations? The answer is clearly negative'.

Daten, fragments of narratives escaping the story's blueprint.[51] Well known examples may be found in words of unknown meaning such as the *strava*, the big feast thrown in occasion of Attila's funeral; the *haliurunnae*, or Gothic *Hexen* who gave birth to the Huns; or the *Winnili*, the name around which the Lombards allegedly gathered in the first days of their migration.[52] What we cannot access is whether the words or fragments were meaningless blather, for us unintelligible puns, local lore, or indeed traditions significant for the *gens* as a whole. In an article of his, Walter Goffart suggested with a rather hilarious and provocative passage, that these names and anecdotes were the product of isolated authorial will.[53] It may have been the case on some occasions, but on others, words seem to have been signifiers of deeper narrative layers, perhaps meaningful for the *gens*' identity.[54] In a few cases we are confident of their importance, such as with the mention of Woden, appearing for the first time in a cluster of seventh century narratives.[55] Here deconstruction and linguistic-turn reach their limits.

2. The same goes for the *Kern*. Expanding the concept to some extent, as Peter Heather suggested, may overcome Wenskus' elitist approach. Even more suggestively, replacing it with the idea of *Träger einer Tradition* as Herwig Wolfram recommended, we would abandon the closeness that the word *Kern* implies.[56] In many narratives of the Migration Ages we

51 *Vor-ethonographische Daten*: Wolfram, 'Terminologisches', pp. 797–799; Id., *Gotische Studien*, pp. 215–221.

52 On *strava*: Jordanes, *Getica* 258 (ed. Mommsen, p. 124); Otto Manchen-Helfen, *The World of the Huns: Studies in Their History and Culture* (Berkeley – Los Angeles, 1973), pp. 274–278; Wolfram, *Gotische Studien*, p. 221. *Haliurunnae*: Jordanes, *Getica* 78 (ed. Mommsen, p. 76), Wolfram, *Gotische Studien*, pp. 255–257. *Winnili*: *Origo gentis Langobarodorum* 1 (ed. Waitz, p. 2); Stefano Gasparri, *La cultura tradizionale dei Longobardi: struttura tribale e resistenze pagane* (Spoleto, 1983), pp. 14–16.

53 Walter Goffart, 'Two Notes on Germanic Antiquity Today', in *Traditio*, 50 (1995), pp. 9–30, at p. 29: 'I like to imagine Cassiodorus dictating his history and pausing to question his *saio* adjutant: "By the way, Giberich, how does one say 'witch' in your language?" "That's 'haliurunna', sir". "Thank you, Giberich. I don't suppose you know whether it takes two n's or one?'. Also: Pohl, 'Von der Ethnogenese', p. 50.

54 Herwig Wolfram, '*Origo et religio*: Ethnic Traditions and Literature in Early Medieval Texts', in *Early Medieval Europe*, 3 (1994), pp. 19–38.

55 Francesco Borri, 'Drinking with Woden: A Re-Examination of Jonas's *Vita Columbani* 1.27', in *Columbanus and the Peoples of post-Roman Europe*, ed. A. O'Hara (Oxford, 2018), pp. 189–203.

56 Wolfram, 'Terminologisches', p. 795, also Id., *Die Goten*, p. 12. Peter Heather, 'The Huns and the End of the Roman Empire in Western Europe', in *English Historical Review*, 110 (1995), pp. 4–41, which gained him the definition 'neo-Romantic': Michael Kulikowski, 'Nation

read how defeated armies joined the victorious side, becoming part of the winning aristocracies' political orders and perhaps, eventually embracing their identity. Others were not given this option because they were too entangled with the losing side: they were exiled, imprisoned, or put to death.

These leading individuals of Barbarian *gentes* surrounded the king and the other major aristocracies resemble the military retinues of kings and major royals where trusted warriors met the children of the powerful for some time and learned their role in society. Indeed, the retinue of lords apparently cover many of the functions that Wenskus attributed to the *Traditionskern*.

4 Talking Tall and Drinking Wine

In scholarly debate, warrior retinues are often called *Gefolgschaften*, a nineteenth century term meant to be a direct translation of the Latin expression *comitatus* used by Tacitus to describe the warrior household of the *Germani* (*Germ.* 14). In historiographic discourse, the *Gefolgschaft* was a group of fighting men bound by a relationship of reciprocity and fidelity where the free swore an oath of allegiance to a leader recognizing his authority. The *fidelitas*, with a too rash essentialism if compared with the English concept of fidelity, was an emotional as well as military and economic bond constituting the glue of aristocratic society.

Like *Traditionskern*, the Wenskus' conception of *Gefolgschaft* owed to Walter Schlesinger as *Stammesbildund und Verfassung*' reviewers had acknowledged.[57] In several work, Schlesinger theorized warrior retinues' German character and the sacred oath binding the warrior to his lord.[58] In his view, the *Gefolgschaft* was by no means a marginal institution to society. It was, rather, the epistyle of the alleged state-building ability of the Germans, and even the *Grundherrschaft*, which is land possession, originated in the relationship between the lord and his follower.[59] These were concepts rooting further in the past, reflecting the work of early twentieth century scholarship, particularly of Heinrich

vs Army a Necessary Contrast', in *On Barbarian Identity*, pp. 69–84, at p. 83. I think 'Goth' would have been more appropriate.

57 Wallace-Hadrill, Rev. of R. Wenskus, p. 138.

58 Walter Schlesinger, 'Herrschaft und Gefolgschaft in der germanisch-deutschen Verfassungsgeschichte', in *Historische Zeitschrift*, 176 (1953), pp. 225–275; Christoph Landolt – Dieter Timpe – Heiko Steuer, 'Gefolgschaft', in *Reallexikon der Germanischen Altertumskunde*, 10 (1998), pp. 553–544, at p. 538.

59 Schlesinger, 'Herrschaft und Gefolgschaft', p. 229.

Dannenbauer, reacting to the eighteenth century reading of ancient Germanic society as democratic and egalitarian. Dannenbauer was a medieval historian and early supporter of the NSDAP (*Nationalsozialistische Deutsche Arbeiterpartei*). He published his reflections on the topic in a 1941 article, printed in the prestigious *Historisches Jahrbuch*.[60] We notice the strong social Darwinism dominating the text, together with the unsettling NS-jargon: 'Freiheit – ja die gab es in Altgermanien, aber für die Starken'.[61] It was an aspect of the dominating *Soldat* fancy of an all-male dominated society, discussed by Klaus Theweleit many years ago in his seminal *Männerphantasien*.[62]

Schlesinger followed this line of research. His characterization of the retinue as Germanic allowed him a largely diachronic approach so that his ideas originated out of a parallel reading of Tacitus' *Germania* and *Beowulf*, which was the approach of the *Germanische Altertumskunde*. It was from the old English poem that the idea of the *Gefolgsmänner* living by their lord probably came.[63] The legacy of Schlesinger's work was twofold. On the one hand, his definition of the *Gefolgschaft* as a long-term relationship of reciprocity between lord and follower – 'auf freiwilliger Unterordnung Freier unter einen Herrn zu kriegerischem Zweck beruht die Gefolgschaft'– is still valuable in its simplicity.[64] On the other, his reading of the Germanic past gave rise to a great debate. The same Wenskus wrote a difficult piece on the long discussion following Schlesinger's work; Jos Bazelmans penned an excellent article on the same topic.[65]

The most highly criticized idea was that of fidelity, *Treue*, sworn by oath. In Schlesinger's view, fidelity for the sake of fidelity was an original characteristic of the Germans. Hans Kuhn condemned the all-German scale and pervasiveness of retinue in Schlesinger's terms, revealing its contiguity with the

60 Heinrich Dannenbauer, 'Adel, Burg und Herrschaft bei den Germanen. Grundlagen der deutschen Verfassungsgeschichte', in *Historisches Jahrbuch*, 61 (1941), pp. 1–50.

61 Dannebauer, 'Adel, Burg und Herrschaft', p. 13.

62 Klaus Theweleit, *Männerphantasien*, 2 vols (Frankfurt am Main – Basel, 1977–1978).

63 Schlesinger, 'Herrschaft und Gefolgschaft', p. 235: 'Der Gefolgsmann unterstellte sich damit dem Schutz des Herrn, er trat gleichsam in sein Haus ein und wurde sein Tischgenosse'.

64 Ibid.

65 Reinhard Wenskus, 'Die neuere Diskussion um Gefolgschaft und Herrschaft in Tacitus Germania', in *Beiträge zum Verständnis der Germania des Tacitus II*, eds. G. Neumann – H. Seemann – H. Jankuhn (Göttingen, 1991), pp. 311–33; Jos Bazelmans, 'Conceptualising Early Germanic Political Structure: A Review of the Use of the Concept of Gefolgschaft', *Images of the Past: Studies on Ancient Societies in North-Western Europe*, eds. N. Roymans – F. Theuws (Amsterdam, 1991), pp. 91–129, at pp. 112–115. I would like to thank the author for sending me a copy of this article.

'conceptual word of the 1930s and 1940s'.[66] This exclusive German character was also rapidly attacked, particularly by František Graus.[67] A further criticism was made of the continuous character of the *Gefolgschaft*. Long enduring continuities, almost archetypal in character, are always fascinating but often omit macroscopic differences in distant contexts.[68] Rosemary Woolf and, more recently, Steven Fanning deconstructed the idea of a long-lasting institution.[69] As I am writing, few scholars maintain a more traditional reading, looking at the warband as a continuous entity throughout time, although not necessarily limited to Germanic speaking barbarians. The most notable example of this view is that of Michael Enright who thoughtfully debated the female role in the armed band.[70] In other works, the idea of warrior retinues was deconstructed together with many features of early medieval society imagined during the gloomy extremes of Romantic and Nationalist pan-Germanism.

Leaving aside the idea of an ahistorical and intrinsically German institution, we know that in ancient Mediterranean and European societies powerful men mustered armed bands ready to use violence in their behalf while receiving prestige, protection, and wealth in return, 'as such it is one particular form of a much more general phenomenon of *clientage*'.[71] Royal and aristocratic households became increasingly common during the decades of the Roman Army's twilight. *Gefolgschaften* may have well originated in this climate.

In recent decades, the alleged Germanic, or even barbaric, originality of many military institutions of the early Middle Ages have undergone significant deconstruction. Weaponry, rituals, and hierarchies, which have been read at length as barbarian in character, have been consistently interpreted as partially or completely Roman, often originating from the middle ground at the imperial frontiers. Even warrior inhumations which had been widely considered quintessentially Germanic, such as the so-called *Föderatengräber* of Northern Gaul, have been reinterpreted as the product of the complicated

66 Hans Kuhn, 'Die Grenzen der germanischen Gefolgschaft', in *Zeitschrift der Savigny-Stiftung für Rechtsgeschichte. Germ. Abt.*, 73 (1956), pp. 1–83; Bazelmans, 'Conceptualising', p. 116.

67 František Graus, 'Über die sogenannte germanische Treue', in *Historica*, 1 (1959), pp. 5–44.

68 Jos Bazelmans, *By Weapons Made Worthy: Lords, Retainers, and their Relationship in Beowulf* (Amsterdam, 1999), pp. 5–9; Steven Fanning, 'Tacitus, Beowulf, and the Comitatus', in *Haskins Society Journal*, 9 (2001), pp. 17–38; Landolt – Tiempe – Steuer, 'Gefolgschaft', pp. 538–539.

69 Rosemary Woolf, 'The Ideal of Men Dying with their Lord in the *Germania* and in *The Battle of Maldon*', in *Anglo-Saxon England*, 5 (1976), pp. 63–81; Fanning, 'Tacitus'.

70 Enright, *Lady with a Mead Cup*.

71 Bazelmans, *By Weapons Made Worthy*, p. 92. On relationships of personal dependency see also: Giuseppe Albertoni, *Vassalli, feudi, feudalesimo* (Roma, 2017[3]).

social changes of the late imperial frontier, dominated by imperial goods and the Roman language of power.[72] Counterarguments which saw them as the product of the army's 'Germanization' are often circular in character.[73] Stefan Esders carefully showed how the Merovingian Franks drew their military hierarchy from the Roman army.[74] Even kingship, which was extensively considered the most original aspect of Barbarian monarchies, had strong roots in the imperial *Herrschaft* as clearly shown by Mathias Becher.[75]

Late Roman men of power had their own retinues of soldiers which constituted among the most efficient tools to enforce authority.[76] From the end of the fourth century we find these groups of bodyguards often being called *buccellarii*.[77] The name, however, may have referred to an older institution which gained significance in late Antiquity. Oliver Schmitt has suggested the presence of German elements in the development of the *buccellarii*: it may have been the case, but with confidence we may ascribe the centrality of these retinues in the politics of those years to what is called the *spätrömische Militäradel*.[78] The very name *buccellarii* reveals its closeness to the ideal of *Gefolgschaft* in that the *buccellatum* was a hard loaf of bread which was given to the soldiers to eat. In this sense, the *buccellarii* were 'hardtack eaters' to borrow Guy Halsall's translation, that is, warriors who shared their lord's food.[79]

72 Guy Halsall, *Cemeteries and Society in Merovingian Gaul: Selected Studies in History and Archaeology, 1992–2009* (Leiden – Boston, 2010), pp. 93–167.

73 Ivi, p. 135: 'this evidence is twisted and distorted to support *a priori* assumptions', see also pp. 150–151.

74 Stefan Esders, 'Treueidleistung und Rechtsveränderung im frühen Mittelalter', in *Rechtsveränderung im politischen und sozialen Kontext mittelalterlicher Rechtsvielfalt*, eds S. Esders – C. Reinle (Münster, 2005), pp. 25–62, pp. 39–40.

75 Mathias Becher, 'Herrschaft im Übergang von der Spätantike zum Frühmittelalter. Von Rom zu den Franken', in *Von der Spätantike zum frühen Mittelalter. Kontinuitäten und Brüche, Konzeptionen und Befunde*, eds, T. Kölzer – R. Schieffer (Ostfildern, 2009), pp. 163–188, pp. 164–75.

76 Ivi, pp. 168–69.

77 Oliver Schmitt, 'Die *Buccellarii*. Eine Studie zum militärischen Gefolgschaftswesen in der Spätantike', in *Tyche*, 9 (1994), pp. 147–174; Oliver Berck, *Riding for the Powerful:* Buccellarii *and the Ideal of Retainer Loyality* (Marburg, 2014); Fernando Carlos Ruchesi, 'Los bucellarii y el Imperio Romano: sus orígenes, empleo y la cuestión de la cohesión social de sus componentes', in *Studia Historica. Historia Antigua*, 10 (2016) pp. 167–188.

78 Schmitt, 'Die *Buccellarii*', p. 153; Alexander Demandt, 'Der spätrömische Militäradel', in *Chiron*, 10 (1980), pp. 609–636.

79 Guy Halsall, *Warfare and Society in the Barbarian West, 450–900* (London, 2003), p. 45.

Thereafter, warrior retinues raised as tax-paid armies declined and the state's monopoly on violence crumbled.[80] They eventually constituted the core of many early medieval armed forces.[81] In contrast to *Traditionskerne*, retinues are very visible in the sources of the age; I will not gather the evidence for this here, although I aim to do it elsewhere. It will suffice to say that narrative and normative sources shed light on their functioning. Warrior retinues did not share a single denomination as *comitatus* mostly represents the authority of a *comes* rather than an aristocratic following, but we do find a rich terminology which, in given contexts, could signify *Gefolgschaft* and *Gefolgsmänner*. Nouns such as *trustis*, *contubernia* and *pueri*, *leudes* or *antrustiones*, *gasindi* are very suggestive of military relationships of dependency or reciprocity.[82] Armed bands were the object of early medieval laws because of the trouble they caused and the responsibility their members shared for their actions.[83] Atop his numerous studies, Heiko Steuer, maintained that *Gefolgschaften* could be traced archaeologically.[84] It has been also suggested that an aristocratic warrior retinue is depicted in one famous miniature found in the Golden Psalter of Saint Gall.[85]

The question is how the retinue of warriors could have fulfilled the functions that Wenskus attributed to the *Traditionskern*. Wenskus saw in the *Gefolgschaft* a capacity for attracting the vast armies which eventually founded kingdoms at the heart of Roman provinces. In one passage, he explicitly stated 'eine Institution, die von Haus aus die Angehörigen verschiedener Völkerschaften vereinigte, war geeignet, sich nahezu unbegrenzte Massen verschiedenster Herkunft einzuverleiben'.[86] This is very similar to the way he described the

80 Jeroen W. P. Wijnendaele, 'Warlordism and the Disintegration of the Western Roman Army', in Circum Mare*: Themes in Ancient Warfare*, ed. J. Armstrong (Leiden, 2016), pp. 185–203.

81 Halsall, *Warfare*, p. 101: 'Royal and aristocratic households made up the core of armies', referring to the Carolingians.

82 Régine Le Jan, 'Satellites et bandes armées dans le monde franc (VIIe–Xe siècles), in *Le combattant au Moyen Age* (= *Actes des congrès de la Société des historiens médiévistes de l'enseignement supérieur public*, 18 [1987]), pp. 97–105.

83 See the article of Fabio Botta in present volume.

84 Heiko Steuer, 'Helm und Ringschwert – Prunkbewaffnung und Rangabzeichen germanischer Krieger: eine Übersicht', in *Studien zur Sachsenforschung*, 6 (1987), pp. 190–236: Heiko Steuer, 'Archäologie der Gefolgschaft', in *2000 Jahren Varusschalcht –Konflikt*, ed. S. Burmeister (Stuttgart, 2009), pp. 309–319.

85 Eric Goldberg, '*More Devoted to the Equipment of Battle than the Splendor of Banquets*: Frontier Kingship, Military Ritual, and Early Knighthood at the Court of Louis the German', in *Viator*, 30 (199), pp. 41–78, at pp. 47–48; Le Jan, 'Satellites et bandes armées', p. 99.

86 Wenskus, *Stammesbildund*, p. 440.

Traditionskern, although the differences in scale are conspicuous. A second aspect is the function of the *Gefolgschaft* in cementing group cohesion and in the transmission of its identity. The major issue here is the long shadow cast by later poetry such as the abovementioned *Beowulf*, but also *Heliand*, *Waltharius*, or the *Battle of Maldon*, dating between the ninth (at the earliest) and the eleventh centuries, which were staged ón the social dynamics of the *Gefolgschaft* while greatly magnifying them. Due to suggestions originating from these narratives, we can imagine the lordly hall lighted by a fire, the lady pouring fine southern wines for the nobler guests and mead for the young ones, together with music, storytelling, and exchanges of gifts.

Nowadays this is to a great extent considered a literary trope where the stories illuminate narrative ideals rather than an accurate description of historical institutions.[87] Still, several behavioural codes and practices of Heroic Poetry find confirmation in histories, chronicles, and annals across kingdoms and the boundaries of language since the seventh century. These enlighten us of a peculiar etiquette and values characteristic of the armed bands of the early Middle Ages, from England to Italy, as well as in Central Europe. The great hall, crowded by warriors, figures in many narratives as a preferential stage for cohesion, identity and the passing-down of tradition. In the second quarter of the eighth century, Bede's famous parable of the sparrow, saw the bird flying from a storm into a warm and lighted royal hallway while the king and his retinue were sitting around the fire.[88] Also, in the stories of king Alboin and Authari, narrated by Paul the Deacon at the end of the eighth century, we find warrior followers and lordly halls as favoured surroundings.[89] The same was true for the story of the *rex Vindenorum* Samo: it is in his halls that the sublime chauvinist rants and the obsession with honour – which escalated into the wars against King Dagobert – were staged.[90] In the halls, joint leisure time and communal dining strengthen the aristocracy's ties, confirming its role in society.[91] Guy Halsall wrote that

87 i.e. Paul Battles, '*Contending Throng*: Scenes and the Comitatus Ideal in Old English Poetry, with special attention to *The Battle of Maldon* 122a', in *Studia Neophilologica*, 83 (2011), pp. 41–53.

88 Bede, *Historia ecclesiatica gentis Anglorum* II, 12 (ed. Colgrave – Mynors, p. 184).

89 Alboin: Paul the Deacon, *Historia Langobardorum* I, 23–24 (ed. Bethmann – Waitz, pp. 61–62); Authari: IVI, III, 30 (pp. 109–110).

90 *Chronicarum quae dicuntur Fredegarii* IV, 68 (ed. Krusch, p. 154).

91 Bettina Arnold, 'Drinking the Feast: Alcohol and the Legitimation of Power in Celtic Europe', in *Cambridge Archaeological Journal*, 9 (1999), pp. 71–93.

> a warrior had to be sure of those who stood to each side of him (especially on his unshielded right-hand side) in the battle-line. This necessity – as we shall see, it was literally a matter of life and death – surely lay behind the communal feasting and drinking, which helped to bond the warrior classes.[92]

Stories were materially embedded in the very drinking vessels displayed whose importance has been highlighted in different occasions.[93] Authors keenly described these cups: we know of the goblet believed to be Cunimond's skull; while the *Waltharius*-poet narrated how Attila's vessel carried the deeds of his people.[94] Fantastic glasses and drinking horns are ubiquitous in archaeological collections; the seventh century vessels found in the mound at Sutton Hoo burial site are iconic examples of these.[95] The insistence on these setting is so consistent over such a wide range of different sources that these dynamics were clearly not only operative on the level of literary discourse, but also informed actual practices. Chadwick had already commented that the deeds narrated in the Heroic Age were received by an audience which aimed at replicating them in real life.[96]

Gefolgschaften were, like Wenskus' *Traditionskern*, a social group within a larger group. Different to the *Traditionskern*, however, these are highly visible. Moreover, the social dynamic of the *Gefolgschaften* apparently became a distinctive characteristic of early Medieval Europe spanning from Wales to the Slavic region east of the Frankish kingdoms.[97] These were places where cohesion among warriors was enforced: also František Graus saw the *Gefolgschat* as the distinctive bearer of tribal traditions.[98] Places of identity, thereafter, albeit not necessarily of ethnicity: it was a matter of *Klasse* rather than *Rasse*.

---

92 Halsall, *Warfare*, p. 34.

93 Bonnie Effros, 'The Ritual Significance of Vessels in the Formation of Merovingian Christian Communities', in *The Construction of Communities in the Early Middle Ages: Texts, Resources and Artifacts*, eds., R. Corradini – M. Diesenberger – H. Reimitz (Leiden – Boston – Köln, 2003), pp. 213–227.

94 Cunimond's skull: Paul the Deacon, *Historia Langobardorum* II, 28 (ed. Bethmann – Waitz, p. 87); Attila's cup: *Waltharius* vv. 307–309 (ed. Strecker, p. 36).

95 Rupert Bruce-Mitford, *The Sutton Hoo Ship Burial III: Late Roman and Byzantine Silver, Hanging-Bowls Drinking Vessels, Cauldrons and Other Containers, Textiles, the Lyre, Pottery Bottle and Other Items*, 2 vols (London, 1983).

96 Chadwick, *The Heroic Age*, pp. 432–463.

97 For Celtic Europe see the somehow controversial: Steven Evans, *Lord of Battle: Image and Reality of the Comitatus in Dark Ages Britain* (Woodbridge, 1997); the Slavic East: Graus, *Die Nationenbildung*, pp. 35–36.

98 Graus, *Die Nationenbildung*, pp. 141–143.

5 Traditionskern, Gefolgschaft: The Punchline

In Reinhard Wenskus' view, *Gefolgschaft* and *Traditionskern* shared many similarities even though an explicit relationship between the two was never made. They were both catalysts to build larger coalitions, sharing a strong allure. However, Wenskus implied a fundamental difference between the two because the *Traditionskern*, as an Iron Age aristocratic group surrounding the sacral kings, predated the warrior retinues of the Migration Age. The *Traditionskern*, formed of women and men from the tribes' highest lineages, was apparently able to attract heterogeneous warriors from different *Stämme* which formed the *Gefolgschaft*. In turn, the retinue of a successful lord was able to draw conspicuous groups which could maintain their own traditions, or 'auch größere Gruppen, die ihre gentile Tradition bewahren konnten, schlossen sich zuweilen diesen Wanderverbänden an.'[99] There were, therefore, differences in both precedence and magnitude. After all, the warriors' retinue appears as an extension of the *Traditionskern* becoming visible in the narratives of the ages. As mentioned, *Traditionskern* is a concept that explains the volatile nature of barbarian tribes together with their traditions' alleged longevity, even if *Traditionen* were probably different in nature as Wenskus' understood them. Although an abstract reading, we agree that traditions must have been perpetuated by some *Träger*. From a plethora of sources, warrior retinues appear as the stage and the preferred audience for the transmission of stories which enforced that cohesion between warriors and their identities.

A tale told by Paul the Deacon in his *Historia Langobardorum* may shed some light on the complex relationships between traditions, their bearers, warrior retinues and identity. It is the story of the *sculdhais* Argait and duke Ferdulf, depicting the frontier life of Friuli's aristocracy and their conflictual existences during the eighth century. In the story, the *sculdhais* was accused by the duke of being an *arga* because of a military failure of his and because the obvious assonance with his name. The word *arga* 'appears to be Paul's version of the Old Norse adjective *argr*, which scholars of that language and society generally agree was an extreme insult that implied more than simply "coward" but rather sexual inadequacy' as Ross Balzaretti suggested. The word *arga* seems, thereafter, a brutal and abusive word, like most early medieval insults recorded in Barbarian laws.[100] *Arga* is mentioned also in the laws of Rothari

99 Ivi, p. 441.

100 Ross Balzaretti, 'Sexuality in Late Lombard Italy, c.700–c.800 AD', in *Medieval Sexuality: A Casebook*, eds. A. Harper – C. Proctor (New York, 2008), pp. 7–31, at p. 16–17; Halsall, *Warfare*, p. 11: 'These are terrible insults in a society, like sixth-century Frankish, which was

where the king imposed the payment of twelve *solidi* for every offender. Further, it features in *Hilderandslied* where its mention finally moved Hildebrand to attack his son Hadubrand.[101] It has been suggested that the name Argait may have originated from a reference to an arrow's point.[102] Regardless, the similarity between Argait and *arga* is evident: we do not know of a further Argait in Lombard history and the name must have immediately sounded awkward and wrong, like Gaylord Focker or Biggus Dickus are to a modern audience acquainted with comedies.[103]

Mortally offended, the *sculdahis* eventually challenged the duke to singlehandedly attack the Slavic marauders once the Lombards were outnumbered. They both charged from the most unfavourable position. The duke's warriors, fearing the loss of honour, followed the two in this reckless move so that the whole of the Friuli aristocracy died that day. Only one man, necessary for the survival of the story, escaped the battlefield. In the aftermath of the military disaster, duke Pemmo gathered the children of the fallen aristocrats and raised them together with his own.

The whole story was perhaps built at some point around the name Argait, which echoed the slur *arga*. The name may have been passed down through oral traditions like many other fragments of forgotten stories. The tale's blueprint may have been the law of Rothari where a fight could have washed away the dishonour of being called an *arga*.[104] As in Herwig Wolframs' interpretation of Reinhard Wenskus' writings, traditions survive in words loaded with meaning, perhaps growing distant from an original context. In some occurrences, were nouns carrying significance for the warrior class and perhaps the

largely constructed around ideas of the 'warrior''. Karl E. Gade, 'Homosexuality and Rape of Males in Old Norse Law and Literature, in *Scandinavian Studies*, 58 (1986), pp. 124–141, at p. 134. Fascinating is also a reference in James Patrick Mallory – Douglas Q. Adams, *The Oxford Introduction to Proto-Indo-European and the Proto-Indo-European World* (Oxford, 2006), p. 188: 'Ht ārki ~ arga "mounts (used with respect to a male animal)"'.

101 *Edictum Rothari* 381 (ed. Bluhme, p. 88); *Hildebrandslied* v. 57 (Engl. transl. Bostock, pp. 76–77).

102 Maria Giovanna Arcamone, 'Die langobardischen Personennamen in Italien', in *Nomen et gens. Zur historischen Aussagekraft frühmittelalterlicher Personennamen*, eds., D. Geuenich – W. Haubrichs – J. Jarnut (Berlin – New York, 1997), pp. 157–175, at p. 167. Thanks to Carla Falluomini for having discussed the topic with me.

103 Gaylord Focker: *Meet the Parents*, dir. by Jay Roach (TriBeCa Productions and Nancy Tenenbaum Productions, 2000); Biggus Dickus: *Monthy Phython's Life of Brian*, dir. by Terry Jones (Python [Monty] Pictures and Orion Pictures, 1979).

104 *Edictum Rothari* 381 (ed. Bluhme, p. 88); Balzaretti, 'Sexuality', p. 17: 'To me it seems highly likely that Paul, instead of recording what the two men said, in fact modeled his story on this law [...]'.

Barbarian *gens*. In the *Historia Langobardorum*, Argait became a fictional hero, a brave and doomed warrior named after the most insulting Lombard word, who could redeem his tragic destiny by dying on the battlefield together with the whole aristocracy of Friuli.

For us, the narrative highlights the existence of aristocratic retinues, such as the one raised by duke Pemmo together with narratives enforcing its members' identity. Such tales may have been proper to a *gens*, but they were often mundane in character and distant from the concept of a sacred history as ideated by Wenskus, being intelligible in different corners of Europe, well beyond the Germanic world, where similar values and norms were shared. Stories as such, perhaps originating from older words loaded with meaning, may have been subject to change and transformation remaining part of a shared tradition. Despite these recalibrations, the story of Argait leads us to a band of warriors living under the same roof sharing stories of their larger-than-life ancestors, which enforced the group's identity and cohesion. As such, early Medieval *Gefolgschaft* finally brings us to a closure near where we started, Reinhard Wenskus' *Traditionskern*, its value for identity and cohesion.

Bibliography

Primary Sources

Bede, *Historia ecclesiatica gentis Anglorum*, ed. B. Colgrave – R. A. B. Mynors, *Bede's Ecclesiastical History of the English People* (Oxford, 1969)

Edictum Rothari, ed. F. Bluhme, MGH, LL, 4 (Hannover, 1863), pp. 3–90

Hildebrandslied, trans. J. K. Bostock, *A Handbook on Old High German Literature*, rev. K. C. King – D. R. McLintock (Oxford, 1976²) pp. 76–77

Jordanes, *Getica*, ed. T. Mommsen, MGH, Auct. ant. 5 (Berlin, 1880), pp. 53–138

Origo gentis Langobarodorum, ed. G. Waitz, MGH, SS rer. Lang. (Hannover, 1878), pp. 1–6

Paul the Deacon, *Historia Langobardorum*, ed. L. Bethmann – G. Waitz, in MGH, SS rer. Lang. (Hannover, 1878), pp. 12–187

Waltharius, ed. K. Strecker, MGH, Poetae, 6.1 (Weimar, 1951) 1–83

Secondary Literature

Albertoni, Giuseppe, 'Intervista a Herwig Wolfram', in *Reti Medievali. Rivista*, 9 (2008), pp. 1–28

Albertoni, Giuseppe, *Vassalli, feudi, feudalesimo* (Roma, 2017³)

Amory, Patrick, *People and Identity in Ostrogothic Italy 489–554* (Cambridge, 1997)

Arcamone, Maria Giovanna, 'Die langobardischen Personennamen in Italien', in D. Geuenich – W. Haubrichs – J. Jarnut, eds., *Nomen et gens. Zur historischen Aussagekraft frühmittelalterlicher Personennamen* (Berlin – New York, 1997), pp. 157–175

Arnold, Bettina, 'Drinking the Feast: Alcohol and the Legitimation of Power in Celtic Europe', in *Cambridge Archaeological Journal*, 9 (1999), pp. 71–93

Arviddson, Stefan, *Aryan Idols: Indo-European Mythology as Ideology and Science* (Chicago, 2006)

Balzaretti, Ross, 'Sexuality in Late Lombard Italy, c.700–c.800 AD', in A. Harper – C. Proctor, eds., *Medieval Sexuality: A Casebook* (New York, 2008), pp. 7–31

Battles, Paul, '*Contending Throng*: Scenes and the Comitatus Ideal in Old English Poetry, with special attention to *The Battle of Maldon* 122a', in *Studia Neophilologica*, 83 (2011), pp. 41–53

Bazelmans, Jos, 'Conceptualising Early Germanic Political Structure: A Review of the Use of the Concept of *Gefolgschaft*', in N. Roymans – F. Theuws, eds., *Images of the Past: Studies on Ancient Societies in North-Western Europe* (Amsterdam, 1991), pp. 91–129

Bazelmans, Jos, *By Weapons Made Worthy: Lords, Retainers, and their Relationship in Beowulf* (Amsterdam, 1999)

Becher, Mathias, '*Non enim habent regem idem Antiqui Saxones* … Verfassung und Ethnogenese in Sachsen während des 8. Jahrhunderts', in H.-J. Häßler – J. Jarnut – U. Lund Hansen, eds., *Sachsen und Franken in Westfalen. Zur Komplexität der ethnischen Deutung und Abgrenzung zweier frühmittelalterlicher Stämme* (= *Studien zur Sachsenforschung*, 12 [1999]), pp. 1–31

Becher, Mathias, 'Herrschaft im Übergang von der Spätantike zum Frühmittelalter. Von Rom zu den Franken', in T. Kölzer – R. Schieffer, eds., *Von der Spätantike zum frühen Mittelalter. Kontinuitäten und Brüche, Konzeptionen und Befunde* (Ostfildern, 2009), pp. 163–188

Beck, Heinrich, 'Wenskus, Reinhard', in *Reallexikon der Germanischen Altertumskunde*, 33 (2006), pp. 454–455

Berck, Oliver, *Riding for the Powerful:* Buccellarii *and the Ideal of Retainer Loyality* (Marburg, 2014)

Borri, Francesco, 'Drinking with Woden: A Re-Examination of Jonas's *Vita Columbani* I.27', in A. O'Hara, ed., *Columbanus and the Peoples of post-Roman Europe* (Oxford, 2018), pp. 189–203

Borri, Francesco, 'Re Clefi (572–74)', in I. Barbiera – F. Borri – A. Pazienza, eds., *I Longobardi a Venezia. Studi per Stefano Gasparri* (Turnhout, 2020), pp. 39–50

Bruce-Mitford, Rupert, *The Sutton Hoo Ship Burial III: Late Roman and Byzantine Silver, Hanging-Bowls Drinking Vessels, Cauldrons and Other Containers, Textiles, the Lyre, Pottery Bottle and Other Items*, 2 vols (London, 1983)

Chadwick, Hector Munro, *The Heroic Age* (Cambridge, 1912)

Chadwick, Hector Munro, *The Nationalities of Europe and the Growth of National Ideologies* (Cambridge, 1945)

Dannenbauer, Heinrich, 'Adel, Burg und Herrschaft bei den Germanen. Grundlagen der deutschen Verfassungsgeschichte', in *Historisches Jahrbuch*, 61 (1941), pp. 1–50

Demandt, Alexander, 'Der spätrömische Militäradel', in *Chiron*, 10 (1980), pp. 609–636

Effros, Bonnie, 'The Ritual Significance of Vessels in the Formation of Merovingian Christian Communities', in R. Corradini – M. Diesenberger – H. Reimitz, eds., *The Construction of Communities in the Early Middle Ages: Texts, Resources and Artifacts* (Leiden – Boston – Köln, 2003), pp. 213–227

Emmerich, Wolfgang, *Zur Kritik der Volkstumsideologie* (Frankfurt am Main, 1971)

Enright, Michael. *Lady with a Mead Cup: Ritual, Prophecy and Lordship in the European Warband from La Tène to the Viking Age* (Dublin, 1996)

Esders, Stefan, 'Treueidleistung und Rechtsveränderung im frühen Mittelalter', in S. Esders – C. Reinle, eds., *Rechtsveränderung im politischen und sozialen Kontext mittelalterlicher Rechtsvielfalt* (Münster, 2005), pp. 25–62

Evans, Steven, *Lord of Battle: Image and Reality of the Comitatus in Dark Ages Britain* (Woodbridge, 1997)

Fabbro, Eduardo, *Warfare and the Making of Early Medieval Italy (568–652)* (Abingdon, 2020)

Fanning, Steven, 'Tacitus, Beowulf, and the Comitatus', in *Haskins Society Journal*, 9 (2001), pp. 17–38

Filotico, Francesco, 'Reinhard Wenskus e il concetto di etnicità nell'età delle invasioni barbariche', in *Nuova rivista storica*, 95 (2011), pp. 786–826

Fritze, Wolfgang H., *Untersuchungen zur frühslawischen und frühfränkischen Geschichte bis ins 7. Jahrhundert, in Europäische Hochschulschriften* (Frankfurt am Main, 1994)

Gade, Karl E., 'Homosexuality and Rape of Males in Old Norse Law and Literature, in *Scandinavian Studies*, 58 (1986), pp. 124–141

Garber, Marjorie, *Loaded Words* (New York, 2012)

Garipzanov, I. H. – P. J. Geary – P. Urbańczyk, eds., *Franks, Northmen, and Slavs: Identity and State Formation in Early Medieval Europe* (Turnhout, 2008)

Gasparri, Stefano, *La cultura tradizionale dei Longobardi: struttura tribale e resistenze pagane* (Spoleto, 1983)

Gasparri, Stefano, 'Tardoantico e alto Medioevo: metodologie di ricerca e modelli interpretativi', in *Il Medioevo (secoli v–xv), viii: Popoli, poteri, dinamiche* (Roma, 2006), pp. 27–61

Geary, Patrick, *Women at the Beginning: Origin Myths from the Amazons to the Virgin Mary* (Princeton, 2006)

A. Gillet, ed., *On Barbarian Identity: Critical Approaches to Ethnicity in the Early Middle Ages* (Turnhout, 2002)

Gillet, Andrew, 'Ethnogenesis: A Contested Model of Early Medieval Europe', in *History Compass*, 4 (2006), pp. 241–260

Gingrich, Andre, 'Ruptures, Schools, and Nontraditions: Reassessing the History of Sociocultural Anthropology in Germany', in F. Barth – A. Gingrich – R. Parkin – S. Silverman, eds., *One Discipline, Four Ways: British, German, French and American Anthropology* (Chicago, 2005), pp. 61–153

Goffart, Walter, 'Two Notes on Germanic Antiquity Today', in *Traditio*, 50 (1995), pp. 9–30

Goldberg, Eric, '*More Devoted to the Equipment of Battle than the Splendor of Banquets*: Frontier Kingship, Military Ritual, and Early Knighthood at the Court of Louis the German', in *Viator*, 30 (199), pp. 41–78

Graus, František, 'Über die sogenannte germanische Treue', in *Historica*, 1 (1959), pp. 5–44

Graus, František, *Die Nationenbildung der Westslawen im Mittelalter* (Sigmaringen, 1980)

Graus, František, Rev. of R. Wenskus, *Stammesbildung und Verfassung: das Werden der Frühmittelalterlichen Gentes*, in *Historia*, 7 (1963), pp. 185–191

Graus, František, *Volk, Herrscher und Heiliger im Reich der Merowinger. Studien zur Hagiographie der Merowingerzeit* (Praha, 1965)

Hachmann, Rolf, Rev. of R. Wenskus, *Stammesbildung und Verfassung. Das Werden der Frühmittelalterlichen Gentes*, in *Historische Zeitschrift*, 198 (1964), pp. 663–674

Hachmann, Rolf, *Die Goten und Skandinavien* (Berlin, 1970)

Halsall, Guy, *Warfare and Society in the Barbarian West, 450–900* (London, 2003)

Halsall, Guy, *Cemeteries and Society in Merovingian Gaul: Selected Studies in History and Archaeology, 1992–2009* (Leiden – Boston, 2010)

Hauck, Karl, 'Von einer spätantiken Randkultur zum karolingischen Europa', in *Frühmittelalterliche Studien*, 1 (1967), pp. 3–93

Heather, Peter, 'The Huns and the End of the Roman Empire in Western Europe', in *English Historical Review*, 110 (1995), pp. 4–41

Heather, Peter, 'Ethnicity, Group Identity, and Social Status in the Migration Period', in Garipzanov – Geary – Urbańczyk, eds., *Franks, Northmen, and Slavs*, pp. 17–49

Jarnut, Jörg. *Agilofingerstudien. Untersuchungen zur Geschichte einer adligen Familie im 6. und 7. Jahrhundert* (Stuttgart, 1986)

Kaspersky, Robert, 'Ethnicity, Ethnogenesis, and the Vandals: Some Remarks on a Theory of Emergence of the Barbarian gens', in *Acta Poloniae Historica*, 112 (2015), pp. 201–242

Kobler, Michael, Rev. of R. Wenskus, *Stammesbildung und Verfassung. Das Werden der Frühmittelalterlichen Gentes*, in *Deutsches Archiv für Erforschung des Mittelalters*, 19 (1963), pp. 253–254

Kuhn, Hans, 'Die Grenzen der germanischen Gefolgschaft', in *Zeitschrift der Savigny-Stiftung für Rechtsgeschichte. Germ. Abt.*, 73 (1956), pp. 1–83

Kuhn, Hans – Wenskus, Reinhard, 'Amaler', in *Reallexikon der Germanischen Altertumskunde*, 1 (1973), pp. 246–248

Kuhn, Hans – Wenskus, Reinhard, 'Balthen', in *Reallexikon der Germanischen Altertumskunde*, 2 (1973), pp. 13–14

Kulikowski, Michael, 'Nation vs Army a Necessary Contrast', in Gillet, ed., *On Barbarian Identity*, pp. 69–84

La Rocca, Cristina, 'Donare, distribuire, spezzare. Pratiche di conservazione della memoria e dello status in Italia fra VIII e IX secolo', in G. P. Brogiolo, ed., *Sepolture tra IV e VIII secolo* (Mantova, 1998), pp. 77–87

Landolt, Christoph – Dieter Timpe – Heiko Steuer, 'Gefolgschaft', in *Reallexikon der Germanischen Altertumskunde*, 10 (1998), pp. 553–544

Le Jan, Régine, 'Satellites et bandes armées dans le monde franc (VIIe–Xe siècles), in *Le combattant au Moyen Age* (= *Actes des congrès de la Société des historiens médiévistes de l'enseignement supérieur public*, 18 [1987]), pp. 97–105

Liebeschuetz, Wolf, 'Debate about the Ethnogenesis of the Germanic Tribes', in H. Amirav – B. ter Haar Romeny, eds., *From Rome to Constantinople: Studies in Honour of Averil Cameron* (Leuven, 2007), pp. 341–356

Lübke, Christian, 'Christianity and Paganism as Elements of Gentile Identities to the East of the Elbe and Saale Rivers', in Garipzanov – Geary – Urbańczyk, eds., *Franks, Northmen and Slavs*, pp. 189–203

Mallory, James Patrick – Douglas Q. Adams, *The Oxford Introduction to Proto-Indo-European and the Proto-Indo-European World* (Oxford, 2006)

Murray, Alexander Callander, 'Reinhard Wenskus on "Ethnogenesis", Ethnicity, and the Origin of the Franks', in Gillet, ed., *On Barbarian Identity*, pp. 339–368

Pohl, Walter, 'Tradition, Ethnogenese und literarische Gestaltung: eine Zwischenbilanz', in K. Brunner – B. Merta, eds., *Ethnogenese und Überlieferung. Angewandte Methoden der Frühmittelalterforschung* (Wien – München, 1994), pp. 9–26

Pohl, Walter, 'Gentilismus', in *Reallexikon der Germanischen Altertumskunde*, 10 (1998) pp. 90–101

Pohl, Walter, 'Ethnicity, Theory, and Tradition: A Response', in Gillet, ed., *On Barbarian Identity*, pp. 221–239

Pohl, Walter, 'Die Namen der Barbaren. Fremdbezeichnung und Identität in Spätantike und Frühmittelalter', in H. Friesinger – A. Stuppner, eds., *Zentrum und Peripherie: Gesellschaftliche Phänomene in der Frühgeschichte* (Wien, 2004), pp. 95–104

Pohl, Walter, 'Von der Ethnogenese zur Identitätsforschung', in M. Diesenberger – W. Pohl – B. Zeller, eds., *Neue Wege der Frühmittelalterforschung* (Wien, 2018), pp. 9–34

Ruchesi, Fernando Carlos, 'Los bucellarii y el Imperio Romano: sus orígenes, empleo y la cuestión de la cohesión social de sus componentes', in *Studia Historica. Historia Antigua*, 10 (2016) pp. 167–188

Schlesinger, Walter, 'Herrschaft und Gefolgschaft in der germanisch-deutschen Verfassungsgeschichte', in *Historische Zeitschrift*, 176 (1953), pp. 225–275

Schlesinger, Walter, 'Über germanisches Heerkönigtum', in T. Mayer, ed., *Das Königtum. Seine geistigen und rechtlichen Grundlagen* (Lindau – Konstanz, 1956), pp. 105–141

Schmitt, Oliver, 'Die *Buccellarii.* Eine Studie zum militärischen Gefolgschaftswesen in der Spätantike', in *Tyche*, 9 (1994), pp. 147–174

Schwarz, Andreas, 'Discussion to Heather', in P. Heather, ed., *The Visigoths from the Migration Period to the Seventh Century: An Ethnographic Perspective* (Woodbridge, 2003), pp. 73–92

Steuer, Heiko, 'Helm und Ringschwert – Prunkbewaffnung und Rangabzeichen germanischer Krieger: eine Übersicht', in *Studien zur Sachsenforschung*, 6 (1987), pp. 190–236

Steuer, Heiko, 'Archäologie der Gefolgschaft', in S. Burmeister, ed., *2000 Jahren Varusschalcht –Konflikt* (Stuttgart, 2009), pp. 309–319

Theweleit, Klaus, *Männerphantasien*, 2 vols (Frankfurt am Main – Basel, 1977–1978)

v. Uslar, Rafael, 'Stämme und Fundgruppen: Bemerkungen zu *Stammesbildung und Verfassung* von R. Wenskus', in *Germania*, 43 (1965), pp. 138–147

v. Uslar, Rafael, Rev. of R. Wenskus, *Stammesbildung und Verfassung. Das Werden der Frühmittelalterlichen Gentes*, in *Nachrichten aus Niedersachsens Urgeschichte*, 34 (1965), pp. 143–144

Vitiello, Massimiliano, *Theodahad: A Platonic King at the Collapse of Ostrogothic Italy* (Toronto - Buffalo – London, 2014)

Wallace-Hadrill, John Michael, Rev. of R. Wenskus. *Stammesbildung und Verfassung: das Werden der Frühmittelalterlichen Gentes*, in *English Historical Review*, 79 (1964), pp. 137–139

Wenskus, Reinhard, *Stammesbildung und Verfassung. Das Werden der Frühmittelalterlichen Gentes* (Köln – Graz, 1961)

Wenskus, Reinhard, 'Über die Möglichkeit eines allgemeinen interdisziplinären Germanenbegriffs', in H. Beck, ed., *Germanenprobleme in heutiger Sicht* (Berlin – New York, 1986), pp. 1–21

Wenskus, Reinhard, 'Die neuere Diskussion um Gefolgschaft und Herrschaft in Tacitus Germania', in G. Neumann – H. Seemann – H. Jankuhn, eds., *Beiträge zum Verständnis der Germania des Tacitus II* (Göttingen, 1991), pp. 311–331

Wijnendaele, Jeroen W. P., 'Warlordism and the Disintegration of the Western Roman Army', in J. Armstrong, ed., Circum Mare*: Themes in Ancient Warfare* (Leiden, 2016), pp. 185–203

Wolfram, Herwig, '*Origo et religio*: Ethnic Traditions and Literature in Early Medieval Texts', in *Early Medieval Europe*, 3 (1994), pp. 19–38

Wolfram, Herwig, *History of Goths* (Berkley – Los Angeles – London, 2000)

Wolfram, Herwig, *Gotische Studien. Volk und Herrschaft im frühen Mittelalter* (München, 2005), pp. 231–33

Wolfram, Herwig, 'Terminologisches', in U. Ludwig – T. Schilp, eds., *Nomen et fraternitas. Festschrift für Dieter Geuenich zum 65. Geburtstag* (Berlin – New York, 2008), pp. 787–802

Wolfram, Herwig, *Die Goten. Von den Anfängen bis zur Mitte des sechsten Jahrhunderts. Entwurf einer historischen Ethnographie* (München, 2009^{5})

Wood, Ian, 'Deconstructing the Merovingian Family', in R. Corradini – M. Diesenberger – H. Reimitz, eds. *The Construction of Communities in the Early Middle Ages: Texts, Resources and Artefacts* (Leiden – Boston – Köln, 2003), pp. 149–171

Wood, Ian, *The Modern Origins of the Early Middle Ages* (Oxford, 2013)

Woolf, Rosemary, 'The Ideal of Men Dying with their Lord in the *Germania* and in *The Battle of Maldon*', in *Anglo-Saxon England*, 5 (1976), pp. 63–81

CHAPTER 13

The Lombard Army Between Myth and Reality: *Farae, Arimanniae, Arimanni*

Stefano Gasparri

1 The Arimannic Theory and the Frontiers

The military issue, along with religion, has always been the centre of attention for historians of the Lombards. In time, both of these questions have been transformed into general frameworks for the interpretation of the history of this people. Traditionally, the religious issue has been held to be controversial to the extent that historiography tended to veer between the two extremes of the presumed 'indifference' of the Lombards to it and the so-called 'drama' of the Lombards, resulting from their – again presumed – uncertain and late adoption of Christianity.[1] In lieu of these now superseded views, a more recent interpretation was advanced, suggesting that the Lombard attitude toward Christianity was defined by a kind of 'deliberate ambiguity'.[2] The military issue, on the other hand, never suffered from such doubts. The Lombards were a bellicose people, whose entire social organization, relationship with the Italian territory, and with the local population, was shaped by their warlike nature.

The military dimension did in fact play a key role in the earliest forms of organization of the Lombard people, whose very identity – like that of other barbarian groups – was forged in the melting pot of the late Roman army and its foederati units.[3] This emerges very clearly from the events that took place in the first decades after their arrival in Italy, also leaving traces in the later

1 The traditional hypotheses are set out in Gian Piero Bognetti, 'S. Maria foris Portas di Castelseprio e la storia religiosa dei Longobardi', in Id., *L'età longobarda*, II (Milan 1967); s. also Ottorino Bertolini, 'Le Chiese longobarde dopo la conversione al cattolicesimo e i loro rapporti con il papato', in *Le Chiese nei regni dell'Europa occidentale e i loro rapporti con Roma sino all'800*, I, Settimane di studio del CISAM, 7 (Spoleto 1960), pp. 491–492 for the thesis on the 'drama' of the Lombards forced to fight the papacy.

2 Walter Pohl, 'Deliberate Ambiguity: The Lombards and Christianity', in *Christianizing People and Converting Individuals*, eds. G. Armstrong – I.N. Wood (Leeds 2002), pp. 47–58. On the problem of Lombard Arianism, to be strongly re-dimensioned: Steven C. Fanning, 'Lombard Arianism Reconsidered', in *Speculum* 6 (1981), pp. 241–258.

3 Guy Halsall, *Barbarian Migrations and the Roman West, 376–568* (Cambridge 2007).

 | DOI:10.1163/9789004698017_014

history of the kingdom. While these conclusions are now widely accepted in international historiography, some aspects of Lombard history are more problematic. They have been raised, in particular, by Italian historians, who have, until recently, and on occasion even now, proposed interpretations which, if accepted, would contribute to making the Lombard kingdom an almost unique case among post-Roman political formations.[4]

The military question revolves around three key areas, all of which are interlinked: the nature of Lombard settlement in the Italian territory; the characteristics of public military obligations; and the very identity of the Lombards, starting from their military characterization. I would like to begin with the caveat that I do not intend here to dwell on the early period of the invasion, touched upon above, and whose interpretation is influenced by the famous passages by Paul the Deacon referring to the Lombard *hospites*, the *tertia*, and the native *populi adgravati*, which have been commented upon on countless occasions by numerous scholars, including myself.[5] I do not think it necessary therefore to explore this aspect any further. When the Lombards occupied the part of Italy later forming their kingdom, they would have taken into account the need for the military control of the territory resulting from the hostilities with the Byzantine forces and from their relations with the local people, which were initially far from easy. This statement, which is so obvious that it does not require any additional explanation, stems from the fact that the Lombard entry into Italy was a genuine invasion, regardless of whether we accept that there was an invitation to them from Narses to do so or not.[6] What interests me is trying to establish to what extent the whole military issue affected the interpretation of the Lombard kingdom in its mature phase, that is, in the final century of its life as an independent kingdom prior to the Frankish conquest in 774.

---

4 For an overview of the issue, Stefano Gasparri, 'La questione degli arimanni', in *Bullettino dell'Istituto Storico Italiano per il Medio Evo e Archivio Muratoriano*, 87 (1978), pp. 121–153.

5 Just two examples: Walter Pohl, '«Per hospites divisi». Wirtschaftliche Grundlagen der Langobardischen Ansiedlung in Italien', in *Römische Historische Mitteilungen*, 43 (2001), pp. 179–226; Stefano Gasparri, 'Le basi economiche del potere pubblico in età longobarda', in *Taxation and Rent. Fiscal problems from Late Antiquity to Early Middle Ages* (*Entre el impuesto y la renta. Problemas de la fiscalidad tardoantigua y altomedieval*), eds. P.C. Díaz and I. Martín Viso (Bari 2011), pp. 71–85.

6 Neil Christie, 'Invasion or Invitation? The Lombard Occupation of Northern Italy', in *Romanobarbarica*, 11 (1991), pp. 79–108; Walter Pohl, 'Premesse e conseguenze della formazione del regno longobardo in Italia', in Id., *Le origini etniche dell'Europa. Barbari e romani tra antichità e medioevo* (Roma 2000), pp. 149–165; Jorg Jarnut, 'Langobarden zwischen Pannonien und Italien', in *Slowenien und die Nachbarländer zwischen Antike und karolingischer Epoche*, ed. R. Bratož (Ljubljana 2000), pp. 73–79.

First, we need to establish how long the initial military requirements would have lasted. This is where one of the most celebrated theories in the historiography on the Lombards comes into play, concerning the *arimanniae*. This hypothesis first emerged over a century ago, fostered by Pier Silverio Leicht, Aldo Checchini and Fedor Schneider, and was later developed in a contradictory manner chiefly by Gian Piero Bognetti. It was enthusiastically embraced by Italian historians who made it one of the pillars of the interpretation of the Lombard age that dominated the discourse until the 1970s, if not even later.[7] Even after this date, the 'arimannic theory' continued to flourish in minor historiographical works. It was very much present at a local level, as well as among archaeologists less familiar with the more recent results of historical research, who had not advanced beyond Bognetti.[8] Most importantly, even in such cases where this theory has not been openly upheld, it has nonetheless given rise to lasting negative consequences. As far as the issue of the *fara* is concerned, I will consider it merely as a kind of appendix to the arimannic question. The relatively recent case of the *fara* postulated by archaeologists with reference to the cemetery found at Collegno, in Piedmont, is paradigmatic of an interpretation of the *fara* as a form of settlement with the function of contributing to the military control of the local area.[9]

The arimannic theory postulated the existence of a particular type of military settlements that were colonies of Lombard warriors, or *arimanni*, stationed in the fiscal lands by the king, concentrating on the boundaries but also in the heartlands, in all areas of strategic importance in terms of protection and defence. Today, various studies – from Giovanni Tabacco's work onwards – have shown that there is no documentary foundation for this theory, based as it was exclusively on a misinterpretation of the sources, and very late ones at that. Moreover, the word '*arimanniae*' is not mentioned in the sources from the Lombard period, and the *arimanni* are never mentioned in the narrative

7 See above, note 4, e Pier Silverio Leicht, 'Ricerche sull'arimannia', in *Atti dell'Accademia di Udine*, ser. III/9 (1902), pp 35–62 e Id., *Studi sulla proprietà fondiaria nel Medio Evo*, I–II (Verona-Padova 1903–1907); Aldo Checchini, 'I fondi militari romano-bizantini considerati in relazione con l'arimannia', in *Archivio giuridico F. Serafini*, 78 (1907), pp. 407–475; Fedor Schneider, *Die Entstehung von Burg und Landgemeinde in Italien* (Berlin 1924).

8 Again, for the sake of brevity, I will give just two examples: Andrea Magno, 'Il "limes" di Serravalle Pistoiese: considerazioni sull'invasione longobarda della Toscana settentrionale', in *Studi Medievali*, XXXIX/2 (1998), pp. 783–807, and Giovanni Stranieri, 'Un limes bizantino nel Salento? La frontiera bizantino-longobarda nella Puglia meridionale. Realtà e mito del 'limitone dei Greci'', in *Archeologia Medievale*, XXVII (2000), pp. 333–355.

9 Aldo A. Settia, 'Una 'fara' a Collegno?', in *Bullettino storico bibliografico subalpino*, 103 (2005), pp. 263–276.

sources, appearing only occasionally in public documents (laws and diplomas), and only once in private documents. Consequently, the *arimanni* and the arimannic lands, known to us mainly from the ninth century onwards, can provide no insight into the military strategies of the Lombard sovereigns: the Carolingian *arimanni* were merely freemen who had escaped from the expansion of the territorial lordship and who had remained associated with the public authority, with the *arimanniae* representing the memory of the lands formerly in their possession.[10]

The arimannic theory, on the other hand, claimed to reconstruct the military strategy of the Lombard kings through the study of toponomastics, which is now considered to be almost completely worthless as a historical source. In a country like Italy, where Lombard and Byzantine lands adjoined each other in numerous places, bisected by a mountainous ridge, incised by numerous rivers, and also traversed by major Roman roads, toponyms with 'arimannic' or 'germanic' origins (like *fara* or other similar names) can be frequently found in places easily classifiable as having military importance. This appeared to lend weight to the arimannic theory, which was the main piece of evidence for the existence of a Lombard settlement of a military nature, entirely separate from Roman settlements. But this so-called 'evidence' is completely misleading. Such false data has multiplied military frontiers within Italy, wherever Lombard and Byzantine territories met, in search of strategic motivations even where these were hardly plausible at all. Consequently, frontiers and borders have been identified everywhere.

From a historical viewpoint, the very concept of border is not neutral, and needs to be examined critically. As Lucien Febvre has demonstrated, the concept of a linear military frontier did not exist prior to the 19th century and was the product of nation states. It was subsequently applied by French and British colonialism to colonial possessions in Asia and Africa over a century ago. The same concept was also applied to the Roman Empire; as lucidly explained by Charles Whittaker, it was thought that the Romans had identified the Rhine and Danube as borders, for the same motives that guided the politics of the European states: because these great rivers were natural frontiers that were both linear and military. Moreover, the Roman *limes* was seen as a symbol of conquest and dominion, a defensive line holding back the indistinct, dangerous flood of barbarism. This perceived Roman ideology was consistent with

10 This is the thesis developed in his seminal book by Giovanni Tabacco, *I liberi del re nell'Italia carolingia e postcarolingia* (Spoleto 1966); on Tabacco's historiography, see Stefano Gasparri, 'Il popolo-esercito degli arimanni. Gli studi longobardi di G. Tabacco', in *Quaderni dell'Accademia delle scienze di Torino*, XIV (2006), pp. 21–36.

the ideology of European colonialism.[11] In reality though, it is very hard to prove that the border always represented a complete discontinuity between two worlds entirely alien to each other.

We owe a new idea of the frontier as an open space, potentially in movement, to the United States, which were literally shaped by the frontier: the reference is to Frederick Jackson Turner's late 19th-century studies into frontier society and to the research carried out by Walter Prescott Webb, who wrote in 1931 that a frontier is not 'a line to stop at, but an area inviting entrance'.[12] This concept of permeable frontier has been applied by historians and archaeologists, from the late 1980s onwards, to the borders of the late Roman empire, where – despite the *limes* theory – relations between Romans and barbarians were intense, bringing together two worlds that were not in the least clearly distinguished from each other. As a result, today the Roman limes is mostly viewed as a channel of communication between populations living on either side, and not as a frontier line.[13]

In order to tackle the issue of Lombard Italy, we need to be very aware of the different ways in which the concept of linear frontier in its various forms (limit of civilization, military barrier, natural geographical element) came into being, even though the area under discussion here is on a far smaller scale. I should also point out that criticizing the arimannic theory does not mean excluding the possibility of the existence of Lombard *castra* in areas considered hot spots in military terms: on the contrary, it would be logical to assume that there would be fortifications in the areas closest to the Byzantine lands.[14] However, this should not lead us to imagine that actual military frontiers existed between the Lombard and Byzantine lands. The only frontier in the classical sense that can be identified – with many provisos – is the Alpine frontier, with the many *clusae* blocking the routes descending from the mountain passes to the valley floor. These barriers are linked to the late Roman defensive system, the *tractus Italiae circa Alpes*, which existed from the 4th century onwards. This was still in place in the Gothic period and, in part, also in the Lombard period,

---

11 Lucien Febvre, *Il Reno. Storia, miti, realtà*, ed. P. Schlotter (Roma 1998) (1st ed. 1931); Charles R. Whittaker, *Les fròntieres de l'Empire romain* (Paris 1989).

12 Frederick J. Turner, *The Frontier in American History* (New York 1920) (Turner's thesis was presented for the first time in 1893); Walter Prescott Webb, *The Great Plains* (Waltham, Massachusetts 1931).

13 See Walter Pohl, Ian Wood and Helmut Reimitz, eds., *The Transformation of Frontiers: from Late Antiquity to the Carolingians* (Leiden-Boston-Köln 2001), and Florin Curta, ed., *Borders, Barriers and Ethnogenesis. Frontiers in Late Antiquity and the Middle Ages*, (Turnhout 2005).

14 See for example Igor Santos Salazar, *Una terra contesa. Spazi, potere e società nell'Emilia orientale dei secoli VI–X* (Firenze 2011).

as shown by the many *castra* referred to by Paul the Deacon in his *Historia Langobardorum*.[15]

Nonetheless, the military role of the Alpine frontier needs to be reassessed, given that it never truly held invaders back: between the end of the fifth century and the end of the eighth century, it let in the Goths of Theodoric, the Lombards of Alboin, the Franks of Childepert, the Avars, and the Frankish armies of Pippin and Charlemagne. The overall impression is that these defence structures were rather modest. Moreover, the limited economic means of the period would have made it impossible for the public authorities to fund the upkeep of imposing fortifications.

Leaving aside the Alpine frontier, we should underline that the suggestions occasionally made in the historiography regarding fortified frontiers between the Lombards and Byzantines – whether in the Venetian plain, the Byzantine Liguria, or in central and southern Italy – are all based on completely improbable data, both archaeologically speaking and in terms of written sources.[16] An actual military frontier-line never existed between Lombard and Byzantine Italy. And no one has ever found evidence that the Lombard public power had created lands – *arimanniae* – in border areas or areas of military interest, stationing large groups of soldiers there for the purpose of defence.

In general, we need to reject the idea that the Lombard settlement had a military character throughout its history. Even more so do we need to reject the idea, connected to the first, that there was a Lombard settlement separate from the 'Roman' settlement (a distinction of which there is no trace in either the written or the archaeological sources). As well as drawing upon Germanic type toponymy (in search of words like *fara, sala, arimanniae*)[17] and searching for the anti-Arian sanctoral dedications so dear to Bognetti – and which he believed to be at the centre of Lombard settlements[18] – these views were also supported by more complex concepts like those expressed in the 1970s by

15 On the Alpine border, Aldo A. Settia, 'Le fortificazioni dei Goti in Italia', in *Teoderico il Grande e i Goti d'Italia Atti del XIII Congresso internazionale di studi sull'Alto Medioevo* (Spoleto 1993), pp. 101–112.

16 See above, note 8; other examples are Gina Fasoli, 'Tracce di insediamenti longobardi nella zona pedemontana tra il Piave e l'Astico e nella pianura tra Vicenza, Treviso e Padova', Atti del I Congresso internazionale di studi sull'Alto Medioevo (Pavia 1952), pp. 303–315, and Stefano Del Lungo, 'La frontiera settentrionale del ducato di Spoleto', in *I Longobardi dei ducati di Spoleto e Benevento*, Atti del XVI Congresso internazionale di studi sull'alto medioevo, II (Spoleto 2003), pp. 1325–1349.

17 This the title of the old book of Adriano Cavanna, *Fara, sala e arimannia nella storia di un vico longobardo* (Milano 1967).

18 The issue recurs in many of the author's works, all collected in Gian Piero Bognetti, *L'età longobarda*, 4 vols., Milano 1966–1967, in particular in volume 2, 'S. Maria foris Portas di Castelseprio e la storia religiosa dei Longobardi'.

Karl Schmid.[19] According to him, Lombard monastic foundations, established by the kings, but also by the aristocracy, had a military function, which was to protect recently conquered lands – in other words, frontier territories – ranging from the ancient case of Bobbio to the more recent one of Nonantola. Such ideas have gained considerable traction and are frequently repeated without further reflection; but these monasteries were not fortresses, they were rather the expression of a desire to emphasize a strong public presence in what was often a newly acquired zone. The latter is clearly a more nuanced concept than the one seeing Lombard rulers as soldier kings only capable of militarily occupying territories.

2 *Arimanni* and *Exercitales*, the True Lombards?

While the arimannic theory has now been definitively disparaged, the same is not true of the idea it engendered, that there was a standing army in the Lombard kingdom. This issue is re-examined in an important essay published recently, significantly entitled 'Ritorno ai Longobardi' (Return to the Lombards), which once again underlines the military connotation of the Lombard identity – an interpretation that can be shared partially, but not in reference to a relatively selected group.[20]

The author, Paolo Delogu, places himself in opposition, albeit politely, to 'decades of deconstructive historiography'. His intention is to prove that the 'true' Lombards were a distinct entity from the general population of freemen, even at the end of the kingdom's history, and that they were defined by warlike memories and traditions, nurtured by an enduring language and by specific legal traditions. Underpinning the entire work is the author's desire 'to correct integrationist conceptions imagining a kingdom by now [*in the* 8th *century*] inhabited only by a mixed population that came under the general heading of Lombard'.[21] In order to achieve these aims, he focuses on two key terms: *exercitalis* and *arimannus*. In short, having cleared the field from the *arimanniae*, we find the *arimanni* re-emerging.

19 Karl Schmid, 'Zur Ablösung der Langobardenherrschaft durch die Franken', in *Quellen und Forschungen aus italienischen Archiven und Bibliotheken*, 52 (1972), pp. 1–36.

20 Paolo Delogu, 'Ritorno ai Longobardi', in *Desiderio. Il progetto politico dell'ultimo re longobardo*, Atti del I Congresso internazionale di studio, ed. G. Archetti (Spoleto 2015), pp. 19–50.

21 Delogu, 'Ritorno ai Longobardi', p. 28.

Actually, Delogu is more concerned with the term *exercitalis*, which appears frequently in the sources, and which he considers to be technical in nature. He claims that this term points to a class of freemen who were obliged to provide continuous military service by joining the royal army, by serving in the armed retinues of dukes and gastalds carrying out local policing functions, and by attending royal assemblies. He argues that in the eighth century, *exercitalis* was the equivalent of *arimannus*, even though the latter never acquired the technical character of the former term and continued to refer generically to freemen under Lombard law. At the same time, the term *arimannus* seemed to be used by kings as a kind of 'ideal representation', indicating a type of freeman who was closer to the public authority, participating more extensively in the kingdom's political and military activities, and identifying more strongly with a specifically Lombard ethnic and military tradition.[22] As is always the case when seeking to identify precise, discrete groups within the population of freemen, there is a tendency to bring about some contradictions: such is the case when Jorg Jarnut linked the appearance of the term *arimannus* in the eighth century to the extension of the status of freeman to the ancient Roman population (which he claimed was still partly separate from the Lombards), suggesting that a Germanic term was introduced ex novo in order to define a population that was mostly Roman![23] Going back to the case we have just examined, if *arimannus* represented the ideal type of Lombard with all of the traditional associations, it is hard to understand why this word would have been used to refer generically to the freemen of the kingdom.

The essay points out unequivocally to the fact that the political participation of freemen within the kingdom rested on military foundations – something with which we can only agree. What I find less convincing are the conclusions that the author draws from this. Delogu believes that there was a segment within the free population, defined in law as *exercitales/arimanni*, or sometimes simply as 'Lombards', that was characterized by a more intense degree of political participation than other freemen, expressed through their permanent military functions. They were the 'true Lombards' who, despite not being biological descendants of the ancient invaders, fully embraced the Lombard identity.[24] The existence of men engaged in permanent military service is not confirmed by any of the Italian sources, unlike those from the Frankish world, where groups performing military activities in the course of the eighth

22 Delogu, 'Ritorno ai Longobardi', pp. 35–36.

23 Jorg Jarnut, 'Beobachtungen zu den langobardischen Arimanni und Exercitales', in *Zeitschrift für Rechtsgeschichte, Germ. Abt.*, 88 (1971), pp. 1–28.

24 Delogu, 'Ritorno ai Longobardi', p. 38.

and ninth centuries tended gradually to emerge as vassals. The figure of the vassal is absent in Lombard Italy and cannot in any way be compared to that of *gasind*, given the absence of any reference to a military function of the latter.[25] It is true that not all freemen are defined as *exercitales*, as shown by the long list of lay witnesses called upon to testify during the inquest ordered by Liutprand in 715 to investigate the heated dispute between the dioceses of Arezzo and Siena.[26] All of this clearly implies that the subjects defined only as *liberi homines* did not participate in the call-up, but this could depend upon a variety of factors such as their physical condition, poverty (making it impossible for them to arm themselves), or age. The qualification of *exercitales senes* does not contradict this because it may be explained by their particularly long service, involving various expeditions in the retinue of the king or of a duke, that would have led to the military definition noted by the writer of the documents. In fact, the classifications contained in such documents rely extensively on the capacity of the individual writers to reflect local situations rather than on ancient notarial conventions. This explains why it is impossible to solve the enigma represented by the definition of so many witnesses as *viri devoti* and why we certainly cannot do so by imagining – as Pier Maria Conti did so long ago – that they were men who took a particular type of oath to the king – of a clearly military nature, if we are to believe the author.[27]

As far as our *exercitales/arimanni* are concerned, it seems hard to argue that they swore a particular type of oath. At most, they would have taken the same oaths that all freemen did. The only rather hermetic evidence of the existence of such an oath is in Liutprand's *Notitia de actoribus regis*, which states that anyone daring to usurp public land was also guilty of perjury by virtue of having sworn an oath of loyalty to the king. By cross-referencing various articles of law, we can see that the *arimanni* mentioned in the *Notitia* are nothing more than freemen of the kingdom, in a general sense.[28]

3 The Structure of the Lombard Army

Establishing who would effectively have been mobilized by the public army is no simple matter. The starting point for every investigation of this kind must

25 Stefano Gasparri, 'Strutture militari e legami di dipendenza in Italia in età longobarda e carolingia', in *Rivista storica italiana*, 98 (1986), pp. 664–726.

26 Luigi Schiaparelli, ed., *Codice diplomatico longobardo*, I, Fonti per la storia d'Italia 62 (Roma 1929), n. 19, pp. 61–78.

27 Pier Maria Conti, *Devotio e viri devoti in Italia da Diocleziano ai Carolingi* (Padova 1971).

28 *Notitia de actoribus regis*, 5, in Claudio Azzara and Stefano Gasparri, eds., *Le leggi dei Longobardi. Storia, memoria e diritto di un popolo germanico* (Roma 2005), pp.254–255.

necessarily be the military laws 2 and 3 enacted by Aistulf.[29] These laws clearly limit military mobilization by the king to those – landowners or merchants – who could afford armour, from the level of foot archers to knights with coats of mail or cuirasses. The divisions ranged from light infantry to heavy cavalry (to the extent that armour could be heavy at that time). Aistulf 2 also goes on to establish that the wealthiest men – that is, those owning more than the seven manses (*casas massarias*) corresponding to the requirement to procure the heaviest armament – should provide additional weapons and horses, again in proportion to their wealth. Who were the recipients of these additional supplies? It is clear that they were intended for those who lacked the wealth to procure a complete armour but were physically best equipped to perform military functions on the battlefield. The fact that this supply of weapons seems to be linked to the order for mobilization, and does not appear to imply a permanent state, makes it impossible to associate it with the existence of a group of professional warriors. On the contrary, it offered an opportunity for social advancement through the use of arms to men in good physical shape. We should also bear in mind Liutprand 62, which establishes that the lowest compensation for the killing of a freeman in self-defence is to be paid for a 'minima persona, qui exercitalis esse invenitur': clearly the full degree of freedom began at this point, which is why the wergeld for those who were not considered freemen-*exercitales* – for the reasons mentioned above – was lower, bringing them closer to the world of those who were not free.[30]

At the opposite extreme, Aistulf 7, enacted in 750 like the previous titles, established heavy penalties for royal officials sending men defined as *homines potentes* back home from the army (*de exercitu*): thus implying that taking part in military actions was not always perceived as a socially prestigious affair, even within the elites of the kingdom. This can be explained if we consider the fact that members of the Lombard aristocracy would have viewed their military identity as self-evident, while participation in single military expeditions could have potentially interfered with their social and political activity.[31] However, this was a new phenomenon, as shown by the fact that, while Liutprand 83 mentions unlawful exemption for the first time, this refers to people of modest means who were exempted by royal officials in order to obtain their services privately.[32] The fact that Aistulf 3 equates landowners to merchants with regard to mobilization by the king is not only proof of the complexity of society in the Lombard kingdom in the mid-eighth century, but also underlines

29 *Ahist.* 2–3, in Azzara-Gasparri, eds., *Le leggi dei Longobardi*, pp. 280–281.

30 *Liut.* 62, in Azzara-Gasparri, eds., *Le leggi dei Longobardi*, pp. 174–175 (anno 724).

31 Ahist. 7, in Azzara-Gasparri, eds., *Le leggi dei Longobardi*, pp. 282–283.

32 *Liut.* 83, in Azzara-Gasparri, eds., *Le leggi dei Longobardi*, pp. 186–187 (anno 726).

the absence of any mention of particular groups of permanent warriors in contemporary laws.[33] The suggestion that these titles of Aistulf refer to the mobilized population in general, but do not extend to the entire military system because there was a group (the *arimanni/exercitales*) performing this activity on a permanent basis, is without any basis in the sources. Naturally, this does not mean that there were no armed retinues belonging to the king or to members of the higher echelons of the aristocracy. Yet again, unlike in the Frankish kingdom, such groups (which were certainly not 'more Lombard' than others) could not have been very numerous given that there is virtually no trace of them in the sources. It is possible that this accounts for the greater weakness of the Lombard aristocracy compared to the Frankish nobility.[34]

Ratchis 4, which describes the military equipment (horse and spear) of the *arimannus* summoned to participate in the *caballicatio* with his *iudex*, is not sufficient to fill this documentary void. In this case too, the text of the law deals with the public duties of freemen and does not focus on a specific group. The testament of a certain Teuderacius, an important landowner with estates in Sabina in the period of Desiderius and Adelchis, who left his property to his wife and daughter, given that 'in ista via dirigimus, in Traspadum de dominatione dominorum nostrorum viam agendo vel facendo', probably shows us one of these freemen-*exercitales–arimanni* who had been summoned to participate in a public mission, possibly together with his duke, rather than a professional warrior in permanent service.[35] Even the *arimannus* riding off together with his *iudex* to carry out policing operations was just a normal freeman – certainly not a *minima persona* – being summoned for this purpose.

Any attempt to identify groups that were 'more Lombard' than others will continue to encounter new contradictions. Can the two men drawing up documents (that look more like testaments) before leaving with the army be considered 'more Lombard'? The first of these two celebrated documents was drawn up by Gaiprandus of Lucca in 755, 'because I have been summoned to march against the Franks', while the second one was drawn up in 769 by the Pisan Domnolinus, who, also called up to the army and 'given that we know not God's judgement', let Austriconda, *dulcissima sorore germana mea*, still living in his house, dispose of her property. Was the Lucchese bishop Walprandus,

33 See n. 29 above.

34 Chris Wickham, 'Aristocratic Power in Eight-Century Lombard Italy', in *After Rome's Fall. Narrators and Sources of Early Medieval History. Essays presented to Walter Goffart*, ed. Alexander Callander Murray (Toronto 1998), pp. 153–170.

35 Ignazio Giorgi and Ugo Balzani, eds., *Il Regesto di Farfa di Gregorio da Catino*, II (Roma 1883), n. 75 (768), pp. 72–73. *Ratch. 4*, in Azzara-Gasparri, eds., *Le leggi dei Longobardi*, pp. 262–263 (anno 745).

son of Duke Walpertus, who made his will before setting off to fight the Franks also a 'true Lombard', to use Delogu's expression again?[36]

As Chris Wickham explained so well, the Lombard warriors were no match for the Franks. Despite the relative frequency of military operations from the time of Liutprand onwards, the Lombard army was modest in military terms, as revealed by numerous episodes, including their defeat, almost without a fight, by the Frankish army of Pippin and Charlemagne, which is only the most recent and striking example.[37]

All the individual cases mentioned above highlight the military obligations of the entire class of freemen, which were performed with a greater or lesser intensity according to the various situations described, but which theoretically applied to all freemen under Lombard law and not just to a specific group in permanent service. The freemen identifying with the traditions of their lineage – many of which were invented or constructed – 'were' Lombard by virtue of this identification and would certainly have 'felt' Lombard. We can accept that these traditions would have been stronger among the aristocratic groups, which probably emphasized them for reasons linked to social prestige, as in the celebrated case of Paul the Deacon's ancestors or that of Rothari's genealogy in the prologue to the laws. This would have been particularly so in the case of the Friulans, who were the kingdom's most warlike aristocracy.[38] Potentially, though, all freemen, and not just a particular group, would have shared these traditions, albeit with different degrees of intensity, but not enough to distinguish between the groups. Going back to Liutprand's inquests of 715, it is significant that when the royal notary Gunteram refers to the witnesses in an abbreviated form, he defines them all as *arimanni*: this after having meticulously described who could be defined as *exercitalis* and who could simply be referred to as a freeman. Similarly, the Sienese *exercitus* mentioned in a document of 730 together with the *sacerdotio* of that city, merely indicated the entire lay population of freemen, like Gunteram's *arimanni*.[39]

Two examples from the ninth century illustrate just how enduring the military tradition of the Lombard kingdom was. The first relates to the monastery of San Bartolomeo, Pistoia, whose abbot was able to avoid military service in

---

36 Stefano Gasparri, *Desiderio* (Roma 2019), pp. 155–156.

37 Wickham, 'Aristocratic Power'.

38 Paulus Diaconus, *Historia Langobardorum*, IV, 37, pp. 131–132; *Roth.* prol., in Azzara-Gasparri, eds., *Le leggi dei Longobardi*, pp. 14–17; on the Friulan arustocracy, Gasparri, *Desiderio*, pp. 161–162, and Id., 'Cividale longobarda', in *Storia di Cividale nel Medioevo. Economia, società, istituzioni*, ed. Bruno Figliuolo (Cividale del Friuli, 2012), pp. 72–76.

39 For Gunteram, s. above nt. 26; the charter of 730 has been published by Schiaparelli, *Codice diplomatico longobardo*, I, n. 50, pp. 164–171.

812 because the heirs of the founder – Gaidoaldus, a physician at the court of Liutprand – on the basis of their ancient land ownership, could perform this service, which had not been cancelled by the foundation of the monastery on the family's estate.[40] Some years later, in 832, at the end of a lengthy legal dispute with the bishop of Piacenza, the descendents of a certain Hermefrit, who lived in the final decades of the eighth century, recognized that they were subject to episcopal law, but at the same time they saw their freedom acknowledged, or at least it was recognized the fact that they were *liberi homines arimanni exercitales*.[41] This definition could not be clearer: the distinctive mark of freedom, even in Carolingian Italy, in the face of a threatening expansion of the aristocracy, was the public function of participation in the royal army. Hermefrit's heirs, like their ancestors from the Lombard period, were free because of their status as *arimanni* or *exercitales*. Clearly, no distinction can be made between these two terms.[42]

This is the true picture of the Lombard army, an army without permanent militias cultivating the traditions of their lineage in a specific manner. And its typical structure based on the mobilization of the class of freemen – landowners and merchants – proved capable of surviving the Frankish conquest. It owed its stability not so much to its efficiency in war – which, as we have seen, was rather modest – but to the fact that military service in the royal army was the backbone of a social organization whose point of reference lay in public power. Through this service, it was able to maintain the link, indispensable for both parties, between king and freemen: Lombards, *arimanni*, *exercitales*, however we wish to define them according to the circumstances, the normative sources and authors involved. They were not so much formidable warriors devoted to war, as the cornerstone of Lombard society, as it was constructed during the course of the seventh century, and carried on towards the Carolingian age.[43]

40 Carlo Manaresi, ed., *I Placiti del Regnum Italiae*, I, Fonti per la storia d'Italia, 92 (Roma 1955), n. 25, pp. 77–80.

41 Paola Galetti, *Le carte private della cattedrale di Piacenza*, I (748–848), (Parma 1978), n. 27, pp. 75–78.

42 To conclude, I just want to mention here one of the most famous essays asserting the difference between *arimanni* and *exercitales*: Ottorino Bertolini, 'Ordinamenti militari e strutture sociali dei Longobardi in Italia', in *Ordinamenti militari in Occidente nell'Alto Medio Evo*, Settimane del CISAM XV, 1 (Spoleto 1968), pp. 429–580.

43 Stefano Gasparri, ''Nobiles et credentes omines liberi arimanni'. Linguaggio, memoria sociale e tradizioni longobarde nel regno italico', in *Bullettino dell'Istituto Storico Italiano per il Medio Evo e Archivio Muratoriano*, 105 (2003), pp. 25–51.

Bibliography

Primary Sources

Ahistulfi leges, eds. C. Azzara – S. Gasparri, *Le leggi dei Longobardi*, pp. 279–295
Codice diplomatico longobardo, ed. L. Schiaparelli, vol. 1 (Fonti per la storia d'Italia, 62) (Roma 1929)
Edictum Rothari, eds. C. Azzara – S. Gasparri, *Le leggi dei Longobardi*, pp. 13–127
I Placiti del Regnum Italiae, ed. C. Manaresi, vol. 1 (Fonti per la storia d'Italia 92) (Roma 1955)
Il Regesto di Farfa di Gregorio da Catino, eds. I. Giorgi and U. Balzani, 3 vols. (Roma 1879–1914)
Liutprandi leges, eds. C. Azzara – S. Gasparri, *Le leggi dei Longobardi*, pp. 137–244
Notitia de actoribus regis, eds. C. Azzara – S. Gasparri, *Le leggi dei Longobardi*, pp. 251–258
Paulus Diaconus, *Historia Langobardorum*, ed. L. Bethmann – G. Waitz, MGH, *Script. rerum Lang.* (Hannoverae, 1878), pp. 12–187

Secondary Literature

Azzara Claudio and Stefano Gasparri, eds., *Le leggi dei Longobardi* (Roma 2005)
Bertolini, Ottorino, 'Le Chiese longobarde dopo la conversione al cattolicesimo e i loro rapporti con il papato', in *Le Chiese nei regni dell'Europa occidentale e i loro rapporti con Roma sino all'800*, I, Settimane di studio del CISAM, 7 (Spoleto 1960), pp. 429–580
Bertolini, Ottorino, 'Ordinamenti militari e strutture sociali dei Longobardi in Italia', in *Ordinamenti militari in Occidente nell'Alto Medio Evo*, Settimane del CISAM XV, I (Spoleto 1968), pp. 429–580
Bognetti, Gian Piero, 'S. Maria foris Portas di Castelseprio e la storia religiosa dei Longobardi', in Id., *L'età longobarda*, vol. 2 (Milan 1967)
Bognetti, Gian Piero, *L'età longobarda*, 4 vols. (Milan 1966–1967)
Cavanna, Adriano, *Fara, sala e arimannia nella storia di un vico longobardo* (Milan 1967)
Checchini, Aldo, 'I fondi militari romano-bizantini considerati in relazione con l'arimannia', in *Archivio giuridico F. Serafini*, 78 (1907), pp. 407–475
Christie, Neil, 'Invasion or Invitation? The Lombard Occupation of Northern Italy', in *Romanobarbarica*, 11 (1991), pp. 79–108
Conti, Pier Maria, *Devotio e viri devoti in Italia da Diocleziano ai Carolingi* (Padova 1971)
Curta, Florin, ed., *Borders, Barriers and Ethnogenesis. Frontiers in Late Antiquity and the Middle Ages*, (Turnhout 2005)
Del Lungo, Stefano, 'La frontiera settentrionale del ducato di Spoleto', in *I Longobardi dei ducati di Spoleto e Benevento*, Atti del XVI Congresso internazionale di studi sull'alto medioevo, II (Spoleto 2003), pp. 1325–1349

Delogu, Paolo, 'Ritorno ai Longobardi', in *Desiderio. Il progetto politico dell'ultimo re longobardo*, Atti del I Congresso internazionale di studio, ed. G. Archetti (Spoleto 2015), pp. 19–50

Fanning, Steven C., 'Lombard Arianism Reconsidered', in *Speculum*, 6 (1981), pp. 241–258

Fasoli, Gina, 'Tracce di insediamenti longobardi nella zona pedemontana tra il Piave e l'Astico e nella pianura tra Vicenza, Treviso e Padova', Atti del I Congresso internazionale di studi sull'Alto Medioevo (Pavia 1952), pp. 303–315

Febvre, Lucien, *Il Reno. Storia, miti, realtà*, ed. P. Schlotter (Roma 1998) (1st ed. 1931)

Galetti, Paola, *Le carte private della cattedrale di Piacenza*, I (748–848) (Parma 1978)

Gasparri, Stefano, '*La questione degli arimanni*', in *Bullettino dell'Istituto Storico Italiano per il Medio Evo e Archivio Muratoriano*, 87 (1978), pp. 121–153

Gasparri, Stefano, 'Strutture militari e legami di dipendenza in Italia in età longobarda e carolingia', in *Rivista storica italiana*, 98 (1986), pp. 664–726

Gasparri, Stefano, 'Le basi economiche del potere pubblico in età longobarda', in *Taxation and Rent. Fiscal problems from Late Antiquity to Early Middle Ages* (*Entre el impuesto y la renta. Problemas de la fiscalidad tardoantigua y altomedieval*), eds. Pablo C. Díaz & Iñaki Martín Viso (Bari 2011), pp. 71–85

Gasparri, Stefano, 'Cividale longobarda', in *Storia di Cividale nel Medioevo. Economia, società, istituzioni*, ed. B. Figliuolo (Cividale del Friuli 2012), pp. 72–76

Gasparri, Stefano, 'Il popolo-esercito degli arimanni. Gli studi longobardi di G. Tabacco', in *Quaderni dell'Accademia delle scienze di Torino*, XIV (2006), pp. 21–36

Gasparri, Stefano, *Desiderio* (Roma 2019)

Halsall, Guy, *Barbarian Migrations and the Roman West, 376–568* (Cambridge 2007)

Jarnut, Jorg, 'Beobachtungen zu den langobardischen Arimanni und Exercitales', in *Zeitschrift für Rechtsgeschichte, Germ. Abt.*, 88 (1971), pp. 1–28

Jarnut, Jorg, 'Langobarden zwischen Pannonien und Italien', in *Slowenien und die Nachbarländer zwischen Antike und karolingischer Epoche*, ed. R. Bratož (Ljubljana 2000), pp. 73–79

Leicht, Pier Silverio, 'Ricerche sull'arimannia', in *Atti dell'Accademia di Udine*, ser. III/9 (1902), pp 35–62

Leicht, Pier Silverio, *Studi sulla proprietà fondiaria nel Medio Evo*, 2 vols. (Verona-Padova 1903–1907)

Magno, Andrea, 'Il "limes" di Serravalle Pistoiese: considerazioni sull'invasione longobarda della Toscana settentrionale', in *Studi Medievali*, XXXIX/2 (1998), pp. 783–807

Pohl, Walter, 'Premesse e conseguenze della formazione del regno longobardo in Italia', in Id., *Le origini etniche dell'Europa. Barbari e romani tra antichità e medioevo* (Roma 2000), pp. 149–165

Pohl, Walter, '«Per hospites divisi». Wirtschaftliche Grundlagen der Langobardischen Ansiedlung in Italien', in *Römische Historische Mitteilungen*, 43 (2001), pp. 179–226

Pohl, Walter, 'Deliberate Ambiguity: The Lombards and Christianity', in *Christianizing People and Converting Individuals*, eds. G. Armstrong – I.N. Wood (Leeds 2002), pp. 47–58

Pohl Walter, Ian Wood and Helmut Reimitz, eds, *The Transformation of Frontiers: from Late Antiquity to the Carolingians* (Leiden-Boston-Köln 2001)

Prescott Webb, Walter, *The Great Plains* (Waltham, Massachusetts 1931)

Santos Salazar, Igor, *Una terra contesa. Spazi, potere e società nell'Emilia orientale dei secoli* VI–X (Firenze 2011)

Schmid, Karl. 'Zur Ablösung der Langobardenherrschaft durch die Franken', in *Quellen und Forschungen aus italienischen Archiven und Bibliotheken*, 52 (1972), pp. 1–36

Schneider, Fedor, *Die Entstehung von Burg und Landgemeinde in Italien* (Berlin 1924)

Settia, Aldo A., 'Le fortificazioni dei Goti in Italia, in Teoderico il Grande e i Goti d'Italia', Atti del XIII Congresso internazionale di studi sull'Alto Medioevo (Spoleto 1993), pp. 101–112

Settia, Aldo A., 'Una 'fara' a Collegno?', in *Bullettino storico bibliografico subalpino*, 103 (2005), pp. 263–276

Stranieri, Giovanni, 'Un limes bizantino nel Salento? La frontiera bizantino-longobarda nella Puglia meridionale. Realtà e mito del 'limitone dei Greci'', in *Archeologia Medievale* XXVII (2000), pp. 333–355

Tabacco, Giovanni, *I liberi del re nell'Italia carolingia e postcarolingia* (Spoleto 1966)

Turner, Frederick Jackson. *The Frontier in American History* (New York 1920)

Wickham, Chris, 'Aristocratic Power in Eight-Century Lombard Italy', in *After Rome's Fall. Narrators and Sources of Early Medieval History. Essays presented to Walter Goffart*, ed. A.C. Murray (Toronto 1998), pp. 153–170

Whittaker, Charles R., *Les fròntieres de l'Empire romain* (Paris 1989)

PART 4

Geometries of the Power and Military Justice

CHAPTER 14

Laeti and *Gentiles*: Military Germanic Settlements in Roman Gaul

Jean-Pierre Poly

ABBREVIATIONS: AE = Sitzmann – Grünzweig, *Die altgermanische Ethnonymie*; ASD = Bosworth – Northcote Toller, *An Anglo-Saxon Dictionary*; CTh = cf. *Codex Theodosianus*; ND = *Notitia Dignitatum*; RGA = *Reallexikon der germanischen Altertumskunde*, eds. H. Jankuhn – H. Beck (Berlin 1973–2008); TWB = Gysseling, *Toponymisch Woordenboek*.

1 Introduction

After the crisis of the third century, the *res publica romana* needed workers for its *agri deserti* and recruits for its army. The Germanic barbarians had stormed the northern imperial borders. They stayed on as slaves or soldiers. Thus began a 'Transformation of the Roman world', a change in the society of orders or classes in western Europe and the beginning of the transition from an empire-state to the future nation-states. In this movement the army played a major part, as was emphasized by Fabio Botta and Luca Loschiavo in their introduction to the congress of Cagliari.[1]

Amongst the questions then discussed was the status of the immigrants settled in the Empire: were they foreigners (*peregrini*) or *de facto* citizens, bondsmen or free men, peasants or soldiers, 'barbarians' or (almost) Romans? The old answers went not without some ideological bias. In Cagliari, Valerio Marotta summarized the studies about the 'imperial barbarians'.[2] We shall try to add some observations about their military settlements in Gaul.

1 Fabio Botta and Luca Loschiavo, *Presentazione*, in *Civitas, Iura, Arma; organizzazioni militari, istituzioni giuridiche e strutture sociali alle origini dell'Europa* (*secc. III–VIII*), (Lecce 2015), p. 5. Jean-Pierre Poly, '*Laeti et gentiles.* Les établissements militaires de la Gaule romaine', in *Ravenna Capitale. L'esercito romano e l'alba dell'Europa. Modelli concettuali e sperimentazioni sociali tra linguaggi, istituzioni e diritto* (*secoli IV–VIII*), ed. G. Bassanelli Sommariva (Santarcangelo di Romagna 2020), p. 15.

2 Alessandro Barbero, *Barbari. Immigrati, profughi, deportati nell'impero romano* (Roma-Bari 2006), p. 177, and *infra* fn. 24. Alain Chauvot, 'Les lètes et leur devenir', in *Les 'barbares' des Romains. Représentations et confrontations*, eds. A. Becker - C. Freu, - O. Huck - H. Huntzinger

 | DOI:10.1163/9789004698017_015

The barbarians' settlements are linked to the recruitment of the military units of foreigners, the *auxilia palatina*.[3] Some historians think that such units had no more to do with the nations whose names they bore, since these nations no longer existed beyond the Rhine. Their recruits would have been common *tirones*. But the existence of military settlements in Gaul allowed the maintenance of ethnic-named units.

True, brevet officers from barbarian origin were generally given Roman citizenship, thus joining 'the wide horizon of imperial policy'.[4] But at the lower level of soldiery, the heredity of the status and the interdiction of marriage with provincial women maintained the cohesion of gentile communities.[5] We may consider these rank-and-file *peregrini* from two different points of view: the Romans', from annals, panegyrics and administrative documents such as the rosters in the *Notitia Dignitatum*,[6] or the barbarians', from the names they used and the emblems they bore.

The ethnonyms transmitted by Latin or Greek texts have been studied by the 19th century philology, condensed in the old RGA and later conveniently modernized by the current edition.[7] Notwithstanding this effort of *aggiornamento*,

---

(Metz 2016), p. 259 (for Seeck, *gentiles* is apposition to *laeti*, but CTh 7.20.12, mal traduit, ND Oc 42.34, 51–70, *Sarmatae gentiles*). Luca Loschiavo, 'Insediamenti barbarici e modelli di coesistenza nell'Italia altomedievale', in *Immigrazione e integrazione*, ed. F. Rimoli, p. 317; Valerio Marotta, 'Il problema dei *laeti*. Fonti e storiografia', in *Civitas, Iura, Arma; organizzazioni militari, istituzioni giuridiche e strutture sociali alle origini dell'Europa* (*secc.* III–VIII), eds. F. Botta - L. Loschiavo (Lecce 2015), p. 117; Jean-Pierre Poly, 'Les Saxons du Bessin et leurs compagnons. Ethnies et acculturation en Gaule du Nord au début du Moyen Age', in *La Normandie, terre de traditions juridiques*, eds. G. Davy and Y. Mausen (Rouen 2016), p. 25 (*laeti batavi, gentiles suevi*); Alain Chauvot, 'Julien, Ammien et les lètes. A propos d'Ammien, XX, 8, 13' and Id., 'La situation juridique des barbares dans l'Empire tardif et la question des unions mixtes', both in *Les 'barbares' des Romains. Représentations et confrontations*, eds. A. Becker, C. Freu, O. Huck, H. Huntzinger (Metz 2016), here respectively at p. 285 and 306 (where he criticizes Mathisen and Moderan).

3 Soazick Kerneis, '*Vox divi, vox populi*', in *Une histoire juridique de l'Occident. Le droit et la coutume* (IIIe–IXe *siècles*), S. Kerneis ed. (Paris 2018), p. 167. Chauvot, *La situation juridique*, p. 278 (importance du *ius gentium*). Ethnicity: Jean-Pierre Poly, '*Leges barbarorum*. La création des lois des nations" in *Une histoire juridique de l'Occident. Le droit et la coutume* (IIIe–IXe *siècles*), ed. S. Kerneis (Paris 2018), pp. 198, 201.

4 Walter Pohl, *Le origini etniche dell'Europa* (Roma 2008), pp. 22, 129.

5 CTh 3.14.1; Soazick Kerneis, '*Francus cives, miles romanus in armis*', in *Droit, religion et société dans le Code Théodosien*, eds. J.J. Aubert and P. Blanchard (Genève 2009), p. 377. Inheritance: CTh 7.20.12.

6 *Notitia Dignitatum* (ed.: Seeck; Neira Faleiro; ed. Maier, *The compilation*, index).

7 Cf. RGA, *his nominibus*, to which may be added Alexander Sitzmann and Friederich E. Grünzweig, *Die altgermanischen Ethnonyme. Ein Handbuch zu ihrer Etymologie* (Wien 2008), who note the composition with *-mann* and *-ware* but omit *-aeht*, *-baere*, *-har*, *-haw*(*an*), *-hun*(*n*)

most of the present historical studies inherited two *a priori* ideas: first, a barbarian group would bear but one name, which, by the way, was already a creed of the Roman administration; secondly, the names, supposedly coming from a faraway past, would be no more understood by their locutors.

As shown in the *Beowulf*, both statements are untrue. Reinhard Wenskus insisted that the *gentes* should adapt incessantly to the hazards of their existence and only succeeded in doing so by clinging to their *Traditionskerne* or, to use Claude Lévi-Strauss words, 'par le moyen d'une mythologie qui ... réajuste constamment sa propre grille pour qu'elle offre la moindre résistance au torrent des évènements'.[8] To this grid the ethnic names had to fit and, like the custom they reflected, they were everlasting, miscellaneous and pliant.

But the 19th century philology of Lachmann or Müllenhoff did not care about the dawning anthropology.[9] It privileged an etymological discourse which inserted the ethnic names in a rigid linguistic reconstitution of 'proto-germanic' and 'indo-european' languages. Vocabulary and meanings were built up like Zeuxis' woman, from diversely attested elements put together, a legitimizing past of Germanic unity.

We may thus understand their disregard for the contingencies having occurred prior to the fixing of the names in a customary Latin form: first of all the conveyance by go-between into the army whose unofficial language has always been hazardous, to say the least; then the unconscious associations, the puns and, most of all, this very soul of oral societies, the oblique speech known in Norse as *stórordhr*, with its *heiti* and its *kenningar*.[10] To the names of the different groups should be added the closely related colours and figures used by them.[11] Those iconographic elements, although mostly left aside by

as, *-hyse*, *-ter/tear*; Corinna Scheungraber and Fiederich Grünzweig, *Die Altgermanischen Toponyme sowie ungermanische Toponyme Germaniens. Ein Handbuch zu ihrer Etymologie* (Wien 2014), for connected toponymie. With the help of *proto-germanic* and *indo-european*, *Cherusci* should be 'warriors', *Sugambri* 'very active', *Suevi* 'soi/self', Marcel Meulder, '*Florus* II.30.24–25: trois peuples germaniques fonctionnels', in *Revue belge de philologie et d'histoire*, 85 (2007), p. 77.

8 Reinhard Wenskus, *Stammesbildung und Verfassung* (Köln 1961), p. 138. «By the mean of a mythology which constantly readjusts its own grid so that it puts up the less resistance to the stream of events»; Claude Levi-Strauss, *Mythologiques*, IV (Paris 1971), p. 546.

9 But already an anthropological perspective in Rudolf Much, *Völkernamen*.

10 Anthropological approach, Pohl, *Le origini etniche*, p. 17. Linguistic rules, Roland Breton, *Géographie des langues* (Paris 1995), pp. 7, 22. The *o.e.* dialects are the only languages of Low-German soon and widely written; forms in ASD.

11 These elements were depicted on the shields of the *auxilia* in the *Notitia Dignitatum*. The text is known from the renaissance copies of a medieval archetype in Speyer (Σ). The shields on the plates and the names in the lists are ill-assorted (ed. Seeck, p. XX):

philological studies, were a major expression in ancient societies.[12] Moreover, the condition of *laeti* and *gentiles* was most of the time considered as one general question, and detailed situations were rarely taken into account. But *Der Teufel steckt im Detail.*[13]

In this perspective, we shall first identify the Germanic settlements in Gaul (I) before trying to estimate their demographic weight (II).

2 Ethnic Districts and Auxiliary Units

After thirty years of Germanic wars, the imperial administration gave up the project of a Roman province between the Rhine and the Elbe. A fortified border was built, a wide glacis was defined. The army crushed the *nationes* living in this area and settled groups of survivors in special frontier districts. Let us observe the evolution of the status from the first settlers to the fourth century, before sketching a panorama of the barbarian settlements and trying a brief overview of their *Weltanschauung*.

2.1 *Dediticii, One Status, Three Stage*

The groups bound to army rule in the first-fourth centuries had the same legal status of *dediticii*, but their situation went through three stages we shall name ranger-, tiller- and roster-*dediticii*.

In Caesar's time lived on the Lower Rhine the warlike *Sugambri* (**SwogAm-Bere*), the Rumbling (like Thor 'the thunderer') bearers of (their maternal) Uncle (the bull Frey). Circa AD 9 Tiberius had them transported to the great island between the northern and the middle arms of the Rhine. In the following years, the name, says Tacitus, was 'almost annihilated', but not the people.

---

the shields-plates followed indeed an honorary order, durable unless sanctions, while the names-lists were the tactical order set for a campaign, grouping the units in divisions (*μοιραι*) of three pairs according to their real strength and abilities. In the original, the shields on the plates would have no underlines but insertion marks; unaware of it, Σ created underlines out of the lists. Following some simple rules (Jean-Pierre Poly, 'Emblèmes et couleurs du Nord. Les écus des auxiliaires francs de l'armée romaine au IVe siècle', in *Bulletin de la Société Nationale des Antiquaires de France* 2019, under publication) we coded the elements of each shield according to the best mss., BNF *lat. 9661* and *Bodleian Canon.lat.misc. 378* (the ms. *Bayerische Staatsbibliothek Clm 10291²* was but an unskillfull step to the theft of Σ; Maier, Cnd, App. 4, 20, 28). An Excel pivot-table sorts out the similars.

12 Michel Pastoureau, *Une histoire symbolique du Moyen Age occidental* (Paris 2004), pp. 13, 17.

13 'The devil is in the details'. Names were given as singular forms, like nowadays 'the German is musician' or 'the French is rational'.

In 16, Germanic contingents who help the army are called *Batavi* (**BatHaw*), Scouts of the Quarrel, referring to their function as *speculatores* and to the dissent which had split them from the other *Chatti*, devotees of the wolf *Hatti*, the Called One.

Near them, other refugees had been transported, their kin (*ex eadem origine*) the *Canninefates*, (**GanNeah NefaAehtas*), The Families of the Nephew Come Near, probable survivors of the *Marsi* (**MaerHyse*), Youth of Maer – also 'an ancient name' for the *Chatti* – slaughtered in 14.[14]

Farther upstream other *Chatti*, the *Mattiaci* of *Mattium* (Metze) who surrendered in 15, were settled north of the Taunus around the *Aquae Mattiacae*. Upper yet, in the *Decumates agri*, were transported some *Suebi*/*Suavi* (**Su*(*h*) *ebbe, Swebbe/Swefe*), Hoist the (Vanic) Sow (their ensign), understood ironically as the homophonic Lullers, that is the Slayers, in fact a northern part of them called *Chasuari* (**HaseWare*) Hold/Inhabit the (river) Hase. With them were two other groups, the *Tencteri*, the Expelled Split (of the *Chasuari*?) with their companions the *Usipi/Usipetes* (**UthSibAehtas*), The Families of the Outside Sippe, both together called the *Tubantes*, Two (clans)-Bond.[15]

All of these people inhabited tracts of land now belonging to the emperor, they had no more kings, they served in *auxilia*. The gift of imperial territory made them 'a part of the Empire' as says Tacitus, but that did not mean they were Roman citizens. All had made a *deditio*. When Gaius wrote his legal handbook in the 160ties, he stated that the main category of *dediticii* were *peregrini* defeated in battle, free men yet but with a peculiar status, the *pessima libertas* which bent them to the will of the *princeps*.[16] The Augustus could grant quasi-rights to them: e.g. the *Batavi* had theirs commanding officers chosen among their noblemen, the *Tencteri* had not; the same *Batavi* did not pay any tribute, the *Frisii* did. But all were, to quote Tacitus, 'tools set apart for use in war'.

During the second century, these six Germanic communities of *dediticii* lived in the eastern fringes of Gaul. Their lands were recruiting districts, not

---

14 Poly, 'Laeti et gentiles', p. 19, fn. 17–18. AE p. 256, though linking *Sugambri* and *Gambrivi*, p. 138, notes *Ambri* amongst *Teutones* and the similitude *Batavi, Chamavi, Frisiavi*; understands 'good isle', p. 115; links with *Canninefates*, p. 75, in Kinhem, but discards *nefat* (though *am*). *Amor*, cf. *infra* fn. 38.

15 Poly, 'Laeti et gentiles', p. 19, fn. 19; slayer = he who puts asleep; for 'asleep' (dull) note nowaday *schwabenstreich*. AE p. 253 links *Suevi* and *Sueones*, the Swedes, devotees of the sow, a *totemtier*, but understands *die Eigenen* or *die Schlafrigen*; pp. 284, 266 links *Tencteri–Usipi/Usipetes* and *Tubantes* but understands Two-Lands ; notes the ending-similitude *Tencteri-Bructeri*, p. 64.

16 He defines the *liberti* of the *lex Aelia Sentia* from the *dediticii peregrini*, not the other way as some modern commentators do (*Gaii Inst.* 1.14 and 25).

defence areas, since beyond the Rhine free Germany kept more or less quiet. But during the next century, empire and border collapsed.

After a while the imperial army counterattacked. In 286–288, the Rhine was cleared, in 293 the Batavian Isle retaken. The high command resumed control over the *dediticii* and captured many prisoners.[17] A panegyric mentions two kinds of people: *laeti* 'restablished *in postliminium*' in the territory of *Treveri*, and *Franci* 'received *in leges*' in the territory of the *Nervii*. Those are the first reliable occurrence of the words *laeti* and *Franci*.

The *laeti* settled near Treves are probably the former *Chatti Mattiaci*; their command will be listed in 395 in the roll of the *praepositurae laetorum et gentilium* of the *Notitia Dignitatum*. Their older district having been swept over by the advancing *barbaricum*, their transfer sets them back into Roman territory with their previous rights. But they remain *deditici* since the citizens' *postliminium* is in their case an extension by *fictio iuris*, the usual device of Roman law, 'judge, do as if ...'. Among the other *laetii*, the *Batavi* are not mentioned; at least partly faithful to the empire, they withdrew to Arras and Noyon where their commands will be mentioned in the list of the *praepositurae*.[18]

To these former groups – let us call them ranger-*dediticii* – are added new people, *Franci* received in Nervian territory under conditions of a treaty (*sub leges*). Those are most probably the *Bructeri minores* of king Gennobaud, an allied Frankish border king who had ordered his people to 'learn to serve as clients (*obsequia discere*)'. But having betrayed their *foedus*, their *perfidia* deserved their degradation to the rank of *dediticii*, though some of the conditions first granted (*leges*) still held on. On the Upper Rhine, some clans of the Suevian league, after crossing the river and ransacking as far as Langres, had been beaten and settled around Troyes.[19]

Other invaders had merely surrendered *à merci*, such as the *Chamavi*, Scouts of the Border, elements from the *Batavi-Canninefates*, a part of those *speculatores* gone to the Saxons; with them were their neighbors *Frisii* (**FreoHyse*),

17 Poly, 'Laeti et gentiles', p. 20, fn. 21. In 286 slaughtered *Heruli* and *Chaibones* (*Panegyrici Latini* II, 5 et III, 7); treaty with Franks (*Bructeri* ?), submission of Alamans (Dioclétien, *Pan. Lat.* II, 5, 8–10; by Constance, *infra* fn. 22); 293, *Frisii* and *Chamavi* (*Pan. Lat.* IV, 9); in 308 prisoners sent to the circus, 'their perfidy making them unfit for military service and their arrogance unsuitable for servitude' (*Pan. Lat.* VII, 12); 313, perfidious Franks beaten (*Bructeri minores* ?), joined by *Cherusci* pushed by the Saxons to the former territory of the *Chatti* (*Chattovari, infra* fn. 31) ; *Lantionae* and *Alamanni* (ex-*Suevi*) ; *Tubantes* beaten (*Pan. Lat.* X, 18).

18 Poly, 'Laeti et gentiles', p. 20, fn. 23, 24. *Falmenna* > Famenne (**FaeleMenn*), Men of the Faithful (people), near the Belgian Yvoy, not Carignan (ND Oc 42.38). Map 1: 6.1, 9.1, 8.1.

19 Poly, 'Laeti et gentiles', p. 20, fn. 25. Map 1: 7.1.

Youth of Freo. Both defeated *nationes* were then scattered in Amiénois and Beauvaisis.[20]

The situation of these new *dediticii* is different from that of their predecessors: they are 'intended for the tilling of the wilderness', scattered in small groups assigned to the great public domains granted to the *honestiores*. They till the land as do the *coloni*, grow cattle, pay taxes and when needed, join the army as the Roman *tirones* given for recruits by the landlords.[21] We might call them tiller-*dediticii*.

The former ranger-*dediticii* in the *Decumates campi*, *Tubantes* and *Chasuarii*, who had unwisely joined the *Suevi/Alamanni* invaders, are reduced to the same condition, they till the wasted region of Langres. Later, they will be gathered again: the *Laterculus Veronensis*, omitting their retrogradation in Langres, note the transfer of their three establishments (quasi-*civitates*) in *Belgica prima*; the *Notitia Dignitatum*, still naming them *laeti lingonenses*, confirms they are 'distributed in several places of the first *Belgica*'.[22]

This second transfer may have occurred in AD 357. Julian's troops were marching from Reims to beat off the Alemannic invaders while another army was coming from Augst: '*laeti* barbarians, smart to snatch at an opportunity of plunder, went between the camps of both armies and fell suddenly on Lyon'. These barbarians styled *laeti* could not come from Treves, Beauvais or Amiens, they are the *dediticii* of Langres-Dijon, nearest to Lyon, who had left field-working for booty. Their transfer in three small districts near Toul – the Blois and both Ornois – must be the outcome of their rash endeavour.[23]

Like the ranger-*dediticii*, the tiller-*dediticii* are named *laeti*. The usurper Magnentius, born in Amiens from a Briton father and a Frankish mother, had learned Latin 'among the *laeti*, a Gallic nation (*ethnos gallikos*)', meaning the *laeti* were something like an *ethnos*, one of the barbaric *nationes*, but settled in Gaul. The former or future soldiers amongst them had to know some Latin. The

20 Poly, 'Laeti et gentiles', p. 20, fn. 26. *C(h)am(h)ari criusiani*, The Repenting (*hrywsenne*) Grey (= wolf) of the Border (*Laterculus Veronensis*, 13.11).

21 Skyres in the East (in 406, CTh 5.6.3), where recruiting systems were different, Constantin Zuckerman, 'Two reforms of the 370s. Recruiting soldiers and senators', in *Revue d'études byzantine*, 56 (1998), p. 80.

22 Poly, 'Laeti et gentiles', p. 21, fn. 28. Not *Lingones*, the citizens. *Bedensis*, **Bed*, Those Who Asked (=*foederati*); *Odornenses*, **OdEarnA*e (> river Orne), The Winnings of the Blest (837 and 870, *Ann.St Bertin*, p. 22 and 173–174). Map 1: 5.1, 5.2.

23 Poly, 'Laeti et gentiles', p. 21, fn. 29.

child Magnentius must have been reared at the home of his Frankish mother's kin, the *Chamavi* scattered in Amienois in AD 296.[24]

But for those called *Suevi*, an old name, these tribes were styled in the fourth century *Franci*.

2.2 Franci, *a Common Name*

Like the word *laeti*, *franci* comes from the popular language of the borders smuggled into Latin by the army. It was not as such the name of an ethnic league, but a term of abuse.

In free *Germania*, *laete* had named bondsmen who paid a tribute. In old-English, the word is glossed 'he who abandons, that is the slow-ones (*dissidens id est tardi*)'; the 'let (alive)' are people who abandoned a fight, therefore supposed not very lively in war or at work. Later, in old-high-German, *laz* became 'lazy, apathetic, silly'. In the Empire the Germanic word sounded like latin *laetus*, 'glad', a pun by those free *Germani* who volunteered in the army: shame to the German *dediticii*, alias *laeti*, they are in bonds, and glad to be![25]

Franks might well be the reverse insult. In old-English dialects, *freceo* meant a glutton, *fraec/frec* was 'greedy, desirous', whence later 'bold', and because of the greed *fraecne/frecne* was not only 'grievous' and 'daring' but 'dangerous, ferocious', even 'wicked'. *Fraecganga* was a *profugus apostata*, an exile because he broke his peace-oath, a *wearg* roaming like a lonely wolf, Frekki thus being a legendary wolf.[26] *Fraecne/Fraence* was a name of abuse given by the *dediticii-laeti* to the free *Germani* beyond the Rhine who plundered them. The *Peutinger's Tabula* calls *F(ranci)* three nations near the coast, *Chauci, Amsivari, Bructuri* and one near the river, *Chamavi*, styled '*Franci* too'.[27] Between *Chauci* and *Chamavi* was later written *Ernepstini* (**AernHeapasTen*), The Ten Former

24 A. Hugo M. Jones, *The Later Roman Empire*, I-II (Baltimore 1986), p. 620; Barbero, *Barbari*, p. 180, the *laeti* would have been Roman prisoners back from Germany, the *dediticii* joining them later; but freed citizens had the *postliminium*, the texts insist they must have their lands back where they had lived (*ibidem*, 199; Zosimos, 3.52). For *laeti gentiles* (ND Oc 42), read *laeti (7=et) gentiles* as in 42.34, '*laeti Batavi et gentiles Suevi*'; the gallic adjectives *Lingonenses* or *Nervii* (37, 39) are used by the text to avoid employing several ethnic names.

25 ASD, p. 613. Barbero, *Barbari*, p. 184. The 'compagnies disciplinaires' of the French army were called 'les joyeux'.

26 '*(H)attica lingua, Francos hoc est feros*' (*Gesta regum Francorum* 1–5); '*a feritate morum ... ferocitas animi*' Isidor, *Etym*. 9.2.101. Gloses o.e.: *gifere vel frec = Ambro* (ASD, p. 331, 475), *Saxones Ambrones* in Northumberland, Ps.-Nennius, 57.

27 *Tabula Peuting*., p. 1988. Pascal Arnaud, 'L'origine, la date de rédaction et la diffusion de l'archétype de la Table de Peutinger', in *Bulletin de la Société Nationale des Antiquaires de France*, 1988, p. 302; 'Franks too' because the *Chamavi* were ancient *Sugambri Istaevones*

Troops/*Centenae*, that is the seven 'hundreds' of the *Frisii* joined by the three of the *Bructeri minores*, all ex-*auxilia* of the army turned hostiles.[28]

Half a century later, the situation changed again. The imperial pretenders needed soldiers. The two status – rangers and tillers – were then unified by the creation of special districts, 'delegations' (*praeposituræ*). The Germanic *dediticii*, the *laeti*, are no longer considered as *gentes*, they form like the Roman soldiers a hereditary corps assigned to the public services, *corpora publicis obsequiis deputata*, says a constitution in 465 which describes the soldier as 'he with an ancestry (given) to Mars, *laetus* or imperial'.[29]

This new policy began in 358. In the winter of 357, Julian had captured two *cunei* of Frankish ex-*auxiliarii* of Magnentius who were ransacking the right bank of the Meuse between Liege and Maastricht. Remembering the revolt of the *laeti lingonenses* some months ago, he got rid of them by sending them in the East to his uncle Constantius (II). But next year, he agrees to the settlement near the former *Franci Nervii* of another Frankish group called *Salii* (**SalEoh*), Halter-Horse, who had marched from the Salland beyond the border to the north-west of Tongres.

Julian received afterwards a part of the *Chamavi* who had crossed over in the Batavian Isle, at this time styled *Quadi* (**CuWaed*) a northern fraction,

while the others, *Bructeri* and *Chauci minores*, were *Ingaevones*, Poly, 'Laeti et gentiles', p. 22, fn. 33.

28 *Getalu vel heapas vel hundredu = centurias*, ASD p. 521. The old hundreds in NW Germany were four cardinal cantons, or in French 'quartiers' (N, W, E, S), named *ende*. Later on, with the diffusion of the bilinear kin, came *gà/ge*, **geAe*. Some *pagi* kept the old name because they had less or more than four hundreds: Twente (*Tuihanti*), 'Two ends', inhabited by two different *gentes, Frisi* and *Marsi* (3th century, with sign of a knife=Frey and of an axe=Thor, *Inscriptions of Roman Britain*, n°234–236), Vriezenveen and Ootmarsum; Drente (*Threanti*), 'Three ends', because the *Bructeri minores* (600 *stadia* from the Rhine, around the *Lupias*/Lauwers parallel to Ems and Weser) had only three 'ends' left, the fourth being probably the *Amsivari* remained on the Ems with the *Chauci*. Ten (ends) (**ErnHeapasTen* alias *Flevi*, 'Those of the Vlie') were the seven *Frisii* (the 7 red spots on their medieval banner, counting the four *pagi* of the *Frisii minores* in their islands and the three of the *maiores* Wester-, Oster-, Sudergà; the fourth formerly in Twente, afterwards in Westerflinge, 'fled to the West' by the *Salii* of the Salland) and the three *Bructeri minores* of Drente. Along the Rhine, expelled from between Ems and Weser by the Saxons *Angrivari*, were the *Bructeri maiores*, **BoRuhTaer*, Split of Both-Rough (the wolves, white *Cimbri* and red *Chatti* ?), later Αρβορυχοι (**ArBoRuh*), Glorious Both-Rough. All this, nowadays a bit tangled, was then well-known. References, Poly, 'Laeti et gentiles', p. 22, fn. 34.

29 CTh Nov. II of Libius Severus, 465: '[....] *laeti et* [....] *aliaque corpora publicis obsequiis deputata*', parallel in *Lex rom. Burg.* '*ex martianitano lito andoronico vel quocumque alio corpore publico*', read '*ex marti avitano, laeto an dominico*' (not *alito* as Poly, 'Leges barbarorum', p. 202); Barbero, *Barbari*, p. 189; the *Salii* are the *laeti* settled 'near the watch-posts' in Toxandria, the forts on the road to Boulogne (Libanios, *ibidem* p. 129).

remained in Saxony, of the homonymous southern goups. He reinforced them with two older Suevian groups, one back from Britany where they had been sent, the other probably withdrawn from Troyes. At last, he subdued some *Cherusci* (**HearuhAesc*), Ash of the Sacrificial Mound, expelled by the Saxons and come to the Rhine in the former land of the *Chatti*, therefore named *Chattovari*.[30]

All these groups now obey a *praefectus* commanding a *praepositura laetorum* or *gentilium* where their families live on a land divided in shares. Like the districts of the first century, each *praepositura* is the reserve (*reliquatio*) whence come the levees for a specific unit. Instead of sending recruiting officers in the *latifundia,* meeting the ill-will of the great *possessores* to part with able workers, the high command needs but to order the *praefectus* to march recruits from the reserve-district of the unit, while the unit sends back the discharged soldiers, wounded or *emeriti.*

The able male population of the *praepositura* is *conscripta,* inscribed on the military roll (*matricula/katalogos*). As stressed by Julian in a letter to Constancius, all the youth may be drafted.[31] Let us call this third evolution roster-*dediticii.* A difference with the ranger-*dediticii* is the presence of Roman neighbours in the *praepositura,* landlords (*possessores*) or bondsmen (*tributarii*). The tax in corn (*annona*) due by those Romans was gathered in local depots (*condita*) for the maintenance of the soldiers' families.[32]

2.3 *Barbarian Panorama*

After the pogrom of Pavia in august 408, many barbarian auxiliaries must have returned to their reserve-districts, bringing back with them their ensign (*vexillum*) named *band,* the leather strip of the *vexillum* wearing the regimental medals. The *praepositurae* will then be called 'bands', that is to say banners. We

30 Poly, 'Laeti et gentiles', p. 24, fn. 42–45. AE p. 236 ignores the halter painted on the shield of the *Notitia Dignitatum.* Map 1: 11.1, 13.1, 7.2 and 13.2, 11.2.

31 *Reliquatio,* 360–380 (CIL 7.137), Soazick Kerneis, *Les Celtiques. Servitude et grandeur des auxiliaires bretons dans l'empire romain* (Clermont 1998), p. 215. *Praepositus laetis* in 369 (CTh 7.20.10) commanding the reserve, not the unit. *Katalogos,* Procopios, 5.12.9–17. Poly, *Leges barbarorum,* p. 200. According to Barbero, *Barbari,* p. 190, the recruitment did not mean a relation *praepositura-auxilium,* though he accepts it for the Sarmatians and the *Salii,* with quite a right analysis pp. 39 and 114.

32 Herwig Wolfram *The Roman Empire and its Germanic Peoples* (1990; engl.transl. Berkeley 1997), p. 113; Poly, 'Les Saxons du Bessin', p. 73.

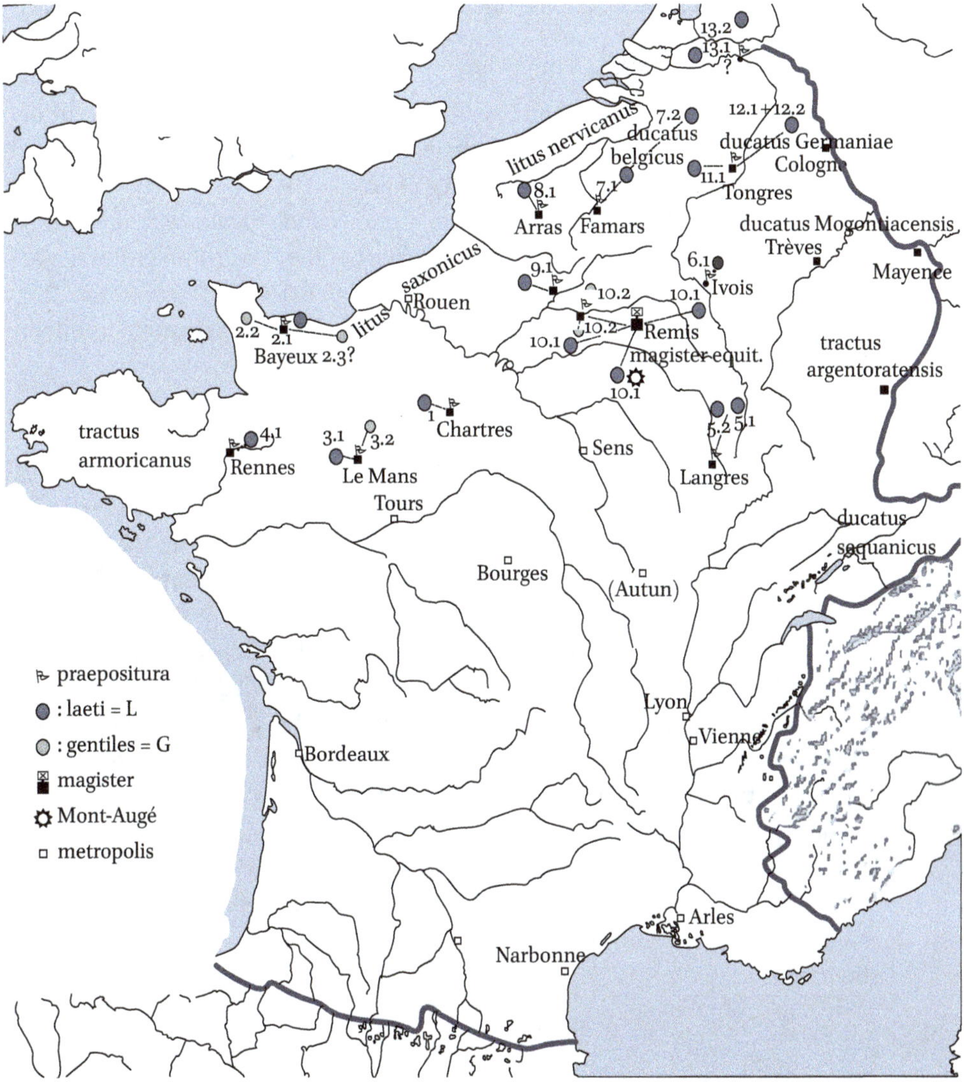

1.1 : L teutomirani (33)	5.2 : L lingones (chasuari?) (37)	10.2 : G (suevi ?) 1/2 + 1/2 (41)
2.1 : L batavi (34)	6.1 : L (h)atti (38)	11.1 : L lagenses (salii) (43)
2.2 : L suevi (34)	7.1 : L nervii (franci+frisii?) (39)	(and in the gap in ND Oc 42.44/45)
3.1 : L suevi (35)	7.2 : L nervii (suevi) (39)	[12.1 : L suevi]
3.2 : G suevi (35)	8.1 : L batavi nemetacenses (40)	[12.2 : L hattovari]
4.1 : L franci (chamavi) (36)	9.1 : L batavi (et) contraginenses (41)	[13.1 : L chamavi (et) quadi?]
5.1 : L lingones(tubantes?) (37)	10.1 : L (franci and suevi?) (42)	[13.2 : L suevi]

MAP 14.1 *Praepositurae laetorum et gentilium* in Gaul c. 400 AD [(33)= in ND Oc 42 ; 12-13=in the gap after 43]

find six of them for the *laeti* beyond the Ardenne and two more for *foederati*, outside the border.[33]

The first reserve-district (*praepositura*) was for the *Salii* in the territory of Tongres, in Lowaige/Law. There was held the general muster at the head of the division of six (*moira*), its name being therefore *Hasbania*, the Ban of the Command.[34] There were levied the units of *Salii*.

To the West, in Nervian territory, there was a second *praepositura*, the *Bra*(*c*)*bant* (**Baeh*(*W*)*racBand*), the Band of the Exile of the (Oath-)Rings. It was the reserve of the unit of the *Brachiati*, a Latinisation of **Beah*(*W*)*racGehat*, The Recall of the (Oaths-) Rings Exile, back from their former assignment as tillers in Amiénois-Beauvaisis.[35]

In the region of *Geel* (Yellow, the suebian color), north of the Keerbergen, Return Hills (*cerr*), there may have been a third reserve-district, that of the *Iovii*, the name of the *Suevian/Alamannic* unit of the young Constantine I. They helped him to proclaim himself *jovius* in York, and 'returned' from Britany with him. During Chlodio's raid to Cambrai, they will leave Geel, go up the Schelde and settle in the area of Lille, renaming their settlement Caribant, the Band of the Return.[36]

A fourth reserve-district was at first probably in the East of the region, and thus called *Osterbant*. But like the *Iovii*, the troops settled there must have followed Chlodio to Cambraisis where they transported their former name, Ostrevant, though it was now situated in the western part of the region. What was left of them in the East may have moved to the *pagus Hattoaria* or *Spe*(*He*)*tueria*, the *Hattuaria* of the Scouts, and the former establishment, void of troops, was then called, by reference to the past, Herve, **HereWaeg*, The Grey One Off. The *Hattoarii* being former *Cherusci* (**HearuhAesc*), their unit was named *Ascarii*, the Grey One (=wolf) of the Ash, bearing a jumping wolf on their shield.[37]

---

33 Reorganized *terrae laeticae* in 399, CTh 13.11.10. Outside: *Ampsivari probably in* Bagband (*beag* = (oath)ring), *Bructeri* in Bursibant (TWB 98 ; **BoRuhHyseBant* ; ND Oc 5.187, 188). Not as Poly, 'Sous les chênes de Salaheim. La loi salique, l'armée romaine et le bilan de la barbarie', in *Civitas, Iura, Arma*, eds. F. Botta - L. Loschiavo p. 202.

34 An **Hasbanband* being not euphonious was reduced to *Hasban*(*ia*) (Hesbaye). ND Oc 5.29 = 177. Wenskus, *Stammesbildung*, pp. 526, 524; Karl Ubl, *Sinnstiftungen eines Rechtsbuchs. Die Lex Salica im Frankenreich* (Ostfildern 2017), p. 71. The relation to the shields is not taken into account.

35 Poly, 'Laeti et gentiles', p. 24, fn. 43 and here, *supra* fn. 19, *infra* fn. 50.

36 Poly, 'Laeti et gentiles', p. 24, fn. 44.

37 Poly, 'Laeti et gentiles', p. 24, fn. 45. Pikes in ash-wood (Tacitus, *Annales* 2.14 and 21). AE p. 95, *herut* (hart) suff. *-isc*, or *horsc*.

In the Batavian Isle there was a fifth reserve-district, whose name was kept by its former place, the Teisterbant (**TehHaestTerBant*), the Band of the Split of Bound-Furious (Fenrir wolf). There could have been levied the unit of the *Defensores*.[38]

Between the Rhin and the Ijssel there was a sixth district, later sunk in the flood of the Aelhmer, the Swifterbant (**SwefTerBant*), the Band of the Suevian Split. They were likely those *Suevi* from Troyes, whose men might form the unit of the yellow *Cornuti*.[39]

Those last four bands amongst the six could be the *praepositurae* in *Germania Ia* cancelled at the end of the list of the *Notitia* and hastily replaced by a new Aquitanian *praepositura*, out of geographical order.

Meanwhile, the system of the *praepositurae* spread on. In 364, when Valentinian made Treves his capital, he had to face three dangers: in the North, the Saxons took over the Frankish piracy and assaulted the coastal regions; in the South-East, the Alamanic league was still aggressive; in the North-West the Bagaude rebelled again. The high command reorganized the defence, sending troops from the Belgian *praepositurae* to the coast and to the West. According to the *Notitia*, a group which bears the name of a Frankish king and Roman officer, Theodemer, – maybe a cavalry unit – was settled near Chartres in Thimerais, their *praepositura* being at the head of the list.[40] Other Frankish *laeti* were near Rennes in the Vendelais (**Wand Deal*), whose name recalls the legendary AurVandil carried in Jylland by Thor.[41] *Laeti Suevi* were in Maine (in Charnie ?).[42] Saxon newcomers of Condren styled *Batavi* were sent in Bessin.[43]

All this military disposition was strengthened by new recruitments. To weaken the league of the Alamans, Valentinian had incited the Burgundians to invade their territories, bringing about a migration of *Suevi* whom he received with a *foedus iniquus*. As *foederati* these Southern Suevians were not called *laeti* but *gentiles*, meaning both clansmen and pagans. They were sent in Bessin and

38 Better than *Tehster*, as in Poly, 'Laeti et gentiles', p. 26, fn. 46. Cf. the countries of Amor, and *Sicambria* in (*H*)*attica lingua*. Possible unit: *Defensores* (not *Armigeri defensores*), ex-guards to the Chamavians Magnentius or Arbogastus ? Later exiled in the East after Arbogast's rebellion? Tale of AurVandil's Toe (*Skaldskaparmàl* 3), François-Xavier Dillmann, *L'Edda. Récits de mythologie nordique par Snorri Sturluson* (Paris 1991), p. 114.

39 Poly, 'Laeti et gentiles', p. 26, fn. 47.

40 Poly, 'Laeti et gentiles', p. 26, fn. 48. *Teutoniciani* read *Teutomiriani* > Timerais.

41 **WandDeal* ou *WandAelle*, celticized *Gwennel*. Jean-Pierre Poly, *Le chemin des amours barbares* (Paris 2003), pp. 114, 118.

42 Poly, 'Laeti et gentiles', p. 26, fn. 50.

43 Poly, 'Laeti et gentiles', p. 26, fn. 51. *Contraginenses* (ND Oc. 42.41) **GundDragen*.

Cotentin and around Reims where in 383 other Suevian, those being *laeti*, the Brisings, were added, later transferred in Aquitania and inscribed as such in a correction of the *Notitia*.[44]

These *praeposituraе* north of the Somme were known as **AlHopRih*, the Command of All Troops/*Centenae*.[45] Cologne and its region, former headquarter of the *Germania Inferior*, was abandoned to the *Bructeri maiores* and *Amsivari* who established their own kingdom, **RihBoHare*, Both Grey Ones of the Command.[46]

2.4 *Gentile Cultures*

The settlers of the *praepositurae* were not a broken barbarian mob but former fractions of several *gentes*, each still carefully distinguished from the others. With their ethnic names went favorite colours and emblems picturing their identity. As Isidorus of Seville observed, the Germanic *gentes* were singled 'by their weapons, the difference of colour in their customary dress (*habitus*) and the various pronunciations of the language'.[47]

Hiw was therefore colour and kin. The colours on the shields were carefully chosen, expressing the custom of the group, a main devotion to a god and a kinship system: red was the colour of the devotees of Donner/Thor, patron of a patrilinear system; yellow was the colour of the Vanic gods and their matrilinear system, joined by white groups; Woden and the bilinear system had blue, associated with red when the 'renchaînement' was patrilinear, with yellow or white when it was matrilinear.

The monolinear unions consecrated by a god, Thor or Inga, were endogamous: the union was preferential, between first cousins – 'frères et soeurs classificatoires' – of a primordial lineage (*Sippe*) specific to each *natio*. Between different 'nations', there could be no consecrated marriage, only second rank

---

44 Poly, 'Laeti et gentiles', p. 27, fn. 52. 'He gathered many youngsters taken from the barbarians inhabiting on the Rhine (the *Suevi*), and rustics from the *gentes* submitted to the Romans (the tiller-*dediticii*) and enlisted them in the army' (Zos. *Hist. Nov.* 4.12.1), 'the *laetus*, the Alaman (=*gentilis*), the Sarmat (=*gentilis*), the *vagus* (vagrant) and the veteran's son' (CTh 7.20.12).

45 Poly, 'Laeti et gentiles', p. 27, fn. 53.

46 Procopios, *supra* fn. 31. *Gallia Belgica Alobrices* (*Cosmogr. Raven.*, 4.24.1). Otherwise, Eugen Ewig, *Die Stellung Ribuariens in der Verfassungsgeschichte des Merowingerreiches* (Bonn 1969), p. 26.

47 Isidorus, *Etym.* 9.2.97, from an older text, in Walter Pohl, 'Telling the difference: signs of ethnic identity', in *Strategies of distinction*, eds. W. Pohl – H. Reimitz (Leiden 1998), p. 17, who note the 'ancestral weapons', the *francisca* and the *saex*. Jean-Pierre Poly, '*Arcana conjugalia*. Le secret des mariages germaniques', in *Histoires de famille. A la convergence du droit penal et des liens de parenté*, ed. L. Otis-Cour (Limoges 2012), p. 21.

unions (*friedelehe*), less binding. But the bilinear system, patronized by Wodan, combined matri- and patrilineages by linking two primordial lineages. First cousins were excluded and since the second-rank unions spread the blood of the two clanic sibs, the consecrated marriage of the bilinear system could be fairly exogamous.[48]

We may have an idea of the nations' faiths by the colours of the shield of the related units in the imperial army. Not many of the red clans were left, where the army levied the *Defensores* and the *Batavi*, possible fraction of the *Sugambri*, and the (Northern) *Heruli*, a part of the *Vandali*. All these were *I(di)stAeWone,* (those who) Stay in the Law of *Idistà*, Dise's Toe, the giant AurVaendil who lost his toe/twig, thus succeeded by his protector Donner/Thor. On the north-sea coast, people were *IngAeWone*, Stay in the Law of Inga, the vanic goddess: the yellow *Gambrivi* (**GamBaereHiw*), Colour/Kin of Gambara, a name of the goddess by the *Langobardi*, ex-**WinnGeolas*, Battle-Yellows. With them were the rallied whites, *Bructeri, Amsivari* and *Frisii*.[49] The main colour was Woden's blue whose worship had spread from Odinsey amongst the *Angli* (**AngAelle*), Fed (=Adopted) by the Unique. The *Langobardi* and other Suevians, formerly yellow or white, as well as many red clans, the *Cherusci* and the *Chatti Mattiaci*, had become blue *Hermin(G)iome*, 'Observe the Lord's Memory', *minne* being the ritual toasts to the supreme god.

The figures on the shields of the units were related to the clans' names. Those of the *Brachiati* were white and bore rings; we have seen that the name of the unit was probably a pun with the latin *brachialis* and a kenning, **Beah(W) racGehat,* Recall of the (Oath) Ring Exile.[50] The shields of the *Salii*/**SalEoh* wore Woden's blue, one with the 'halter', the binding rope (*sal*) of the god, another with the wolf, the witches' 'horse'.[51]

To make a long story short: in the fourth century, the reserves of the *auxilia* and some other gentile units were spread all over northern Gaul. Their shields tell us about a *Weltanschauung* rather far from that of the Christian Empire, even to the seventh or eighth century when bishops still strove to fight it. No wonder. The demographic weight of the peregrine communities had been far more important than is nowadays supposed.

---

48 Poly, *Le chemin des amours*, p. 76; Id., 'Arcana conjugalia', p. 223. Of course, the strict observance of these different kinship-systems was practised only for the noblest unions.

49 Poly, 'Laeti et gentiles', p. 28, fn. 58.

50 Bagbant *supra* fn. 33; *wrac* better than *raec* as in Poly, 'Laeti et gentiles', p. 30, fn. 66; *aeht*, ASD, p. 369.

51 Wenskus, *Stammesbildung*, pp. 520, 549. Bond: Tacitus, *Germ.* 39; *Ibn Fadlan* (ed. Canard, p. 63); Peter Vilhelm Blog, *The Bog People* (Ithaca 1988), fig. 59.

3 Imperial Barbarians: 'Count Up Your Men'

Let us try to compare the numbers of the military peregrine groups, both fighters and non-fighters, and those of the civil population, free men, allegedly Roman, and slaves, whose origins and culture are more dubious.

A regular army is always counting up its men – each morning in fact – since the numbers vary all the time: some in mission, others on leave or reported sick, others deserters or dead. All units were and still are affected by this trend and when there was a lack of new levies, the actual numbers may have been well below the regular ones. To keep up the number between two levies, the commanding officer might constrain some stray elements, vagrants or *manu militari* rounded-up 'volunteers. He needed to muster as a full unit 500 men, at least a minimum of 300.[52]

For the registration of numbers for the civilians, we are going to meet with more difficulty. It was set by the fiscal census of each territory of a 'city'. But in Gaul, nothing is left of the tax-rolls and the account for the civilian population is thus still more hypothetical than for the military.

3.1 *Demography of the Reserves*

Tribal demography is rarely very large. During their migration in 60 BC, the Gallic *Helvetii* and their allies were by their own account 368.000 persons, of whom 92.000 were able to fight.[53] But in quieter times, not all fit men would be warriors. Let us catch a sight of the recruitment ratios from the texts, then look at a small graveyard in Champagne.

In the first century, each *pagus* of the Germans – the Germanic hundred – sent to the chief around 120 youths. In the North-West, the customary division would be four cardinal districts (*end*), giving a theoretical number of 480 young warriors.[54] When the imperial army asked for reinforcements from the allied clans, they must have sent their youngsters. As the Germanic *auxilia* became more numerous, the army shifted its former *cohors* with 6 *centuriae* of 80 men for a *numerus* with 4 *centuriae* of 120 men.

The transported *Sugambri* and *Canninefates* had been an approximate 40.000 persons, men and women, bar the children not taken in account. Let us suppose four customary 'ends'/hundreds in both *nationes*, that is 5.000 persons for each one. Without the cavalry *ala* formed by the noblemen, the army levied eight quingenary cohorts, probably one per each hundred. For all the

52 Warren Treadgold, *Byzantium and Its Army* (Stanford 1995), p. 88.

53 Poly, 'Laeti et gentiles', p. 31 fn. 68.

54 Tacitus, *Germ.* 6 and 14. The Germanic hundred is 120. 'Ends', *supra* fn. 28.

Batavian Isle that meant a recruitment of 4.000 men, 10% of the adult population, 40% of the maximum number of fighters.[55]

Under Constans, son of Constantine, the structure of the elite units was modified to make the service easier. In 356, the *Ioviani* of *Cornacum*, a detachment (*vexillum*) drawn from a Danubian legion, counted two levies, ancients (*seniores*) and cadets (*juniores*).[56] The *Notitia* shows that the cadets, likely bachelors, were generally sent farther away than the ancients who, unless they were punished, served nearer, and were thus able to visit their wives on short leaves, providing children for the relief; once married, a young soldier would shift from the cadets to the ancients. In the 'Briton' *auxilia*, as probably afterwards in others, a tripling of the levies was ordered by Magnentius, then lieutenant of Constans.[57] Some *auxilia* in the *Notitia* do have indeed three levies in the field; but the third levy – youngsters, future recruits officered by some *emeriti* veterans, their drill-sergeants –, would remain in the *praepositura*, though it could sometimes be sent to the army.

To catch sight of the working of the system, we have an archaeological testimony: the graveyard of Mont-Augé.

In the district of Vertus in Champagne, at Mont-Augé near the village of Vert, there was an ancient derelict funerary site: a small *fanum* and two cremation enclosures. Then, sometimes in the second half of the fourth century, some thirty deceased were buried there[58]: in two of their graves, at either end of the graveyard, two 'oboles à Charron' were deposited, left-overs from former *donativa* or soldiers' pay: a *denarius* of Constantius II – received in 355 ? – at the very beginning of the graveyard in the East, and a *solidus* of Valens – received in 378 ? – at the end of it, in the West.[59]

Six of the deceased were soldiers or ex-soldiers, one possibly a non-commissioned officer known by his weapons, the others by their *caligae*, the

55 Poly, 'Laeti et gentiles', p. 31, fn. 70.

56 Thomas Drew Bear, 'A Fourth Century Latin Soldier's Epitaph from Nakola', in *Harvard Studies in Classical Philo*logy, 81 (1977), p. 257 (but *Cornacum* in Pannonie, not the *Cornuti*). The system was therefore earlier than 364, year given in Dietrich Hoffman *Das Spätrömische Bewegungsheer und die Notitia Dignitatum* (Düsseldorf 1969), p. 118.

57 Kerneis, *Les Celtiques*, p. 181.

58 Raymond Lantier, 'Un cimetière du IVe siècle au Mont-Augé', in *L'Antiquité Classique*, 17 (1948), p. 373; Raphaëlle Chossenot, *Carte Archéologique de la Gaule. La Marne 51/1* (Paris 2004), p. 722. In year 1949 a rather good excavation for the period, but the bones, after examination, were thrown away; nowadays we could have known much more from them, cf. Patrick Geary and Krishna Veeramah. 'Mapping European Population through Genomic Research', in *Medieval Worlds* (2016), p. 65.

59 Poly, 'Laeti et gentiles', p. 32, fn. 74.

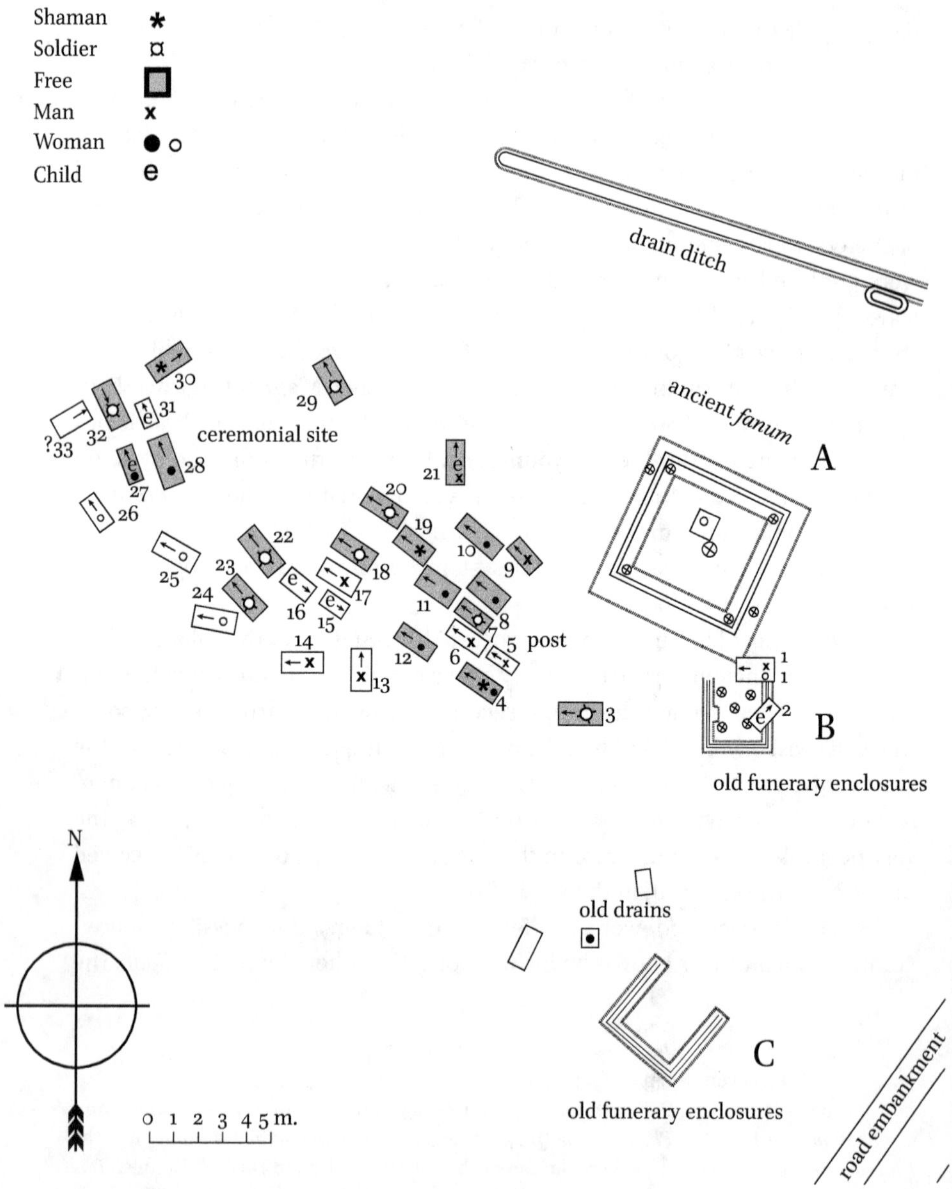

MAP 2 Mont-Augé Graveyard

military nailed shoes, characteristic for Procopius, the weapons being kept in depot.[60] Two generations later, this barbarian graveyard was abandoned.

The small kin group to which the graveyard belonged was situated between two *fara,* the migrant' groups of Suevians. The nearer one bore an illustrious name, *Fara Brisinga* (Férébriange). There was an old tale of a legendary Brisings' collar worn by Freyja and lent to her husband Woden/Odin. The name allows us to date more precisely the graveyard. The *fara* sent its recruits to the units of the *Brisi*(*n*)*gavi,* ancients and cadets, formed in 377 when Gratian marched his troops to the East.[61] Twenty years before, Constantius II had struck a treaty with the *Lentienses Suevi,* but later on they had turned hostile and Gratian defeated them near Horburg, in front of their country of Brisgau, 'They delivered their strong youth to be set among our recruit'. In 380 Gratian was back in Italy before returning to Gaul in the summer 383 to fight and die.[62] The *Brisigavi* then stayed in Champagne alongside other *Suevi* federates (*gentiles*), refugees of 369. All of them went under the command of the 'prefect of the *laeti* and *gentiles* of Reims and Soissons in *Belgica secunda*'.[63]

Eighteen years later, Constantius, general of Honorius, accepts the settlements of the Goths in southern Aquitaine. To watch them, he transfers in Angoumois the *laeti* Brisings and a part of their neighbors, *gentiles Suevi.* The list of the *praepositurae* is rectified and on an erasure, out of the geographic order of the list, the *notarius* inserts a '*praepositura* of the *laeti* (and) *gentiles Suevi* at Angoulême in *Aquitania secunda*'.[64] Honoured with the emperor's name, *Honoriani,* they build a Brisembourg (**BrisingBeorg*) not far from an Aumagne (*Alemania*). The cadets are in Italy, the ancients in Spain guarding

60 Poly, 'Laeti et gentiles', p. 32, fn. 75.

61 Ammien Marcellin, Hist. 31.10.1–5, 8–12, 17. Brisings, according to AE 63, would be a hybrid creation from an ex-celtic Brisgau. *Feria Breisangia,* près d'Etoges, *Stogia,* **StodTihHea,* 'Hill of the Pillar Enclosure', Poly, 'Laeti et gentiles', p. 34, fn. 76. The **LehnTihEnSess* (understood as lat. suff. -*enses*), would be Remains of the One's Loan Enclosure (the Brisings' collar lent by Freyja to Odhinn and the circular wall of the Hexentanz in the Harz where had been the ancient Suevian gathering?).

62 Poly, 'Laeti et gentiles', p. 34, n. 78–79.

63 ND Oc 42.42, *praefectus laetorum* (*et*) *gentilium Remo et Silvanectas* (Senlis) corr. **Susionecias* (Soissons); the lack of the designation *Suevi* may imply the presence of another *natio,* the Frankish *equites Brachiati* in Thimerais (*Teutomiriani*); the *equites Cornuti* in Omois (*Ot*(*ho*)*mensis*) would be Suevians? Poly, 'Laeti et gentiles', p. 34, n. 80.

64 ND Oc 42.44; *praefectus laetorum* (*et*) *gentilium suevorum Arumbernos Aquitaniae secundae* (ms. V), corr. **Anculemos*; cf. the graveyard of Herpes, where are found the spoils from the Saxons of the Loire. Poly, 'Laeti et gentiles', p. 35, fn. 81.

the Pyrenees, gathered before the campaign in Pamplona in 412.[65] The group of Mont-Augé had then left the Champagne.

The graveyard is thus the reflection of a 'letic' hamlet established in 378 and abandoned in 411. It numbers 33 graves: 17, the graves of the free people, are in a coffin, facing the North-West, the sunset of the summer solstice. The 15 others, the graves of the dependents, have no coffin and no special direction.[66] In both categories there are more men than women. There are few children's burials: nearly one half of those children must have died in the first years, without having a grave. Once aged 15, boys would have been sent to the army, while girls went to other Suebian districts for marriage. All would have been buried elsewhere.

The positions of the graves tell us more. The first graves, 1 and 2, are those of poor people, a couple and a child buried in the ditch of the cremation enclosure, perhaps a Gallic family buried before the coming of the Brisings. Next comes the grave n° 3 facing West, an old man with a bad leg buried with a silver 'obole à Charron'. The coin dates back to the time when Constantius II forced the *Lentienses* into submission; the deceased could be one of them-later joining the new unit of the Brising as NCO – thus with soft leather half-boots, not *caligae* – and honourably discharged for a wound.

A wooden pillar marked the area near the *fanum* for the first funerals of the Brisings, a group we shall call Group A. First was buried a woman and at her side an empty grave was dug with offerings, n° 7 et 8. Not far away lay a woman in a hollowed tree trunk, n°4, a traditional burial without any offering.[67] This trunk-burial might be that of the seer, mistress of the old rituals.[68] Then come three other free women's graves, n° 10, 11, 12, each of them with an empty place at its left, for a husband who never came back. Not far away are the graves of four bondsmen's, n° 5, 6, 9, 13.

To the West was another group, the Group B: an old man in a trunk-coffin without offerings, n° 19, a possible successor of the seer woman, and four veterans, n° 18, 20, 22, 23; one of them, n° 22, perhaps another NCO, has been permitted to keep his weapons; no nails of *caligae*, probably leather boots. Three of them have an empty place on the right of their tomb, for a wife still living in 411; the fourth (25) has a bondswoman at his left. The graves of the bondsmen

---

65 *Honoriani*, Christophe Archan, 'L'Epistula Honorii', in *El ejercito, la paz y la Guerra*, eds. J.J. de los Mozos - I. Szaszdi (Valladolid 2009), p. 153.

66 The *contubernia* led astray the *mancipia* of the *possessores* (Sid. Apoll., *Lettres* 3.9).

67 *Totenbaüme* frequent in *Suabia*, Poly, 'Laeti et gentiles', p. 35, fn. 85.

68 'Walburg sibylla of the (Suevian) Semnons' in the 2^{d} century at Assouan, Poly, 'Laeti et gentiles', p. 34, fn. 86.

are on the southern margin, n° 14, 15, 16, one of them perhaps a smith, the other with two children.

Beyond a new area left open for the more recent funerals come the last graves: a young woman and a small girl, n° 28 and 27, with an empty place on their left. A veteran, n° 32 and, facing him, a poor person's rather large grave, perhaps that of a bondswoman and her child, which had been plundered before the excavation. Another veteran, n° 29, and a young man in a trunk-coffin without offerings, n° 30; the direction of his grave is unorthodox, facing the morning summer sun. Two bondswomen are on the southern margin, n° 26 and 23, an old woman with a straw shroud.

We may thus have an insight on the demography of the group and the working of the 'letic' system. The graveyard of the hamlet counted six soldiers – four rather old –, to whom we may add the two 'tree-men', perhaps shamans, the absence of offerings in their graves concealing their military status. The grave of a young boy with a waist belt, n° 21, facing the North, could be that of a young Frank reared to be a soldier; still growing up, he had received no *caligae* yet. The number of the deceased military points to a 'tent-fellowship' (*contubernium*), as in the *Lex Salica*: the two squads of a combat group, three and four men with two corporals (*circitores*: *tetrarchos* and *pentarchos*) and the sergeant (*decurion, biarchos*).[69]

As shown before, each territorial hundred of a *praepositura* sent its recruits to a regimental hundred in both levies (*seniors/iuniores*) of the unit (*numerus*); more precisely, to the levy of the *iuniores* when the men of this levy, getting older and married, go over to the *seniores,* replacing the losses and the discharges. The number of about ten old soldiers in a hamlet seems to show that the recruits were sent by *contubernia*, ten for each hamlet. Each of his four *centenarii* would take in their schooling retinue – the *sacebarones*, 'militia born' – approximately ten lads, each *decania* from some 48 hamlets. The *praefectus*/graf, 'the catcher', thus avoided discussing the recruitment with the entire group. Batches of recruits would be sent to the army when ordered. The maximum service was 25 years, but many, worn out, were discharged before that time. The dead would not be repatriated as is shown by the empty grave n° 8.

The *praepositura* stayed in Champagne some thirty years, from the end of 378 when they came back from the East, to 411 when they left for the Angoumois. When the graveyard began, all men were young, even the oldest of them (*seniores*) married to new wives, young too. Then, the men endured losses in 378 at Andrinople, and perhaps some in 383 when Gratian fought near Paris

69 Treadgold, *Byzantium*, pp. 89–90, 93. *Contubernium*, PLS 14.6, 42.1 and 3, 43.3.

or in 388 when Maxim was defeated on the Sava. During these ten years, no recruits came from the *praepositura*, the children were too young. The levies had to be at their lower effective, 300 men each, may be including some vagrants enlisted as 'volunteers'.

We are told that the losses of the *Rivus Frigidus*, in 394, were heavy among the 'barbarians', let us suppose, for the sake of argument, half the effective (which by the way makes a regular death rate hard to tell). In the hamlet, four married men (*seniores*) are missing (the husbands of n° 8, 10, 11, 12), five old or wounded men were discharged not much later, as the first recruits would come (n° 18, 19, 29, 22, 23). The half-effective of the surviving *milites iuniores* (150 men?) must have been allowed to marry, thus joining the (150?) surviving milites *seniores*, to replace their losses for a lower effective of 300 men. But their own previous levy of *iuniores* would then be almost fallen to nothing. New recruits from the *praepositura* had to restore it at least at the 300 men level.

Admitting the conditions in the 'natural' demography – a medium fecundity rate of 7 and a medium descent rate of 3,4 –, in each hamlet the likely ten wives of the married *contubernium* of *seniores* had to beget in three years an average of 6 or 7 male fifteen-agers, that is 12 or 14 children, boys and girls, in fact much more because of miscarriages – half of them –, bar some unfits.[70] It means that in 395–396, all the women had to send all fit boys born in 379, 380, 381. During these eighteen years, four women out of ten died. We may suppose that the Suebian federates could provide some replacement wives, and some bondswomen could help, but the *graf* would have to wait.

In 395 the Roman western army still had to retreat before Alaric. In 396, the coming of the army on the Rhine was more a prestige performance than a campaign. Probably in 397 and surely in 401–404 the troops could have decent effectives again. This came at the price of heavy death rates amongst the women.[71] The *praepositurae* were conscripts-producing farms where the women had to replace or, if possible, increase the number of auxiliary soldiers.

70 The low percentage of children's graves in the earliest medieval graveyards is noted by Emilie Perez, 'La place de l'enfant dans l'espace funéraire du haut Moyen Age', in *Cahiers d'Histoire* 29 (2010), p. 120. Due to the fear of malevolent deads, they were buried *in aliquo secreto loco* sometimes with the mother if she was dead too (*Burch.Worm., Decretum*, lib. 19 qu. 180 (Gagnon, p. 147) col. 974; Vogel, 'Pratiques superstitieuses ...', p. 754; Gagnon, p. 52 ; Poly, *Le chemin des amours*, p. 165) or sometimes burnt outside the graveyard (Jean-Pierre Poly, Soazick Kerneis, 'Enfances barbares. Les communautés péregrines de Gaule entre loi et coutume', in *L'infanzia nell'alto medioevo, Settimane del* CISAM, LXVIII (Spoleto, 2021), p. 421).

71 Since marriage was forbidden with provincial women (CTh. 3.14.1), a solution was the abduction of *ancillae* (CTh Nov. 11 of Libius Severus).

The *annona* from the *praefectus* and the work of the bondsmen and bondswomen were badly needed.

The graveyard had been used for some thirty years. During this period 31 people were buried, 1 death per year. But this figure accounts neither for the terrible infantile death rate or the miscarriage death of young mothers – dangerous dead buried in the wild, outside the graveyard, therefore leaving no grave – nor for the death rate of unmarried young soldiers dead abroad, nor for young girls married outside the *praepositura* to compensate the first Suebian marriages. We thus get an estimate of 108 individuals, 3,6 deaths per year. With a death rate of 35 °/°°, that would mean a population of some 100 living individuals. Those hundred people of an average hamlet had to replace at intervals the *contubernium* of *iuniores*, whose older soldiers would replace the losses of the *seniores*. The recruitment rate would be one in ten people, an easy rule for the *praefectus* to enforce, a heavy burden for the community to endure.

3.2 *Civilian and Military Populations*

The estimate of civilian population in Gaul is even more difficult.[72] The only definite testimony is a panegyric of 312. Let us draw from this text a general estimate of the civilian population before trying a comparison between civilian and military numbers.

In 312, a Gallic orator thanked the emperor for granting a tax abatement to his native city, Autun: 7.000 heads (*capita*) from 32.000 due, with a relief of the arrears for the past five years.[73]

The *capitatio* bore upon fit men and women from 14 to 65. An imperial constitution in 386 stated that before this date the *capitis norma* was 1 for the men and 1/2 for the women. With this *ratio* and a *sex-ratio* of 1.1, we may suppose a population in Autunois of 42.000 adults.[74] To this number we must add the children under 14, about a third more, say 21.000 individuals. That gives a

72 Josiah Russell, 'Population in Europe', in *The Fontana Economical History of Europe*, I, ed. C. Cipolla (Glasgow 1972), p. 25 (in 500: 5 million). Robert Étienne, 'La Gaule romaine', in *Hist. de la population française*, I, ed. J. Dupâquier (Paris 1988), p. 91 (in 200: 7,5 million; in 400: 5,75). Jean-Nicolas Corvisier and Wieslaw Suder, *La population de l'Antiquité classique* (Paris 2000), p. 86 (in 100: 6 million; in 200: 9,2).

73 *Pan. Lat.* VIII, 5.5, 6.1, 10.5, 11.1–4. *Caput,* personal and distinct from *iugum*, the area ploughed by a team of oxen (André Piganiol, 'La capitation de Dioclétien', in *Revue Historique*, 176 (1935), p. 1; Jones, *The Later Roman Empire*, p. 454.

74 CTh 13.11.2 in 386 extends to four eastern cities a general decree stating that the *capitis norma* had just been shifted to 1/2 for fit men (*ac ternis viris* read *attends viris*) and to 1/4 for women. Discussion Faure and Jones, in Poly, 'Laeti et gentiles', p. 38, fn. 92.

total of 63.000 *humiliores*.[75] Let us add the exempts, that is the *familiae* of 2 or 3 absentee *senatores* and some veterans,[76] this gives a general total of about 65.000 individuals.

The town of Autun had been the capital of the greater Autunois, and its amphitheatre had been built for an audience of 30.000. The greatest cities in Gaul had urban amphitheatres where could be gathered 20.000 to 30.000 persons.[77] But Autun had been destroyed in 270 and in the fourth century the Autunois, after losing the Auxerrois, the Nivernais, the Mâconnais and the Chalonnais, was but a middle-size *civitas*.[78] Let us estimate the population of a great *civitas* 130.000 inhabitants, of a middle one 60.000 and of a small one 20.000. There were at that time in Gaul 17 provinces with 123 *civitates*:[79] the territories of the 13 great ones would gather about 1.690.000 inhabitants, those of the 44 middle ones 2.800.000, and those of the 66 small ones about 1.320.000. That would be a total population of about 5.800.000 habitants, that is 2.800.000 for the 60 *civitates* of the *Gallia ulterior* in the North, and 3.000.000 for the 63 *civitates* of the *VII provinciae* in the South.

But in the North the *tractus maritimi* had been ransacked by Frankish and Saxon pirates at least as far as Amiens, and Alamanic raids had done the same in *Germania superior* and the *civitates* behind as far as Sens and Langres, whence the 'letic' tillers. We may doubt that those regions had an estimated 830.000 civilians. For northern Gaul, 2.500.000 may be a more likely approximation, even too optimistic.[80]

We can compare these numbers with the military Germanic population. In 395 there was a dozen *praeposituræ*, some of them having several districts, for a total of 22. Fifteen would provide two elite units each (ancient and cadets), the others three units each (ancients, cadets, cadets *gallicani*), four districts of *gentiles foederati* providing only one unit, a less heavy recruitment. For an

---

75 Polyptyque of Marseille in 814: 740 individuals, 24% children under 13 (Jean-Pierre Poly, 'Régime domanial et rapports de production 'féodalistes' dans le Midi de la France', in *Structures féodales et féodalisme dans l'Occident méditerranéen* (Roma 1980), p. 9; Jean-Pierre Devroey, 'Elaboration et usage des polyptyques. Quelques éléments de réflexion à partir de l'exemple des descriptions de l'Eglise de Marseille' in *Akkulturation. Probleme einer germanisch-romanischen Kultursynthese in Spätantike und frühen Mittelalter*, eds. D. Hägermann – W. Haubrichs – J. Jarnut (Berlin 2004), p. 436).

76 Landlords, ERAE p. 38, n. 93 and 95.

77 Étienne, 'La Gaule romaine', p. 103.

78 Maurice Chaume, *Les origines du duché de Bourgogne* II, 3 (Dijon 1931 = Aalen 1977), pp. 818, 983, 1014.

79 From the *Notitia Galliarum*, (ed. Seeck, *Notitia*, p. 262).

80 Jones considered Autun as a great town but estimated the population of Gaul less than 2,5 million, after a loss of 10/15 % (*The Later Roman Empire*, II, p. 1039).

average effective of 400 men for each unit, that would mean about 18.000 recruits. With a recruitment of 1 to 10, the Germanic population – culturally speaking –, free people and dependents, would count about 180.000 individuals. We get roughly the same results using the list of the Germanic units recruited in Gaul: three *auxilia palatina* with three levies each, nine with two levies each, three cavalry *vexillationes* with two levies each and some single levies of *gentiles* among the *pseudo-comitatenses*.[81] That would be about 7,5 % of the civilian population in northern Gaul.

But this number falls rather short of the total military population. In the infantry, there were Celtic units – Northern Britons, Caledonians, Irishmen came to Wales –, nearly as many as the Germanic ones. In the cavalry, were Iranian units – Sarmats, Taïfals, Parths, Alans – and other Easterners – Armenians, Syrians –, settled on the Loire and in the future Lorraine and Burgundy, as well as Moorish units in the West, the total amounting to 65 infantry *auxilia palatina* and 18 cavalry *vexillationes*. The *Gallia ulterior* had thus a military peregrine population of about 300.000 'barbarians', at least 12 % of the civilian 'roman' population. Still, if we accept Jones' low estimate of the civilians – 2.500.000 for the entire Gaul, not much more of a million for the North –, the peregrine military population would amount to 30%. Whatever the right number was, we have to admit that the 'barbarian' settlers were not a negligible part of the inhabitants of northern Gaul.[82]

4 Conclusion

The peregrine soldiers and their families of the *praepositurae* were settled in the empire long before the last invaders, Goths, Vandals or Lombards, came as unavoidable federates. The system – 'barbarian' settlements used as recruitment reserves for elite units and directly supported by tax-payers – existed only in the West: in Gaul, Northern Italy, Britain, Raetia and Pannonia prima.[83]

Under Merovingian rule, the peregrines of those settlements became part of the population of northern Gaul. The soldiers and their families became squireens; their neighbors, the *coloni*, payed the taxes to the ex-*praefectus/tribunus* companion (*comes*) to the king; the *possessores* kept to the towns and got what was left of the taxes. All free men were mobilizable, but the king

81 Districts and units taken in account, Poly, 'Laeti et gentiles', p. 39, fn. 99 and 100.

82 Kerneis, *Les Celtiques*, p. 320. Poly, 'Laeti et gentiles', p. 40, fn. 102.

83 *Amm.Marc.* 29.5.15. Poly 2017. ND Oc.34.24, 35.31, 42.51–63. Compare with CTh 5.6.3 in 409, where Skyres are still considered as *coloni*.

would rely upon the old military districts, the *centenae*. *Fraenc* now meant 'free', an old barbarian teaching: no fighting, no freedom.

The system of the *praepositurae* was not a negligible institutional by-product but the militarization of whole regions, a remedy to the political collapse of the empire in the West. It was a compromise worked through the army between two worlds, a part of the *Barbaricum* – the variegated European kin societies (*gentes*) – come in imperial territory and the *Romania* – *nobiliores* and *humiliores* subjects of an empire-state (*res publica*).

It did bring a social transformation. A wide bureaucratic empire and its heavy *Amtsaristokratie* were replaced by small kingdoms, scanty public administrations and local kin-lords.[84] For the slaves and the *coloni*, the change of masters would make no great difference. For the common 'barbarians' insiders and even for some *humiliores*, it must have been a liberation, to take Salvian's word for it. He remembered the siege of Trier, the selfish behaviour of the worthies and the starvation of the commoners.[85] *Quos vult perdere, Jupiter dementat.*

Bibliography

Primary Sources

Ammiani Marcellini Rerum gestarum libri, ed. J.C. Rolfe, 3 vols. (Cambridge, Mass., 1935–1939); *Histoires* (ed. and French transl.) XIV–XVI by E. Galletier (Paris 1968), XVII–XIX by G. Sabbah (Paris 1970), XX–XXII by J. Fontaine (Paris 1996), XXIII–XXV by J. Fontaine (Paris 1977), XXVI–XXVIII by M.-A. Marié (Paris 1984), XXIX–XXXI by G. Sabbah (Paris 1999)

Burchardus Wormatiensis Decretum, ed. F. Gagnon, *Le* Corrector sive Medicus *de Burchard de Worms* (Th. Montréal 2011, online)

CIL = *Corpus Inscriptionum Latinarum, consilio et auctoritate Academiae litterarum Borussicae editum* (Berlin 1863)

Codex Theodosianus: *Theodosiani libri XVI cum constitutionibus Sirmondianis et leges Novellae ad Theodosianum pertinentes*, ed. Th. Mommsen and P.M. Meyer, 2 vols. (Berlin, 1905); Engl. transl. by C. Pharr, *The Theodosian Code and Novels and the Sirmondian Constitutions. A Translation with commentary, glossary, and bibliography*, [Princeton, 1952], 10th Print (Union, N.J., 2001)

Gai Institutiones, ed. (and French transl.) ed. J. Reinach, *Gaius, Institutes* (1951, Paris 2019); ed. (and Engl. transl.) W.M. Gordon and O.F. Robinson, *The Institutes of Gaius* (London 1988)

84 B. Dumézil, Servir l'Etat barbare, p. 403.
85 Salvianus, 6.72–74, 82–84, 88–89.

Ibn Fadlan, ed. M. Canard, *Ibn Fadlan, voyage chez les Bulgares de la Volga* (Mayenne 1983)
Inscriptions of Roman Britain, ed. V.A. Maxfield and B. Dobson (London 2006)
Isidori Hispalensis Etymologiae, liber IX, ed. M. Reydellet (Paris 1984) [ed. Lindsay]
Laterculus Veronensis, ed. in O. Seeck, *Notitia dignitatum*, p. 247
Notitia Dignitatum omnium tam ciuilium quam militarium, ed. O. Seeck (1876, Frankfurt 1983); ed. C. Neira Faleiro, *La Notitia Dignitatum. Nueva edicion critica y comentario historico* (Madrid 2005); best ed., still in progress, I. Maier, *The compilation 'notitia dignitatum', Cnd* (//www.notitiadignitatum.org).
Pactus Legis Salicae, ed. K.A. Eckhardt [MGH, *Leg. Nat. Germ.*, 4.1] (Hannoverae 1962)
Panegyrici Latini, ed. R.A.B. Mynors, *XII Panegyrici Latini* (Oxford 1964); ed. (and French transl.) E. Galletier, *Panégyriques latins*, 3 vol. (Paris 1949, 1952, 1955)
Procopiou Kaisareôs Uper tôn Polemôn Logoi (*De bellis*) ed. (and Engl. transl.) H.B. Dewing, *History of the Wars* (Cambridge Mass. 1919)
(Pseudo-Nennius) *Historia Brittonum*, ed. J. Morris in *Arthurian Period Sources*, vol. 8 *Nennius: British History and the Welsh Annals* (London - Chichester 1980)
Salviani De gubernatione Dei, ed. G. Lagarrigue, *Salvien, Oeuvres II* (Paris 1975)
Skaldskaparmàl, French transl. by François-Xavier Dillmann, *L'Edda. Récits de mythologie nordique par Snorri Sturluson* (Paris 1991), p. 105.
Sidonii Apollinaris Epistulae et Carmina, ed. (and French transl.) A. Loyen, *Sidoine Apollinaire, Œuvres*, 3 vols. (Paris 1960–70)
Tabula Peutingeriana, ed. K. Miller, *Itineraria romana* (Stuttgart 1916); *Bibliotheca Augustana* (online : www.fh-augsburg.de)
Taciti Annales I–III and IV–VI, ed. (and French transl.) P. Wuillemier (1923,1949; Paris 2013, 2019)
Taciti Germania, ed. (and French transl.) J. Perret (1949, Paris 1983)
Zôsimou Istoriai neai, ed. (and French translation) F. Paschoud, *Zosime, Histoire Nouvelle*, I–VI (Paris 1971–1989)

Secondary Literature

Archan, Christophe. 'L'Epistula Honorii', in J.J. de los Mozos and I. Szaszdi eds. *El ejercito, la paz y la Guerra* (Valladolid 2009), pp. 153–164
Arnaud, Pascal. 'L'origine, la date de rédaction et la diffusion de l'archétype de la Table de Peutinger', in *Bulletin de la Société Nationale des Antiquaires de France*, 1988, pp. 302–320
Barbero, Alessandro. *Barbari. Immigrati, profughi, deportati nell'impero romano* (Roma-Bari 2006)
Beck, Heinrich, Geuenich, Dieter and Steuer, Heiko eds. *Reallexikon der germanischen Altertumskunde* (Berlin – New York, 1972–2008)
Blog, Peter Vilhelm. *The Bog People* (1964), (Ithaca 1988)

Bosworth, Joseph and Northcote Toller, Thomas eds., *An Anglo-Saxon Dictionary* (1898; Oxford 1964)

Botta, Fabio and Loschiavo, Luca eds., *Civitas, Iura, Arma; organizzazioni militari, istituzioni giuridiche e strutture sociali all'origine dell'Europa* (*secc.* III–VIII), (Lecce 2015)

Breton, Roland. *Géographie des langues* (Paris 1995)

Chaume, Maurice. *Les origines du duché de Bourgogne* II, 3 (Dijon 1931 = Aalen 1977)

Chauvot, Alain. 'Les lètes et leur devenir', in Becker, Audrey, Huntzinger, Hervé *et alii* eds *Les 'barbares' des Romains. Représentations et confrontations*, (Metz 2016), pp. 259–280

Chauvot, Alain. 'Julien, Ammien et les lètes. A propos d'Ammien, XX, 8, 13', *Ibidem*, pp. 281–299

Chauvot, Alain. 'La situation juridique des barbares dans l'Empire tardif et la question des unions mixtes', *Ibidem*, pp. 301–325

Chossenot, Raphaëlle. *Carte Archéologique de la Gaule. La Marne 51/1* (Paris 2004)

Corvisier, Jean-Nicolas and Suder, Wieslaw. *La population de l'Antiquité classique* (Paris 2000)

Devroey, Jean-Pierre. 'Elaboration et usage des polyptyques. Quelques éléments de réflexion à partir de l'exemple des descriptions de l'Eglise de Marseille' in Hägermann, Dieter, Haubrichs, Wolfgang and Jarnut, Jörg eds *Akkulturation. Probleme einer germanisch-romanischen Kultursynthese in Spätantike und frühen Mittelatler* (Berlin 2004), pp. 436–472

Drew Bear, Thomas. 'A Fourth Century Latin Soldier's Epitaph from Nakola', in *Harvard Studies in Classical Philo*logy, 81 (1977), pp. 257–274

Dumézil, Bruno. *Servir l'Etat barbare dans la Gaule franque* (*IV*[e]*–IX*[e] *siècle*) (Paris 2013)

Étienne, Robert. 'La Gaule romaine', in Dupâquier, Jacques ed. *Hist. de la population française*, I (Paris 1988), pp. 65–118

Ewig, Eugen. *Die Stellung Ribuariens in der Verfassungsgeschichte des Merowingerreiches* (Bonn 1969)

Geary, Patrick and Veeramah, Krishna. 'Mapping European Population through Genomic Research', in *Medieval Worlds* (2016), pp. 65–78

Hoffmann, Dietrich. *Das Spätrömische Bewegungsheer und die Notitia Dignitatum* (Düsseldorf 1969–1970)

Jones, Arnold Hugh Martin. *The Later Roman Empire* I–II (Baltimore 1986)

Kerneis, Soazick. *Les Celtiques. Servitude et grandeur des auxiliaires bretons dans l'empire romain* (Clermont 1998)

Kerneis, Soazick. '*Francus cives, miles romanus in armis*', in Aubert, Jean-Jacques and Blanchard, Pascal eds. *Droit, religion et société dans le Code Théodosien*, (Genève 2009), pp. 377–399

Kerneis, Soazick. '*Vox divi, vox populi*', in Kerneis Soazick ed., *Une histoire juridique de l'Occident. Le droit et la coutume* (*IIIe–IXe siècles*), (Paris 2018), pp. 129–186

Lantier, Raymond. 'Un cimetière du IVe siècle au Mont-Augé', in *L'Antiquité Classique*, 17 (1948), pp. 373–402

Lévi-Strauss, Claude. *Mythologiques*, IV (Paris 1971)

Loschiavo, Luca. 'Insediamenti barbarici e modelli di coesistenza nell'Italia altomedievale', in Rimoli Francesco ed. *Immigrazione e integrazione*, I (Napoli 2014), pp. 317–348

Maier, Ingo. *The compilation 'notitia dignitatum', Cnd* (//www.notitiadignitatum.org)

Marotta, Valerio. 'Il problema dei *laeti*. Fonti e storiografia', in Botta, Fabio and Loschiavo, Luca eds. *Civitas, Iura, Arma; organizzazioni militari, istituzioni giuridiche e strutture sociali alle origini dell'Europa* (*secc. III–VIII*), (Lecce 2015), pp. 117–157

Meulder, Marcel. '*Florus* II.30.24–25: trois peuples germaniques fonctionnels', in *Revue belge de philologie et d'histoire*, 85 (2007), p. 77–92

Much, Rudolf. *Völkernamen*, in Hoops, Johannes ed. *Reallexikon der germanischen Altertumskunde* IV, (Strasburg 1918–1919), p. 425

Pastoureau, Michel. *Une histoire symbolique du Moyen Age occidental* (Paris 2004)

Pérez, Emilie. 'La place de l'enfant dans l'espace funéraire du haut Moyen Age', in *Cahiers d'Histoire* 29 (2010), pp. 119–136

Piganiol, André. 'La capitation de Dioclétien', in *Revue Historique*, 176 (1935), pp. 1–13

Pohl, Walter. 'Telling the difference: signs of ethnic identity', in Pohl, Walter and Reimitz Helmut eds. *Strategies of distinction* (Leiden 1998), pp. 17–69

Pohl, Walter. *Le origini etniche dell'Europa* (Roma 2008)

Poly, Jean-Pierre. 'Régime domanial et rapports de production 'féodalistes' dans le Midi de la France', in *Structures féodales et féodalisme dans l'Occident méditerranéen* (Roma 1980), p. 90

Poly, Jean-Pierre. *Le chemin des amours barbares* (Paris 2003)

Poly, Jean-Pierre. '*Arcana conjugalia*. Le secret des mariages germaniques', in Otis-Cour, Lea ed. *Histoires de famille. A la convergence du droit penal et des liens de parenté*, (Limoges 2012), p. 219–230

Poly, Jean-Pierre. 'Sous les chênes de Salaheim. La loi salique, l'armée romaine et le bilan de la barbarie', in Botta, Fabio and Loschiavo, Luca eds. *Civitas, Iura, Arma* (Lecce 2015), p. 193–219

Poly, Jean-Pierre. ' Les Saxons du Bessin et leurs compagnons. Ethnies et acculturation en Gaule du Nord au début du Moyen Age', in Davy, Gilduin and Mausen, Yves eds. *La Normandie, terre de traditions juridiques*, (Rouen 2016), p. 25–88

Poly, Jean-Pierre. '*Leges barbarorum*. La création des lois des nations' in Kerneis, Soazick ed. *Une histoire juridique de l'Occident. Le droit et la coutume* (*IIIe–IXe siècles*) (Paris 2018), p. 187–247

Poly, Jean-Pierre. 'Emblèmes et couleurs du Nord. Les écus des auxiliaires francs de l'armée romaine au IVe siècle', in *Bulletin de la Société Nationale des Antiquaires de France* 2019 (under publication)

Poly, Jean-Pierre. '*Laeti et gentiles*. Les établissements militaires de la Gaule romaine', in G. Bassanelli Sommariva ed. *Ravenna Capitale. L'esercito romano e l'alba dell'Europa. Modelli concettuali e sperimentazioni sociali tra linguaggi, istituzioni e diritto* (*secoli IV–VIII*) (Santarcangelo di Romagna 2020), pp. 15–40

Poly, Jean-Pierre and Kerneis, Soazick. 'Enfances barbares. Les communautés péregrines de Gaule entre loi et coutume', in *L'infanzia nell'alto medioevo, Settimane del CISAM* LXVIII (Spoleto, 2021), pp. 421–454

Russell, Josiah. 'Population in Europe', in C. Cipolla ed, *The Fontana Economical History of Europe*, I, (Glasgow 1972)

Scheungraber, Corinna and Grünzweig, Friedrich. *Die Altgermanischen Toponyme sowie ungermanische Toponyme Germaniens. Ein Handbuch zu ihrer Etymologie* (Wien 2014)

Sitzmann, Alexander and Grünzweig, Friedrich. *Die altgermanischen Ethnonyme. Ein Handbuch zu ihrer Etymologie* (Wien 2008)

Treadgold, Warren. *Byzantium and Its Army* (Stanford 1995)

Ubl, Karl. *Sinnstiftungen eines Rechtsbuchs. Die Lex Salica im Frankenreich* (Ostfildern 2017)

Vogel, Cyrille. 'Pratiques superstitieuses au début du XIe siècle d'après le *Corrector sive Medicus* de Burchard évêque de Worms', in *Mélanges E.R. Labande* (Poitiers 1974), pp. 751–761

Wenskus, Reinhard. *Stammesbildung und Verfassung* (Köln 1961)

Wolfram, Herwig. *The Roman Empire and its Germanic Peoples* (1990; Engl.transl. Berkeley 1997)

Zuckerman, Constantin. 'Two reforms of the 370s. Recruiting soldiers and senators', in *Revue d'études byzantine*, 56 (1998), p. 79–139

CHAPTER 15

Personality of Law or *Ius Speciale Militum*? Around the Origins of the *Leges Barbarorum*

Luca Loschiavo

1 Romano-Barbarian *Regna* and the Principle of the Personality of Law

One of the most complex – and most fascinating – issues when studying the shift from Late Antiquity to the early Middle Ages is the forced coexistence of different ethnic groups in areas where Romano-Barbarian kingdoms were forming. These kingdoms were characterised by new political regimes expressing, in most cases, wholly original social and legal structures, shaped specially to manage this coexistence. For a long time, almost unquestioningly, historiography has recognised in the so-called 'principle of the personality of law' the mechanism which made this coexistence possible, and which greatly facilitated the formation of new kingdoms.[1]

The principle of the personality of law – its theory, and its practice – does not only interest researchers of the past. It is also of great interest for legal scholars, who recognise in this principle one of the historical roots of modern international private law.[2] However, historians and legal scholars do not always

---

1 The literature on the issue is vast. Here are some particularly enlightening works on the different historiographical seasons: Friederich Karl von Savigny, *Geschichte des römischen Rechts im Mittelalter*, 2nd ed. (Heidelberg, 1834), pp. 30–36 and 115–130; Louis Stouff, *Étude sur le principe de la personnalité de des lois depuis les invasions barbares jusqu'au XII siècle* (Paris, 1894); Heinrich Brunner, *Deutsche Rechtsgeschichte*, I, 2nd ed. (Berlin, 1906), pp. 383–399; Theodor Mayer, 'Die Entstehung des 'modernen' Staates im Mittelalter und die freien Bauern', in *Zeitsch. der Sav. Stift. – Germ. Abt.*, 57 (1937), pp. 210–288; Guido Astuti, 'Personalità del diritto', in *Novissimo Digesto Italiano*, XII (1965), pp. 1088–1092; Maurizio Lupoi, *The Origins of European Legal Order*, (orig. ed. 1994) Engl. transl. (Cambridge, 2000), pp. 388–405; Ennio Cortese, *Il diritto nella storia medievale* I. *L'alto medioevo* (Rome, 1995), pp. 58–61, 79, 226–232.

2 Karl Neumeyer, *Die gemeinrechtliche Entwieckelung des internationalen Privat- und Strafrechts*, I–II (Munich, 1901–1916); Eduard Maurits Meijers, *L'histoire des principes fondamentaux de droit international privé à partir du Moyen Age* (Paris, 1935); Bruno Paradisi, *Storia del diritto internazionale nel medio evo* (Milan, 1940), pp. 439–519; Simeon L. Guterman, 'The Principle of the Personality of Law in the Early Middle Ages', in *Univ. Miami Law Review*, 21 (1966), pp. 261–348.

 | DOI:10.1163/9789004698017_016

agree. It is therefore best to proceed with caution, defining the topic as clearly as possible.

From a theoretical point of view, the meaning of 'personality of law' is easily identifiable when we view it alongside, for instance, the better-known 'territoriality of law'. We speak of 'territoriality' when all the people living under a certain legal system are affected by it equally. We have the 'personality' of law when, within a given legal system, each individual can live according to the laws of their country of origin. This means that within that legal system, in conditions of parity, it is possible to accept as many laws as there are ethnic groups living in that territory, regardless of whether the ethnicity is represented by a small or large group of individuals. The law therefore adheres to the individual at birth, like a second skin, and follows him wherever he may go. Full application of the principle of personality therefore implies that an individual in a foreign country can regulate family relationships, undertake obligations, trade, and even respond to criminal convictions according to the law of the country or lineage of origin.

It was nineteenth-century law scholars – engaged in critical editions of Medieval legal sources, the *leges barbarorum* – who suggested that the principle of the personality of law played a key role in the situations outlined above. Although none of the ancient texts which have reached us mentions it clearly, or considers it to be generally applicable, these scholars became convinced that the principle of personality determined the correlation between the different national *leges*, as well as – and especially – Roman law which, in several and nuanced ways, lived on in the former provinces of the Empire, in spite of the absence of the Emperors' formal authority.

As was customary amongst nineteenth- and early-twentieth-century scholars, the main concern was to discover the origins of the principle: that is to say, if it were possible to recognise it as being a typical outcome of the Germanic legal mentality (the 'Germanic' theory), or whether it were possible to identify a model for it in the Roman world which may have inspired it (the 'Roman' theory), and hence a sign of continuity with its recent past. Those who endorsed its Germanic origin stressed how the concept of law in terms of an inherent quality of an individual was a consequence of the fact that many ancient Germanic peoples (in common with those societies that are generally the expression of more primitive civilisations) originated from the grouping together of people irrespective of the territory which hosted them; these peoples changed constantly due to the nomadic life they led.[3] Although the observation appears

3 The first to suggest this hypothesis may have been Montesquieu (Charles Louis de Secondat baron de Montesquieu, *De l'Esprit des lois* [Geneva, 1748], XXVIII.2). The idea was then

sound from an anthropological point of view, it fails to explain the origin of the system of the personality of law applied amongst different populations. It does not clarify how the law of origin was preserved when a group entered another political jurisdiction, nor how it was possible for the individuals who left their community to bring their legal rights with them. Furthermore, from a historical point of view, there is evidence to suggest that the Germanic peoples were not aware of that system before they settled within the confines of the Empire.[4]

Scholars who hypothesised a Roman origin, on the other hand, stressed how the Romans would often allow subjected populations to maintain their own laws, and that in Rome the *peregrini* used a different legal system. However, this case does not exactly describe the above-mentioned principle: Rome did not in fact recognise equal dignity to the laws and rights of all subjected populations. Compared with the system popular in the early Middle Ages there were therefore substantial differences, both in terms of the implementation (the validity of those rights derived from an explicit concession), and of mutual relationships (Roman law always took precedence, and foreigners were always legally inferior to the citizens).

With the origin not being clearly defined by either tradition, a further theory emerged and eventually prevailed. According to this third theory, the principle of the personality of law was a 'natural result of the invasions'.[5] The principle became established in Europe precisely when the new Romano-Barbarian *regna* progressively substituted the – by now exhausted – Western Empire. The period was 'characterised by the existence of nations without territories and of laws without states'[6] – and the principle established itself fully only with the creation of the Carolingian Empire.

adopted by many 'Germanists' (like Waitz and Dahn) and taken up in Schröder's celebrated handbook (Richard Schröder, *Lehrbuch der deutschen Rechtsgeschichte* [Leipzig, 1907]).

4 We can recall the case of those Saxons who, descending together with the Lombards in the wake of Alboin's rule, were nevertheless forced to leave Italy in order not to suffer the imposition of abandoning their customs to adopt those of the numerically superior Lombards (Pauli Diaconi, *Hist. Langobardorum* 3.6). It may therefore be inferred that the groups of Gepids, Heruli and Bulgarians, who like the Saxons had joined Alboin, had instead suffered that fate. Equally, the oldest Salic law (*Pactus legis Salicae*, 41.1) also confirms the existence, alongside the *ingenuus Francus*, of the 'barbarian who lives according to the Salic law', certainly alluding to warriors belonging to another barbarian *natio*, incorporated into the army and equal in status to fellow Franks.

5 So Adhémar Esmein, *Cours élémentaire d'histoire du droit français* (Paris, 1906), pp. 54–55 and fn. 3.

6 Guterman, 'The Principle of the Personality', p. 261.

The issue of origins was not the only question surrounding the principle of the personality of law. In fact, the transition from a purely theoretical level to a more concrete historical basis presented scholars with a series of complex problems to solve.[7] Considering its practical application, the principle was inevitably subject to heavy conditioning. The attestations were also often fragmentary, if not downright contradictory.[8] Most scholars were then convinced that the principle of the personality of law was gradually implemented over time. At first it may have been used by Barbarian individuals exclusively with regard to the Roman population, once subjected. As it was not possible to impose a much more primitive legal tradition, the Germanic peoples accepted that the Romans should continue to follow their own. It was only after some time, when the necessity presented itself of incorporating other Barbarian groups under the same reign, that the sovereigns allowed for the same treatment reserved to the Romans to be extended to the newcomers. With the creation of the Carolingian Empire the principle finally found widespread application, and the various systems established in the meantime to solve the objective difficulties of its practical usage (from the presence of *scabini* to the *professiones legum* and so forth)[9] were institutionalised.

2 Criticism of the Traditional Approach

At the heart of the common conviction that the principle of personality governed the shift from the Roman legal system to that enforced by the new peoples, there were two basic assumptions. On the one hand, it was believed that the arrival of the Barbarians had triggered a genuine 'clash of civilisations', and that the clash could only be resolved with a clear-cut dualism of legal cultures: the Roman law of the conquered was thus juxtaposed with the Germanic

7 Francesco Calasso, *Medioevo del diritto* I. *Le fonti* (Milan, 1954), pp. 110–118 and 183–188.

8 See Stefania Pietrini, *I sovrani goti e il* ius commune *nelle* Variae *di Cassiodoro Senatore*', in *Atti dell'Acc. Romanistica Costantiniana*, XXI (2016), pp. 355–372. To give an example, we can refer here to *Form. Andecav.* 54 (ed. Zeumer, *Formulae*, p. 23) where the *lex romana* and the *consuetudo pagi* (whose 'territorial' meaning is undoubted) or to the even more explicit *Marculfi form.* 1.37 (ed. Zeumer, *Formulae*, p. 67) '… quicquid lex loci vestri de tali causa edocet …'.

9 Guido Astuti, 'Note critiche sul sistema delle fonti giuridiche nei regni romano-barbarici dell'Occidente', (1970) now in G.A., *Tradizione romanistica e civiltà giuridica europea*, (G. Diurni, ed.), I (Rome, 1984), pp. 15–18. Even during Charlemagne's time, the principle was applied in a rather peculiar way; see for example Ennio Cortese, 'Nostalgie di romanità: leggi e legislatori nell'alto medioevo barbarico', in *Ideologie e pratiche del reimpiego nell'alto Medioevo* (Settimane CISAM, XLVI), (Spoleto, 1999) pp. 485–510, here esp. pp. 502–503.

law of the conquerors. On the other hand, there was the conviction that the newcomers firmly opposed the otherwise natural tendency to merge with the populations living stably on the same territory. Therefore, the principle of the personality of law was adopted precisely in order to maintain this dualism. In fact, it safeguarded the individuality of the ethnic groups, and their position of privilege on a legal-formal level.

It is well-known that this way of regarding the penetration and settling of Barbarian peoples in the Roman world underwent a thorough review, starting from the 1970s. It was then that the theory of the clash was juxtaposed by scholars with the theory of the 'transformation of the Roman world'. Today, half a century later, most scholars are convinced that the Barbarians were not *tout court* enemies of *romanitas*, and that they did not wish to keep themselves separate from the resident population. Rather than overthrowing the existing socio-economic structures, they aimed at being part of them, making the most of all the possible advantages. The objective for the Barbarian warriors was to take the place of the previous élites, or to intermingle with them whenever taking their place proved too arduous an endeavour. They basically wanted to become a part of the existing system, and take maximum advantage from it.

But before this important historiographical change was completed, the old view had already received serious criticism from scholars who were themselves legal historians. Thus – following the steps of Ernst Levy and Paulo Merea's research on so-called 'Roman vulgar law' – Alfonso García Gallo in the '40s[10] and his pupil Álvaro d'Ors in the '50s[11] had already begun to criticise the application of the principle of the personality of law under Visigothic rule. They tried to demonstrate the territorial rather than the personal value of the *Codex Euricianus* and Alaric's *Breviarium*. Although to date there are still doubts, as there were then, about this theory,[12] the Spanish scholars were not alone in their considerations. Other legal historians found it very difficult to

10 Alfonso García Gallo, 'Nacionalidad y territorialitad del derecho en la época visigoda', in *Anuario de historia del dercho español*, XIII (1936–1941), pp. 91–124.

11 Àlvaro d'Ors, 'La territorialidad del derecho de los Visigodos', in *I Goti in Occidente* (Settimane CISAM, III), (Spoleto, 1956), pp. 363–408; Id., *El Codigo de Eurico. Edición, Palingenesia, Indices* (Estudios Visigoticos, II), (Rome-Madrid, 1960 = Madrid, 2014), pp. 1–12.

12 See Astuti, *Note critiche*, p. 15; Harald Siems, 'Lex Romana Visigothorum', in *Handwörterbuch zur deutschen Rechtsgeschichte*, II (Berlin, 1978), pp. 1940–1949, here p. 1941; Paul David King, 'King Chindasvind and the First Territorial Law-code of the Visigothic Kingdom', in *Visigothic Spain: New Approaches*, ed. E. James (Oxford, 1980) pp. 131–157; Id., 'The Alleged Territoriality of Visigothic Law', in *Authority and Power*, eds. B. Tierney – P. Linehan (Cambridge, 1980) pp. 1–11. See also Jean-Pierre Poly, '*Leges Barbarorum*. La création des lois des nations', in *Une histoire juridique de l'Occident*, ed. S. Kerneis (Paris 2018), pp. 187–247, here p. 233.

recognise the full implementation of the principle of personality when looking at Ostrogothic and Burgundian rulers, not to mention the reigns created by the Franks, the Longobards, and the early Anglo-Saxons. Just as there is no trace of 'brotherhood' connecting even neighbouring Barbarian peoples, it is equally impossible to find a general "principle that protected the laws of other people, or which allowed foreigners to bring their own laws with them and to be governed by them".[13]

The fact that it is difficult to recognise a full differentiation on the basis of the personality of the laws in force within the new *regna* does not invalidate the evidence that those legal systems were characterised by a certain dualism, or rather plurality. The Gothic kingdom of Toledo for around a century and a half maintained two distinct bodies of laws: the repeatedly-updated *Leges Visigothorum*, and the *Breviarium Alaricianum*. Likewise, the Ostrogoth Theodoric and the Burgundian Gundobad allowed different legal systems to be adopted for the Romans and for the Barbarians coexisting in their respective kingdoms. Except for mixed cases, different judges were supposed therefore to apply different legal systems. Finally, the Franks and the Longobards allowed the Romance population to retain their own legal tradition undisturbed.

We are thus back to the initial issue. If there is no systematic application of the principle of the personality of law during the first centuries of the Middle Ages, how the pluralism of law was effectively carried out amongst Romano-Barbarian kingdoms still needs to be explained. Over the past decades there have been new attempts to provide explanations, some more convincing than others.[14] The issue, however, does not seem to have found a solution as yet. I would like to enter the debate by suggesting a different hypothesis, which stems from two preliminary considerations that take into account the new discoveries in historical research surrounding the arrival of Barbarian populations in Roman territory.

3 A Different Interpretation

The first – if not new – consideration derives from the observation that the encounter of the Barbarians with the Roman world started and continued in

13 So Lupoi, *The Origins*, p. 393. See also Patrick Amory. 'The meaning and purpose of ethnic terminology in the Burgundian laws', in *Early Medieval Europe*, 2.1 (1993), pp. 19–23.

14 Karol Modzelewski, *Barbarian Europe*, (orig. ed. 2004) Eng. transl. (Frankfurt am Main 2015), ch. 2 § 3; Nicholas Everett, 'How territorial was Lombard law?', in *Die Langobarden: Herrschaft und Identität*, eds. W. Pohl – P. Erhart, (Vienna, 2005), pp. 345–360 and Poly, *Leges Barbarorum*, pp. 212–215.

the name of militias and mercenary war. For a long time the Barbarian warriors measured themselves against the Imperial Army. In particular, when they were recruited by the legions, together with Roman fighting techniques they also learned Latin (the necessary medium to understand orders, and the *lingua franca* in the melting pot that was the late-Roman Army), and military discipline, i.e. above all respect for hierarchy and the duty of obedience to non-tribal leaders.[15] But Barbarian warriors also ended up coming in contact with legal establishments hitherto unknown. For instance: regular pay, and furlough at the end of service; the possibility to name an heir through a will; the obligation to resort to the commander in case of offences, damages or injuries received from a comrade in arms[16] and to consign to the commander any valuables found in or around the encampment.[17] More importantly, they learned the principle according to which soldiers benefited from a special, privileged legal condition, quite different from that bestowed on the rest of the population.

No less significant for these groups was the experience of *militia foederata*. Again, the mere duty – one should say 'by contract' – of protecting a section of frontier, of presiding over a portion of imperial territory (a *tractus*), represented for those warriors (and for their families, who often came with them or were newly-formed) the submission to discipline. Even if it did not coincide entirely with that imposed by regular troops, it was nevertheless a military discipline, and inevitably included much of the Roman military order.[18]

15 Some cases of disobedience to superiors are considered in chapters 2, 3, 11, 19, 21, 25 of the so-called *Leges ex Ruffo*, a short criminal law collection also known as *Nómoi stratiotikoí* and dating from between the second half of the fourth and the seventh centuries (ed. G. Famiglietti, *Ex Ruffo leges militares* [Milan 1980], pp. 18–25). Similarities can be noted with some passages of *De re militari* of the Digest (*e.g.* D. 49.16.3.4; 49.16.10; 49.16.3.13).

16 See *Ex Ruffo leges* ch. 4 (ed. Famiglietti, pp. 18–19); also D. 49.1612.2. Regarding soldiers turning to their superiors for the decision of their disputes see Ramsay MacMullen, *Soldier and Civilian in the Later Roman Empire* (London, 1963), p. 54. This Roman rule should be compared with *Lex Alamannorum* LXXXIV (ed. Eckhardt, p. 148).

17 *Ex Ruffo leges* ch. 10 (ed. Famiglietti, pp. 20–21), and D. 47.2.43.4 and 7 should be compared respectively with the chapters 260 and 16 of the *Edictum Rothari* (ed. Bluhme, pp. 15 and 63).

18 We can consider the case of the Rugian Frideric, reproved by Theodoric in 491 for not showing sufficient respect for the people of Pavia who had been entrusted to him; cf. Herwig Wolfram, *History of the Goths*, Eng. transl. (Berkely 1990) p. 282. Other cases in which Roman officers imposed military discipline on Barbarian warriors recruited as federates or allies (or require Barbarians to respect precepts that are alien to them) are mentioned in Fausto Goria, 'Romani, cittadinanza ed estensione della legislazione imperiale', (1984) now in F.G., *Diritto romano d'Oriente*, eds. P. Garbarino *et all.* (Alessandria 2016), pp. 76–77.

For years several scholars – such as Alexander C. Murray,[19] Thomas Anderson,[20] Wulf Eckart Voß,[21] Jean-Pierre Poly,[22] and Soazick Kerneis[23] – have highlighted the presence of elements of Roman military discipline in the *Pactus legis Salicae*: a document for a long time considered one of the most exemplary of the 'genuine' Germanic legal tradition. Even before them, Gian Piero Bognetti[24] and Ottorino Bertolini[25] had made steps in that direction. Both had recognised in the Longobard system – beyond the titles taken from the Roman Army[26] – clear traces of rules and principles of the Roman military law. It was not so long ago that I myself tried to point towards a few further examples of rules of the *leges barbarorum* which clearly derived from the Roman army.[27]

The second observation concerns the political and legal culture of the Barbarians; in particular, we have to consider the way in which the newcomers

---

19 Alexander Callander Murray, 'From Roman to Frankish Gaul: 'centenarii' and 'centenae' in the administration of the Merovingian Kingdom', in *Traditio*, 44 (1988), pp. 59–100.

20 Thomas Anderson, 'Roman Military Colonies in Gaul, Salian Ethnogenesis and the Forgotten Meaning of Pactus Legis Salicae 59.5', in *Early Medieval Europe*, 4(2) (1995), pp. 129–144.

21 Wulf Eckardt Voß, 'Vom römischen Provinzialprozeß der Spätantike zum Rechtsgang des frühen Mittelalters', in *Recht im frühmittelalterlichen Gallien. Spätantike Tradition und germanische Wertvorstellungen*, eds. H. Siems – K. Nehlsen-von Stryk – D. Strauch (Cologne-Weimar-Vienna, 1995) pp. 73–108.

22 Jean-Pierre Poly, 'La corde au cou. Les Francs, la France et la loi salique', in *Genèse del l'Etat moderne en Méditerranée*, (École française de Rome) (Rome, 1993), pp. 287–320 and Id., 'Freedom, warrior's bond, legal book. The 'Lex Salica' between Barbarian custom and Roman law', in *Clio@Themis*, 10 (2016), pp. 1–25.

23 E.g. Soazick Kerneis, '*Francus civis, miles Romanus*: les barbares de l'Empire dans le Code Théodosien', in *Droit, religion et société dans le Code Theodosien*, eds. J.-J. Aubert – P. Blanchard (Geneva, 2009), pp. 377–399; Ead., 'Vox divi et vox populi: la pluralité des droits', in *Une histoire juridique de l'Occident. Le droit et la coutume (IIIᵉ–IXᵉ siècle)*, ed. S. Kerneis (Paris, 2018), pp. 173–186.

24 Gian Piero Bognetti, 'L'influsso delle istituzioni militari romane sulle istituzioni longobarde del secolo VI e la natura della "fara"', (1953) now in G.P.B., *L'età longobarda*, vol. 3 (Milan, 1967) pp. 1–46.
Bognetti, *L'influsso*.

25 Ottorino Bertolini, 'Ordinamenti militari e strutture sociali dei Longobardi in Italia', in *Ordinamenti militari in Occidente nell'alto medioevo*, (Settimane CISAM, XV) (Spoleto, 1968), I, pp. 429–607.

26 Lombard leaders continue to be called *duces*, *comites*, *centenarii* and *decani*; see Bertolini, 'Ordinamenti militari', pp. 482–493.

27 Better than with the fragments collected under the Digest's title *De re militari* (D. 49.16) – however, it is noticeable how close the D. 49.16.3.17 and *Edictum Roth.* 244 are – the proximity can be noticed by comparing the rules present in the *Leges militares ex Ruffo* (*supra* n. 15); see Luca Loschiavo, *Figure di testimoni e modelli processuali tra antichità e primo medioevo* (Milan, 2004), pp. 82–103, 159–176, 195–213.

settled within the Empire. Even when the Barbarian warriors settled by force, their establishment within the Roman territory was always, somehow, effected under legal parameters. At least formally, the Barbarian leaders and the Roman Emperors always found a way to give a 'normalising' interpretation of the new situation.[28] Officially, the access and settlement of the new peoples was never presented as conflicting with the imperial will. The Barbarian sovereigns recognised (at least formally) the higher authority of the Emperor. In exchange, they received formal titles of legitimacy valid amongst the resident population.[29] Whenever possible, they retained their pre-existing administrative systems. Therefore, the rise of new regimes could be presented as being carried out within the Roman state (and the *societas christianorum* with which it was identified) rather than against it. Everything happened within a framework of *continuity* rather than the *breakdown* or subversion of pre-existing structures.

From these premises stems the above-mentioned hypothesis. It is a question of trying to establish whether the foundation of the legal pluralism that is so characteristic of post-Roman times was the problematic principle of personality, or – at least in its initial phase with the rise of new political figures – the well-known dichotomy *ius speciale / ius commune*. In other words, once they had gained control over the former Roman provinces, the new settlers could have started to establish their relationship with the resident population by simply applying the same form of interaction between the Roman Army and civilians, which ensured a special, more advantageous legal status compared to those who did not participate in the military field.

On a theoretical level, the possibility of benefiting from a different treatment from the 'common' one can be explained by the fact of belonging to a specific category. Shifting from the general legal system is not justified on the basis of some special ethnic connotation (as is the case of the personality of law), but by the importance bestowed on the function connected to the

28 See (for the Burgundian world) Ian Wood, 'The Burgundians and Byzantium', in *Western Perspectives on the Mediterranean*, eds. A. Fischer – I.N. Wood (London, 2014) pp. 1–22 and (for the Visigoths) Christine Delaplace, *La fin de l'empire romain d'Occident. Rome et les Wisigoths de 382 à 531* (Rennes, 2015). The case of the Lombards is different, although also in this case the narrative today can no longer develop in the traditional terms of the clash of civilizations; see Stefano Gasparri, 'La migrazione longobarda in Italia tra mito e realtà', in *Le migrazioni nell'alto medioevo* (Settimane CISAM, LXVI), (Spoleto, 2019), pp. 375–393.

29 In this context we can remember the imperial recognition long invoked and finally obtained by Theodoric the Amal: the title of *praeses provinciae*, with which bishop Remigio of Reims greeted Clovis, the double role of Gundobad and, finally, the title of *flavius* which the Lombard sovereigns boasted starting from king Autari.

individual category.[30] The military function is the one most frequently seen as worthy of justifying the existence of special or extraordinary rules. The *ius militare* almost represents the prototype of the *ius speciale* (another, later example is represented by the church, with the special forum reserved to clerics … but here too we should remember the role of the very ancient idea of *militia Christi*).

4 The Ostrogothic Kingdom

Today we know that the question of the post-Roman Western world must take into account the specific differences across regions. Any verification should therefore consider the new *regna* individually. I will here focus on two: the Ostrogothic kingdom, and the Burgundian kingdom.

The kingdom in which Theodoric the Amal transformed the diocese of Italy was short-lived (lasting barely half a century). However, it had a deep effect on the history of the Romano-Barbarian West. It is well-known that Theodoric had a dual institutional profile. On the one hand he presented himself as a high magistrate of the Empire: a Western *imitatio* of the Emperor of Constantinople. On the other hand, Theodoric was the sovereign of an army-population in which the Gothic component was predominant, but not absolute.[31] His politics wavered occasionally through the decades, but it was based on a few clear principles.

Firstly, he strengthened his monarchy, thus ensuring the loyalty of those men with whom he had unseated Odoacer – in spite of the heavy price in terms of effort and bloodshed. In order to preserve their loyalty it was imperative to turn their dream into reality: give them a homeland (a *Gothìa*) and holdings. This way, Theodoric turned his men from mercenaries into landowners.

30 Cortese, *Il diritto nella storia*, pp. 79–80. See also Stefano Gasparri, 'Identità etnica e identità politica nei regni barbarici postromani: il problema delle fonti', in *Civis/civitas. Cittadinanza politico-istituzionale e identità socio-culturale da Roma alla prima età moderna*, eds. C. Tristano – S. Allegria (Montepulciano, 2009), at p. 196. Here one must also consider what Kerneis writes in this same volume about justice in the late Roman provinces (p. ++): '… in the case of mixed trials …, the search for the truth depends on the origin and the condition of the parties, as if already, anticipating the times to come, the trial begins with the question: 'Under which law do you live?''.

31 During the long journey from the Black Sea to Italy the *exercitus Gothorum* grew, by including other Barbarian groups and even Roman provincials. Furthermore, it is likely that several soldiers of the defeated Odoacer also chose to enlist under the banner of the winner.

However, he did not wish them to cease being soldiers.[32] In order for them not to lose their warlike attitude, and for them to keep the army strong and tight, it was important to preserve their traditions, their language, their religion in order to prevent them from blending completely with the Roman civil population, which was numerically superior. At the same time, Theodoric wanted to gain the favour of the Romans. To this aim, he intended to preserve the social, economic, cultural, and therefore religious and legal structures which existed before the arrival of the Goths.[33]

In Theodoric's vision, the two different realities were meant to coexist and collaborate without mixing, each with their separate roles and operating 'in a spirit of friendship'. The Goths and other Barbarian populations dealt with the military sphere (and therefore the defence of the borders, and preserving internal order); the Romans with administration, agriculture, commerce, and all manner of civilian activity. The policy was dualistic, though not downright 'separatist', as has been hypothesised.[34] To return to the issue at hand: if therefore the function of the Goths was military, their specific status gave them the possibility of living in the new political organisation in accordance with the special laws of the army. In practical terms this meant resolving their controversies before a military judge, the *comes Gothorum*. This figure was originally supposed to incarnate tribal authority and military command. However, with the establishment of the new kingdom it morphed into an emanation of royal power.[35] The role ceased to be that of ethnic judge and turned into an itinerant official (and not necessarily one of Gothic origin, for that matter)[36] who answered only to the king and whom the king sent wherever he was needed.

---

32 Roman military rules also prohibited soldiers from cultivating fields and keeping animals: *Ex Ruffo leges militares*, ch. 29 (ed. Famiglietti, pp. 26–26); C.I. 12.35(36).15.

33 Pierfrancesco Porena, *L'insediamento degli Ostrogoti in Italia* (Rome, 2012), here esp. pp. 18–181.

34 So Adriano Cavanna, 'Diritto e società nei regni ostrogoto e longobardo', (1984) now in A.C., *Scritti (1962–2002)* (Naples, 2007) I, p. 524. Stefano Gasparri and Cristina La Rocca, *Tempi barbarici. L'Europa occidentale tra antichità e medioevo (300–900)* (Rome, 2012), pp. 100–101, view the Ostrogothic kingdom in terms of 'dualism'. See also Patrick Geary, *The Myth of Nations. The Medieval Origins of Europe* (Princeton – Oxford, 2003), pp. 107–108.

35 Kayoko Tabata, 'I comites Gothorum e l'amministrazione municipale in epoca ostrogota', in *'Humana sapit'. Études d'antiquité tardive offerts à Lellia Cracco Ruggini*, eds. J.-M. Carrié – R. Lizzi Testa (Turnhout, 2002) pp. 67–78; Gideon Maier, *Amtsträger und Herrscher in der Romania Gothica. Vergleichende Untersuchungen zu den Institutionen der ostgermanischen Völkerwanderungsreiche* (Stuttgart, 2005), pp. 207–222.

36 For example, considering his name, the *comes Colosseus* sent by Theoderic to govern the *Pannonia Sirmiensis* would appear to be Roman: Cassiodorus, *Var.* 3.23 (ed. Mommsen, p. 91); Sean D.W Lafferty, *Law and Society in the Age of Theoderic the Great. A Study of the* Edictum Theoderici (Cambridge, 2013), p. 103.

A well-known *varia* by Cassiodorus – the *Formula comitivae Gothorum per singulas civitates* – outlined his duties.[37] In case of a dispute, it confirmed that the Romans had the right to resort to their own natural judges (*cognitores*) and their complex legal tradition. The *comes* was meant to solve conflicts among the Goths, or between Goths and Romans. Cassiodorus' choice of words, always very polished and refined, further testifies to the military nature of his jurisdiction. It should be noted that extending the military jurisdiction to civilians in 'mixed' trials was not altogether a new concept. Considering late-imperial legislation it is clear that the authority and expertise of military judges had expanded significantly during the 4th and especially the 5th centuries.[38] And after all, even the circumstance noted by Amory – in which Athalaric allowed two Roman litigants to be judged by a Gothic *comes* (and therefore by a military court) – was not entirely original.[39] The only (albeit substantial) new aspect on a practical level was, in these cases, the imposition of a *prudens* (an expert on Roman law) alongside the *comes*. The former had the double function of suggesting, in case of necessity, the use of certain Roman laws the latter would not know about, and most of all to check that the Gothic *comes* remained impartial and did not penalise the Roman litigant.[40]

37 Cass., *Var.* 7.3.1: 'Since we know that, by God's blessing, the Goths live intermingled with you, lest any trouble should arise among partners, as is accustomed to happen, we deemed it necessary to send you a *comes* who, being a man lofty and hitherto comported with good character, ought to decide litigation between Gothic parties according to our edicts (*secundum edicta nostra inter duos Gothos litem debeat amputare*), and even if any kind of dispute will have arisen between a Goth and a freeborn Roman (*inter Gothom et Romanum natum*), by summoning to himself a wise Roman advisor, will he be able to disarm the dispute with fair reason (*certamen possit aequabili ratione discingere*). Between two Romans, however (*Inter duos autem Romanos Romani*), let those Romans who we have sent to the provinces as legal representatives hear the case, so that each may be served by his own laws and a single justice may embrace all people under a diversity of judges (*ut unicuique sua iura serventur et sub diversitate iudicum una iustitia complectatur universos*)' (my translation); see for a comment Luca Loschiavo, *L'età del passaggio. All'alba del diritto comune europeo (secoli III–VII)*, 2nd ed. (Turin, 2019), pp. 162–169.

38 After Voß, *Vom römischen Provinzialprozeß* and Loschiavo, *Figure di testimoni*, pp. 82–88, see Fausto Goria, 'Giudici civili e giudici militari nell'età giustinianea', (1995) now in F.G., *Diritto romano d'oriente*, pp. 225–239.

39 Patrick Amory, *People and Identity in Ostrogothic Italy*, 489–554 (Cambridge, 1997), p. 72.

40 According to Giulio Vismara, 'Il diritto nel regno dei Goti', (1992) now in G.V., *Scritti di storia giuridica*, VIII (Milan, 1996), p. 20, Roman citizens were subjected to the judgment 'of a military commander, moreover a foreigner'. Nonetheless, it should not be forgotten that – unlike normal military jurisdiction – in this case Theodoric had precisely added the mandatory assistance of the *prudens* in order to limit the potential abuses of such a judge.

As regards the application of the law, when solving a dispute between Goths the *comes* had to refer to the royal *edicta*. No documents illustrating Ostrogothic customs have reached us; however, one cannot rule out the possibility that some existed. In this regard, two considerations should be made: on the one hand, Theodoric's *exercitus* did not constitute an ancient, or even a very compact ethnic unit. On the other hand, the customs developed during the previous historical phases (all characterised by a tribal structure) could not very well be adapted to the living conditions created by the new kingdom. In any case, the sovereign would have had no interest in maintaining them. There are therefore several elements suggesting that between the 5th and the 6th century the Ostrogothic *belagines* only partially reflected the ancient legal traditions (to be found above all in family law). Mostly, they must have expressed the customs (connected especially with the repression of crimes) established during the exercise of *militia foederata*, and handed down into military encampments (e.g. the coming of age when demonstrating the ability to fight; the adoption by weapon; the penal system based on compositions; the *mundium* over women).

When dealing with disputes between Goths and Romans, the *comes* necessarily had to apply Roman law (specified several times by Cassiodorus as being the *ius commune* for Goths and Romans).[41] It is a question of understanding which 'Roman law' the *comes* was meant to use in practice, and what sort of *aequabilis ratio* – as pointed out in the *formula* – he should apply. At this point, the discussion must be extended to the so-called *Edictum Theoderici*, which may have been compiled to facilitate the task of the Ostrogothic *comes* (I use the conditional because it is impossible to completely remove all doubts as to whether the authorship of the text can be attributed to the Ostrogothic king). What is certain is that, besides the style of the prologue and the epilogue, the 154 chapters of which it is made up seem to correspond perfectly to this aim. The Edict basically regulates particular cases, potential sources of contrast between Romans and Barbarians. It is not, therefore, a personal law and, after all, its content mostly proposes vulgarised Roman laws.

An awareness of the territorial reach of the Edict is of no help in understanding its relationship with 'official' Roman law, which remained in force. Such a relationship acquires more clearly defined connotations once we interpret the text as an instrument working alongside and in juxtaposition with common law (the Imperial Roman Law) according to the exception/rule dichotomy. In other words, the Edict responds to the needs of a special (military) judge in

41 Calasso, *Medioevo del diritto*, p. 77; Pietrini, *I sovrani goti*, pp. 363–367.

solving special controversies (special because civilians are temporarily bound up in the military legal system). It is precisely because of this special status that he is granted prerogatives of freedom and elasticity in the interpretation and adaptation of the law in force, which would have been debarred for ordinary judges. The terminology chosen by Cassiodorus in the *Formula comitivae Gothorum* is enlightening. While, in fact, for the disputes between two Goths it is said that the judge would have to *litem amputare*, when the controversy sees a contrast between a soldier (i.e. a Goth) and a civilian (i.e. a Roman), the *comes* (with the support of the *prudens*) has to 'certamen … aequabili ratione discingere': the verb *discingere* (literally 'unfasten') clearly alludes to the military belt – the *cingulum* – which in these cases the *comes* is invited to loose.

If one were to abandon the technical-legal framework of personal law in favour of the juxtaposition *ius commune* / *ius speciale*, it seems to me that the reality of the legal system during Theodoric's kingdom becomes much clearer.

5 The Co-existence of Barbarians and Romans: The *Edictum Theodorici*

If we read the Edict through this new lens, we find that its laws could actually reveal the intent of the sovereign of the *Regnum Italiae* and magistrate of the Empire. Theodoric, desirous to achieve a spirit of collaboration between Goths and Romans so often evoked in the *Variae*, firstly tries to reassure the Romans (the 'civilian' population) about the rightness and impartiality of the military judge. The opening series of chapters of the Edict defines the discipline which should guide the actions of the judge: any detrimental or harmful behaviour should be severely punished (chapters 1–14).

Immediately following that, there are rules aimed at eliminating or correcting the Gothic warriors' attitudes deriving from their tribal and mercenary past, and incompatible with the objective of achieving peaceful coexistence. With this in mind, it is easier to comprehend the several chapters, which basically reproduce late-imperial rules in heavily sanctioning banditry or violence against women. In particular, chapters 15 and 16,[42] by re-proposing ancient principles of Roman criminal law, mainly aimed at countering the acts of banditry that characterised daily countryside life in late-antiquity, and that belonged to the *habitus* of the Barbarian tribes even when they were bound by a *foedus* with the Romans (a habit that Theodoric's warriors had long shared

42 *Edictum Theoderici* (ed. Bluhme, p. 254).

before arriving in the West and which they were evidently struggling to lose). Similarly, to give another example, the insistence and seriousness with which the abduction of women was punished (chapters 17–22)[43] could be due to the king's will to repress a widespread behaviour among Gothic soldiers, and a common source of hatred between them and the resident population, rather than to fighting the so-called 'bride kidnapping' (a practice certainly disliked by the church). These are behaviours into which the soldiers of the *exercitus gothorum* easily lapsed – much like soldiery across all centuries who, in occupied territories, considered raids and plunder against the population, and violence against the women of that territory, as a sort of *ius predae*.

It is also worth noting that much later on, in chapter 59,[44] the legislator returns to the theme of violence against young women of an *ingenua* condition. This time the chapter has no immediate precedent among the Roman sources: it is therefore possible that this is an innovative intervention on the part of the legislator which may have derived from the resolution of a specific court case. In this case, the rule does not differentiate on the basis of ethnicity, but of census. The first part outlines the cases in which are involved the rich and noble, granting them a chance of escaping capital punishment by marrying the violated woman (a 'shotgun marriage', therefore), or in the case of individuals who are already married, on the condition of giving adequate compensation enough to put the woman in a position of 'marrying well' in any case.

It is worth wondering to whom this rule was actually aimed. It is hard to imagine Roman aristocrats going around molesting Gothic women. Such behaviour would have exposed them inevitably to the violent wrath of Barbarian warriors. It is easier to think, therefore, that the rule was mainly addressed to the exponents of the most important Gothic families, who certainly occupied important roles within the *exercitus gothorum*. By allowing in certain

43 *Ivi*, p. 154 (Engl. transl. Lafferty, *Law and Society*, pp. 254–256).

44 *Ivi*, p. 158 (Engl. transl. Lafferty, *Law and Society*, p. 267): 'Anyone who forcibly defiles a freeborn virgin, should he be supported by a suitable patrimony and is of noble rank, shall be compelled to accept her as his wife: provided that he is of the understanding that a fifth of his patrimony is to be conferred to her under the title of a dower. But if he commits such acts while already having a wife, he shall be compelled under worthy and solemn (written) guarantee to give a third of his patrimony to that one whom he forcibly corrupted; so that she, who lost her modesty on account of him, may find a respectable husband. But the oppressor and violator of freeborn modesty shall he afflicted with the most extreme punishment if he is not supported either by a suitable patrimony or nobility'.

cases to waive capital punishment, substituting a lighter punishment, it is not difficult to recognise in it a logic often adopted in military penal law.[45]

The regulation of inheritance law is even more interesting. Chapter 32 explicitly refers to those Barbarians who fight for the state, and to them it grants the *libertas testandi*.[46] This is clearly an example of *ius speciale*. And we can truly say 'special' in a double sense: firstly, in the usual meaning, in that soldiers had greater freedom compared with civilians,[47] and secondly, because of the divergence from the 'Germanic' tradition which, as is well-known, allows only for intestate succession. In fact, chapter 33 reinforces the previous concession, since it punishes anyone who tries to restrain the recently conceded freedom to write a will, possibly by referring to tribal tradition.[48]

The Barbarian element – in the Edict the word 'Barbarian' undoubtedly represents a 'Barbarian soldier at the service of the Empire' – is evoked again with reference to court trials. Chapter 145, in fact, refers back to the content of chapter 5, which calls for a judgment *in absentia* for those who fail three times to appear in court after the judge's formal summons. The rule is now explicitly extended to the 'Barbarian' subjects.[49] We are once again dealing with a rule which clashes with the Germanic legal tradition, which does not admit for a trial to be held when one of the parties is absent.[50] The trial against the

---

45 See *Leges ex Ruffo*, chapters 39 and 40 (ed. Famiglietti, pp. 28–29).

46 *Edictum Theoderici* ch. 32 (ed. Bluhme, p. 155; Engl. transl. Lafferty, *Law and Society*, p. 258): 'We grant the freedom of making wills to barbarians who are known to have performed military service for the state, in any way they wish and are able, whether they are at home or at a military camp'. See Gasparri, *Identità etnica*, p. 196.

47 See in this sense D. 29.1.1.pr. On the topic see now Ruggiero in this volume.

48 *Edictum Theoderici* ch. 33 (ed. Bluhme, p. 155; Engl. transl. Lafferty, *Law and Society*, p. 259): 'Anyone who expects an inheritance from another who has died intestate, should he have prevented the testator from making a will as he wished, let his inheritance be confiscated just as if he was unworthy (*indignus*)'.

49 *Edictum Theoderici* ch. 145 (ed. Bluhme, p. 167; Engl. transl. Lafferty, *Law and Society*, p. 291): 'If any barbarian, upon being summoned for a third time by the authority of a judge exercising the appropriate jurisdiction, and duly called upon by judicial directives refuses to come to the judge by whose command he has been summoned, he shall deservedly receive a sentence of contempt after examination of the case; and moreover he shall be judged to have lost the case concerning which he was summoned. There must these conditions: that either the reliability of the guarantee issued by the defendant shows, or the testimony of freeborn and honourable men confirms, that he who was summoned for the third time was one of the *capillati*; and that it is clearly obvious to those giving testimony that he who lost by judicial decree disdained to give response and had no wish to come before the court'.

50 Heinrich Brunner and Claudius Freiherrn von Schwerin, *Deutsche Rechtsgeschichte*, II, 2nd ed. (Berlin, 1928 = 1958), pp. 441–453, 604–606; Franca Sinatti d'Amico, *Le prove*

absentee, however, is present in military law: contumacy is an example of disobedience, therefore it should be appropriately punished.[51] In this case, too, we can notice how the intervention of the legislator clearly sacrificed ethnic tradition in favour of the (Roman) military legal structure.

Therefore, if the Edict can be attributed to Theodoric the Amal, we can observe how really the sovereign and high imperial magistrate used the legislative power he had at his disposal. First, in this context it is worth recalling how the edicts of the Roman magistrates – being based on the imperium – should have been applied to all the people included in the territorial area within which the magistrate himself exercised his power. In particular, the edict of the peregrine praetor necessarily concerned all relations between Romans and foreigners, and between foreigners only (whatever the relationship was between Rome and the legal system to which the foreigner belonged). But also the provisions eventually issued by a *magister militum* for the region placed under his military command had the same extension. In case of disputes between foreigners (i.e. non-Romans) of the same origin, the magistrate generally ordered that they be judged according to their ethnic law. Gaius mentions the consideration of similar 'national laws' in several passages (*Inst.*, 1.92; 1.193; 3.96; 3.120–121). In some cases the magistrate allowed litigants from the same city to be judged by compatriot judges. In disputes, however, between Romans and foreigners, and between foreigners from different cities, the magistrate applied the law they had in common (if there was a rule of *ius gentium*); if, however, a common rule did not exist, he then applied the norm that seemed most suitable to the case (Gai *Inst.*, 4.37). So Theodoric had two means to use his power. On the one hand, he adapted Roman law to the skills of Gothic judges; on the other, he intervened in Barbarian traditions to modify them wherever possible, adapting them to the *ius speciale* that is military law. In so doing – it should be noted – the author of the Edict put into practice the statements in the prologue and epilogue, according to which Barbarians and Romans (and therefore military and civilians), each in their own different ways, preserved the 'iuris publici reverentia' and the 'legibus omnibus devotio'.[52]

---

giudiziarie nel diritto longobardo. Legislazione e prassi da Rotari ad Astolfo (Milan, 1968), pp. 60–96, 180–189.

51 See *Pauli Sententiae* 5.5a.6(7) and D. 42.1.53.1 *et* 3; D. 48.3.9; D. 49.16.6.2.

52 *Edictum Theoderici, Prolog.* (ed. Bluhme, p. 152; Engl. transl. Lafferty, *Law and Society*, p. 248): '... so that both barbarians and Romans, while maintaining the respect due to the public law and dutifully preserving in their entirely the rights of everyone, may clearly know what they are obliged to follow concerning the items specified in the present edicts'. *Epil.* (ed. Bluhme, p. 168; Engl. transl. p. 294): '... all loyal barbarians and Romans must abide by them ...'.

6 The Burgundian Legislation

What has been said about the Ostrogothic legal system is broadly applicable also to the longer-lasting Burgundian kingdom. After several decades from their settlement in Sapaudia, the Burgundian sovereigns still considered themselves as the representatives, in a western province, of the imperial power of Constantinople. Just like Theodoric, they too needed an ethnically-composite minority of Barbarian soldiers to coexist with a numerically superior resident Roman (or rather Romance) population.[53] The formal legitimisation of their power over the imperial province perhaps differed from that of the Ostrogothic sovereign in that it had an even stronger military connotation. Officially, they were the *magistri militum per Gallias*, and they were in charge of territorial defence. Since, however, the transalpine province had long been militarised, they were also in charge of civil government.

The Burgundian military regime was one in which the old civil judiciaries had been mostly absorbed into the military hierarchy. The problem of legal pluralism – a civil structure with a dominant multi-ethnic army – was therefore similar to the issues faced by Theodoric. The Burgundians also favoured a practical approach. The sovereign-*magister militum* certainly had limited regulatory powers; nevertheless, he constituted the apex of jurisdiction, both for the army and for the civil population (since he had absorbed the powers of the governor of the province).

In any case, not even Burgundian sovereigns were exempted from respecting – and demanding respect for – Roman legal tradition: in order to preserve formal legitimacy, Romans had to be judged in accordance with Roman law.[54] This notion should be contextualised in order to be fully appreciated. Almost certainly, Roman legal texts and civil (therefore Roman) judges were available, and they constituted a reference point.[55] Furthermore, there is more

53 Ian Wood, 'The political structure of the Burgundian kingdom', in *Chlodwigs Welt. Organisation von Herrschaft um 500*, eds. M. Meier – S. Patzold (Stuttgart, 2014) pp. 383–396.

54 See *Liber constitutionum, prima constitutio*, § 8: 'Inter Romanos vero, ... sicut a parentibus nostris statutum est, Romanis legibus praecipimus iudicari ...' and *const. extrav.* 20 issued by Sigismund in 516: '... ut inter Romanos etiam de hoc negotio legis Romanorum ordo servatur' (ed. De Salis, pp. 32 and 119; Engl. transl. K. Fisher Drew, *The Burgundian Code*, pp. 20 and 92). The same Sigismund, writing to emperor Anastasius in 516 (*Alcimi Ecdicii Aviti opera*, ed. Peiper, p. 100), underlined how he had inherited from his ancestors the *romana devotio* which inspired his own government; on the ideological meaning of these expressions, see Santo Mazzarino, 'Aetio, la 'Notitia dignitatum' e i Burgundi di Worms', in S.M., *Antico, tardoantico ed èra costantiniana* (Bari, 1980) II, pp.139–140.

55 Roman judges are mentioned in §§ 5, 11 and 13 of *Lib. Const., prima constitutio* (ed. De Salis, pp. 31–33; Engl. transl. Fischer Drew, pp. 18–21) but we can imagine that they were all military magistrates (see the words *omnes etiam militantes* in § 5).

than one argument to think that Roman legal culture in the Sapaudia (or at least at Lyon) was of a high standard (probably not inferior of that of the contemporary Rome and Ravenna).[56] Over and above the theoretical knowledge available, it is plausible to think that, in the case of a dispute between two Romans, they would prefer more often than not to resort to a *potens* (Roman or Burgundian)[57] or to the local bishop. For Roman *provinciales* too, the interest in the effectiveness of the decision – that is to say, in the actual capacity to carry out the decision – must have prevailed over the formal respect for the late-imperial legal system.

When the dispute was between a Roman and a Barbarian (Burgundian or belonging to other babarian tribes), if what has hitherto been said is valid, they were tried by a military judge, who would have inevitably applied the special law so dear to soldiers. We have no certain data about the previous period, but § 13 of the so-called *prima constitutio* – the one that opens the *Liber Contitutionum* and which, according to Ian Wood, in the version that has been handed down to us, can be dated to Easter 517[58] – although its tone lacks clarity, seems to imply that in these cases the custom was to have a court made up of two *comites*, one Burgundian and the other Roman.[59] This in a way replicated the solution encountered in Cassiodorus' *formula comitive Gothorum*: although the *prudens* is substituted by a second *comes* (albeit 'Roman'), the aim remains that of remedying the ignorance of the *ordo legum* (*… se de legum ordine incerti esse non possint*).

In order to give as much certainty as possible to the law the *comites* needed to apply (and also to limit any potential breach), the Burgundian sovereign-generals thought it necessary to *constituere* – i.e. to 'fix' – that law. They did so by appropriately simplifying and modifying 'official' Roman law (or, if one

56 On legal culture in late antique Gaul: Detlef Liebs, *Römische Jurisprudenz in Gallien* (*2. bis 8. Jahrhundert*) (Berlin, 2002), esp. pp. 41–62.

57 The preference for the assistance of a Burgundian protector in the settlement of a dispute is implicit in *Lib. Const.*, *const.* XXII (ed. De Salis, p. 60; Engl. transl. Fischer Drew, p. 39). Rules that try to limit the patronage of *potentes* – whether respected or not – are also contained in the so-called *Codex Euricianus* (ch. 310) and in the Edict of Theodoric (ch. 43, 44, 46); see Cortese, *Il diritto nella storia*, pp. 87–88, 93–96 and Lafferty, *Law and Society*, pp. 119–120.

58 Ian Wood, 'Burgundian law-making, 451–534', in *Italian Review of Legal History*, 3 (2017), n. 15 (https://irlh.unimi.it/), here p. 7.

59 *Lib. Const.*, *prima constitutio* § 13 (ed. De Salis, p. 33; Engl. transl. Fischer Drew, p. 21): 'Nullam causam absente altero iudice vel Romanus comes vel Burgundio iudicare praesumat, quatenus studeant, ut saepius expetentes se de legum ordine incerti esse non possint' (*Let no Roman or Burgundian count, in absence of the other judge, presume to decide any case however often they may desire it, so that consulting frequently they may not in doubt concerning the provisions of the laws*).

prefers, 'common' law), as well as by taking into consideration the deep-rooted customs of the Barbarian military. To fulfil this duty, they sought the aid of the high command of their army, made up of the principal *comites*; they undoubtedly also consulted other local figures (lay or ecclesiastical authorities) who were knowledgeable in Roman law.[60]

Already Gundioc after 443, then Chilperic, and then Gundobad, must have made individual legal interventions in this sense. When around 501 Gundobad unified all power in his hands, he probably had those measures collected for the first time. They often solved specific cases, and had a strong 'vulgar' connotation.[61] They formed the first nucleus of the *Lex Burgundionum* – or rather *Liber Constitutionum*, as it is termed in the manuscripts which have been handed down to us – corresponding roughly to the first 41 chapters of the text that has been preserved. This first nucleus was later revised and expanded by Gundobad himself, and in 517 and again later, by his son Sigismund.

Like Theodoric, the Burgundian sovereigns pursued several aims. On the one hand, in fact, it was a case of securing the privileged legal status of their soldiers and therefore also their traditions and customs, as long as they did not clash with Roman public order (and the need for cohesion within the new political structure). On the other hand, however, they also had the duty to shield the population of the provinces – and the Catholic Church – from the potential threat of judges with little knowledge in the legal field,[62] and little inclined to equanimity (or even to just listening[63]), especially when dealing with disputes involving Romans (and therefore civilians) and Burgundians (and therefore soldiers like them) together.

In order to achieve this second aim, a second text was compiled. To this day it is known by the name of *Lex romana Burgundionum*, but its original name is almost certainly *Forma et expositio legum*.[64] In accordance with the supposed general acceptance amongst the Germanic peoples of the principle of personality, in the past it was hypothesised that the two collections clashed; it should, however, be stressed that this was not in fact the case. There is certainly no conflict in terms of content (both the first and second collections contain mostly vulgarised Roman laws); nor in terms of recipients (Romans

60 Hermann Nehlsen, 'Lex Burgundionum', in *Handwörterbuch zur deutschen Rechtsgeschichte*, II, (Berlin, 1978), p. 1911; Wood, *Burgundian law-making*, pp. 6, 13–16.

61 For a different opinion see Modzelewski, *Barbarian Europe*.

62 References to the possible crudeness (*simplicitas*) of such judges can be found, e.g., in *Lib. Const.* § 11 (ed. De Salis, p. 33).

63 *Lib. Const., prima constitutio* § 12 (ed. De Salis, p. 33).

64 See *Lib. Const., prima constitutio* § 8 (ed. De Salis, p. 32) and Wood, *Burgundian law-making*, p. 11.

and Barbarians – whether Burgundian or not – when they enter into legal relationships of a certain importance). *Forma et expositio legum* (FEL) represents a completion of the *Liber Constitutionum* (but perhaps the text already existed before 517), of which it even tends to copy the order. Just like the *Edictum Theoderici*, it is mainly addressed to judges (whose knowledge of Roman law was evidently doubted at court).[65] This is demonstrated by the fact that, if the rules of 'common' Roman law are reduced to mere *summaria*, they often come with references to their official texts (which means that their respective rules did not stop being – at least in theory – the law in force). There are also several references to parallel chapters of the *Liber Constitutionum*.

Forma et expositio is therefore made up of short summaries (hence rightly called *summaria*), of precise indications of corresponding legal texts (*transmissiones*), and references to parallel passages (*paratitla*): they are nothing less than the instruments that teachers used daily in late antiquity law schools.[66] This does not mean that *Forma et expositio* was an educational text. Simply – as the title suggests – the text has a clearly illustrative intention, with adaptations determined by particular economic and social conditions,[67] and therefore perfectly functional in practical use (even for non-experts).

7 Between (Ancient and Modern) Customs and Royal Innovations: Solving Disputes

Since, as has been pointed out, both texts were aimed equally at Burgundians and Romans, it is plausible to expect them to testify to the vulgarisation of Roman law, as well as to a certain reception of Burgundian customs. Ancient

65 So Hermann Nehlsen, 'Lex Romana Burgundionum', in *Handwörterbuch zur deutschen Rechtsgeschichte*, II (Berlin, 1978) p. 1930 and Wood, *Burgundian law-making*, p. 6. According to Liebs, *Römische Jurisprudenz*, pp. 176–179, the *Lex Romana Burgundionum* (i.e. the *Forma et expositio legum*) was not an official text, but merely the work of a private individual (*schlichte Privatarbeit*).

66 As well as the classic work of Herman Jan Scheltema, *L'enseignement de droit des antecécesseurs*, (1970) now in H.J.S., *Opera minora: ad iuris historiam pertinentia*, eds. N. van der Wal *et all.* (Groningen, 2004) pp. 58–110, see Detlef Liebs, 'Juristenausbildung in der Spätantike', in *Juristenausbildung in Europa zwischen Tradition und Reform*, eds. Ch. Baldus – T. Finkenauer – T. Rüfner (Tübingen, 2008) pp. 31–45.

67 A very informative case is that of *manumissio in ecclesia* recently analysed by Stefan Esders, 'Early medieval use of late antique legal texts: the case of the 'manumissio in ecclesia'', in *Configuration du texte en histoire. Proceedings of the 12th International Conference on Studies for the Integrated Text Science*, ed. O. Kano (Nagoya, 2012), here pp. 60–61 and 65.

traditions in fact emerge here and there across both texts. It is not, however, an especially noteworthy presence. On the contrary, space for popular tradition is in truth minimal,[68] perhaps even less than has been stated in the past. More evident than the desire to preserve tradition is the innovative role of the legislator-sovereign. And, once again, solutions already present in late-Roman military customs may have served as a source of inspiration.

There is no point in this context repeating what has been said about the *Edictum Theoderici* and the trials *in absentia*[69] and the special *libertas testandi* recognised specifically to soldiers (and not to Barbarians in general).[70]

Anyone wishing to find a few examples of figures traditionally considered as being 'Germanic' could start with the so-called *Haussuchung*: the special procedure used for searching a property where stolen goods are suspected of being stored.[71] Such a procedure is referred to in both texts (in LC 16.1[72] and in FEL 12.1[73]). This cannot, however, be considered as exclusive to the Burgundian tradition, nor is it even broadly Germanic. The *Haussuchung* is in fact quite

---

68 See e.g. *Lib. Const.* 77.3 (ed. De Salis, p. 102; Engl. transl. Fischer Drew, p. 74): '... Ideoque rectius visum est ut consuetudinis huius superfluitate remota' (... *And therefore it seems right that when the existing inapplicable law has been removed*) and than Nehlsen, 'Lex Burgundionum', pp. 1909–1910 and Wood, 'Burgundian law-making', p. 21.

69 *Lib. Const.*, 8.5–6, see also 17.4–5 (ed. De Salis, pp. 50 and 55–56).

70 In *Forma et expositio legum Romanarum* (FEL) ch. 45 the dichotomy *ius commune / ius speciale* appears even more explicit: §§ 2 and 4 addresses the Roman citizens (*cives Romani*) to which § 3 opposes the soldiers (*milites*) and not, generically, the Burgundians or barbarians (ed. De Salis, pp. 161–162): [3] 'Militibus vero aliter si vellint testamentum facere permissum est; quibus licet et sine conpetenti testium numero testari et filios praeterire; nec per inofficiosi querillam eorum constat testamentum posse disrumpi' (to compare with Gai *Inst.* 2.109). The norm is clearly related to *Lib. Const.* 1.1 which refers indistinctly to all subjects but, in turn, refers to an even more ancient rule (*ordo prioris legis*) acknowledged in *Lib. Const.* 24.5 (see ed. De Salis, p. 41 nt. 3) which refers to the Burgundian warriors only.

71 Heinz Holzhauer, 'Haussuchung', in *Handwörterbuch zur deutschen Rechtsgeschichte*, I (Berlin, 1971) pp. 2040–2046; Loschiavo, *Figure di testimoni*, pp. 156–158.

72 *Lib. Const.* 16.1 (ed De Salis, p. 54; Engl. transl. Fischer Drew, p. 33): 'Quicumque vestigium de quolibet animali secutus fuerit et ad domum alterius vestigio ducente pervenerit, ac si eum ad cuius domum venerit prohibuerit domum suam intrare ad res suas requirendas, pro hoc, quod reposcit, is, qui eum de domo sua ab inquisitione repulerit, pro fure teneatur obnoxius; ita ut inquisitionem nec mulieri liceat denegare' (*If anyone has followed the tracks of an animal, and following those tracks comes to another's house, and if he to whose house he comes prohibits his entering the house to seek back his property, let him who drives him away from his house when he is making inquiry about that which he seeks back be held for punishment as a thief, with the further provision that it is not permitted a woman to deny questioning* [*to refuse to reply to an inquiry*]).

73 FEL 12.1 (ed De Salis, pp. 136–137): 'Si quis ingenuus querentem animalia aut res suas, domum ad inquirendum intrantem prohibuerit, pro fure teneatur, ita ut rem, quae

closely related to the Roman *quaestio lance licioque*,[74] and it is also so widespread that it may even have Indoeuropean roots. Furthermore – although there is no way to ascertain this – it cannot be ruled out that traces of the ancient procedure mentioned by Gaius had been maintained in the rules (written, or orally transmitted) used in military encampments when a soldier who had been robbed demanded to search the tent of the fellow soldier suspected of theft.

A similar argument can be made about the system of pecuniary settlements (*Geldstrafe*; *blood money*) which appears both in LC (several times) and in FEL (ex. 2.6). Somewhat 'traditionally' considered to this day as a Germanic specificity,[75] the system – especially well-suited to solving blood feuds or at any rate abusive behaviour – was certainly not unknown in the late-Roman world, and it could easily be enumerated amongst the customs that were widespread in military encampments.[76]

A much more interesting topic is the way disputes were settled, seen for a long time as one of the most characteristic traits of the entire Germanic legal

---

requiritur, quadruplum solvat, ea tamen ratione, ut cum tribus ingenuis testibus, ubi suspicionem inveniendi furti habet, ingrediatur'.

74 Described in Gai. *Inst.* 3.192–193 and Aulus Gell., *Noctes Att.*, 11.18.9; see also Macrobius, *Saturn.*, 6.1 and than Loschiavo, *Figure di testimoni*, pp. 8–9 and n. 16 (with bibliography).

75 See e.g. Patrick Wormald, 'Law and Ethnicity in the post-Roman West', in *Regna amd Gentes. The Relationship between Late Antique and Early Medieval Peoples and Kingdoms in the Transformation of the Roman World*, eds. H.-W. Goetz – J. Jarnut – W. Pohl (Leiden-Boston-Cologne, 2003), p. 30 and Gasparri, *Identità etnica*, p. 195.

76 Settling offences (*delicta*) was always possible in Roman law. As for the crimes (*crimina*), the faculty granted to the parties to settle a public criminal trial by settlement was certainly admitted in the classical era and right up to the years of Diocletian (see C. 2.4.18, which is certainly genuine in its first part). Significant limitations were introduced only starting with Constantine, when the persecution of criminal acts became the exclusive prerogative of the state; cf. Maria E. Peterlongo, *La transazione nel diritto romano* (Milan, 1936), pp. 145–156, 230–244. However, it is likely that, also from this point of view, Constantine's will to reform met with strong resistance, especially in provincial contexts (where it was easier for the *potentes* to crush the rights of the *debiliores*). In the army, in any case, when it came to punishing soldiers who had caused damage to the people or to the property of others, the military judges certainly preferred settlement solutions (both between fellow soldiers and the civilian population). Above all – considering how important it was not to lose military staff – they will have much preferred to impose fines and even corporal punishments (as long as they did not involve the impairment or physical elimination of the soldier). It is easy to imagine that there were 'rate cards' for this purpose which the officers could refer to; see for example Vegetius, *Epitoma rei militaris*, III.4.9–10, and Kerneis, *Vox divi*, p. 166.

tradition, often in clear juxtaposition with the principles of Roman law.[77] The typically Germanic way of settling disputes was not on the basis of judicial ascertainment of the truth and subsequent ascription of who was right and who was wrong, but by relying on divine judgement by means of instruments aimed at triggering divine intervention (trial by ordeal), and/or capable of immediately terminating the controversy, such as a collective oath or trial by combat.

While Burgundian legislation bears no trace of actual ordeals, among the titles of the *Liber Constitutionum* we find almost immediately the relevance of oaths in trials. In title 8 the suspect is obliged to make an oath of innocence with his kinspeople (thus having the function of *sacramentales* or *coniuratores*). The claimant is granted the right, however, not to settle for the accused's so-called 'purgative' oath, and the claimant can try to *convincere* the latter (i.e. demonstrate their guilt) by judicial combat.[78] The rule is expressly aimed at all subjects, both Burgundian and Roman, and it therefore has an undoubtedly territorial significance. The Romans would appear then to have been forced to adopt a procedure alien to their legal culture. In actual fact, it is important to consider two factors. Firstly, the rule is only applied in mixed controversies (it was not applied when both parties to the dispute were Roman). Secondly, neither the oath nor the combat were new to the Roman mentality. They are both well-documented in the military world. In the case of the duel, the connection is only too apparent. As for the oath – in all its expressions – it is not only one of the main features in trials of late-antiquity, but in the form of *sacramentum*

77 A critical examination of the historiography on the subject is in Luca Loschiavo, 'La risoluzione dei conflitti in età altomedievale. Un excursus storiografico', in *Il diritto per la storia. Gli studi storico giuridici nella ricerca medievistica*, eds. E. Conte – G. Miglio (Rome, 2010), pp. 91–111.

78 *Lib. Const.* ch. 8 (ed. De Salis, p. 49; Engl. transl. Fischer Drew, pp. 29–30): '1. Si ingenuus per suspicionem vocatur in culpa, tam barbarus quam Romanus, sacramenta prebeat, cum uxore et filiis et propinquis sibi duodecimus iuret. … 2. Quod si ei sacramentum de manu is, cui iurandum est, tollere voluerit, antequam ecclesiam ingrediantur illi, qui sacramentum audire iussi sunt, quod a iudicibus ternos semper ad sacramentum audiendum praecipimus delegari, contestantur se nolle sacramenta percipere, et non permittitur qui iuraturus erat post hanc vocem sacramenta praestare; sed ad nos illico diriguntur Dei iudicio committendi' (1. *If a native freeman, either barbarian or Roman, is accused of a crime through suspicion, let him render oath, and let him swear with his wife and sons and twelve relatives … 2. But if he who must take oath wishes to take it with raised hand* [de manu], *and if those who are ordered to hear the oath-those three whom we always command to be delegated by the judges for hearing an oath-before they enter the church declare they do not wish to receive the oath, then he who was about to take oath is not permitted to do so after this statement, but they (the judges) are hereby directed by us to commit the matter to the judgment of God* [*i.e., to ordeal*]).

militiae to the commander it identifies, in a way, military status itself. The life of a Roman soldier was littered with oaths – whether individual or collective, promissory, assertive, or purgative.[79]

It should also be noted that in 502 Gundobad brought back the rule – adding a couple of new, interesting points (LC 45)[80] – with the aim of thwarting its fraudulent or superficial use by the men under his command (*homines nostros*). First of all, the party who refused the oath was not to fight in a duel with the

79 The *sacramentum militiae* is described in Vegetius, *Epit. rei milit.*, 2.5. We know from Procopius (*De bello vandalico*, 2.18.6–7) that at the time of enlisting, even the *buccellarii* took an oath that also involved the emperor. On the extensive use of the oath in justice administered by military judges (especially in the provinces), see Murray, *From Roman to Frankish Gaul*, p. 86 and Noel Lenski, 'Schiavi armati e formazione di eserciti privati nel mondo tardoantico', in *Ordine e sovversione nel mondo greco e romano*, ed. G. Urso (Pisa, 2009), p. 160. More generally, on the characteristics of this military justice and on the general tendency to expand the use of oaths in the trial context, see Loschiavo, *Figure di testimoni*, pp. 82–95.

80 LC 45 (ed. De Salis, pp. 75–76; Engl. transl. Fischer Drew, p. 52): 'Multos in populo nostro ... ita cognovimus depravari, ut de rebus incertis sacramenta plerumque offerre non dubitent et de cognitis iugiter periurare. Cuius sceleris consuetudinem submoventes praesenti lege decernimus, ut, quotiens inter homines nostros causa surrexerit, et is, qui pulsatus fuerit, non deberi a se quod requiritur, aut non factum quod obicitur, sacramentorum oblatione negaverit, hac ratione litigio eorum finem oportebit inponi: ut, si pars eius, cui oblatum fuerit iusiurandum, noluerit sacramenta suscipere, sed adversarium suum veritatis fiducia armis dixerit posse convinci, et pars diversa non cesserit, pugnandi licentia non negetur; ita ut unus de eisdem testibus, qui ad danda convenerint sacramenta, Deo iudicante confligat; quoniam iustum est, ut si quis veritatem rei incunctanter scire se dixerit et obtulerit sacramentum, pugnare non dubitet. Quod si testis partis eius, cui obtulerit sacramentum, in eo certamine fuerit superatus, omnes testes, qui se promiserant iuraturos, trecenos solidos multae nomine absque ulla induciarum praestatione cogantur exolvere. ... ut veritate potius quam periuriis delectentur.' (*We know that many of our people are corrupted ... so that they do not hesitate frequently to offer oaths about uncertain matters and likewise to perjure themselves about known matters. To break up this criminal practice, we decree by the present law that as often as a case shall arise among our people and he who has been accused denies by offering oaths that that which is sought is owed by him, and that that has been done which is charged, it is fitting that an end be made to their litigation in this manner: if the party to whom oath has been offered does not wish to receive the oath, but shall say that the truthfulness of his adversary can be demonstrated only by resort to arms, and the second party* [*the one accused*] *shall not yield* [*the case charged*], *let the right of combat not be refused; with the further provision that one of the same witnesses who came to give oath shall fight, God being the judge. For it is just that if anyone shall say without delay that he knows the truth of the matter and shall offer to take oath, he should not hesitate to fight. But if the witness of him who offered oath was overcome in that combat, let all witnesses who promised that they would take oath be compelled to pay a fine of three hundred solidi without any grant of delay. ... so that as a result, one may delight in truth rather than falsehood*).

person who was to purge himself of the accusation, but with one of those who offered to pledge the same oath. In this way, at least one of the 'oath-helpers' would be in the position of risking his own physical well-being.[81] Secondly, in the case of defeat on the part of the party refusing the oath, all 'oath-helpers' were obliged to pay a very large fine (three hundred *solidi*).

Several aspects of this particular intervention on the part of the legislator should be analysed further. Firstly, the addressees should be taken into account. While chapter 8, as has been noted, addresses Barbarians and Romans collectively, the rule in question here is explicitly aimed at Gundobad's 'men', i.e. his soldiers. This detail can help us better understand the true aim of the legislator. It is not a question of fighting the reprehensible phenomenon of 'paid' witnesses. Rather, what is implied here is the custom of fellow soldiers making sworn oaths with the aim of granting each other (favourable) witness statements should there be a trial. This undoubtedly pernicious custom was widespread amongst Roman soldiers[82] and re-emerges in post-Roman legislations among populations well-accustomed to *militia foederata* and other forms of collaboration with the late-Roman army.[83] Essentially, the Burgundian sovereign gives the soldier who, in a trial, is up against the sworn testimony of his adversary's closest comrades, the possibility of fighting one of them in a duel, thus putting the dispute on a more congenial, less unfavourable level for him.

Another noteworthy factor is the repeated emphasis on the necessity for the oath to be truthful. We are therefore dealing with proper witnesses (whether actual or presumed), rather than 'Germanic' *coniuratores*. Finally, one of the legislator's reasons for introducing this measure may have been the intention of weakening the system of solidarity among tribes (a kind of logic which must have been retained in armies formed by uniting entire family groups of Barbarians). It is easy to sense the kind of threat that such a solution could constitute for the cohesion of the clans: it made it very difficult to find relatives and friends willing to act as oath-helpers. Once again, in the author's opinion, an 'ethnic' reading of the rule would be misleading.

8 Conclusions

The hypothesis proposed in the present article is corroborated in the analysis of two different situations constituted by the Ostrogothic kingdom and the

81 Not his life: normally, the fight stopped with the loser surrendering.

82 See e.g. Tacitus, *Ann.*, 3.10.2.

83 Loschiavo, *Figure di testimoni*, pp. 87–88.

Burgundian kingdom. Of course, the scenario of the kingdoms which substituted the imperial government in the West is much broader than this. Therefore, the significance of the hypothesis cannot be extended automatically to other political situations, such as those of the Visigoths, the Franks, the Langobards, and the Anglo-Saxons. It would be necessary to analyse the specificities of the legal frameworks of each kingdom to arrive at valid conclusions.

For the time being, may it suffice to have demonstrated how useful it can be to approach the topic by taking into account the mental and cultural paradigm within which the transition from Roman to post-Roman times took place. In spite of the differences witnessed by this shift across the imperial provinces, one common trait seems to emerge: the strong militarisation of the protagonists, and of the contexts in which they operated. As historians have acknowledged, between the fourth and sixth centuries it is the Roman aristocratic class – especially from the provinces, but also to some extent from the Italian peninsula as a whole – that changed its characteristics, assuming an increasingly military trait.[84] The organisational models, which the long experience of the military apparatus had allowed to emerge through time, were certainly present in the minds of those who now occupied positions of power. Those models, characterised by a large degree of elasticity and adaptability, now proved to be perfect as a way to facilitate the co-existence of disparate ethnic groups. The newcomers, though motivated by different – if not opposed – needs compared with the residents', could not completely subjugate the latter, nor could they think of completely subverting the pre-existing order.

Bibliography

Primary Sources

Alcimi Ecdicii Aviti opera, ed. R. Peiper [MGH, *Auct. ant.*, VI.2] (Berolini, 1883)

Aulus Gellius, *Noctes Atticae*, ed. P.K. Marshall (Oxford, 1968)

Cassiodorus Senatoris, *Variae*, ed. Th. Mommsen [MGH, *Scriptores – Auct. ant.* XII] (Berlin, 1894) *etiam* Flavio Magno Aurelio Cassiodoro Senatore, *Varie*, (dir.

84 Giovanni Tabacco, *Egemonie sociali e strutture del potere nel medioevo italiano* (Turin, 1974 = 2000), pp. 49–136; Alexander Demandt, 'The Osmosis of the Late Roman and Germanic Aristocracies', in *Das Reich und die Barbaren*, eds. E.C. Chrysos – A. Schwarcz (Vienna, 1989), pp. 75–86; Ralph Mathisen, *Roman Aristocrats in Barbarian Gaul: Strategies for Survival in an Age of Transition* (Austin, Texas, 1993); Karl Ferdinand Werner, *Naissance de la noblesse : l'essor des élites politiques en Europe*, 2nd ed. (Paris, 1999). Delaplace, *La fin de l'empire*, p. 275, has recently underlined the presence of officials coming from the Roman provincial aristocracy in the army of the Visigoths already in the fifth century.

A. Giardina), eds. A. Giardina – G.A. Cecconi – I. Tantillo – F. Oppedisano, **6 vols.** – Engl. transl. by M.S. Bjornlie, *The Variae. The Complete Translation.* (Oakland, CA, 2019)

Codex Euricianus, ed. A. D'Ors, *El Código de Eurico*, Madrid 2014[2]

Digest, v. Iustinianus, *Digesta*

Edictum Rothari, ed. F. Bluhme, *Edictus Langobardorum*, in *Leges Langobardorum* [MGH *Leges* IV] (Hannoverae, 1868), pp. 3–90

Edictum Theoderici regis, ed. F. Bluhme [MGH *Leges* V] (Hannoverae, 1875–1889); Engl. transl. S. Lafferty, *Law and Society in the Age of Theoderic the Great. A Study of the* Edictum Theoderici (Cambridge, 2013), pp. 243–294

Forma et expositio legum Romanarum, in *Leges Burgundionum*

Formulae Andecavenses, v. *Formulae Merowingici et Karolini aevi*

Formulae Merowingici et Karolini aevi, ed. K. Zeumer [MGH, *Formulae*] (Hannover 1886)

Gaius, *Institutiones*, ed. (and Engl. transl.) W.M. Gordon – O.F. Robinson, *The Institutes of Gaius*, London 1988

Iustinianus, *Digesta*, ed. T. Mommsen – P. Krüger, *Corpus Iuris Civilis*, I (Berlin, 1963[17]) – Engl. transl.: *The Digest of Justinian*, English translation edited by Alan Watson, 4 vols. (Philadelphia, Pennsylvania, 1985)

Leges ex Ruffo, ed. G. Famiglietti, *Ex Ruffo leges militares* (Milano 1980)

Lex Alamannorum, ed., K.A. Eckhardt [MGH, *Leges nat. Germ.*, V,1] (Hannoverae, 1966) – Engl. transl. by Th.J. Rivers, *Laws of the Alamans and Bavarians* (Philadelphia, 1977)

Leges Burgundionum, ed. L.R. de Salis [MGH, *Leges nat. Germ.* II.1] (Hannoverae, 1892)

Liber Constitutionum, v. *Leges Burgundionum*, Engl. transl. K. Fisher Drew, *The Burgundian Code. Book of Constitutions or Law of Gundobad* (Philadelphia 1972)

Macrobius (Ambrosius Theodosius), *Saturnalia*, ed. N. Marinone (Torino 1987)

Marculfi Formulae, v. *Formulae Merowingici et Karolini aevi*

Nómoi stratiotikoí, see *Leges ex Ruffo*

Pactus Legis Salicae, ed. K.A. Eckhardt [MGH, *Leg. Nat. Germ.*, 4.1] (Hannoverae, 1962)

Pauli Sententiae, ed. M. Bianchi Fossati Vanzetti, *Pauli Sententiae. Testo e interpretatio* (Padova 1995)

Paulus Diaconus, *Historia Langobardorum*, *Historia Langobardorum*, ed. L. Bethmann – G. Waitz, MGH, *Script. rerum Lang.* (Hannoverae, 1878), pp. 12–187

Procopius Caesariensis, *De bello vandalico* in P.C., *De bellis – Tôn Polemôn Logoi*, ed. (and Engl. transl.) H.B. Dewing, *History of the Wars* (Cambridge, Mass., 1919)

Tacitus (Publius Cornelius), *Annales* I–III and IV–VI, ed. P. Wuillemier (1923,1949; Paris 2013, 2019)

Vegetius, *Epitoma rei militaris*, ed. A. Önnerfors (Stuttgart, 1995)

Secondary Literature

Amory, Patrick, 'The meaning and purpose of ethnic terminology in the Burgundian laws', in *Early Medieval Europe*, 2.1 (1993), pp. 19–23

Amory, Patrick, *People and Identity in Ostrogothic Italy*, 489–554 (Cambridge, 1997)

Anderson, Thomas, 'Roman Military Colonies in Gaul, Salian Ethnogenesis and the Forgotten Meaning of Pactus Legis Salicae 59.5', in *Early Medieval Europe*, 4(2) (1995), pp. 129–144

Astuti, Guido, 'Personalità del diritto', in *Novissimo Digesto Italiano*, XII (1965), pp. 1088–1092

Astuti, Guido, 'Note critiche sul sistema delle fonti giuridiche nei regni romano-barbarici dell'Occidente', (1970) now in G.A., *Tradizione romanistica e civiltà giuridica europea*, (G. Diurni, ed.), I (Rome, 1984), pp. 15–18

Bertolini, Ottorino, 'Ordinamenti militari e strutture sociali dei Longobardi in Italia', in *Ordinamenti militari in Occidente nell'alto medioevo*, (Settimane CISAM, XV) (Spoleto, 1968), I, pp. 429–607

Bognetti, Gian Piero, 'L'influsso delle istituzioni militari romane sulle istituzioni longobarde del secolo VI e la natura della "fara"', (1953) now in Id., *L'età longobarda*, vol. 3 (Milan, 1967) pp. 1–46.

Brunner, Heinrich, *Deutsche Rechtsgeschichte*, I, 2nd ed. (Berlin, 1906)

Brunner, Heinrich and Claudius Freiherrn von Schwerin, *Deutsche Rechtsgeschichte*, II, 2nd ed. (Berlin, 1928 = 1958)

Calasso, Francesco, *Medioevo del diritto*. I. *Le fonti* (Milan, 1954)

Cavanna, Adriano, 'Diritto e società nei regni ostrogoto e longobardo', (1984) now in Id., *Scritti (1962–2002)* (Naples, 2007) I, pp. 513–574

Cortese, Ennio, *Il diritto nella storia medievale* I. *L'alto medioevo* (Rome, 1995)

Cortese, Ennio, 'Nostalgie di romanità: leggi e legislatori nell'alto medioevo barbarico', in *Ideologie e pratiche del reimpiego nell'alto Medioevo* (Settimane CISAM, XLVI), (Spoleto, 1999) pp. 485–510

D'Ors, Àlvaro, 'La territorialidad del derecho de los Visigodos', in *I Goti in Occidente* (Settimane CISAM, III), (Spoleto, 1956), pp. 363–408

D'Ors, Àlvaro, *El Codigo de Eurico. Edición, Palingenesia, Indices* (Estudios Visigoticos, II), (Rome-Madrid, 1960 = Madrid, 2014)

Delaplace, Christine, *La fin de l'empire romain d'Occident. Rome et les Wisigoths de 382 à 531* (Rennes, 2015)

Demandt, Alexander, 'The Osmosis of the Late Roman and Germanic Aristocracies', in *Das Reich und die Barbaren*, eds. E.C. Chrysos – A. Schwarcz (Vienna, 1989) pp. 75–86

Esders, Stefan, 'Early medieval use of late antique legal texts: the case of the 'manumissio in ecclesia'', in *Configuration du texte en histoire. Proceedings of the 12th International Conference on Studies for the Integrated Text Science*, ed. O. Kano (Nagoya, 2012) pp. 55–66

Esmein, Adhémar, *Cours élémentaire d'histoire du droit français* (Paris, 1906)

Everett, Nicholas, 'How territorial was Lombard law?', in *Die Langobarden: Herrschaft und Identität*, eds. W. Pohl – P. Erhart (Vienna, 2005), pp. 345–360

García Gallo, Alfonso, 'Nacionalidad y territorialitad del derecho en la época visigoda', in *Anuario de historia del dercho español*, XIII (1936–1941), pp. 91–124

Gasparri, Stefano, 'Identità etnica e identità politica nei regni barbarici postromani: il problema delle fonti', in *Civis/civitas. Cittadinanza politico-istituzionale e identità socio-culturale da Roma alla prima età moderna*, eds. C. Tristano – S. Allegria (Montepulciano, 2009), pp. 193–204

Gasparri, Stefano, 'La migrazione longobarda in Italia tra mito e realtà', in *Le migrazioni nell'alto medioevo* (Settimane CISAM, LXVI), (Spoleto, 2019), pp. 375–393

Gasparri, Stefano and Cristina La Rocca, *Tempi barbarici. L'Europa occidentale tra antichità e medioevo (300–900)* (Rome, 2012)

Geary, Patrick, *The Myth of Nations. The Medieval Origins of Europe* (Princeton – Oxford, 2003)

Goria, Fausto, 'Romani, cittadinanza ed estensione della legislazione imperiale', (1984) now in F.G., *Diritto romano d'Oriente*, pp. 21–86

Goria, Fausto, 'Giudici civili e giudici militari nell'età giustinianea', (1995) now in F.G., *Diritto romano d'oriente*, pp. 225–239

Goria, Fausto, *Diritto romano d'Oriente*, eds. P. Garbarino *et all.* (Alessandria, 2016)

Guterman, Simeon L., 'The Principle of the Personality of Law in the Early Middle Ages: A Chapter in the Evolution of Western Legal Institutions and Ideas', in *Univ. Miami Law Review*, 21 (1966), pp. 261–348

Holzhauer, Heinz, 'Haussuchung', in *Handwörterbuch zur deutschen Rechtsgeschichte*, I (Berlin, 1971) pp. 2040–2046

Kerneis, Soazick, '*Francus civis, miles Romanus*: les barbares de l'Empire dans le Code Théodosien', in J.-J. Aubert – P. Blanchard (eds.), *Droit, religion et société dans le Code Theodosien* (Geneva, 2009) pp. 377–399

Kerneis, Soazick, ed., *Une histoire juridique de l'Occident. Le droit et la coutume (IIIe–IXe siècle)* (Paris, 2018)

Kerneis, Soazick, 'Vox divi et vox populi: la pluralité des droits', in Ead., *Une histoire juridique de l'Occident*, pp. 173–186

King Paul David, 'King Chindasvind and the First Territorial Law-code of the Visigothic Kingdom', in E. James (ed.), *Visigothic Spain: New Approaches* (Oxford, 1980), pp. 131–157

King Paul David, 'The Alleged Territoriality of Visigothic Law', in B. Tierney – P. Linehan (eds.), *Authority and Power* (Cambridge, 1980), pp. 1–11

Lafferty, Sean D.W., *Law and Society in the Age of Theoderic the Great. A Study of the* Edictum Theoderici (Cambridge, 2013)

Lenski, Noel, 'Schiavi armati e formazione di eserciti privati nel mondo tardoantico', in G. Urso (ed.), *Ordine e sovversione nel mondo greco e romano* (Pisa, 2009), pp. 145–175

Liebs, Detlef, *Römische Jurisprudenz in Gallien (2. bis 8. Jahrhundert)* (Berlin, 2002)

Liebs, Detlef, 'Juristenausbildung in der Spätantike', in Ch. Baldus – T. Finkenauer – T. Rüfner (eds.), *Juristenausbildung in Europa zwischen Tradition und Reform* (Tübingen, 2008) pp. 31–45

Loschiavo, Luca, *Figure di testimoni e modelli processuali tra antichità e primo medioevo* (Milan, 2004)

Loschiavo, Luca, 'La risoluzione dei conflitti in età altomedievale. Un excursus storiografico', in E. Conte – G. Miglio (eds.), *Il diritto per la storia. Gli studi storico giuridici nella ricerca medievistica* (Rome, 2010), pp. 91–111

Loschiavo, Luca, *L'età del passaggio. All'alba del diritto comune europeo (secoli III–VII)*, 2nd ed. (Turin, 2019)

Lupoi, Maurizio, *The Origins of European Legal Order*, (orig. ed. 1994) Engl. transl. (Cambridge, 2000)

MacMullen, Ramsay, *Soldier and Civilian in the Later Roman Empire* (London, 1963)

Maier, Gideon, *Amtsträger und Herrscher in der Romania Gothica. Vergleichende Untersuchungen zu den Institutionen der ostgermanischen Völkerwanderungsreiche* (Stuttgart, 2005)

Mathisen, Ralph W., *Roman Aristocrats in Barbarian Gaul: Strategies for Survival in an Age of Transition* (Austin, Texas, 1993)

Mayer, Theodor, 'Die Entstehung des "modernen" Staates im Mittelalter und die freien Bauern', in *Zeitsch. der Sav. Stift. – Germ. Abt.*, 57 (1937), pp. 210–288

Mazzarino, Santo, 'Aetio, la 'Notitia dignitatum' e i Burgundi di Worms', in Id., *Antico, tardoantico ed èra costantiniana* (Bari, 1980) II, pp.132–160

Meijers, Eduard Maurits, *L'histoire des principes fondamentaux de droit international privé à partir du Moyen Age* (Paris, 1935)

Modzelewski, Karol, *Barbarian Europe*, (orig. ed. 2004) Eng. transl. (Frankfurt am Main 2015)

Montesquieu, Charles Louis de Secondat baron de, *De l'Esprit des lois* (Geneva, 1748)

Murray, Alexander Callander, 'From Roman to Frankish Gaul: 'centenarii' and 'centenae' in the administration of the Merovingian Kingdom', in *Traditio*, 44 (1988), pp. 59–100

Nehlsen, Hermann, 'Lex Burgundionum', in *Handwörterbuch zur deutschen Rechtsgeschichte*, II, (Berlin, 1978), pp. 1901–1915

Nehlsen, Hermann, 'Lex Romana Burgundionum', in *Handwörterbuch zur deutschen Rechtsgeschichte*, II (Berlin, 1978) p. 1927–1933

Neumeyer, Karl, *Die gemeinrechtliche Entwieckelung des internationalen Privat- und Strafrechts*, I–II (Munich, 1901–1916)

Paradisi, Bruno, *Storia del diritto internazionale nel medio evo* (Milan, 1940)

Peterlongo, Maria Emilia, *La transazione nel diritto romano* (Milan, 1936)

Pietrini, Stefania, 'I sovrani goti e il *ius commune* nelle *Variae* di Cassiodoro Senatore', in *Atti dell'Acc. Romanistica Costantiniana*, XXI (2016), pp. 355–372

Poly, Jean-Pierre, 'La corde au cou. Les Francs, la France et la loi salique', in *Genèse del l'Etat moderne en Méditerranée*, (École française de Rome) (Rome, 1993), pp. 287–320

Poly, Jean-Pierre, 'Freedom, warrior's bond, legal book. The 'Lex Salica' between Barbarian custom and Roman law', in *Clio@Themis*, 10 (2016), pp. 1–25

Poly, Jean-Pierre, '*Leges Barbarorum*. La création des lois des nations', in S. Kerneis, ed., *Une histoire juridique de l'Occident*, pp. 187–247

Porena, Pierfrancesco, *L'insediamento degli Ostrogoti in Italia* (Rome, 2012)

von Savigny, Friederich Karl, *Geschichte des römischen Rechts im Mittelalter*, 2nd ed. (Heidelberg, 1834)

Scheltema, Herman Jan, *L'enseignement de droit des antecécesseurs*, (1970) now in H.J.S., *Opera minora: ad iuris historiam pertinentia*, eds. N. van der Wal *et all.* (Groningen, 2004) pp. 58–110

Schröder, Richard, *Lehrbuch der deutschen Rechtsgeschichte* (Leipzig, 1907)

Siems, Harald. 'Lex Romana Visigothorum', in *Handwörterbuch zur deutschen Rechtsgeschichte*, II (Berlin, 1978), pp. 1940–1949

Sinatti d'Amico, Franca, *Le prove giudiziarie nel diritto longobardo. Legislazione e prassi da Rotari ad Astolfo* (Milan, 1968)

Stouff, Louis, Étude sur le principe de la personnalité de des lois depuis les invasions barbares jusqu'au *XII siècle* (Paris, 1894)

Tabacco, Giovanni, *Egemonie sociali e strutture del potere nel medioevo italiano* (Turin, 1974 = 2000)

Tabata, Kayoko, 'I comites Gothorum e l'amministrazione municipale in epoca ostrogota', in J.-M. Carrié – R. Lizzi Testa (eds.), *'Humana sapit'. Études d'antiquité tardive offerts à Lellia Cracco Ruggini*, (Turnhout, 2002) pp. 67–78

Vismara, Giulio, 'Il diritto nel regno dei Goti', (1992) now in G.V., *Scritti di storia giuridica*, VIII (Milan, 1996)

Voß, Wulf Eckardt, 'Vom römischen Provinzialprozeß der Spätantike zum Rechtsgang des frühen Mittelalters', in H. Siems – K. Nehlsen-von Stryk – D. Strauch (eds.), *Recht im frühmittelalterlichen Gallien. Spätantike Tradition und germanische Wertvorstellungen*, (Cologne-Weimar-Vienna, 1995) pp. 73–108

Werner, Karl Ferdinand, *Naissance de la noblesse : l'essor des élites politiques en Europe*, 2nd ed. (Paris, 1999)

Wolfram, Herwig, *History of the Goths*, Eng. transl. (Berkely *et all.*, 1990)

Wood, Ian, 'The Burgundians and Byzantium', in *Western Perspectives on the Mediterranean*, eds. A. Fischer – I.N. Wood (London, 2014) pp. 1–22

Wood, Ian, 'The political structure of the Burgundian kingdom', in *Chlodwigs Welt. Organisation von Herrschaft um 500*, eds. M. Meier – S. Patzold (Stuttgart, 2014) pp. 383–396

Wood, Ian, 'Burgundian law-making, 451–534', in *Italian Review of Legal History*, 3 (2017), n. 15, pp. 1–27

Wormald, Patrick, 'Law and Ethnicity in the post-Roman West', in *Regna amd Gentes. The Relationship between Late Antique and Early Medieval Peoples and Kingdoms in the Transformation of the Roman World*, eds. H.-W. Goetz – J. Jarnut – W. Pohl (Leiden-Boston-Cologne, 2003), pp. 21–53

CHAPTER 16

Late Roman Military Justice and the Birth of Ordeal

Soazick Kerneis

1 Introduction

In the seventies, a team of archaeologists made an important discovery at Chesterholm/Vindolanda. In a heap of garbage, it exhumed a set of 1.000 tablets dated from the years 85 to 130.[1] The corpus is quite exceptional because of the support of the writing since the Vindolanda tablets are thin sheets of wood 1 to 3 millimetres thick, about the same size as a modern postcard.[2] Wood material is brittle, and the amazing preservation of the remains is due to a happy combination of circumstances. In 105 AD, the garrison is moved to the Danube. When the time came to evacuate the fort, the archives were destroyed. But the rain began to fall and stopped the fire. A lot of shelves have thus been saved; a first-rate administrative documentation able to provide information about the soldiers' daily life.[3]

The documentation has fascinated the specialists of the Roman army. Legal historians have been also interested since several tablets concern the judicial

1 Alan Bowman and David Thomas, eds., *Vindolanda : The latin Writing-Tablets* (*Tabulae Vindolandenses I*) (London 1983); Id., *The Vindolanda Writing-Tablets* (*Tabulae Vindolandenses II*) (London 1994); Id. (with contributions by J. Pearce), *The Vindolanda Writing-Tablets* (*Tabulae Vindolandenses III*) (London 2003). Electronic link with translation and comments, http://vindolanda.csad.ox.ac.uk/tablets. References to tablets are given to Bowman and Thomas edition (*Tab. Vind.*). See also Alan Bowman, *Life and Letters on the Roman Frontier* (London 1994) and Antony Birley, *Garrison Life at Vindolanda. A Band of Brothers* (Charleston 2002).

2 Bowman – Thomas, *Tabulae Vindolandenses I*, pp. 51–71. Birley, *Garrison Life*, pp. 31–40. D. 32.52.1: librorum appellatione continentur omnia uolumina, siue in charta siue in membrana sint siue in quauis alia materia: sed et si in philyra aut in tilia (ut nonnulli conficiunt) aut in quo alio corio, idem erit dicendum. *Under the designation of 'books' all volumes are included, whether they are made of papyrus, parchment, or any other material whatsoever; even if they are written on bark (as is sometimes done), or upon any kind of prepared skins, they come under the same appellation* (transl.: https://droitromain.univ-grenoble-alpes.fr/Anglica/D32_Scott.htm).

3 Other tablets at Carleon, Roger Tomlin, 'Roman Britain in 1985', in *Britannia* 17 (1986), pp. 450–452 and Carlisle, Roger Tomlin, 'The Twentieth Legion at Wroxeter and Carlisle in the first Century: the Epigraphic Evidence', in *Britannia* 23 (1992), pp. 150–153.

 | DOI:10.1163/9789004698017_017

activity of the camp.[4] They show the severity of the military discipline, the prefect acting as a coercive power. They also reveal the invasive influence of the discipline extending to the civilians surrounding the fort. Above all, this documentation issued from castral practice emphasizes a new representation of the very idea of justice, a justice dominated by the personal power of the hierarchy, far away from the values of the former *Res publica*.

At the end of the first century, Vindolanda is already following an administrative procedure. Written records are usual in the army and legal cases are registered like the others in the *tabulae publicae*. Since the prefect is a judge, the *cognitio* is applied and the litigant may appeal to a higher authority, following the hierarchy even until the emperor. Several letters request the indulgence of the prefect up to the fort.[5] *Exoro*, I implore your benevolence.[6] In the Army, the practice of going over the heads of the supervising officers, unusual in classical Roman law, is spreading.[7]

It is well known that the ancient procedure of the *ordo iudiciorum priuatorum* has receded under the Empire. The trial takes place before a judge, delegated by the emperor who has the *cognitio*. Scholars underline the pragmatism of this development which had grown on a case-by-case basis.[8] The role played

4 Michael Peachin, *Five Vindolanda Tablets, Soldiers and the Law*, in *Tyche* 14 (1999), underlines the possibility to compare with similar judicial activities elsewhere in the Empire. A Carlisle tablet (H 476. Inv. N° 14, see Roger Tomlin, 'Roman manuscripts from Carlisle: the ink-written tablets', in *Britannia* 29 (1998), pp. 55–68) mentions a letter addressed to the prefect who asked about the names of the soldiers having not their spears. Maybe it must be related to the capital sanction punishing the soldier who have lost his weapons, D. 49.16.3.13: Miles, qui in bello arma amisit vel alienavit, capite punitur: humane militiam mutat. (*A soldier who has lost his arms in time of war, or has sold them, is punished with death, and it is only through indulgence that he may be transferred to another branch of the service* (transl. https://droitromain.univ-grenoble-alpes.fr/Anglica/D49_Scott.htm#XVI)

5 *Tab. Vind* III, 719 : *indulgentia tua* [...*bonitas quae*]. *Tab.Vind.* II, 284 : *agas si uom* [/*re et eam clem*[/ *lo exsigas id* .... The editor notes that «There is a space after *ẹt*; *ẹam* following it is plausible if we assume that the mark before ẹ is not ink. Following that, the reading *clem* looks good and suggests a reference to the recipient's clementia» (https://romaninscriptionsofbritain.org/inscriptions/TabVindol284)

6 *Tab. Vind.* II, 307 : *a uobis omn*[/*exoro d*[..../]*er*/]*scribti*/]*e ad accus*/*a*]*ntium* ...

7 Soazick Kerneis, 'Armée romaine et procédure administrative. L'apport de l'épigraphie britannique', in *Revue des Etudes Militaires Antiques*, 4 (2007), pp. 93–122.

8 Giuseppe I. Luzzato, *Il problema d'origine del processo extra ordinem. I: Premesse di metodo. I cosidetti rimedi pretoir* (Bologne 1965); Id., 'In tema di origine nel processo «extra ordinem»', in *Studi Volterra*, II (Milan 1971); Jean-Philippe Lévy, 'La "*litis denuntiatio*" et sa place dans l'évolution de la procédure extraordinaire' and Maxime Lemosse, 'Le procès provincial classique', both in *Mélanges à la mémoire de André Magdelain*, eds. M. Humbert – Y. Thomas (Paris 1998); William Turpin, 'Formula, *cognitio* and proceedings *extra ordinem*', in *Revue Internationale des Droits de l'Antiquité*, 46.3 (1999).

by the provincial governor has often been highlighted. I think that the judicial practice of the military-supervised populations on the borders of the Empire also contributed to the transformation of the procedure. In the army, as justice is framed by hierarchy, the judgment can be reviewed by a higher authority. It is quite fascinating to note that in this remote part of the Empire, new judicial conducts, later usual in the rest of the Empire, have been experimented very early. The influence of the military practice in the evolution of the late procedure must therefore be emphasised.

In the late Empire, administrative procedure gives an important role to the judge who increasingly resorted to torture. *Res iudicata pro veritate habetur*. In the name of truth, the judge grasps the body of the suspect in order to extract a confession. Despite the severity of military discipline, soldiers cannot be tortured. But the obsession of truth remains and in the fourth and fifth centuries, new forms of torture arise. Popular sources provide us with information and disclose the existence of a marginal practice, still exceptional, but which will become the rule later on. It is now the civilian practice that influences the military trial by making it harder.

I will first look at the sources and then try to understand what they tell us.

2 The Documentation

In the 19th century, a deposit of curse tablets was found in the Roman amphitheatre of Trier.[9] *Defixiones* are very common in the ancient Mediterranean world.[10] They can be described as thin inscribed sheets of lead addressed to the infernal gods and asking them to perform an action against a person or an object. They are usually rolled up or folded and pierced with a nail which is intended to symbolically fix the fate. They are often placed in graves, sometimes in wells. They are used for competition, love spells, but sometimes also in legal cases. In this case, they are judicial prayers and have to be considered as legal materials.[11]

---

9 CIL XIII, pars IV, *Addenda,* (Berlin, 1916) n° 11340. Amina Kropp, *Defixiones: Ein aktuelles Corpus lateinischer Fluchtafeln* (*dfx*) (Speyer 2008); Richard Wünsch, 'Die laminae litteratae aus dem Keller des Trierer Amphitheaters', in *Bonner Jahrbücher*, 119 (1910).

10 Auguste Audollent, *Defixionum tabellae* (Paris 1904), pp. 63–65. Celia Sanchez Natalias, 'Le *defixiones* durante la tarda antichità e la loro iconografia', in *Chaos e Kosmos*, 14 (2013), p. 1.

11 The merite of the so-called *defixiones* for legal history has been emphasised. Michel Huvelin, 'Les tablettes magiques et le droit romain', in *Etudes d'histoire du droit commercial romain* (*histoire externe-droit maritime*), eds. P. Huvelin – H. Lévy-Bruhl (Paris 1929); Henk

These strange procedures are attested in the Roman West, in Gaul for example, with a few tablets written in Gaulish.[12] But the largest deposits are located in Britain, mainly in the Severn Estuary, around three sanctuaries, Sul Minerva in Bath, Mercury in Uley and Nodons in Lydney Park.[13] These inscriptions, dating from the second to the fourth centuries, are written in vulgar Latin, and almost exclusively deal with cases of theft.[14] They result from the interaction of norms and intend to accommodate different representations about dispute resolution, mixing Roman trial and tribal revenge.[15]

The political context for the fifteen curse tablets discovered in Trier is nevertheless quite different.[16] The great imperial city has been fortified. In the middle of the third century, the place had served as a refuge for the population around fleeing from the Germanic tribes. In the following centuries, this military vocation gets stronger. The amphitheatre is integrated into the surrounding wall. The binding spells have been found in its basement.[17] Some of them are too fragmentary to be interpreted. What remains is fascinating by the plurality of cultures. Most of the tablets are written in Latin characters but that does not mean they used Latin language, others are in Greek, one uses runes.[18]

The amphitheatre had become a place of pagan worship, sources name an altar dedicated to Mars and Ana: an entity assimilated to Anna Perenna

---

Versnel, 'Les imprécations et le droit', in *Revue historique du droit français et étranger*, 65 (1987).

12 At Aumagne (Saint-Hilaire-de-Villefranche, Charente-Maritime), two lead tablets in latin in a judicial context, Louis Maurin, *Inscriptions Latines d'Aquitaine; Santons* (Bordeaux 1994), p. 104. At Bad-Kreuznach (*vicus* et *castellum*), ten lead tablets dating back to Vespasian One of them: 'sic non possit repo(nde)re quas(tioni/tionibus)' (CIL XIII, 7550–7555 IV, sp. 7555 II). *Defixiones* in gaulish at Chamalières and La Vayssière, Pierre-Yves Lambert, *La langue gauloise* (Paris 1995), pp. 149–172.

13 Bath curse tablets, Barry Cunliffe, *The Temple of Sulis Minerva at Bath*, vol. 2: *The Finds from the Sacred Spring* (Oxford 1988), (quoted Tab. Sul). Uley ones, Ann Woodward and Peter Leach, *The Uley Shrines: Excavation of a ritual complex on West Hill, Gloucestershire, 1977–9* (London 1993); Kropp, *Defixiones*, pp. 1–24; Hélène Ménard, 'Le vol dans les tablettes de la Bretagne romaine (*Britannia*)', in *Revue historique du droit français et étranger*, 78 (2000); Soazick Kerneis, 'La question enchantée. Les jugements des dieux dans l'île de Bretagne (IIe–IVe siècles)', in *Revue historique du droit français et étranger*, 88/4 (2010).

14 About the development of new legal representations and the parallels between curse tablets of Roman Britain and the *Latin leges barbarorum*, James Adams, 'British Latin: The Text, Interpretation and Language of the Bath Curse Tablets', in *Britannia*, 23 (1922).

15 Kerneis, 'La question enchantée'.

16 Edith Mary Wightman *Roman Trier and the Treveri* (London 1970), pp. 92–98.

17 *CIL* XIII additamenta 11340-III-XIII. Wünsch, *Die laminae litteratae*; Kropp, *Defixiones*, pp. 1–15.

18 About sarmatic context, Jean-Pierre Poly, 'Les trois flèches des Pakhthouas. Ordalie et acculturation en Gaule à la fin du IVe siècle', in *Droit et Cultures*, 65 (2013).

or Diana. It seems difficult that, after the edict of Thessalonica, a pagan altar would be allowed in a city which was both a former capital of the Western Empire and an ecclesiastical metropolis. We may assume that the altar goes back to the government of the pagan Arbogast. The short pagan resurgence during his rule would explain the low number of tablets, while the dispersal of the collection suggests that the pagan oratory had been destroyed after his fall.[19]

As for the spells themselves, four of them take place in a judicial context, one of them talking about torture by nails, usual in Majesty trials.[20] They do not deal with trivial matters of love or horse-racing, but they are exceptional legal sources able to give information about alternative justice in the late Roman Empire. The Roman army was then a cultural melting pot, of different populations, and this led to legal difficulties. Which rules to apply in a trial opposing parties with different legal traditions?

3 The Birth of Ordeal

One of the inscriptions deserves a special analysis because it provides information on the birth of ordeal in the late Roman trial. The judicial prayer is framed around three paragraphs. The first one is an invocation to the gods, difficult to interpret and written in a Celtic language, probably old Irish. Then follows the request in Latin, divided into two parts. First: 'At your altar at the entrance, Ana and Mars linkers, Revenge me after the cauldron', then: 'Bind Eusebius in the nails, and revenge me. On the reverse side in a kind of title: 'Deposed against Eusebius.[21]

The Latin part of the text is the easiest to understand for a legal historian. The so-called nails evoke the practice of torture, usual in the late Roman Empire. Yan Thomas has pointed out significant changes occurred in the imperial criminal procedure. While the *Res publica* emphasized the freedom

19 Wünsch, *Die laminae litteratae*, points out the spreading of archaeological remains.

20 The citations are to Kropp, *Defixiones*: Nr. 4.1.3/7, 8, 9, 11.

21 CIL XIII, 11340, III, AE 1911, 150: In aditus ara uestra/Annam et Martem/uinculares ut me/ uindicetis de cucuma Eusebium in ungulas/obligetis et me/uindicetis.Depositum in Eusebium. Wünsch, *Die laminae litteratae*, Nr. 24; Kropp, *Defixiones*, Nr. 4.1.3/9 argues that *cucuma* is a surname. Soazick Kerneis, 'Les ongles et le chaudron. Pratiques judiciaires et mentalités magiques en Gaule romaine', in *Revue historique du droit français et étranger*, 83 (2005), p. 178.

of the citizens, in the Empire the trial is devoted to the search for truth and this exigency allows torture during the investigation by the judge.[22]

The aim of the trial is to establish the *rei veritas*, and the judge, delegated by the emperor, has the power, the *plena inquisitio*, to get in-depth understanding through an investigation.[23] In this context, the regime of the *confessio* is changing.

In the classical tradition, the confession occurred in civil matters. *Confessus pro iudicato*, the confession recorded an acknowledgement of debt, making the continuation of the trial useless.[24] In 314, a constitution of Constantine states that the extorted confession, i.e. the confession obtained under torture (*ex-torquere*), is the regular way of investigation in criminal matters, because it is in accordance with the truth.[25]

The regime of this confession in criminal law is nevertheless strange. It is not used for the search of evidence but rather to establish the subordination of the accused to the judge. Whereas in civil matters, the contractual nature of the *confessio* left the parties on an equal footing, the criminal confession recognizes a fault and creates a position of dominance that allows torture. The *confessio* obtained under torture shapes the relation of domination between the judge and the parties. It is a performative speech that can be seen as an

22 Yan Thomas, 'La procédure de la majesté. La torture et l'enquête depuis les Julio-Claudiens', in *Mélanges à la mémoire d'André Magdelain. Droit, histoire et religion de Rome*, eds. M. Humbert – Y. Thomas (Paris 1998).

23 Cosimo Cascione, 'La verdad juridica en el derecho de la Antigüedad Tardia. Siglos III–VI d.C. (derecho postclasico y justinianeo)', in *Derecho y verdad II. Genelogía(s)* (Valencia 2015), p. 435. Gian Gualberto Archi, 'Les preuves dans le droit du Bas-Empire', in *Recueil de la Société Jean Bodin*, XVI (1965), focuses attention on the exigence of truth, mainly through the struggle against falsehood. Pasqualina Farina, *La querela civile di falso. I. Origini e vicende storiche* (Roma 2017).

24 Yan Thomas, '*Confessus pro iudicato.* L'aveu civil et l'aveu pénal à Rome', in *L'Aveu*, ed. J.-C. Maire Vigueur (EFR 88) (Rome 1986), p. 94.

25 *Codex Theodosianus*, 9.40.1: Qui sententiam laturus est, temperamentum hoc teneat, ut non prius capitalem in quempiam promat severamque sententiam, quam in adulterii vel homicidii vel maleficii crimine aut sua confessione aut certe omnium, qui tormentis vel interrogationibus fuerint dediti, in unum conspirantem concordantemque rei finem convictus sit et sic in obiecto flagitio deprehensus, ut vix etiam ipse ea, quae commiserit, negare sufficiat. (*Emperor Constantine Augustus to Catullinus. If any person is about to pronounce sentence, he shall maintain such moderation that he shall not pronounce a capital sentence or a severe sentence against any person until such person has been convicted of the crime of adultery, homicide, or magic, either by his own confession, or at any rate by the testimony of all witnesses who have been subjected to torture or to questioning when such testimony is concordant and in agreement, pointing to the same end of the matter. Thus the accused shall be so detected in the crime charged that even he himself can scarcely deny the crime which he committed*).

acknowledgement of the sovereignty of the judge. By confessing, the accused surrenders to the mercy of the judge. According to Yan Thomas, the *confessio* is first and foremost a kind of appeal to the judge's clemency, an act of submission that makes grace possible.[26] The nails in the text refer to a torture usual in the late criminal trial. The applicant claims that Eusebius has wrongly accused him and must endure the nails. But the first step of the procedure concerns himself and, as we shall see, it is an odd one.

The text gives two mentions of the cauldron, one in Latin with the word *cucuma*, the other in Celtic with the term *coire*. Of course, in the Roman tradition, there is no cauldron in the imperial courts. However, the cauldron did play a crucial role among the ancient Celts and it is striking to note that the inscription begins with few lines in a Celtic language. As we know, writings in this language are very rare, ancient Celts did not write, preferring the oral tradition, so that this source is of a great value.

How to explain Celtic language in a Roman inscription? A mention of Saint Jerome testifies to the presence of *Atecotti* around Trier and describes them as rough and dangerous people because they used to slice the buttocks of shepherds and the breasts of women.[27] They were people from Ireland who had been established around Trier in the 360s. They are one of the barbarian military communities attested by the *Notitia Dignitatum*, a document recording the units of the Roman army at the end of the fourth century. Among them, it mentions the *Atecotti* as auxiliaries established in Gaul.[28]

The trial implied by the tablet of Trier opposes Eusebius, the accuser, and a barbarian soldier, coming from the rude *Atecotti* of the Trier area. For the legal historian, the inscription is a fascinating record documenting a mixed trial. Eusebius has charged an Irish soldier and brought him before the judge, a military one as the defendant belongs to the army. If the accuser fails, he has to endure the nails. But the first stage of the proceedings concerning the native soldier raises a difficulty. How to test the truth of his words since he has the privilege of not being tortured? The question is all the more difficult since the need for truth is crucial regarding the statements of a barbarian litigant.[29]

26 Thomas, 'Confessus pro iudicato', p. 109.

27 Hieronymus, *Aduersus Jouinianum* (ed. Migne, col. 296). Philip Rance, '*Attacotti, Déisi* and Magnus Maximus: the Case for Irish Federates in Late Roman Britain', in *Britannia* 32 (2001).

28 *Notitia Dignitatum* : *Atecotti iunores Gallica* (ND OC 5.70 = Oc 7, 78) et *Honoriani Atecotti seniores* (ND OC. 5 197 = Oc 7. 74).

29 *Codex Iustinianus* 9.41.8: Imperatores Diocletianus, Maximianus: Milites neque tormentis neque plebeiorum poenis in causis criminum subiungi concedimus (*A soldier who has lost his arms in time of war, or has sold them, is punished with death, and it is only through*

As their gods are foreign to Rome, it is difficult to trust their oath. The problem is addressed in one of the letters of Saint Augustine in which the Holy Father tries to reassure his correspondent Publicola, worried about the dangers that the oath of the barbarian employees may represent.[30] At the end of the fourth century, in this part of North Africa, Berber nomads, here the Arzugis, often ask to cross the border, offering to convoy goods, guarding granaries, guiding travellers ... In this case, the officer delivering the safe-conduct asked them to take an oath regarding their actions within the Empire and they must repeat it in front of the private individuals who hire them. But what value have these words? As they invoke barbarian gods, are they not likely to pollute the water in the pools, the grain of the crops, or even the air we breathe? Augustine pleads for appeasement, perhaps in vain.[31]

In daily life, barbarians can cause difficulties to those who employ them. In criminal trials, the difficulty is further complicated. Not only Roman law has to be accommodated with native traditions, but also the truth must be revealed with a distrust about the oath but without the support of the torture. An alternative solution has to be found.

For the ancient Irishmen, truth is also a core concept and for them, the vector of truth is the cauldron. Truth is *Fir* and it is consubstantial with the idea of justice, mainly the justice of the king. The role of the king is to render justice, to give true judgments, and he can do it through the cauldron, which is also called *fir*, because it is deeply linked with knowledge.[32] We therefore can understand the presence of the cauldron in a judicial source. What the nails and the cauldron have in common is their quest for the truth and both are

---

indulgence that he may be transferred to another branch of the service. D. 49.16.3.1: Poenae militum huiuscemodi sunt: castigatio ... nam in metallum aut in opus metalli non dabuntur nec torquentur (*Military punishments are of the following kinds: namely, castigation, fines, the imposition of additional duties, transfer to another branch of the service, degradation from rank, and dishonorable discharge; for soldiers are neither condemned to labor in the mines nor subjected to torture.*

30 Claude Lepelley, 'Le serment païen malédiction démoniaque: Augustin devant une angoisse des chrétiens de son temps', in *Le serment*, II. *Théories et devenir*, ed. R. Verdier (Paris 1991); Kevin Uhalde, 'Barbarian Traffic, Demon Oaths and Christian Scruples', in *Romans, Barbarians and the Transformation of the Roman World. Cultural Interaction and the Creation of Identity in Late Antiquity*, eds. R.W. Mathisen – D. Shanzer (Farnham, Surrey and Burlington, VT, 2011), p. 253.

31 On the posterity of his point of view, Corinne Leveleux-Teixeira, 'Prêter serment au Moyen-Age. La *virtus verborum* au risque du droit', in *Le pouvoir des mots au Moyen-Age*, eds. N. *Bériou* – J.-P. *Boudet* – I. *Rosier-Catach* (Turnhout 2014), pp. 186–187.

32 Christophe Archan, *Les chemins du jugement. Procédure et science du droit dans l'Irlande médiévale* (Paris 2007), pp. 263–272.

useful in ascertaining the truth. Integrating the cauldron into the late Roman procedure is a means of performing the truth when « regular » torture cannot be applied. But as it became the equivalent of the nails, the very meaning of the Celtic cauldron has evolved. The nails must tear off the confession. The text lets us assume that the cauldron replaces torture. We do not know the precise process of the ritual, but because the cauldron is given as the equivalent of nails, a Roman-style torture, we can assume it works as a test of truth required of the barbarian.

On the basis of what is known about the later medieval judicial use of the cauldron, it may be assumed that the barbarian litigant has to put his hand into the boiling water in the cauldron, perhaps the hand with which he has sworn the oath. Because soldiers cannot be exposed to torture, the solution adopted is to use the cauldron as a test of truth, a sort of equivalent of the nails. But while the cauldron was traditionally used for divination, in the late Empire it becomes a tool used for checking the patient's word. In the fourth century, the idea is that the truth has to be extracted from the body of the accused. There is a profound link between body and truth and, in this context and for this reason, the cauldron becomes an instrument of torture. Body and cauldron are associated, as well the elements of Nature, fire and water, are called upon. By accepting the judgment of his god, the barbarian soldier submitted to his authority, so that the ordeal, like the confession extorted under torture, helped to frame the subordination of the accused to his judge and to perform the criminal trial as a political tool of domination.

In principle, torture is applied in serious cases and this small piece of lead relates perhaps to a major case, a *lèse-majesté* trial. It reveals that in the case of mixed trials and for severe cases, the search for the truth depends on the origin and the condition of the parties, as if already, anticipating the times to come, the trial begins with the question: 'Under which law do you live?'. For the judge, an indigenous ritual may well be enough; the evidence is free. But for the defendant, the ordeal means asking his God to be exposed to the justice of the Nature, perhaps the most dangerous. The integration of indigenous rituals in the Roman procedure and their distortion in order to shape them into the frame of the criminal trial helped to solve judicial difficulties. It impacted in fact on the whole representation of the judicial truth.

4 Conclusions

I believe that the development of ordeal in Europe in the early Middle Ages must be interpreted in conjunction with the emergence in the late Empire of the representation of a judgment saying the truth and the related use of torture. Rather than a native institution of the Germans, the birth of the European ordeal may be found in the border area of Rome, and it stems from the need in a military context to reach the judicial truth in the difficult cases of mixed trials.[33] It is the genius of Rome to make use of ancient native rituals and beliefs in order to test the value of the words. In a trial between a provincial civilian and a peregrine soldier, judicial truth derives from different practices and these various ways enrich the concept with further meanings. With the god's intervention, the trial becomes a holy time able to create new forms of subjugation.

Michel Foucault pointed out to judicial practices as a kind of observatory able to 'locate the emergence of new forms of subjectivity'.[34] By paying attention to popular sources issued from the margins of the Empire, other forms of trial appear involving various populations attached to values different from those of the *Res publica romana*. Framing dispute resolution is not so easy when religious beliefs and requirements of reparation are strongly opposing each other. At the heart of the process of accommodation, there is the rite, a sort of lowest common denominator, an elementary form of the trial able to combine various traditions. In the late Empire, this trial became brutal, a drama that accentuated the disparity between the judge and the litigants, under the gaze of God.

Bibliography

Epigraphic Sources

CIL = *Corpus Inscriptionum Latinarum*, consilio et auctoritate Academiae litterarum Borussicae editum, Berlin, 1863-

Tabulae Vindolandenses I–II, eds. A. Bowman – D. Thomas, *Vindolanda : The latin Writing-Tablets*, (London 1983 and 1994)

Tabulae Vindolandenses III, eds. A. Bowman – D. Thomas with contributions by J. Pearce, *The Vindolanda Writing-Tablets* (London 2003)

33 Michel Foucault, *Leçons sur la volonté de savoir. Cours au Collège de France (1970–1971). Le savoir d'Œdipe* (Paris 2011), p. 82.

34 Michel Foucault, *La vérité et les formes juridiques, Dits et écrits*, t. 1, eds. D. Defert – F. Ewald (Paris 1954), pp. 1406–1408.

Primary Sources

Codex Theodosianus: *Theodosiani libri* XVI *cum constitutionibus Sirmondianis et leges Novellae ad Theodosianum pertinentes*, ed. Th. Mommsen – P.M. Meyer, 2 vols. (Berlin, 1905) – Engl. transl. by C. Pharr *The Theodosian Code and Novels and the Sirmondian Constitutions. A Translation with commentary, glossary, and bibliography*, [Princeton, 1952], 10th Print (Union, N.J., 2001)

Hieronymus, *Aduersus Jouinianum libri duo*, ed. J.P. Migne, Partrol. Lat. 23, coll. 205–384

Iustinianus, *Codex*, ed. P. Krüger in T. Mommsen – P. Krüger, *Corpus Iuris Civilis*, II (Berlin, 1954[11]) – Engl. transl. B.W. Frier, *The Codex of Justinian. A New Annotated Translation*, I–III, (Cambridge, 2016)

Iustinianus, *Digesta*, ed. T. Mommsen – P. Krüger, *Corpus Iuris Civilis*, I (Berlin, 1963[17]) - From: *The Digest of Justinian*, English translation edited by Alan Watson, 4 vols. (Philadelphia, Pennsylvania, 1985)

Notitia Dignitatum: accedunt Notitia urbis Constantinopolitanae et laterculi prouinciarum, ed. O. Seeck (Berlin 1876 = Frankfurt, 1983); ed., C. Neira Faleiro, *La Notitia Dignitatum. Nueva edicion critica y comentario historico* (Madrid, 2005); ed. in progress, I. Maier, *The compilation 'notitia dignitatum', Cnd* (//www.notitiadignitatum .org). Units: Hoffmann, 1969–1970)

Secondary Literature

Adams, James N. 'British Latin: The Text, Interpretation and Language of the Bath Curse Tablets', in *Britannia*, 23 (1922), pp. 1–26

Archan, Christophe. *Les chemins du jugement. Procédure et science du droit dans l'Irlande médiévale* (Paris 2007)

Archi, Gian Gualberto. 'Les preuves dans le droit du Bas-Empire', in *Recueil de la Société Jean Bodin*, XVI (1965), pp. 415–594

Audollent, Auguste, *Defixionum tabellae* (Paris 1904)

Birley, Anthony, *Garrison Life at Vindolanda. A Band of Brothers* (Charleston 2002)

Bowman, Alan K., *Life and Letters on the Roman Frontier* (London 1994)

Bowman, Alan K. and David J. Thomas, eds., *Vindolanda : The latin Writing-Tablets* (*Tabulae Vindolandenses I*) (Britannia Monographs 4) (London 1983)

Bowman, Alan K. and David J. Thomas, eds., *The Vindolanda Writing-Tablets* (*Tabulae Vindolandenses II*) (London 1994)

Bowman, Alan K. and David J. Thomas, eds., (with contributions by John Pearce). *The Vindolanda Writing-Tablets* (*Tabulae Vindolandenses III*) (London 2003)

Cascione, Cosimo. 'La verdad juridica en el derecho de la Antigüedad Tardia. Siglos III-VI d.C. (derecho postclasico y justinianeo)', in *Derecho y verdad II. Genelogía(s)* (Valencia 2015), pp. 389–494

Cunliffe, Barry (ed.). *The Temple of Sulis Minerva at Bath*, vol. 2: *The Finds from the Sacred Spring* (Oxford 1988)

Farina, Pasqualina. *La querela civile di falso. 1. Origini e vicende storiche* (Roma 2017) https://romatrepress.uniroma3.it/libro/la-querela-civile-di-falso-i-origini-e-vicende-storiche/

Foucault, Michel. *La vérité et les formes juridiques*, (D. *Defert* – F. *Ewald, eds.*), *Dits et écrits*, t. 1 (Paris, 1954)

Foucault, Michel. *Leçons sur la volonté de savoir. Cours au Collège de France (1970–1971). Le savoir d'Œdipe* (Paris, 2011)

Huvelin, Paul. 'Les tablettes magiques et le droit romain', in *Etudes d'histoire du droit commercial romain (histoire externe-droit maritime)*, eds. P. Huvelin – H. Lévy-Bruhl (Paris 1929) pp. 219–271

Kerneis, Soazick. 'Les ongles et le chaudron. Pratiques judiciaires et mentalités magiques en Gaule romaine', in *Revue historique du droit français et étranger*, 83 (2005), pp. 155–181

Kerneis, Soazick. 'Armée romaine et procédure administrative. L'apport de l'épigraphie britannique', in *Revue des Etudes Militaires Antiques*, 4 (2007), pp. 93–122

Kerneis, Soazick. 'La question enchantée. Les jugements des dieux dans l'île de Bretagne (IIe–IVe siècles)', in *Revue historique du droit français et étranger*, 88/4 (2010), pp. 483–498

Kropp, Amina. *Defixiones: Ein aktuelles Corpus lateinischer Fluchtafeln* (*dfx*) (Speyer 2008)

Lambert, Pierre-Yves. *La langue gauloise* (Paris 1995)

Lemosse, Maxime. 'Le procès provincial classique', in *Mélanges à la mémoire de André Magdelain*, eds. M. Humbert – Y. Thomas (Paris 1998), pp. 239–246

Lepelley, Claude. 'Le serment païen malédiction démoniaque: Augustin devant une angoisse des chrétiens de son temps', in *Le serment*, II. *Théories et devenir*, ed. R. Verdier (Paris 1991), pp. 53–61

Leveleux-Teixeira, Corinne. 'Prêter serment au Moyen-Age. La *virtus verborum* au risque du droit', in *Le pouvoir des mots au Moyen-Age*, eds. N. *Bériou* – J.-P. *Boudet* – I. *Rosier-Catach* (Turnhout 2014), pp. 171–188

Lévy, Jean-Philippe. 'La "*litis denuntiatio*" et sa place dans l'évolution de la procédure extraordinaire', in *Mélanges à la mémoire de André Magdelain*, eds. M. Humbert – Y. Thomas (Paris 1998), pp. 247–257

Luzzato, Giuseppe Ignazio. *Il problema d'origine del processo extra ordinem. I: Premesse di metodo. I cosidetti rimedi pretoir* (Bologne 1965)

Luzzato, Giuseppe Ignazio. 'In tema di origine nel processo «extra ordinem»', in *Studi Volterra*, II (Milan 1971), pp. 665–757

Maurin, Louis. *Inscriptions Latines d'Aquitaine; Santons* (Bordeaux 1994)

Ménard, Hélène. 'Le vol dans les tablettes de la Bretagne romaine (*Britannia*)', in *Revue historique du droit français et étranger* (RHDFE), 78 (2000), pp. 289–299

Peachin, Michael, *Five Vindolanda Tablets, Soldiers and the Law*, in *Tyche* 14 (1999), pp. 223–235

Poly, Jean-Pierre. 'Les trois flèches des Pakhthouas. Ordalie et acculturation en Gaule à la fin du IVe siècle', in *Droit et Cultures*, 65 (2013), pp. 15–77. Published on line, October 01, 2013, consulted on October 05, 2017: http://droitcultures.revues.org/2992

Rance, Philip. '*Attacotti, Déisi* and Magnus Maximus: the Case for Irish Federates in Late Roman Britain', in *Britannia* 32 (2001), pp. 243–270

Sanchez Natalias, Celia. 'Le *defixiones* durante la tarda antichità e la loro iconografia', in *Chaos e Kosmos*, 14 (2013), pp. 1–19

Thomas, Yan. '*Confessus pro iudicato*. L'aveu civil et l'aveu pénal à Rome', in *L'Aveu*, ed. J.-C. Maire Vigueur (EFR 88) (Rome 1986), pp. 89–117

Thomas, Yan. 'La procédure de la majesté. La torture et l'enquête depuis les Julio-Claudiens', in *Mélanges à la mémoire d'André Magdelain. Droit, histoire et religion de Rome*, eds. M. Humbert – Y. Thomas (Paris 1998), pp. 477–501

Tomlin, Roger S.O. 'Roman Britain in 1985', in *Britannia* 17 (1986), pp. 363–454

Tomlin, Roger S.O. 'The Twentieth Legion at Wroxeter and Carlisle in the first Century: the Epigraphic Evidence', in *Britannia* 23 (1992), pp. 141–158

Tomlin Roger S.O. 'Roman manuscripts from Carlisle: the ink-written tablets', in *Britannia* 29 (1998), pp. 38–84

Turpin, William. 'Formula, *cognitio* and proceedings *extra ordinem*', in *Revue Internationale des Droits de l'Antiquité*, 46.3 (1999), pp. 499–574

Uhalde, Kevin. 'Barbarian Traffic, Demon Oaths and Christian Scruples', in *Romans, Barbarians and the Transformation of the Roman World. Cultural Interaction and the Creation of Identity in Late Antiquity*, eds. R.W. Mathisen – D. Shanzer (Farnham, Surrey and Burlington, VT, 2011), pp. 253–262

Versnel, Henk S. 'Les imprécations et le droit', in *Revue historique du droit français et étranger* (*RHDFE*), 65 (1987), pp. 5–22

Wightman, Edith Mary. *Roman Trier and the Treveri* (London 1970)

Woodward, Ann and Peter Leach. *The Uley Shrines: Excavation of a ritual complex on West Hill, Gloucestershire, 1977–9* (London 1993)

Wünsch, Richard. 'Die laminae litteratae aus dem Keller des Trierer Amphitheaters', in *Bonner Jahrbücher*, 119 (1910), pp. 1–12

CHAPTER 17

Collective Criminal Responsibility and Comrades' Solidarity: From Roman Military Formations to Barbarian Armed Bands

Fabio Botta

1 Prologue

While perhaps counter-intuitive, combining the legal notion of 'solidarity' with criminal law can be very useful in the field of military justice systems. In the subject of military organizations and their law, solidarity has a crucial role in the relationship between the individual and the community (i.e. the *corpus*) to which the person structurally belongs, both in a 'functional' and in a 'moral' sense. Such a solidarity is the expression of a bond of reciprocal aid among individuals connected to each other by mutual fidelity and loyalty. This bond is in its turn often rooted in pre-legal or meta-juridical values that the law simply acknowledges and, at times, directs.

From this perspective, it is particularly interesting to consider the numerous provisions found in the various 'barbarian' or 'Romano-barbaric' legal structures in Europe between the fifth and the ninth centuries. Those sources seem to indicate a set of provisions pointing to joint liability ('*in solidum teneri*', denoting a repressive function) in case of murders, thefts and robberies by subjects belonging to military structures, such as *trustis, hlóđ, schilde*.[1] Such an interest is magnified when comparing that repressive dispositions with other rules that can be found in the most ancient Roman law which, depending on the point of view we want to adopt as an interpretative key, could also be taken as their common *ratio*, their foundation.

In this perspective, obviously, one must be aware of the specific features of the different legal systems, distant from each other both on a historical-social level and on an anthropological-cultural one. Indeed, only too often, the similarities between those Germanic legal orders that made use of those repressive measures and the Roman legal tradition are only apparent. Thus, for instance, nouns such as *contubernium* or *manus* (*collecta*), clearly derived from the

1 Examples and sources in Wilhelm E. Wilda, *Geschichte des deutschen Strafrechts* (Halle 1842), pp. 612 *ff.*; Heinrich Brunner, *Deutsche Rechtsgeschichte*, II (München-Leipzig 1928), pp. 739 *ff.*

 | DOI:10.1163/9789004698017_018

Roman military organization, are also found in the 'barbaric' sources, but there they are used as synonyms of the Germanic terms mentioned above (*trustis, hlóđ, schilde*).[2]

In each legal system that we shall consider, all the provisions related to the repression of 'group' violence must be studied against the background of the way of perpetrating offences which today we would define as conspiracy. This is true both in order to assess them as legal rules within a 'military justice system' and to associate the meaning of 'solidarity' used in the current language with the technical-legal language of Roman origin ('*in solidum teneri*').

Despite the variety of forms in which the repression of those behaviors takes place in the various Germanic legal orders, it is in the conspiracy element that the legal concept of 'cumulative liability' (a sort of '*in solidum teneri*') emerges – both when the offence is a private *delictum* and when it is a crime punishable by public penalty.

2 Associative Crime and Armed Band in Roman Law: From Edict of Lucullus to the '*Lex Quisquis*'

In the Roman legal experience, 'cumulative liability' meant the imposition of the same punishment – be it pecuniary or afflictive – to all those participating in the commission of the crime, whatever their actual role in bringing about the injurious event,[3] as long as the conduct was both intentional and causally-oriented towards that event.[4] This approach results in the assimilation of any

2 Wilda, *Geschichte*, p. 613 *ff.*; Brunner, *Deutsche Rechtsgeschichte*, p. 745 *f.*

3 In the system of *publica iudicia*, complicity, however defined, involves the same punishment for the participants, since all of them are individually considered to have infringed the *lex publica* prohibiting the committed *crimen*. See Theodor Mommsen, *Römisches Strafrecht* (Leipzig 1899), p. 101 f.; Contardo Ferrini, *Diritto penale romano, Esposizione storica e dottrinale* (repr. Rome 1976), pp. 109 *ff.*; Laurent Chevailler, 'Contribution à l'étude de la complicité en droit pénal romain', in *Revue historique de droit français et étranger*, 31 (1953), pp. 201 *ff.*; Giannetto Longo, 'La complicità in diritto penale romano', in *Bullettino dell'Istituto di Diritto Romano*, 61 (1958), pp. 108 *ff.*; Ugo Brasiello, 'Concorso di persone nel reato (dir. rom.)', in *Enciclopedia del Diritto*, VIII (Milan 1961), pp. 561 *ff.*; Victor Manuel Amaya-Garcia, *Coautoria y complicidad: estudio historico y jurisprudencial* (Madrid 1993), pp. 87 *ff.*; and more recently Dennis Bock, *Römischrechtliche Ausgangspunkte der strafrechtlichen Beteiligungslehre. Täterschaft und Teilnahme im römischen Strafrech* (Berlin, 2006), pp. 51 *ff.*

4 See Ulp. (71 *ad ed.*) D. 43.24.15.2: '[…] Labeo scribit, et si communi consilio plurium id factum sit, licere vel cum uno vel cum singulis experiri: opus enim, quod a pluribus pro indiviso factum est, singulos in solidum obligare. si tamen proprio quis eorum consilio hoc fecerit, cum omnibus esse agendum, scilicet in solidum: itaque alter conventus alterum non liberabit

accomplice to the crime into its actual perpetrator.[5] Each offender deserves the same punishment, which could be modified exclusively in consideration – and only in repression *extra ordinem pro modo admissi* – of the circumstances subjectively attributable to each one.[6]

It suffices to recall that the gathering together of armed men in order to commit a wrongdoing (especially against property) with violence finds a first repressive model in the Republican age with the edict of Lucullus of 76 BCE *de hominibus armatis coactisve.*[7] The edict increased the penalty provided for the case of *fur nec manifestum* (punished at *duplum*) elevating it to the *quadruplum* for damage or theft caused by an armed band.

Hence, in the late-Republic legal regime of *delicta privata,* armed band takes the form of a non-associative tort, because the responsibility borne by the members of the band is therefore that of the cumulative liability arising from the offence and the only bond common to each participant in the *rapina* consisting in a fine (a multiple of the damage) payable individually by each participant (so that the victim could seek compensation for the harm from

---

et rell'. From a different perspective of *societas sceleris,* also Paolo Garbarino, 'Brevi note su '*societas criminis*'', in *Diritto@Storia,* 11 (2013).

5 Vincenzo Giuffrè, *La repressione criminale nell'esperienza romana. Profili* (Napoli 1997⁴), p. 152.

6 Thus Laura Solidoro, 'La repressione della criminalità organizzata. Tra diritto romano e categorie moderne', in *Esperienze giuridiche a confronto. Aspetti del diritto pubblico e privato dall'età romana alle configurazioni moderne. Lezioni* (Napoli 2001), p. 3: 'la mancata previsione, nel diritto romano più risalente, della costante repressione degli accordi delittuosi in quanto [...] nell'ambito di reati collettivi comuni [...] l'aggregato criminoso costituiva esclusivamente una circostanza aggravante o attenuante, non una fattispecie autonoma'. On the (possible and case-based: see Wilhelm Langenbeck, *Die Lehre von der Theilnahme am Verbrechen* (Jena 1868), pp. 84 *ff.*) variation of the punishment *extra ordinem,* an essential bibliography in Bernardo Santalucia, *Diritto e processo penale nell'antica Roma* (Milano 1998²), p. 247 nt. 208.

7 Cic. *Pro Tull.* 7: 'Iudicium vestrum est recuperatores quantae pecuniae paret dolo malo familiae P. Fabi vi hominibus armatis coactisve damnum datum esse M. Tullio'. For all, I refer the reader to the now classic works of Udo Ebert, *Die Geschichte des Edikts de hominibus armatis coactisve* (Heidelberg 1968), pp. 3 *ff.*; Marco Balzarini, *Ricerche in tema di danno violento e rapina nel diritto romano* (Padova 1969), pp. 1 *ff.*; Letizia Vacca, *Ricerche in tema di actio vi bonorum raptorum* (Milan 1972), pp. 47 *ff.*; Ead., 'Delitti privati e azioni penali nel principato', in *Aufstieg und Niedergang der Römischen Welt* (*ANRW*), II.14 (Berlin-New York 1982), pp. 684 *ff.*; Laura Solidoro, 'La «familia» nell'editto di Lucullo', in *Atti dell'Accademia di Scienze morali e politiche,* 92 (1981) (Napoli 1982), pp. 7 ff. In the sense of what is stated in the text, with particular regard to bands of slaves and freemen hired for the pursuit of personal aims of the political leaders of the moment, and on the relationship between the edict of Lucullus and the so-called *Formula Octaviana,* see now Giovanni Cossa, *Attorno ad alcuni aspetti della lex Iulia de vi publica et privata* (Rome 2007), pp. 5 f. with previous bibliography.

any one of those responsible for it). This repressive model then tends to be reproduced even when the character of *crimen publicum* is attributed to those same cases, such as in the *lex Plautia de vi*[8] (and in the subsequent *leges de vi,*[9] even if bearing arms in public – without however specifying whether in a band or not – for specific purposes is punished both in the *lex Cornelia de sicariis*[10] and in the *lex Iulia de maiestate*[11]). These are always crimes which, albeit committed by the group, involve the punishment of each single perpetrator, if the sanctioned conduct is effectively attributable to him, with the same penalty (the only one invariably imposed by the *lex publica* for the infraction) due to every other participant.

All this is confirmed in

> D.47.8.2 (Ulp. 56 *ad ed.*) Praetor ait: 'si cui dolo malo hominibus coactis damni quid factum esse dicetur sive cuius bona rapta esse dicentur, in eum, qui id fecisse dicetur, iudicium dabo'. [1] Hoc edicto contra ea, quae vi committuntur, consuluit praetor. nam si quis se vim passum docere possit, publico iudicio de vi potest experiri, neque debet publico iudicio privata actione praeiudicari quidam putant: sed utilius visum est, quamvis praeiudicium legi Iuliae de vi privata fiat, nihilo minus tamen non esse denegandam actionem eligentibus privatam persecutionem.
>
> The praetor says: 'If any loss be said to have been inflicted with deliberate wrongfulness by armed men on someone or if his goods be said to have been forcibly taken, I will grant an action against the person alleged to have done this. And if a slave be said to have done it, I will grant a noxal action against his master.' [1] By this edict, the praetor makes provision against what is done by force. For if someone can show that he has been subjected to force, he can initiate a public prosecution for force; and

8 Andrew W. Lintott, *Violence in Republican Rome* (Oxford 1968), pp. 107 *ff.*; Luigi Labruna, *Vim fieri veto* (Napoli 1971), pp. 24 *ff.* Also see Werner Vitzthum, *Untersuchungen zum materiellen Inhalt der 'lex Plautia' und 'lex Julia de vi'* (München 1966), pp. 82 *ff.*; Balzarini, *Ricerche*, pp. 181 *ff.*; Id., 'Violenza (dir. rom.)', in *Enciclopedia del Diritto*, XLVI (Milan 1993), pp. 830 *ff.* Other bibliography, questions of dating and of (problematic) connection with the *lex Lutatia*, in Santalucia, *Diritto e processo*, pp. 155 ff., and more recently in Cossa, *Attorno ad alcuni aspetti*, p. 7 and note 21 (with previous literature).

9 *PS*. 5.3.3 'Hi, qui aedes alienas uillasue expilauerint effregerint expugnauerint, si quidem in turba cum telis coacta fecerint, capite puniuntur'; *Coll.* 1.13.1 'Qui telum tutandae salutis causa gerit, non uidetur hominis occidendi causa portare' (= D. 48.6.11 pr. e 2).

10 D. 48.8.1 (Marcian. 14 *inst.*); C. 9.16.6 Diocl. et Maxim.; *PS*. 5.23.1.

11 D. 48.4.1 (Ulp. 7 *de off. proc.*). See Alexander Nogrady, *Römisches Strafrecht Strafrecht nach Ulpian. Buch 7 bis 9 De officio proconsulis* (Berlin 2006), p. 146 *f.*

> there are those who think that a public prosecution should not be prejudiced by a private action. It seemed more advantageous, however, that although it be prejudicial to the lex Julia de vi, nonetheless, the action should not be refused to those preferring to seek private redress.[12]

That is, the band ('armed', if the violence is already in the *rapere*) is only the reason, the context, of the 'solidarity' in the punishment, a way in which the 'complicity' in the crime is affirmed.

As previously mentioned, the band is not punished *per se*: the liability of the member cannot be established purely on the basis of his belonging to it. Unlike present-day criminal law codes, therefore, Roman law knew no form of 'associative crime', because of the principle of individual responsibility clearly visible in the Roman criminal justice system (we will see its applications in the military legal order later on). This is also true when, as in the case of the *lex de collegiis* and its applications, the law intends to repress the membership of *factiones illicitae* by presumptively attributing to each member the commission of the offences connected with the associations to which he belongs, because they have been prohibited the legislator in the qualification itself of the *secta* as *illicita*.[13]

On the other hand, since 'objective' criminal liability has generally little to do with Roman law, we can appreciate the contrast – perhaps only apparent – between two constitutions of Arcadius sent to the same p.p. Eutychian, less than two years apart.

The more recent of them (and its Visigothic *interpretatio* in the *Breviarum Alaricianum*), put the principle of individual responsibility, perhaps as a 'humanitarian' correction of a possible application perceived as abusive and excessive, somehow differently with respect to the prior *lex Quisquis*:

> CTh. 9.40.18 [= Brev. 9.30.4] Impp. Arcad. et Honor. AA. Eutychiano p.p. Sancimus, ibi esse poenam, ubi et noxa est. Propinquos, notos, familiares procul a calumnia summovemus, quos reos sceleris societas non facit; nec enim affinitas vel amicitia nefarium crimen admittunt. Peccata igitur suos teneant auctores, nec ulterius progrediatur metus, quam reperitur

12 Transl. by Watson, *The Digest*, IV, p. 279.

13 Francesco De Robertis, *Il diritto associativo romano* (Bari 1938), pp. 301 *ff.* Previously Theodor Mommsen, *De collegiis et sodaliciis Romanorum* (Kiliae 1843 = Napoli 2006), pp. 90 *ff.* Further observations in Fabio Botta, '*Pro legibus facit inquisitio.* Processo provinciale e *nomen christianum* nelle opere apologetiche di Tertulliano', in Κοινωνία, 44/I (2020), pp. 166 *ff.*

delictum. Hoc singulis quibusque iudicibus intimetur. dat. viii. kal. Aug. Constantinopoli, Theodoro v. c. cos. (399 Iul. 25)

We sanction that there shall be punishment where there is also guilt. We remove from fear of calumny close kinsmen, acquaintances, and members of the household who have not been rendered guilty by participation in crime; for kinship or friendship does not make one guilty of a nefarious crime. Only the authors of crimes, therefore, shall be held liable, and fear shall advance no further than crime is found. This regulation shall be made known to each and every judge.

Interpretatio. Poena illum tantum sequatur, qui crimen admisit. propinqui vero, affines vel amici, familiares vel noti, si conscii criminis non sunt, non teneantur obnoxii. nemo de propinquitate criminosi aut de amicitiis timeat, nisi qui scelus admiserit.

Interpretation: Punishment shall pursue only the person who committed the crime. Near kinsmen, kinsmen by marriage, friends, members of the household, or acquaintances, if they are not accomplices of the crime, shall not be held guilty. No one of the kinsmen or friends of a criminal, except the one who has committed a crime, shall have fear.[14]

The *lex Quisquis*[15] punished the *factio* (*cum militibus vel privatis, barbaris etiam*) and its membership (he who *'factionis ipsius susceperit sacramenta vel dederit'*) for a specific case that posed a considerable danger (*de nece virorum illustrium, qui consiliis vel consistorio nostra intersunt etc.*):

CTh. 9.14.3 pr. Arcad. et Honor. AA. Eutychiano pp. Quisquis cum militibus vel privatis, barbaris etiam scelestam inierit factionem aut factionis ipsius susceperit sacramenta vel dederit, de nece etiam virorum illustrium, qui consiliis et consistorio nostro intersunt, senatorum etiam, nam et ipsi pars corporis nostri sunt, cuiuslibet postremo qui nobis militat cogitarit, eadem enim severitate voluntatem sceleris qua effectum puniri

14 Translation by Pharr, *The Theodosian Code*, p. 258.

15 See Richard A. Bauman, 'Some problems of the *Lex Quisquis*', in *Antichton*, 1 (1967), pp. 49 *ff.*, but especially a broad-ranging analysis in Marco Sperandio, *Dolus pro facto. Alle radici del problema giuridico del tentativo* (Napoli 1998), pp. 197 *ff.* (with complete previous bibliography) and, comprehensively, Laura Solidoro, *Profili storici del delitto politico* (Napoli 2002), pp. 52 *ff.* A definitive overview of the issues surrounding the two constitutions of Arcadius in Paolo Garbarino, 'Appunti sulla *lex Quisquis* (CTh. 9,14,3)', in *Bullettino dell'Istituto di Diritto Romano*, 107 (2013), pp. 147 *ff.*

iura voluerunt: ipse quidem utpote maiestatis reus gladio feriatur bonis eius omnibus fisco nostro addictis, filii vero eius, quibus vitam imperatoria specialiter lenitate concedimus, paterno enim deberent perire supplicio, in quibus paterni, hoc est hereditarii criminis exempla metuantur, a materna vel avita, omnium etiam proximorum hereditate ac successione habeantur alieni, testamentis extraneorum nihil capiant, sint perpetuo egentes et pauperes, infamia eos paterna semper comitetur, ad nullos umquam honores, nulla prorsus sacramenta perveniant, sint postremo tales, ut is perpetua egestate sordentibus sit et mors solacio et vita supplicio ... Dat. prid. Non Sept. Ancyrae Caesario et Attico conss. (397 Sept. 4).

If any person should enter into a criminal conspiracy with soldiers or civilians, or even with barbarians, or should plan for the death of men of Illustrious rank who participate in Our counsels and Our Consistory, or for the death of Senators, who are also part of Our body, or, finally, for the death of anyone who is in Our imperial service, he shall be struck down with the sword as one guilty of high treason, and all his goods shall be assigned to Our fisc. For the laws have willed that the intent to commit crime shall be punished with the same severity as the actual commission of crime. His sons, indeed, to whom We have granted their lives by especial imperial leniency (for they ought to perish by the same punishment as their father's, since in them must be feared examples of paternal, that is, hereditary, crime) shall be held as persons extraneous to the inheritance and succession of their mother or grandmother, and also of all their nearest kinsmen. They shall receive nothing by the wills of extraneous persons; they shall be needy and poor perpetually; their father's infamy shall accompany them always; they shall never be admitted to any honors and to any oaths of service; finally, they shall be in such sordidness of perpetual want that death shall be to them a solace and life a punishment ...[16]

In the more recent constitution, on the contrary, the same principle is set out in more general terms, referring to each *societas sceleris*.[17] The punishment of the *socius* depends on his actual involvement in the conspiracy to commit the crime, thus excluding the possibility to extend the punishment to relatives and friends of the perpetrators of the crime. This way, however, those personal relationships, by their nature oriented towards solidarity, could have represented

16 Id., p. 236.

17 Garbarino, *Brevi note*, cpv. 2.

(according to ancient Roman tradition[18]), and probably had already represented, the premises for extension of responsibility in the crime to subjects belonging to those circles;[19] and this was not only on account of their presumed aiding and abetting of the perpetrators, but objectively liable also when they were even unaware of the *crimen*.

3 Cases in Roman Military Justice System

Examining the sources concerning the military justice system (the non-legal ones, as we will now see), we might look with special interest at the constitution of Arcadius of 399. This constitution seeks to exclude from its application the *propinqui, adfines, amici* and *familiares* from the *auctores* of the crime, and it may help in shedding light on a previous episode of the *Life of Pescennius Niger*, as reported by Aelius Spartianus in *Historia Augusta*:

> Spartian. *Vita Pescennii*, 10.5 Ob unius gallinacei direptionem decem commanipulones, qui raptum ab uno comederant, securi percuti iussit; et fecisset, nisi ab omni exercitu prope usque ad metum seditionis esset rogatus. [6] et cum pepercisset, iussit ut denorum gallinaceorum pretia provinciali redderent decem, qui simul furto convixerant, addito eo ut tota in expeditione in commanipulatione nemo focum faceret, ne umquam recens coctum cibum sumerent, sed pane ac frigida vescerentur, adpositis speculatoribus, qui id curarent.
>
> For the theft of a single cock, furthermore, he gave an order that the ten comrades who had shared the bird which one of them had stolen, should all be beheaded; and he would have carried out the sentence, had not the entire army importuned him to such a degree that there was reason to fear a mutiny. And when he had spared them, he ordered that each

18 On group solidarity from the origins of the Roman legal tradition, see the bibliography and considerations expressed in Fabio Botta, 'La vendetta come *officium pietatis*', in *Antropologia della vendetta*, eds. G. Lorini – M. Masia (Napoli 2015), pp. 11 *ff.* and in Yann Rivière, 'Solidarité de la famille et responsabilité pénale dans le droit criminel romain', in *L'Homme*, 223/224 (2017), pp. 33–62.

19 On which see Biondo Biondi, *Il diritto romano cristiano*, III (Milan 1954), p. 428 f.; Clemence Dupont, 'Peine et relations pécuniaires entre fiancés et conjoints dans les constitutions rendues de 312 à 565 après Jésus-Christ', in *Revue Internationale des Droits de l'Antiquitè*, 23 (1976), pp. 134; 138 and footnotes; Ferdinando Zuccotti, *"Furor Haereticorum"* (Milan 1992), p. 162.

> of the ten who had feasted on the stolen bird should pay the provincial who owned it the price of ten cocks. At this same time, he ordered that no one during the whole period of the campaign should build a hearth in his company-quarters, and that they should never eat freshly-cooked food, but should live on bread and cold water. And he set spies to see that this was done.[20]

Ten *commanipulones* (perhaps a *contubernium*, since this fraction of the Roman legion normally consisted of ten non-commissioned officers and soldiers)[21] who had eaten together a rooster stolen by only one of them, are condemned to death directly by the emperor,[22] who is particularly strict about military discipline. Saved by the strong dissent against the imperial command by the entire army, the *milites* are sentenced to compensate 'jointly' the owner ten times the value of the stolen property (on top of going without fire in their tent for the entire campaign and eating only raw food).[23]

This, I believe, is a vivid example of collective responsibility which is similar – as regarding the possible bases of the accusation and the methods of imposing the penalty – to specific military-group offences mentioned several times in Roman literary sources first and then also in Byzantine ones (but not found in the early-medieval Germanic ones), such as those that entailed the penalty of decimation.[24]

For the moment I would like to underline the particular form of solidarity in the punishment established by Pescennius Niger, since – despite the peculiarities of the *disciplina militaris* – the penalty that can be imposed on the *fur*

20 Transl. by Magie, *The Scriptores Historiae Augustae*, I, p. 453.

21 Veget., *de re milit.*, 2.8: 'Erant decani, denis militibus praepositi, qui nunc caput contubernii uocantur'. See Erich Sander, 'Zur Rangordnung des römischen Heeres: Der *Duplicarius*', in *Historia* 8 (1959), pp. 239 *ff.*; Ramsay MacMullen, 'The Legion as a Society', in *Historia*, 33 (1984), pp. 440 *ff.*; Jon E. Lendon, '*Contubernalis, Commanipularis*, and *Commilito* in Roman Soldiers' Epigraphy: Drawing the Distinction', in *Zeitschr. Für Papyrologie und Epigraphik*, 157 (2006), pp. 270 *ff.*

22 A precedent in App. *B.C.* 1.109 (an entire cohort executed by Sertorius in Spain after one of its soldiers attempted to rape a woman). See Charles Goldberg, 'Decimation in the Roman Republic', in *The Classical Journal*, 111.2 (2015), p. 154 f.

23 Michele Carcani, *Dei reati delle pene e dei giudizi militari presso i Romani* (1st ed. 1928) (Napoli 1981), p. 98.

24 Cic. *Pro Cluent.* 46.128; App. 2.63; Pol. 6.38.1; Plut. *Ant.* 44.3; *Ex Ruffo leges militares* (ed. Famiglietti) 12. Reviewers on the subject, Sara Phang, *Roman Military Service. Ideologies of Discipline in the Late Republic and Early Principate* (Cambridge 2009), pp. 123 *ff.*; Goldberg, 'Decimation', pp. 141 *ff.* Now, on this topic, see Botta, 'Note in tema di 'decimazione': pena militare, pena 'collettiva', pena 'sacrale'', in *Quaderni Lupiensi di Storia e Diritto*, 11 (2021), pp. 27 *ff.*

is also extended to his closest comrades (who were blameless for the offence,[25] although they would profit from it later on). Those comrades were all united by the common feature of living under the same tent[26] – the *contubernium*, which means, among other things, that they call each other (and are thus perceived by outsiders as) *fratres*.[27]

Even so, it must be noted that the episode told by Historia Augusta remains an isolated case among those available to us. In fact, in the 'special' military justice system known to us through the jurisprudential and authoritative sources of the 'classical' Roman age, there is no trace (up to the Byzantine period[28]) of 'associative' crimes in the narrow sense of the term – that is, attributable to the individual on the simple basis of his belonging to a *societas* oriented towards criminal activities. In truth, there is not even a trace of cases of collective responsibility (which we could define as 'of the *corpus*', using a collective *nomen*[29]), i.e., one that inflicts the punishment to each individual participant because of his simple membership of a band (armed or not) 'collectively' responsible for the offence (since the band is, 'collectively', the perpetrator of the offence). I am referring to those cases which, as mentioned above, have nonetheless been defined by an ancient tradition as the most effective forms of repression and deterrence of properly military crimes.

25 Unless that comrades were thus charged with the crime of 'receiving stolen goods', punished with the same penalty as the *latro*: D.47.16.1 (Marc. 2 *publ. iud.*); *PS* 5.3.4.

26 MacMullen, *The Legion*, p. 443 (notes 22 and 23 with other bibliography): 'They dined as '*contubernales*' of common plate, they marched off to their posts as a group, and as such they prepared for battle and fought. Expanded the term '*contubernalis*' comes to mean not only the normal eight men of a barracks-block but whatever group a person was assigned to for service'.

27 Witnesses in this regard in Cassius Dio 47.22.1 (see Heinz Schulz-Falkenthal, 'Die Unterstützungstätigkeit in einem Militärkollegium der *legio III Augusta* in Lambaesis und das Problem der Sozialleistungen im römischen Vereinswesen', in *Afrika und Rom in der Antike*, eds. H.-J. Diesner – H. Barth – H.-D. Zimmermann (Halle 1968), pp. 157; 163), but also in the text of some imperial constitutions: see in fact the rescript of Caracalla *Aurelio Herculano et aliis militibus* in *Coll.* 1.8.1. A specific study on the epigraphic results by Jana Kepartovà, 'Frater in Militärinschriften – Bruder oder Freund?', in *Listy filologické / Folia philologica*, 109 (1986), pp. 11 *ff.*

28 In which instead there is an explicit provision (*Ex Ruffo leges militares* [ed. Famiglietti], 12) aimed at repressing a 'group' crime with a 'collective' punishment.

29 On the model of the *corpora ex distantibus*: Riccardo Orestano, *Il «problema delle persone giuridiche» in diritto romano* (Torino 1968), pp. 115 *ff.* Comments aimed at discussing a 'responsabilità penale collettiva', but in the light of the brocardo '*societas delinquere non potest*', now by Pietro Paolo Onida, 'La responsabilità penale degli enti collettivi fra diritto romano e diritto moderno', in *Regole e garanzie nel processo criminale romano*, ed. L. Solidoro (Torino 2016), esp. pp. 77 *ff.*, following Leo Peppe, 'Un altro brocardo se ne va', in *Labeo* 48 (2002), pp. 370 *ff.* and Id., 'La responsabilità penale della persona giuridica tra attualità e passato', in *Studi in onore di A. Metro*, IV (Milan 2010), pp. 491 *ff.*

Indeed, Arrius Menander (1 *de re milit.*) states that

> D. 49.16.2 pr. Militum delicta sive admissa aut propria sunt aut cum ceteris communia: unde et persecutio aut propria aut communis est. proprium militare est delictum, quod quis uti miles admittit.
>
> Soldier's crimes or offences are either peculiar or common to other men; and accordingly, their prosecution is either peculiar or common. A peculiar military crime is one which a man commits in his capacity as a soldier[30]

and then he adds that

> D. 49.16.6 pr. Omne delictum est militis, quod aliter, quam disciplina communis exigit, committitur.
>
> Every disorder committed to the prejudice of the common discipline is a military offence.[31]

That is, the soldier can perpetrate a *delictum commune cum coeteris* or a *delictum proprium militare* which is committed *uti miles*: the latter is characteristically the one carried out in contravention of the rules governing his life as a soldier and not as a civilian.[32]

As it is not included in the Roman military justice system described in the 'classical' sources, the associative crime (armed men coming together for criminal purposes), can only remain confined to the 'common' justice system, where is used as a framework to describe the features of conspiracy or, at the most, is used as an aggravating circumstance of crimes which, by their nature,

30 Transl. by Watson, *The Digest*, IV, p. 407.

31 Id., p. 409.

32 See, for all, Erich Sander, 'Das römische Militärstrafrecht', in *Rheinisches Museum für Philologie*, 103 (1960), p. 289; Vincenzo Giuffré, 'Arrio Menandro e la letteratura «de re militari»', in *Labeo*, 20 (1974), pp. 27 *ff.*; Sabino Perea Yébenes, 'El soldado romano, la ley militar y las cárceles *in castris*', in *Castigo y reclusión en el mundo antiguo*, eds. S. Torallas Tovar – I. Pérez Martín (Madrid 2003), pp. 115 *ff.*; Christoph Schmetterer, *Die rechtliche Stellung römischer Soldaten im Prinzipat*, (Wiesbaden 2012), pp. 25 *ff.* More recently, Valerio Massimo Minale, 'Per uno studio sui frammenti *de re militari* di Macro', in *Teoria e Storia del Diritto privato*, 6 (2013) pp. 1 *ff.*; Dario Annunziata, *Temi e problemi della giurisprudenza severiana. Annotazioni su Tertulliano e Menandro* (Napoli 2019), pp. 57 *ff.*, esp. (on the fragments in the text) pp. 71 *ff.* and p. 141 (with other bibliography).

could be committed even by a single person.[33] In such instances, whatever the role a soldier might play in the crime (as in the above-mentioned case of CTh. 9.14.3), so long as there is a joint participation, he would respond *uti singulus*, because he is in any case the perpetrator of the punishable conduct, according to the model – typically Roman – of cumulative liability in the offence.[34]

4 Associative Crime and Armed Band in Ancient Germanic Law and Society

Our understanding of the subject tends to become even more complicated when we turn to the above-mentioned examples of Western early-medieval regulation of joint participation in the crime committed by an armed band.[35]

As mentioned from the beginning, it is in some of these sources that the term *contubernium* is used, as well as *exercitus* or *manus collecta*,[36] apparently derived from the Roman military lexicon.[37] This derivation might suggest that they could be references to actual military structures[38] and hence that soldiers were effectively involved in the cases described in those regulatory provisions.

For example, *contubernium* appears with particular frequency in the Frankish legal experience. Thus, in the *Pactus* (already at the beginning of the sixth century) there are two norms (later reproduced in *L. Sal.* 69.1 and 71.3[39] respectively):

---

33 As an example for all, see Macer (1 *publ. iud.*) D. 47.14.2, but also see *Coll.* 11.2 (= PS. 5.18.2) where the aggravating circumstance for the *abigei* is connected to the *id saepius aut ferro aut conducta manu faciunt*. Here armed group of men is expressed with the meaningful locution *conducta manus*.

34 See e.g., the treatment provided for the bands of *veterani* in CTh. 7.20.7 (= C.12.46.3) on which *infra* note 50.

35 Now see Margareth Helfer, *Il concorso di più persone nel reato. Problemi aperti del sistema unitario italiano* (Torino 2013), p. 23.

36 See Stefan Esders, 'Spätrömisches Militärrecht in der Lex Baiuvariorum', in *Civitas, Iura, Arma. Organizzazioni militari, istituzioni giuridiche e strutture sociali alle origini dell'Europa (secc. III–VII)*, eds. F. Botta – L. Loschiavo (Lecce, 2015), p. 69.

37 See, *generaliter*, Hans Kuhn, 'Das römische Kriegswesen im germanischen Wortschatz', in *Zeitschr. für deutsches Altertum und deutsche Literatur*, 101 (1972), pp. 13 *ff.*

38 Esders, 'Spätrömisches Militärrecht', p. 69: 'hier insbesondere an reguläre militärische Einheiten zu denken ist'.

39 *Lex Sal.* 69.1 (pp. 117–118): 'Si quis collecto contubernio hominem ingenuum antrustionem in domo sua adsallierit et eum occiderit, soledos mille DCCC culpabilis iudicetur. ... 3. Si vero corpus occisi hominis tres vel amplius habuerit plagas, tres quibus inculpatur, qui in eo contubernio fuissent se conprobatur, legem superius conprehensam cogantur exsolvere. 4. Alii vero tres de eo contubernio ... solidus XC componant, et tres adhuc in tercio

> *PLS* 42.1. Si quis collecto contubernio hominem ingenuum in domo sua adsallierit et ibi<dem> eum occiderit .... solidos DC culpabilis iudicetur. ... [3] Si vero corpus occisi hominis <usque ad> tres vel amplius habuerit plagas, tres quibus inculpatur, qui in eo contubernio fuisse probantur, legem superius conprehensam <singillatim> cogantur exsolvere. Alii vero tres de eo contubernio ... XC solidos singuli eorum solvant. Et tres adhuc in tertio loco de eo contubernio ... solidus XLV culpabiles iudicentur ...[40]

> *PLS* 43.3. Si quis vero foris casa sive iter agens sive in agro positus a contubernio fuerit occisius et tres vel amplius habuerit plagas, mallobergo druchteclidio, hoc est tunc tres de eo contubernio, qui adprobati fuerint, singillatim mortem illius coniactent. Et tres, si plures fuerint de eo contubernio <qui adprobati non fuerint> ... solidos XXX solvant <unusquisque illorum>,[41]

while another norm referring to *contubernium* is found in the *Capitula addita* of the beginning of the ninth century

> *Capitula Legi Salicae addita* 72. Si quis ingenuam feminam a contubernio facto aut puellam in itinere aut quolibet loco inferre praesumpserit, quam unus tam plurimi, qui ipsum scelus admisisse fuerint probati, CC solidos culpabilis iudicentur. Et de illo contubernio, si adhuc remanserit, qui ipsum scelus non admiser(un)t et ibi fuisse noscuntur, si plures a(ut) minore numero fuerit tres, et ipsi quadragenus quinos solidos culpabilis iudicetur.[42]

---

loco de ipso contubernio ... solidus XLV solvant. ...'; *L. Sal.* 71. 3 (p. 120): 'Si quis vero foris casa vel iter agens sive in agro a contubernio fuerit interfectus et tres vel amplius habuerit plagas, mallobergo drochlidio, hoc est ut tres de eo contubernio, qui adprobati fuerint, singuli mortem illius componant; tres adhuc, si fuerint de ipso contubernio, XV solidus componant'. See Hans-Rudolf Haegemann, 'Vom Verbrechenskatalog des altdeutschen Strafrechts', in *Zeitschr. der Savigny-Stiftung für Rechtsgeschichte. Germ. Abt.*, 91 (1974), pp. 13 *ff.* Previously, Franz v. Woringen, *Beiträge zur Geschichte des deutschen Strafrechts* (Berlin 1836), pp. 62 *ff.* A broad analysis of the regulations in Jean-Joseph Thonissen, *L'organisation judiciaire, le droit pénal et la procédure pénale de la Loi Salique*, 2nd ed. (Bruxelles-Paris 1882), p. 284 f. and notes.

40 Ed. Eckhardt, pp. 162–164.

41 Ed. Eckhardt, pp. 166–167. On the Frankish sources, see Rolf Sprandel, 'Struktur und Geschichte des merowingischen Adels', in *Historische Zeischrift*, 193 (1961), p. 38.

42 Ed. Eckhardt, p. 242. Thonissen, *L'organisation judiciaire*, pp. 305 *ff.*

Exercitus (together with *manus*) appears in the Edict of Rothari[43]

> *Roth.* 19: Si quis pro iniuria sua vindecanda, super quemcumque cum mano armata cocurrerit, aut exercitum usque ad quattuor homines in vico intraverit, ille prior pro inlecita praesumptione moriatur, aut certe conponat solidos noningentos, medietatem regi et medietatem cui iniuria inlata fuerit. Et illi, qui cum ipso fuerint, si liberi sunt, unusquisque octugenta solidus conponat, medietatem regi et medietatem iniuriam passo, excepto si in ipso vico casas incenderint aut hominem occiserint, secundum qualiter adpreciatum fuerit, ita conponatur illi, cuius casas incensas aut parentes aut servus occisus fuerit;[44]

manus collecta in

> *Lex Thuringorum* 57. Qui domum alterius collecta manu hostiliter circumdederit, trium primorum, qui fuerint, unusquisque solidos LX conponat; et rei similiter; de ceteris, qui eos secuti sunt, solidos X unusquisque et in bannum regis solidos LX[45]

and in

> *Lex Frisionum* 17.4: Qui manu collecta hostiliter villam vel domum alterius circumdederit, ille, qui caeteros collegit et adduxit, weregildum ad partem Regis componat, et qui eum secuti sunt, unusquisque solidos XII, et ei cui damnum, si etiam damnum illatum est, in duplo emendetur ... .[46]

However, the reading of this anthology of sources suggests that in these early-medieval norms some Latin military terms were borrowed to indicate with

---

43 Ed. eds. Azzara – Gasparri, p. 20. A completely different meaning here 'da quello militare specifico', because it means only 'gruppo d'armati' for Ottorino Bertolini, 'Ordinamenti militari e struttura sociale dei Longobardi in Italia', in *Ordinamenti militari in Occidente nell'Alto Medioevo*, I (Spoleto 1968), p. 448 f., with a comprehensive commentary on the rule.

44 Partial exegesis of the clause of the Edict, in relation to the others on the subject of damage to private property (146; 149; 279; 280) and in the face of parallel places in the Anglo-Saxon laws, in Rebecca Colman, "Hamsocn': Its Meaning and Significance in Early English Law', in *The American Journal of Legal History*, 25 (1981), p. 97 f. (with previous literature).

45 Ed. von Richthofen, p. 141.

46 Ed. von Richthofen, p. 671.

a general *nomen*, immediately understandable by the reader, the number of participants in the group, a number required to define the armed group (and thus to specify the method of commission of the offence) and, therefore, the aggravated responsibility of each perpetrator. In the opinion of Wilda,[47] *collectum contubernium* would indicate a band consisting of (at least?) ten active subjects (coinciding with the number of *contubernales* normally attested in the Roman sources, as we have seen); with *manus collecta* evidently five (even though *manus* is the root of *manipulus* and Spartianus's *commanipulones* were ten). For *exercitus*, the Lombard edict gives us the minimum number of participants (four) to assess the existence of 'conspiracy'.[48]

As mentioned in the introductory section, in various contexts other terms evidently not borrowed from the military Latin are used for the same purpose, such as *trustis*,[49] *hlóđ*, *Rädlein*, *Ring* or *Kreis*.[50]

Sometimes the numbers change, five, six, seven, nine components being sufficient for the definition of armed band,[51] to which is sometimes added a *Rädleinführer*,[52] a *prior* or several *priores*.

Now, although for the reasons just mentioned it cannot be excluded (as H. Brunner did[53]) that these armed bands were objectively organized in military forms (and this would explain the use of the terms *hariscild*, *harizuht*, *hariraida* or *heriraida*[54] as in *Lex Ribuaria* 64,[55] which moreover presents a

---

47 Wilda, *Geschichte*, p. 621.

48 Bertolini, *Ordinamenti militari*, p. 450: 'cifra minima, non massima'.

49 Thonissen, *L'organisation judiciaire*, p. 284 fn. 1.

50 Brunner, *Deutsche Rechtsgeschichte*, II, pp. 744 *ff.*

51 And more, obviously. See, for all, Thonissen, *L'organisation judiciaire*, p. 284 fn. 1, and lastly the essays of Wood and Poly in the present volume.

52 On the meaning of *Rädlein* and Rädleinführer, see Ernst Göpfert, 'Beiträge zum obersächsischen Wortschatz (Fortsetzung)', in *Zeitschr. für deutsche Mundarten*, 4 (1909), p. 37 f.

53 Brunner, *Deutsche Rechtsgeschichte*, p. 745.

54 Bertolini, *Ordinamenti militari*, p. 449, who recalls the parallel (High Danish) term *hoerwoerki/herwerky* and the explanation of Klaus von See, *Das Jütsche Recht* (Weimar 1960), p. 182: 'überfall in Jemandes Haus mit bewaffenetem Gefolge, also ein sog. Bandenverbrechen'.

55 *Lex Ribuaria* 64 (66) (ed. Beyerle – Buchner, pp. 117–118): [De homine in domo propria occiso] 'Si quis hominem ad domum propriam cum hariraida interficeret, auctor facti ut triplicem weregildum multetur, et tres priores XC solidos culpabiles iudicentur. Et quanti ei sanguinem fuderent, unusquisque weregeldum conponant. Et quanticumque post auctorem et sanguinem effusuris vel post tres prioris fuerint, unusquisque ter quinos solidos culpabilis multetur, et quicquid ibidem talaverent, restituant'. See Esders, 'Spätrömisches Militärrecht', pp. 68 *ff.* Parallel with the much later *Leges Henrici Primi*, 80,9 in Colman, 'Hamsocn', p. 101 *ff.* Also see Felix Liebermann, *Die Gesetze der Angelsachsen*, I (Halle 1903), p. 597.

case modeled on the above-mentioned *PLS* 42.1), if we wish to understand them as part of the ranks of the army[56] this could be done only with regard to the 'new' structures of military formations in the barbarian kingdoms that took place between the fourth and seventh centuries – adhering to the hypothesis of a «radical break» of those structures from the Roman military system.[57]

The fact that cases of offences constructed in this way are not attested in the Roman military criminal law, but are rather found in the common criminal law first of the Republic and then of the Principate (since in the Roman territory the use of weapons was forbidden to anyone who was not part of the *militia*[58]) derives directly, in Rome and for the entire duration of the Empire, from the 'existence of a permanent army of professional soldiers, maintained by regular taxation in kind and money at the expense of the rest of the population [...] by using a continuous line of defence around Roman territory', expecting 'that peace would reign within the Empire and that there would be no reason for bearing arms, that violence would be illegal'.[59] Conversely, in the early medieval

---

56 Elisabeth Magnou-Nortier, 'Remarques sur la genèse du *Pactus Legis Salicae*', in *Clovis, histoire et* mémoire, I, *Clovis et son temps: l'événement*, ed. M. Rouche (Paris, 1997), p. 502: the *Pactus* would have been inspired by the military discipline of the Roman army.

57 Philippe Contamine, *War in the Middle Ages*, (orig. ed. Paris 1980) (Oxford 1986), p. 15.

58 And this is valid *a fortiori* for the ex-*milites*. In fact, see CTh. 7.20.7. [Imp. Constantius] A. Evagrio p.u. 'Quosdam veteranos hoc nomine indignos latrocinia perpetrare comperimus, ideoque praecipimus, ut veterani, quibus bona mens est, aut arva versent aut certe in optimis negotiis pecuniam tractent et mercimonia agitent. Qui vero neque rus colunt neque utilem vitam commerciis exhibent, mox in ipsos capitaliter exsurgendum. exui enim omnibus privilegiis debent, per quos publica quies perturbatur, ita ut, si levissimum aliquid deliquerint, omnibus subiciantur poenis.' Dat. III id. Aug. Constantio A. VI et Constante II conss. (353? Aug. 11). Veterans engaged in *latrocinium* (hence most probably part of an armed band) punished for disturbance of the *quies publica* whatever the causal contribution (provided that there has been one, even if *levissimum*) of the single participant in the commission of the crimes attributable to the band. On the constitution, see lastly and for all Gianluca Zarro, '*Homines militares* nel tardo antico. Scollamento di società civile e compagine militare. Immigrati, briganti, mercenari', in *Atti dell'Accademia Romanistica Costantiniana*, 23 (2019), pp. 591 *ff.*

59 Contamine, *War*, p. 15. See *supra* note 9. The exception may be seen for example in the concession by Valentinian III to the Italic owners allowing them to defend themselves against raids by the Vandals by establishing their own private armed formations (*Nov. Valent.* 9.1; 13.14), or in the diffusion throughout the fifth century and into the sixth up until Justinian of 'private militias' linked by *sacramentum* to the military leader also in support of the regular army (perhaps already the Constantinian *comitatus*, see *infra* note 60), clearly remains such in the framework of Roman (criminal) law, as seen by the reaction to this institute (which, on the other hand, thus proves to be very widespread and ancient, see Sines. *de regno*, 14–15 and cf. Dick Whittaker, 'Landlords and Warlords in the Later Roman Empire', in *War and Society in the Roman World*, eds. J. Rich – G. Shipley (London 1993), pp. 277 *ff.*; Gastone Breccia, '«*Salus Orientis*». Il nuovo sistema militare

Germanic societies, every free man was usually a soldier and naturally part of an armed band where the connection between the participants could be easily ascertained[60] both by the belonging to the clan (established so as to control the territory) and by the *comitatus* (or, for others, by *Gefolgschaft*),[61] i.e. by that particular bond, perhaps already described by Tacitus,[62] of 'fedeltà e affetto reciproci che esisteva tra un capo guerriero e il suo seguito scelto di compagni fidati',[63] 'bound to him by bonds of loyalty, expressed by service on the battlefield and rewarded by gifts resulting from successful exploits there'.[64]

In the case of tribal or family-based clans, the organized use of arms can also be attributed to the economic phases characterized by a 'natural' economy

---

romano alla prova, 379–400', in *Rivista di Studi Bizantini e Neoellenici*, 41 (2004), p. 11; Alessandro Barbero, *Barbari. Immigrati, profughi, deportati nell'impero romano* (Rome-Bari 2006), pp. 223*f.*, 288) already by Honorius (CTh. 8.1.15), Theodosius I (CTh. 8.1.17), Leo I (C. 9.12.10, *sub tit. ad legem Iuliam de vi publica seu privata*), and Justinian in Novv. 30.5.1 and 116.1, all constitutions aimed at punishing those who keep, host or make use of private militias. An accurate description in Noel Lenski, 'Schiavi armati e formazione di eserciti privati nel mondo tardoantico', in *Ordine e sovversione nel mondo greco e romano*, ed. G. Urso (Pisa 2009), pp. 145*ff.* (with a wide bibliography).

60 Alternately, in the opinion of Malcom Todd, *The Early Germans* (Oxford 1992), p. 31*f.*, but, in one case and another, still giving rise to the internal hierarchy. Todd, referring to the comitatus («retinue»), argues on the one hand, that it represents a 'remarkably stable and stabilizing factor in such societies' and, on the other, that 'it did [...] cut across the bounds of the tribe and the family', so that its members 'for the organization of the tribe [...] were destructive. Young warriors were devoted to the cause of a leader, not the well-being of their tribe'. On *comitatus*, see Peter Heather, *Empires and Barbarians. The Fall of Rome and the Birth of Europe* (Oxford 2012), pp. 43*ff.*; Marco Battaglia, *I Germani. Genesi di una cultura europea*, (Rome 2013), p. 119*ff.* On the relationship between *centena*, *pagus*, *trustis*, *comitatus*, a wide-ranging analysis in Karol Modzelewski, *Barbarian Europe* (Frankfurt am Main 2015), pp. 241*ff.*

61 On *Gefolgschaft*, see Gianfranco Poggi, *La vicenda dello Stato moderno. Profilo sociologico* (Bologna 1988), p. 42. Cf. Walter Schlesinger, 'Lord and Follower in Germanic Institutional History', in *Lordship and Community in Medieval Europe*, ed. F. Cheyette (New York 1968), pp. 64*ff.*, who refers to the classic work by Otto Brunner, *Land und Herrschaft. Grundfragen der territorialen Verfassungsgeschichte Österreichs im Mittelater* (Vienna 1965[5]), in which also *Herrschaft* is opposed to *Genossenschaft*, on which see also Maurizio Lupoi, *Alle radici del mondo giuridico europeo* (Roma 1994), p. 445*f.*, who defines (p. 446) the notion of *Gefolgschaft* 'storiograficamente più pericolosa'.

62 Tac. *Germ.* 13–14. Thus, also Battaglia, *I Germani*, p. 120 (with previous bibliography), who attributes to him the derivation of all forms of patronage based on military brotherhood, fidelity and reciprocity of duties, variously found and named in Germanic societies. However, for an interpretation of the passage in a radically opposite sense, see Lupoi, *Alle radici*, p. 337 and fn. 162.

63 Poggi, *La vicenda*, p. 42.

64 Todd, *The Early Germans*, p. 31.

based on self-sufficiency of the production units. In such an economy the use of arms responded to the requirement that each social and family group 'had to look to their own security, to defend their rights and interest by arms'.[65] In contrast, it was when or where (in the spatial and temporal proximity to the Empire) the economy improved, becoming money- and exchange-based,[66] that the institution of *comitatus* developed more decisively. Hence, we are faced with «a new kind of leader, who owed the strength of his position to having at his beck and call a permanent body of warriors»[67] in order to be able to establish competitive relationships with other leaders in the accumulation (or preservation) of wealth, forcing the lower strata of the population to produce.[68] Although debated and debatable categories (because sometimes indicated as 'dangerous' in the historiography that studied and contrasted different historical phases), we can briefly refer first to the (*Sippen*) *Genossenschaft* and then to the *Gefolgschaft* to outline the essence of the system of domination over the territory (*Herrschaft*) in the barbarian kingdoms.[69]

Regarding the bond of *fides* supporting the institute of *comitatus* (to differentiate it from the notion of *Gefolgschaft* which, when excessively generalized, risks to blur the diversity of the early medieval institutional experiences[70]), it should not be excluded that underlying the formation of a retinue of *satellites*

65 Contamine, *War*, p. 15.

66 For Otto Brunner, *Sozialgeschichte Europas im Mittelalter* (Göttingen, 1978) (Ital. transl., p. 52 *f.*), however, it is a new Western-style economy of trading which arose from the remnants of the late antique state.

67 Heather, *Empires and Barbarians*, p. 46.

68 Id., pp. 117 *ff.*; 398 *ff.*

69 Poggi, *La vicenda*, p. 41 *ff.* See *supra* fn. 54.

70 Lupoi, *Alle radici*, p. 338. *Contra*, Battaglia, *I Germani*, p. 120: «[*comitatus*] difficilmente identificabile con le clientele galliche e romane». On the non-Germanic root, however, of a 'new *fidelitas*', see now especially Luca Loschiavo, *L'età del passaggio. All'alba del diritto comune europeo* (*secoli III–VII*) (Torino 2019²), pp. 144 *ff.*, who strongly emphasises the Roman origin of the *comitatus* (placing in parallel the Frankish *trustis*), taking into account especially its transformation by Constantine into a 'patronage' army, conforming to the oath made directly to the prince, in which a high percentage of elements of barbarian origin found a place. On the oath/*sacramentum* made to a 'leader' to establish a relationship of 'private' loyalty, for example in relation to the *doriferi*/*buccellarii* in the Byzantine army, see Procop. *bell.* 4.18.6 (where, nonetheless, for Walther Kienast, 'Gefolgswesen und *Patrocinium* im spanischen Westgotenreich', *Historische Zeitschrift*, 239 (1984), pp. 29 *ff.*, there is the echo of the *sacramentum* underlying the old Germanic *Gefolgschaft* referred to in Tac. *Germ.* 14.1). Again, on the *satellites* in the juridical and non-juridical sources of imperial late antiquity, see the witnesses of CTh. 9.40.19; 21; 22; Procop. *Anekd.* 4.13 (on Belisarius and his *doriferi*). An exhaustive examination of the witnesses on the subject in Hans-Joachim Diesner, 'Das Buccellariertum von Stilicho und Sarus bis auf Aetius (454/455)', in *Klio* 54 (1972), pp. 331 *ff.*; 336 *ff.*

there could have been the interpretation of the Roman notion of *patrocinium*[71] regarding 'not just military leaders, but also aristocrats who raised warbands'.[72] *Patrocinium* could have been generated directly by the disintegration of the legal system of the Late Antique Roman army – based on the oath (*sacramentum*) of 'personal' loyalty of the single soldier to the 'military leader' (so that Belisarius had his own personal army and the landowners measured their power in terms not of land area but the number of men under their protection[73]). To our purposes, this means that the resulting structure of control of the territory and its resources in the barbarian kingdoms, extending ever more 'vertically' (i.e. marking both the relationship between the local lord and his dependents and the relationship between the local lord and the territorial ruler), seems to have produced a dual effect on the military organization.

On the one hand, this would explain why 'taken together, the Barbarian monarchies did not have a regular army at their disposal',[74] also taking into account that, depending on the different structures assumed by the economy and the taxation during the period, they were generally unable to have sufficient funds to assemble and maintain a stable army dependent on the center.[75] Indeed, the armies were constituted not by local conscriptions but mainly by the call to arms,[76] for the needs of the *regnum*, of the military formations under the control [77] of the *primates*, of the *seniores*, with respect to the *gasindi* (as in the Visigothic experience),[78] of the *thrutin* (in the Scandinavian one),[79] of the *faramanni*

71 Lupoi, *Alle radici*, pp. 415 *ff.*; p. 423, concerning the military necessities.

72 With the words of Ian Wood, 'Transformation of the military in the Late Antique West', in this volume at p. 6 who quoted Amm. Marc. 29.5.34–36. See Whittaker, 'Landlords and Warlords', pp. 288 *f.*

73 Lupoi, *Alle radici*, p. 338. On the same reconstructive line, also see exhaustively Oliver Schmitt, 'Die *Buccellarii*. Eine Studie zum militärische Gefolgschaftswesen in der Spätantike', in *Tyche* 9 (1994), pp. 147 *ff.* (with a wide bibliography).

74 Contamine, *War*, p. 15.

75 Thus, as stated by Wood, 'Transformation', p. ++ there would have been 'the evaporation of the Roman army and its replacement with a force of farmer-soldiers'. On the precedents within the post-Adrianople Empire, see *supra* fn. 49.

76 'In tutte le terre europee prevale l'impiego dei legami di dipendenza per realizzare le finalità militari': Lupoi, *Alle radici*, p. 424 *ff.* (who indicates the Scandinavian political communities as exceptions), with other bibliography.

77 Id., pp. 338 *f.*; 419 *ff.*

78 Ramon D'Abadal y de Vinyals, 'A propos du legs visigothique en Espagne', in *Caratteri del secolo VII in Occidente*, II (Spoleto 1958), pp. 541 *ff.*

79 See, Dennis Howard Green, *The Carolingian Lord. Semantic studies on four old high German words 'Balder', 'Frô', 'Thrutin', 'Hêrro'* (Cambridge 1965), p. 331: 'leader of the *comitatus*'; Karl Kroeschell, *Haus und Herrschaft im frühen deutschen Recht* (Göttingen 1968), pp. 25 *ff.*; Heather, *Empires and Barbarians*, p. 46: 'leader of a warband'.

(in the Lombard and Burgundian ones)[80] or of the *leudi*/*antrustiones*,[81] with respect to the *buccellarii*/*satellites*[82] and the *fideles* making up the *trustis* (as in the development of the Frankish experience, in which the Frankish-Salic name *trust*, as we have seen, can also mean an armed formation).[83]

On the other hand, the form of domination of the territory by means of 'group armed control', being related (both determining it and being determined by it) to an almost complete eclipse of the notion of *limes*,[84] had the effect of inverting the 'classical-Roman' relationship between army, war and 'group' armed violence. It is in this logic that we must interpret the reduction of the 'differences [...] between public warfare and private violence, between the feud or vendetta and a conflict waged by the king in the name of his people'.[85] In this perspective, considering the testimonies of conflicts between local lords relying on their armed retinues or between groups in control of the territories, soon translated into feuds (hence reciprocal acts of armed violence),[86] the legal provisions we have examined thus far could be understood not only

80 Edward A. Thompson, 'The Visigoths from Fritigern to Euric', in *Historia. Zeitschrift für Alte Geschichte*, 12 (1963), p. 120; Alexander Callander Murray, *Germanic Kinship Structure: Studies in Law and Society in Antiquity and the Early Middle Ages* (Toronto, 1983), pp. 89 *ff.*; Matthew Innes, 'Land, Freedom and the Making of the Medieval West', in *Transactions of the Royal Historical Society*, 16 (2006), p. 72 *f.*

81 Laury Sarti, *Perceiving War and the Military in Early Christian Gaul (ca. 400–700 AD)* (Leiden-Boston 2013), p. 28.

82 According to the definition given by Wood, 'Transformation', p. ++: *buccellarii*, as '*satellites* of individual commanders and subsequently of major landowners'. On their Roman 'origin', see Diesner, *Das Buccellariertum*, p. 321 *ff.*; Schmitt, *Die Buccellarii*, p. 148 *ff.*

83 John Beeler, *Warfare in Feudal Europe, 730–1200* (Ithaca-London 1971), p. 4. Notes in this regard in Bernard S. Bachrach, 'Military Organization in Aquitaine under the Early Carolingians', in *Speculum*, 49 (1974), p. 4; Timothy Reuter, 'Plunder and Tribute in the Carolingian Empire', in *Transactions of the Royal Historical Society*, 35 (1985), p. 82 and fn. 42; and especially Lupoi, *Alle radici*, p. 339 fn. 171, with more previous literature. Also see the call to arms of all *leudi* by Dagobert in the *Chronicle* of Fredegar (ed. and transl. Wallace Hadrill, pp. 47 and 73).

84 Contamine, *War*, p. 15.

85 *Ibidem.*

86 Gabriel Monod, 'Les aventures de Sichaire. Commentaire des chapitres XLVII du livre VII et XIX du livre IX de l'*Histoire des Francs* de Grégoire de Tours', in *Revue Historique*, 31 (1886), pp. 268 *f.*, describes, recalling *PLS* 42,1 and the notion of *trueste dominica*, the background and reasons for the feud described therein in this way: 'La société mérovingienne était déja fortement hiérarchisée. Les grands propriétaires groupaient autour d'eux non seulement des esclaves et des colons, mais aussi des hommes libres qui se trouvaient vis-à-vis dans un rapport de dépendance et de protection. Ils sont appelés dans les textes de noms fort divers, *homines, amici, gasindi, suscepti, pares*. Sichaire paraît avoir été le chef de cette espèce de colonie agricole; il recrute parmi eux des hommes armés pour ses entreprises [...] Il avait dans le Poitou un autre domaine organisé comme celui de

as the reaction of the central power to the abuse of 'armed' power[87] aimed at territorial domination by those powerful men or those groups,[88] but also as the rules of the settlement of such 'private wars'.[89] Therefore, the pecuniary punishment normally provided for in those cases (often as wergild), 'designed to prevent a continuing train of violence, appeared in some respects like a ransom, a war indemnity paid to injured party'.[90]

5 Difference between Roman (and Byzantine) Repressive Model and Early Germanic Legal Systems

I believe that there is sufficient information to be able to definitively set out the differences we have observed, in terms of 'armed band' and collective liability, between the Roman legal order (and those derived directly from it) and the regulatory systems of the early-medieval Western experiences. In the provisions regarding the latter, which we have seen thus far, an assembly of armed men is both a requirement of the definition of conspiracy, and also an aggravating factor of offences which, by their nature, can also be committed by a single person. In this context, the punishment of group delinquency should be distinguished depending on whether a legal system normally prohibits the use of arms (being the sole prerogative of the 'centralized' military institution), or it allows that in tolerating (or even encouraging) the presence of 'private' military structures scattered throughout the territory.

I believe that the different regulatory and punitive impact of sanctioning group delinquency can be demonstrated comparing two cases. The first is found in *Capitula Legi Salicae addita* 72 (a text already mentioned, which is worth recalling):

> *Capitula Legi Salicae addita* 72. Si quis ingenuam feminam a contubernio facto aut puellam in itinere aut quolibet loco inferre praesumpserit, quam unus tam plurimi, qui ipsum scelus admisisse fuerint probati, CC

Manthelan'. Other examples in *Decem Libri Historiarum* 2.27; 6.31; 8.30; 10.3. were cited by Wood, 'Transformation', pp. ++.

87 Esders, 'Spätrömisches Militärrecht', p. 69: 'Derartige *collectae* betrieben neben ihren eigentlich militärischen Aufgaben Handlungen fehdeartiger Justiz, die die fränkischen Könige unter Tatbestand der *harizucht* zu bekampfen suchten'.

88 Brunner, *Terra e potere*, p. 25*f.*; 39*ff.*

89 See Giulio Vismara, 'Problemi storici e istituti giuridici della guerra altomedievale', in *Ordinamenti militari in Occidente nell'Alto Medioevo*, II (Spoleto 1968), p. 1141.

90 Contamine, *War*, p. 15.

> solidos culpabilis iudicentur. Et de illo contubernio, si adhuc remanserit, qui ipsum scelus non admiser(un)t et ibi fuisse noscuntur, si plures aut minore numero fuerit tres, et ipsi XLV solidos culpabilis iudicentur,

The second is found in

> Πεῖρα 63.5 (= sch. 11 *ad Bas.* 60.58.1)[91]: Ὅτι ὅτε μετὰ πλήθους ἀφίκεταί τις εἰς οἰκίαν τινὸς καὶ εἴτε ἄκοῦσαν εἴτε ἑκοῦσαν ἁρπάσει γυναῖκα, ἁρπαγὴν πλημμελεῖ καὶ τιμωρεῖται. ὅτε δὲ μόνος ἀπέλθῃ, μή τινας ἔχων τοὺς συνυπουργηόκτας, καὶ ἀφανῶς τὴν γυναῖκα ἑαυτὴν προδεδωκυῖαν ἀφελέσθαι θαῤῥήσει, οὐχ ἁρπαγήν, ἀλλὰ φθορὰν παρθένου ἁμαρτάνει.[92]
>
> If someone came accompanied by a crowd of men to someone else's house and abducted a women, whether she wanted it or not, he committed abduction and is punished. But if he came alone without accomplices and dared to take secretly a consenting woman with him, he did not commit abduction but rape of a virgin.[93]

As late as in the eleventh century, in the context of a legal order which, having 'naturally' adopted the principles of Roman law, prohibited and punished the use of arms by private individuals, the element of conspiring in the abduction of someone (a conduct which could be carried out also by a single person) was represented, for the judges of the Hippodrome, by the generic term πλῆθος, a term at first sight not attributable to the military lexicon.[94] The case, realized μετὰ πλήθους, is abduction; but the *nomen criminis* (φθορά παρθένου) changes if this crime takes place by a single person, thereby implying that acting in a group *per se* expresses the element of *vis*: abduction, in the Byzantine sources, explicitly consists of *stuprum* and violence.[95]

The Frankish legislator did not share the same attitude. For him, faced with a social-anthropological context in which both the assembly of armed

91 Eds. Scheltema – Holwerda – van der Wal, B/IX, p. 3931.

92 Eds. Simon – Reinsch, p. 567.

93 My translation.

94 See Fabio Botta, *Per vim inferre, Studi su 'stuprum' violento e 'raptus' nel diritto romano e bizantino* (Cagliari 2004), pp. 261 *ff.* The term can often be interpreted in the Greek sources as a synonym of the Latin *turba*. However, Wood, 'Transformation', p. 4+, recalled Olympiodorus of Thebes (fr. 38 Blockley) who used the Greek lemma to designate the barbarian troops supporting Galla Placidia ('described as *buccellarii*'). However, the Egyptian writer himself, among the most useful sources for fifth-century military history, provides very interesting *excursus* on the term '*buccellarii*' (frr.7.4; 12 Blockley).

95 See *Bas.* 60.18.5.2 (eds. Scheltema – Holwerda – van der Wal, A/VIII, p. 2882).

men and the use of arms was profoundly different, the *crimen* was the same whether committed by *unus* or by *plurimi*, although the sanctioned conduct was always described as implicitly collective and naturally violent and 'armed', since carried out by the *contubernium*.

This last comment, I believe, also indicates the peculiarities of the formulation of the norms relating to the armed band in the Germanic legal orders and their difference to both the coeval Byzantine experience and – with regard to the regime of responsibility – the Roman model of cumulative liability in the offence. This model is perpetuated without particular variations in the Eastern legal orders, since, as it clearly appears from the decision reported in the *Peira*, the circumstance of the plurality of wrongdoers does not determine any change in the punishment, either with respect to the 'non-aggravated' crime or among the joint participants. Rather, all participants are punished for a crime more serious than the one that each of them would have committed had he acted alone.

Being an accomplice in a crime committed jointly and with arms is considered in a completely different manner in the Western regulations where, unsurprisingly, the pecuniary penalty[96] is proportional to the contribution of each perpetrator to the commission of the offence.[97]

Some explained this difference on the basis of a profound change in the provisions of conspiracy precisely on the subject of 'armed band', a subject that, previously, could not find a definition in the Early Germanic law. There, the only person punished was the perpetrator of the offence, understood in his immediate offensive materiality, and never the abettor or accomplice who had not actually contributed to carrying out the harmful action.[98] In crimes done by 'armed band', instead, the changes in the roles and punishments of the joint participants is described by attributing the position of *Täter* (perpetrator) to the group leader while the contribution of the accomplice, who was called *Folger* (follower), consisted in following the group and being present at the scene of the crime. Thus, the mere making oneself available for the

96 A multiple of the value of the damages done, cumulatively borne by each of the participants *singillatim* (a probable legacy of the *delictum* of Roman tradition?): See François Ricci, 'Note sur les tarifs de la Loi Salique', in *Revue Historique*, 100 (1909), p. 318 f.

97 See *supra*, among those taken into consideration: *Lex Rib.* 67(64); *Lex Fris.* 17.4; *Lex Thuring.* 10.9. Cf. Wilda, *Geschichte*, pp. 611*ff.*; Scipio Sighele, *La teorica positiva della complicità* (Torino 1894), p. 28; Brunner, *Deutsche Rechtsgeschichte*, pp. 746*ff.*

98 Rudolf His, *Das Strafrecht des deutschen Mittelalters, I. Die Verbrechen und ihre Folgen im allgemeinen* (1920 – repr. Aalen 1964), pp. 111*ff.* Briefly, now, Helfer, *Il concorso*, pp. 21*ff.* See *Leges Henrici Primi* 49.7 '... *ex una culpa non debet plures emendationes habere*'.

commission of the crime was already sufficient for the definition of criminal responsibility.[99]

Therefore, it seems to me that, precisely because of the above-mentioned genesis and function of the formations of armed men in early medieval societies, because of their natural hierarchical structure and the bonds of clan and blood or of *fides* that usually underpinned them,[100] it is possible to understand not only the innovative importance of these regulations as argued by scholars of the Germanic penal systems, but also the difference in the treatment of the 'armed band' between the one now seen and what can be attributed to the Roman (and Byzantine) model. In this way, I believe that greater clarity is given to the reason why, on the one hand, the amount of (pecuniary) penalty is based in the 'barbarian' laws on the significance of the individual's contribution in the commission of the collective crime[101] and, on the other hand (as in the conclusion of the provision of the *Capitula* now reproduced), cases of responsibility deriving from mere membership in the criminal conspiracy are defined, given that the punishment (albeit reduced) for the offence committed by the *contubernium collectum* is also extended to those who are only associated with it, despite not having contributed to the realization of the offence itself.

Bibliography

Primary Sources

Aelius Spartianus, *Vita Pescennii*, ed. and Engl. transl. D. Magie, *Scriptores Historiae Augustae*, I, (London – New York 1922)

Appianus Alexandrinus, *Historia Romana – Romaikà*, ed. and Engl. transl. H. White – J. D. Denniston – E. Iliff Robson, *Appian's Roman history*, vols. 1–4 (Cambridge, Mass. – London 1972–1979)

Basilicorum libri LX. Series A, Textus librorum I–LX, voll. I–VIII, eds. H.J. Scheltema – D. Holwerda – N. van der Wal (Groningen 1955–1988); Series B, Scholia I–LX, voll. I–IX, eds. H.J. Scheltema – D. Holwerda – N. van der Wal (Groningen, 1953–1985).

99 Helfer, *Il concorso*, p. 23. Also see previously His, *Das Strafrecht*, p. 128*f.*; Hermann Conrad, *Deutsche Rechtsgeschichte*, I, *Frühzeit und Mittelalter* (Karlsruhe 1962), p. 173.

100 However, see the perplexities, albeit in the comparison with the Roman precedent of the *lex Quisquis*, exposited in this regard by Harald Maihold, 'Die Sippenhaft: Begründete Zweifel an einem Grundsatz des "deutschen Rechts"', in *Mediaevistik*, 18 (2005), esp. p. 126*f.*

101 On *PLS* 42 and 43, see broadly Thonissen, *L'organisation judiciaire*, pp. 285 *ff.* and fns.; Sprandel, 'Struktur', p. 38; Esders, 'Spätrömisches Militärrecht', p. 69.

Capitula Legi Salicae addita, ed. K.A. Eckhardt, in *Pactus legis Salicae* [MGH *Leges nat. Germ.* IV.1] (Hannoverae, 1962), pp. 237–273

Cicero (Marcus Tullius), *Pro Cluentio*, ed. A.C. Clark, *M. Tulli Ciceronis Orationes*, I (Oxford 1905)

Cicero (Marcus Tullius), *Pro Tullio*, ed. A.C. Clark, *M. Tulli Ciceronis Orationes*, VI (Oxford 1911)

Codex Theodosianus: *Theodosiani libri XVI cum constitutionibus Sirmondianis et leges Novellae ad Theodosianum pertinentes*, ed. Th. Mommsen – P.M. Meyer, 2 vols. (Berlin 1905) – Engl. transl. by C. Pharr *The Theodosian Code and Novels and the Sirmondian Constitutions. A Translation with commentary, glossary, and bibliography*, [Princeton, 1952], 10th Print (Union N.J., 2001)

Collatio legum Mosaicarum et Romanarum, eds. P. Krüger, T. Mommsen, G. Studemund, in *Collectio librorum iuris anteiustiniani*, III, (Berolini 1890)

Edictum Rothari, eds. C. Azzara and S. Gasparri, *Le leggi dei Longobardi. Storia, memorie e diritto di un popolo germanico* (Rome, 2005) al. *Edictus Langobardorum*, ed. F. Bluhme, in *Leges Langobardorum* [MGH *Leges* IV] (Hannoverae, 1868), pp. 3–90

Ex Ruffo leges militares, ed. G. Famiglietti (Milano 1980)

Fredegarius, *Chronicon*, ed. and transl. W. Hadrill and J. Michael, *The Fourth Book of the Chronicle of Fredegar* (Oxford 1960)

Gregorius Turonensis, *Libri Historiarum X*, ed. B. Krusch – W. Levison [MGH, *Script. Rer. Meroving.* I.1] (Hannover, 1951) *etiam* [*Historiarum libri decem*, ed. Rudolf Buchner, *Ausgewählte Quellen zur Geschichte des Mittelalters*, 2 (Darmstadt, 1974)] – Engl. transl. by L. Thorpe, *Gregory of Tour, The history of Franks* (London 1974)

Iustinianus, *Digesta*, eds. T. Mommsen – P. Krüger, *Corpus Iuris Civilis*, I (Berlin 1963[17]) – Engl. transl. by A. Watson, *The Digest of Justinian* (Philadelphia 1998)

Leges Henrici Primi, ed. L.J. Downer, *Leges Henrici Primi: Edited with Translation and Commentary* (Clarendon 1972)

Lex Frisonum, ed. K. von Richthofen, [MGH *Leges*, III] (Hannoverae 1863), pp. 631–700

Lex Ribvaria, ed. F. Beyerle – R. Buchner [MGH, *Leges nat. Germ.* III.2] (Hannoverae 1954)

Lex Salica, ed. K.A. Eckhardt, in MGH *Leg. Nat. Germ.*, 4.2 (Hannoverae 1969)

Lex Thuringorum, ed. K. von Richthofen [MGH *Leges*, V] (Hannoverae, 1875–1889), pp. 103–142

Mauricius, *Strategikon*, ed. and Engl. transl. G.T. *Dennis, Maurice's Strategikon. Handbook of Byzantine Military Strategy* (*Philadelphia* 984)

Novellae Valentiniani, v. *Codex Theodosianus*

Olympiodorus, ed. R.C. Blockley, *The Fragmentary Classicising Historians of the Later Roman Empire, Eunapius, Olympiodorus, Priscus and Malchus*, vol. 2 (Liverpool 1983)

Pactus Legis Salicae, ed. K.A. Eckhardt [MGH, *Leg. Nat. Germ.*, 4.1] (Hannoverae 1962)

Pauli Sententiae, ed. M. Bianchi Fossati Vanzetti, *Pauli Sententiae. Testo e interpretatio* (Padova 1995)

Πεῖρα (*Practica ex actis Eustathii Romani*), D.R. Simon – D.R. Reinsch, 'Η Πεῖρα, *Die Peira, Ein juristisches Lehrbuch des 11. Jahrhunderts aus Konstantinopel – Text, Übersetzung, Kommentar, Glossar*, vols. 1–2, in *Forschungen zur byzantinischen Rechtsgeschichte*, Bd. 4, Berlin/Boston 2023, [al. ed. C.E. Zachariae von Lingenthal, in J. Zepi – P. Zepi, *Jus graecoromanum*, IV (Athenai 1931 = Aalen 1962)].

Plutarchus, *Vita Antonii*, ed. C.B. Pelling, *Life of Antony* (Cambridge 1988).

Polybius, *Historiae – Historíai*, ed. and Engl. transl. W. Paton, *The Histories*, vols. 1–6 (Cambridge, Mass. – London 1922–1927)

Procopius Caesariensis, *Anecdota sive Historia arcana*, ed. and Engl. transl. H.B. Dewing, *Anecdota or Secret history* (London – New York 1914)

Procopius Caesariensis, *De bellis – Tôn Polemôn Lógoi*, ed. and Engl. transl. H.B. Dewing, *History of the Wars* (Cambridge, Mass. 1919)

Synesius, *De Regno*; Id., *De Providentia*, ed. J. Lamoureux – N. Aujoulat, *Synésios de Cyrène, Opuscules*. vols. 2–3 (Paris 2008)

Tacitus (Publius Cornelius), *Annales* I–III and IV–VI, ed. P. Wuillemier (1923,1949; Paris 2013, 2019)

Vegetius, *Epitoma rei militaris*, ed. A. Önnerfors (Stuttgart 1995)

Secondary Literature

Amaya-Garcia, Victor Manuel, *Coautoria y complicidad: estudio historico y jurisprudencial* (Madrid 1993)

Annunziata, Dario, *Temi e problemi della giurisprudenza severiana. Annotazioni su Tertulliano e Menandro* (Napoli 2019)

Bachrach, Bernard S., 'Military Organization in Aquitaine under the Early Carolingians', in *Speculum*, 49 (1974), pp. 1–33

Balzarini, Marco, *Ricerche in tema di danno violento e rapina nel diritto romano* (Padova 1969)

Balzarini, Marco, 'Violenza (dir. rom.)', in *Enciclopedia del Diritto*, XLVI (Milan 1993), pp. 830–843

Barbero, Alessandro, *Barbari. Immigrati, profughi, deportati nell'impero romano* (Rome-Bari 2006)

Battaglia, Marco, *I Germani. Genesi di una cultura europea* (Rome 2013)

Bauman, Richard A., 'Some problems of the *Lex Quisquis*', in *Antichton*, 1 (1967), pp. 49–59

Beeler, John, *Warfare in Feudal Europe, 730–1200* (Ithaca-London 1971)

Bertolini, Ottorino, 'Ordinamenti militari e struttura sociale dei Longobardi in Italia', in *Ordinamenti militari in Occidente nell'Alto Medioevo*, I (Spoleto 1968), pp. 429–608

Biondi, Biondo, *Il diritto romano cristiano*, III (Milan 1954)

Bock, Dennis, *Römischrechtliche Ausgangspunkte der strafrechtlichen Beteiligungslehre. Täterschaft und Teilnahme im römischen Strafrech* (Berlin, 2006)

Botta, Fabio, *Per vim inferre, Studi su 'stuprum' violento e 'raptus' nel diritto romano e bizantino* (Cagliari 2004)

Botta, Fabio, 'La vendetta come *officium pietatis*', in *Antropologia della vendetta*, eds. G. Lorini – M. Masia (Napoli 2015), pp. 11–38

Botta, Fabio, '*Pro legibus facit inquisitio.* Processo provinciale e *nomen christianum* nelle opere apologetiche di Tertulliano', in Κοινωνία, 44/I (2020), pp. 147–174

Botta, Fabio, 'Note in tema di 'decimazione': pena militare, pena 'collettiva', pena 'sacrale'', in *Quaderni Lupiensi di Storia e Diritto*, 11 (2021), pp. 27–56.

Brasiello, Ugo, 'Concorso di persone nel reato (dir. rom.)', in *Enciclopedia del Diritto*, VIII (Milan 1961), pp. 561–563

Breccia, Gastone, '*«Salus Orientis». Il nuovo sistema militare romano alla prova, 379*–400', in *Rivista di Studi Bizantini e Neoellenici*, 41 (2004), pp. 3–72

Brunner, Heinrich, *Deutsche Rechtsgeschichte*, vol. 2, München-Leipzig 1928

Brunner, Otto, *Land und Herrschaft. Grundfragen der territorialen Verfassungsgeschichte Österreichs im Mittelater* (Vienna 1965[5])

Brunner, Otto, *Sozialgeschichte Europas im Mittelalter* (Göttingen, 1978) (Ital. transl. *Storia sociale dell'Europa nel Medioevo*, Bologna 1978)

Carcani, Michele, *Dei reati delle pene e dei giudizi militari presso i Romani* (1st ed. 1928) (Napoli 1981)

Chevailler, Laurent, 'Contribution à l'étude de la complicité en droit pénal romain', in *Revue historique de droit français et étranger*, 31 (1953), pp. 200–243

Colman, Rebecca V., 'Hamsocn: Its Meaning and Significance in Early English Law', in *The American Journal of Legal History*, 25 (1981), pp. 95–110

Conrad, Hermann, *Deutsche Rechtsgeschichte, Bd. 1, Frühzeit und Mittelalter* (Karlsruhe 1962)

Contamine, Philippe, *War in the Middle Ages*, (orig. ed. Paris 1980) (Oxford 1986)

Cossa, Giovanni, *Attorno ad alcuni aspetti della lex Iulia de vi publica et privata* (Rome 2007)

D'Abadal y de Vinyals, Ramon, 'A propos du legs visigothique en Espagne', in *Caratteri del secolo VII in Occidente*, II (Spoleto 1958), pp. 541–586

De Robertis, Francesco M., *Il diritto associativo romano* (Bari 1938)

Diesner, Hans-Joachim, 'Das Buccellariertum von Stilicho und Sarus bis auf Aetius (454/455)', in *Klio* 54 (1972), pp. 321–350

Dupont, Clemence, 'Peine et relations pécuniaires entre fiancés et conjoints dans les constitutions rendues de 312 à 565 après Jésus-Christ', in *Revue Internationale des Droits de l'Antiquitè*, 23 (1976), pp. 119–139

Ebert, Udo, *Die Geschichte des Edikts de hominibus armatis coactisve* (Heidelberg 1968)

Esders, Stefan, 'Spätrömisches Militärrecht in der *Lex Baiuvariorum*', in *Civitas, Iura, Arma. Organizzazioni militari, istituzioni giuridiche e strutture sociali alle origini dell'Europa* (*secc. III–VII*), eds. F. Botta – L. Loschiavo (Lecce, 2015), pp. 43–78

Ferrini, Contardo, *Diritto penale romano, Esposizione storica e dottrinale* (repr. Rome 1976)

Garbarino, Paolo, 'Brevi note su 'societas criminis'', in *Diritto@Storia*, 11 (2013)

Garbarino, Paolo, 'Appunti sulla *lex Quisquis* (CTh. 9,14,3)', in *Bullettino dell'Istituto di Diritto Romano*, 107 (2013), pp. 137–165

Giuffré, Vincenzo, 'Arrio Menandro e la letteratura «de re militari»', in *Labeo*, 20 (1974), pp. 27–63

Giuffrè, Vincenzo, *La repressione criminale nell'esperienza romana. Profili* (Napoli 1997[4])

Goldberg, Charles, 'Decimation in the Roman Republic', in *The Classical Journal*, 111.2 (2015), pp. 141–164

Göpfert, Ernst, 'Beiträge zum obersächsischen Wortschatz (Fortsetzung)', in *Zeitschrift für deutsche Mundarten*, 4 (1909), pp. 26–55

Green, Dennis Howard, *The Carolingian Lord. Semantic studies on four old high German words 'Balder', 'Frô', 'Thrutin', 'Hêrro'* (Cambridge 1965)

Haegemann, Hans-Rudolf, 'Vom Verbrechenskatalog des altdeutschen Strafrechts', in *Zeitschr. der Savigny-Stiftung für Rechtsgeschichte. Germ. Abt.*, 91 (1974), pp. 1–72

Heather, Peter, *Empires and Barbarians. The Fall of Rome and the Birth of Europe* (Oxford 2012)

Helfer, Margareth, *Il concorso di più persone nel reato. Problemi aperti del sistema unitario italiano* (Torino 2013)

His, Rudolf, *Das Strafrecht des deutschen Mittelalters, I. Die Verbrechen und ihre Folgen im allgemeinen* (1920 – repr. Aalen 1964)

Innes, Matthew, 'Land, Freedom and the Making of the Medieval West', in *Transactions of the Royal Historical Society*, 16 (2006), pp. 39–74

Kepartovà, Jana, 'Frater in Militärinschriften – Bruder oder Freund?', in *Listy filologické / Folia philologica*, 109 (1986), pp. 11–14

Kienast, Walther, 'Gefolgswesen und *Patrocinium* im spanischen Westgotenreich', *Historische Zeitschrift*, 239 (1984), pp. 23–75

Kroeschell, Karl, *Haus und Herrschaft im frühen deutschen Recht* (Göttingen 1968)

Kuhn, Hans, 'Das römische Kriegswesen im germanischen Wortschatz', in *Zeitschrift für deutsches Altertum und deutsche Literatur*, 101 (1972) pp. 13–53

Labruna, Luigi, *Vim fieri veto* (Napoli 1971)

Langenbeck, Wilhelm, *Die Lehre von der Theilnahme am Verbrechen* (Jena 1868)

Lendon, Jon E., '*Contubernalis*, *Commanipularis*, and *Commilito* in Roman Soldiers' Epigraphy: Drawing the Distinction', in *Zeitschr. Für Papyrologie und Epigraphik*, 157 (2006), pp. 270–276

Lenski, Noel, 'Schiavi armati e formazione di eserciti privati nel mondo tardoantico', in *Ordine e sovversione nel mondo greco e romano*, ed. G. Urso (Pisa 2009), pp. 145–175 : http://fondazionecanussio.org/atti2008/09_Lenski.pdf.

Liebermann, Felix, *Die Gesetze der Angelsachsen*, I (Halle 1903)

Lintott, Andrew W., *Violence in Republican Rome* (Oxford 1968)

Longo, Giannetto, 'La complicità in diritto penale romano', in *Bullettino dell'Istituto di Diritto Romano,* 61 (1958), pp. 103–207

Loschiavo, Luca, *L'età del passaggio. All'alba del diritto comune europeo (secoli III–VII)* (Torino 2019²)

Lupoi, Maurizio, *Alle radici del mondo giuridico europeo* (Roma 1994)

MacMullen, Ramsay, 'The Legion as a Society', in *Historia,* 33 (1984), pp. 440–456

Magnou-Nortier, Elisabeth, 'Remarques sur la genèse du *Pactus Legis Salicae*', in *Clovis, histoire et* mémoire, I, *Clovis et son temps: l'événement*, ed. M. Rouche (Paris, 1997)

Maihold, Harald, 'Die Sippenhaft: Begründete Zweifel an einem Grundsatz des "deutschen Rechts"', in *Mediaevistik,* 18 (2005) pp. 117–144

Minale, Valerio Massimo. 'Per uno studio sui frammenti *de re militari* di Macro', in *Teoria e Storia del Diritto privato,* 6 (2013) pp. 1–31

Modzelewski, Karol, *Barbarian Europe* (Frankfurt am Main 2015) (orig. ed. *Barbarzyn-ska Europa*, Warszawa 2004)

Mommsen, Theodor, *De collegiis et sodaliciis Romanorum* (Kiliae 1843 = Napoli 2006)

Mommsen, Theodor, *Römisches Strafrecht* (Leipzig 1899)

Monod, Gabriel, 'Les aventures de Sichaire. Commentaire des chapitres XLVII du livre VII et XIX du livre IX de l'*Histoire des Francs* de Grégoire de Tours', in *Revue Historique*, 31 (1886), pp. 259–290

Murray, Alexander Callander, *Germanic Kinship Structure: Studies in Law and Society in Antiquity and the Early Middle Ages* (Toronto, 1983)

Nogrady, Alexander, *Römisches Strafrecht nach Ulpian. Buch 7 bis 9 De officio proconsulis* (Berlin 2006)

Onida, Pietro Paolo, 'La responsabilità penale degli enti collettivi fra diritto romano e diritto moderno', in *Regole e garanzie nel processo criminale romano*, ed. L. Solidoro (Torino 2016), pp. 59–81

Orestano, Riccardo, *Il «problema delle persone giuridiche» in diritto romano* (Torino 1968)

Peppe, Leo, 'Un altro brocardo se ne va', in *Labeo* 48 (2002), pp. 370–381

Peppe, Leo, 'La responsabilità penale della persona giuridica tra attualità e passato', in *Studi in onore di A. Metro*, IV (Milan 2010), pp. 491–525

Perea Yébenes, Sabino, 'El soldado romano, la ley militar y las cárceles *in castris*', in *Castigo y reclusión en el mundo antiguo*, eds. S. Torallas Tovar – I. Pérez Martín (Madrid 2003), pp. 115–152

Phang, Sara Elise, *Roman Military Service. Ideologies of Discipline in the Late Republic and Early Principate* (Cambridge 2009)

Poggi, Gianfranco, *La vicenda dello stato moderno. Profilo sociologico* (Bologna 1978)

Reuter, Timothy, 'Plunder and Tribute in the Carolingian Empire', in *Transactions of the Royal Historical Society*, 35 (1985), pp. 75–94

Ricci, François, 'Note sur les tarifs de la Loi Salique', in *Revue Historique*, 100 (1909) pp. 311–324

Rivière, Yann, 'Solidarité de la famille et responsabilité pénale dans le droit criminel romain', in *L'Homme*, 223/224 (2017), pp. 33–62

Sander, Erich, 'Zur Rangordnung des römischen Heeres: Der *Duplicarius*', in *Historia* 8 (1959) pp. 239–247

Sander, Erich, 'Das römische Militärstrafrecht', *Rheinisches Museum für Philologie*, 103 (1960), pp. 289–319

Santalucia, Bernardo, *Diritto e processo penale nell'antica Roma* (Milano 1998²)

Sarti, Laury, *Perceiving War and the Military in Early Christian Gaul (ca. 400–700 AD)* (Leiden-Boston 2013)

Schlesinger, Walter, 'Lord and Follower in Germanic Institutional History', in *Lordship and Community in Medieval Europe*, ed. F. Cheyette (New York 1968), pp. 64–99

Schmetterer, Christoph, *Die rechtliche Stellung römischer Soldaten im Prinzipat,* (Wiesbaden 2012)

Schmitt, Oliver, 'Die *Buccellarii.* Eine Studie zum militärische Gefolgschaftswesen in der Spätantike', in *Tyche* 9 (1994), pp. 147–174

Schulz-Falkenthal, Heinz, 'Die Unterstützungstätigkeit in einem Militärkollegium der *legio III Augusta* in Lambaesis und das Problem der Sozialleistungen im römischen Vereinswesen', in *Afrika und Rom in der Antike*, eds. H.-J. Diesner – H. Barth – H.-D. Zimmermann (Halle 1968), pp. 155–172

Sighele, Scipio, *La teorica positiva della complicità* (Torino 1894)

Solidoro, Laura, 'La «familia» nell'editto di Lucullo', in *Atti dell'Accademia di Scienze morali e politiche*, 92 (1981) (Napoli 1982), pp. 1–46

Solidoro, Laura, 'La repressione della criminalità organizzata. Tra diritto romano e categorie moderne', in *Esperienze giuridiche a confronto. Aspetti del diritto pubblico e privato dall'età romana alle configurazioni moderne. Lezioni* (Napoli 2001)

Solidoro, Laura, *Profili storici del delitto politico* (Napoli 2002)

Sperandio Marco, *Dolus pro facto. Alle radici del problema giuridico del tentativo* (Napoli 1998)

Sprandel, Rolf, 'Struktur und Geschichte des Merowingischen Adels', in *Historische Zeitschrift*, 193 (1961), pp. 33–71

Thompson, Edward Arthur, 'The Visigoths from Fritigern to Euric', in *Historia. Zeitschrift für Alte Geschichte, 12 (1963)*, pp. 105–126

Thonissen, Jean-Joseph, *L'organisation judiciaire, le droit pénal et la procédure pénale de la Loi Salique*, 2nd ed. (Bruxelles-Paris 1882)

Todd, Malcom, *The Early Germans* (Oxford 1992)

Vacca, Letizia, *Ricerche in tema di actio vi bonorum raptorum* (Milan 1972)

Vacca, Letizia, 'Delitti privati e azioni penali nel principato', in *Aufstieg und Niedergang der Römischen Welt* (ANRW), II.14 (Berlin-New York 1982), pp. 682–721.

Vismara, Giulio, 'Problemi storici e istituti giuridici della guerra altomedievale', in *Ordinamenti militari in Occidente nell'Alto Medioevo*, II (Spoleto 1968), pp. 1127–1200

Vitzthum, Werner, *Untersuchungen zum materiellen Inhalt der 'lex Plautia' und 'lex Julia de vi'* (München 1966)

von See, Klaus, *Das Jütsche Recht* (Weimar 1960)

von Woringen, Franz, *Beiträge zur Geschichte des deutschen Strafrechts* (Berlin 1836)

Whittaker, Dick, 'Landlords and Warlords in the Later Roman Empire', in *War and Society in the Roman World*, eds. J. Rich – G. Shipley (London 1993), pp. 277–302

Wilda, Wilhelm Eduard, *Geschichte des deutschen Strafrechts* (Halle 1842)

Wood, Ian, 'Transformation of the military in the Late Antique West' (in this volume pp. 1–27)

Zarro, Gianluca, '*Homines militares* nel tardo antico. Scollamento di società civile e compagine militare. Immigrati, briganti, mercenari', in *Atti dell'Accademia Romanistica Costantiniana*, 23 (2019), pp. 575–607

Zuccotti, Ferdinando, *"Furor Haereticorum"* (Milan 1992)

CHAPTER 18

From the Roman Army to the Laws of the Kingdoms: Concluding Remarks

Walter Pohl

1 Between Paradoxes and New Approaches

The chapters of this volume follow one of the most productive new roads through the jungle of debates about an old historical problem, the replacement of the Western Roman Empire by the kingdoms of the Goths, Franks, Longobards and others. Generations of scholars have discussed why the Western Empire fell, whether it declined because of inner problems or was destroyed by Germanic warriors, and whether that caused an 'end of civilisation', a 'transformation of the Roman world', or was just a rather superficial political shift in the rather continuous development of the Medieval West.[1] Gradually, many scholars move away from the binary models that juxtaposed Romans and Germans, the Empire and the 'barbarians', the civil society that could thrive in the *pax Romana* and the 'communities of violence'[2] that replaced it, and, not least, Roman and Germanic law.

The latter binary constitutes the field in which this book brings significant progress. It builds on ground-breaking work of a number of legal historians such as Stefan Esders, who has dealt with many facets of the problem in numerous publications (to which I will return below), and Luca Loschiavo, the editor of this collection.[3] The main argument unfolded here is that the legal systems

1 Rupture: Peter J. Heather, *The Fall of the Roman Empire: A New History* (London/Oxford 2005); Transformation: *Kingdoms of the Empire. The Integration of Barbarians in Late Antiquity*, ed. Walter Pohl (*The Transformation of the Roman World* 1, Leiden/New York/Köln 1997); Continuity: Walter A. Goffart, *Barbarian Tides: The Migration Age and the Later Roman Empire* (Philadelphia 2006); short overview: Walter Pohl, 'Rome and the Barbarians in the Fifth Century', in *Antiquité Tardive* 16 (2008), pp. 93–101.

2 Binaries: Walter Pohl, 'Die Anfänge des Mittelalters – alte Probleme, neue Perspektiven', in *Mediävistik im 21. Jahrhundert. Stand und Perspektiven der internationalen und interdisziplinären Mittelalterforschung*, eds. H.-W. Goetz and J. Jarnut (Munich 2003) pp. 361–378; Communities of Violence: Ulrich Wiemer, 'Die Goten in Italien. Wandlungen und Zerfall einer Gewaltgemeinschaft', in *Historische Zeitschrift* 296 (2013), pp. 593–628.

3 Luca Loschiavo, *L'età del passaggio. All'alba del diritto comune europeo (secoli III–VII)* (Turin 2nd ed. 2019).

 | DOI:10.1163/9789004698017_019

and institutional models of the new kingdoms built in many ways on norms and practices current in the Roman army. This is a line of research that has received little attention in traditional legal history: First, because apart from Book VII *De re militari* of the *Codex* Theodosianus,[4] there is no coherent body of Roman military law, but it must be pieced together from scattered evidence; and second, because it seemed so obvious that in the West, the Roman army had been ousted by barbarians in the course of the fifth century.

The disappearance of the Roman army from the West was in fact a twofold process, synthesised in Ian Wood's elegant overview of the 'transformation of the military in the Late Antique West' in the introduction to this volume: On the one hand, the loss of a centralised military force was at the core of the dissolution of the Western Roman Empire; it was replaced by the military forces of the successor kingdoms. On the other hand, a standing army funded by a rather sophisticated tax system gradually ceased to exist altogether. Instead of receiving regular pay financed by taxes, soldiers were now endowed with land, and landed property came to be linked to an obligation for military service.[5] It is generally acknowledged that this shift was a key element in the 'transformation of the Roman World'.

Chris Wickham has argued that the disappearance of a standing army marked the transition from a 'strong state', the Roman Empire 'based on taxation and a paid army as an independent resource of political power', and 'weak states', 'with a landed army but also a strong sense of public power acting as a focus for political legitimation, inherited from the Roman World'.[6] An ancient empire was thus replaced by a medieval system in which rulers had to obtain, or to negotiate the armed support of their free subjects, instead of simply putting a standing army to the field. Taxation was reduced to tolls,

4 *Theodosiani libri XVI cum constitutionibus Sirmondianis et Leges Novellae ad Theodosianum pertinentes*, eds. T. Mommsen and P. M. Meyer (Berlin 1905), 1, pp. 309–359; Vincenzo Giuffrè, *"Iura" e "arma". Intorno al VII libro del Codice Teodosiano* (Napoli 1979).

5 For an overview over the process, see, for instance, Guy Halsall, *Barbarian migrations and the Roman West, 376–568* (Cambridge 2007); Mischa Meier, *Geschichte der Völkerwanderung. Europa, Asien und Afrika vom 3. bis zum 8. Jahrhundert n. Chr.* (Munich 2019); Walter Pohl, 'The Emergence of New Polities in the Break-up of the Western Roman Empire', in *Empires and Communities in the Post-Roman and Islamic World, c. 400–1000*, eds. W. Pohl and R. Kramer (New York/Oxford 2021), pp. 28–63; Pierfrancesco Porena, 'La posizione dell'elemento militare nell'impero romano e i 'regni romano-barbarici'', in *Civitas, Iura, Arma. Organizzazioni militari, istituzioni giuridiche e strutture sociali alle origini dell'Europa* (secc. III–VIII), eds. F. Botta and L. Loschiavo (Lecce 2015), pp. 221–257.

6 Chris Wickham, *Framing the Early Middle Ages. Europe and the Mediterranean, 400–800* (Oxford 2006), p. 56. See also John Haldon, *The State and the Tributary Mode of Production* (London 1993).

fees, and similar sources of income, and consequently, medieval state finances remained precarious.[7]

This model is certainly useful for understanding the difference between the political options open to a Roman emperor and an early medieval king, respectively. It also helps to explain why what modern historiography has often seen as the deplorable end of ancient civilisation may have appeared attractive to many contemporaries, for instance landowners who ceased to be taxed, soldiers who were not garrisoned anymore, or clerics who could now attract investments that had so far gone to the sophisticated Roman military system.[8] Yet, as many contributions to this volume show, things were more complicated than the transition model implies. As many large-scale social transformations, it evolved rather differently in different regions,[9] and not without paradox and contradictions.

One paradox lies in the changing role of war and violence. In many respects, the emergence of new kingdoms brought a militarisation of society; the senatorial and civil elites lost influence, their options became limited to ecclesiastic careers, or they disappeared altogether, while a new military elite came into power.[10] Unlike Roman senators, the new warrior class would also demonstrate their status by carrying swords in public. That might well increase small-scale violence, and Gregory of Tours, at the end of the sixth century, has a number of drastic stories to tell in that respect. On the other hand, as Ian Wood convincingly argues in his chapter in this book, in terms of large-scale and interstate

7 Many German medievalists even maintain that medieval kingdoms, and not least, the Holy Roman Empire, were no states at all, just 'Herrschaft ohne Staat', rule without a state, e.g. Gerd Althoff, *Die Ottonen. Königsherrschaft ohne Staat* (Stuttgart / Berlin / Cologne 2000); discussion in: *Der frühmittelalterliche Staat – Europäische Perspektiven*, eds. W. Pohl/V. Wieser (Forschungen zur Geschichte des Mittelalters 16, Vienna 2009). In Italy, we can recognise the same discussion among legal historians. Absence of the state in Middle Ages: Paolo Grossi, *L'ordine giuridico medievale* (Roma / Bari 1995) pp. 52–56 and 130–135, and Id., 'Un diritto senza Stato (La nozione di autonomia come fondamento della costituzione giuridica medievale)', in *Quaderni fiorentini per la storia del pensiero giuridico moderno* 25 (1996), pp. 267–284, *contra* Mario Ascheri, 'Un ordine giuridico senza Stato? Tra miti e disconoscimenti', in *Rechtshistorisches Journal* 15 (1996) pp. 360–372, and Ennio Cortese, 'Nostalgie di romanità: leggi e legislatori nell'alto medioevo barbarico', in *Ideologie e pratiche del reimpiego nell'alto Medioevo* (Settimane CISAM, 46 – Spoleto 1999), pp. 485–510.

8 Ian Wood, *The Transformation of the Roman West* (Leeds 2018).

9 See Wickham, *Framing the Early Middle Ages*.

10 *Early medieval militarisation*, eds. E. Bennett, G. M. Berndt, S. Esders, L. Sarti (Manchester 2021); Laury Sarti, *Perceiving War and the Military in Early Christian Gaul (ca. 400–700 A.D.)* (Leiden / Boston 2013).

violence, the post-imperial societies were mostly more peaceful than the *pax Romana*, especially in the seventh century, not least because it had become more difficult to assemble a large army and motivate it to march into far-away lands. It should also be noted that in the last 70 years of the Western Empire, 22 emperors or pretenders were raised, which led to numerous military conflicts, while in the following 50 years, only two kings ruled in Italy, Odoacer and Theodoric.[11]

Another paradox is that in Late Antiquity, the Roman army was in many respects not a 'Roman' army anymore. There had always been auxiliaries and contingents of barbarian origin, often distinguished by ethnic labels – many of them are listed in the early-fifth-century *Notitia Dignitatum*, although in several cases, these names were little more than reminiscences of past identities. In the fifth century, the Roman army was composed to a considerable extent of soldiers of barbarian origin, up to the highest ranks of *patricii* and *magistri militum*. In most cases these barbarians remained loyal to the Roman cause for all of their lives. Yet they remained distinct, and even after generations, they could still be identified as Goths, Heruls or Huns, and several authors of the period complain about the barbarian origin and often also barbaric behaviour of these 'Roman' soldiers. Some of them made splendid careers, and contributed to changing the balance within late-Roman society. A decisive shift was the arrival of large Gothic groups in Roman provinces in 375, which led to the settlement of federate troops under their own leaders on Roman territory.[12] For people of barbarian origin, there were many ways to opt into the late-antique world which did not necessarily imply becoming 'Romans'. Christianity provided a platform on which integration was possible, although distinctions between diverging factions ('Arians' and other 'heretics') could be maintained. In spite of all the wars and power struggles that accompanied the last century of Roman rule in the West, in the end military control and political power slipped out of Roman hands surprisingly swiftly. In itself, the transition did not cause much damage to the infrastructure.[13]

---

11 List of emperors: Alexander Demandt, *Die Spätantike. Römische Geschichte von Diocletian bis Justinian 284–565 n. Chr.* (Handbuch der Altertumswissenschaft 3/6, München 2007), pp. 611–612.

12 Massimo G. Palazzi, *Foedus cum natione. I trattati alle radici dell'Europa. Aspetti delle relazioni fra Impero romano e popolazioni barbariche fra II e V secolo* (Gallarate 2014).

13 Walter Pohl, 'Christian and barbarian identities in the early medieval West: introduction', in: *Post-Roman Transitions: Christian and Barbarian Identities in the Early Medieval West*, eds. W. Pohl/G. Heydemann (Turnhout 2013), pp. 1–46.

2 From Tax to Land: Conditions of Settlement

Thus, region by region, barbarian armies replaced the Roman *exercitus*. Since the 1980s, the exact procedures of their settlement have spurred heated debates. The traditional opinion was formulated in the 1840s by Theodor Gaupp, who assumed that the barbarians were initially billeted in private houses and estates according to the Roman system of *hospitalitas*, but soon received a share of one third, *tertiae*, of the estate as permanent possession.[14] In the 1980s, Walter Goffart argued that these *tertiae* were not shares of land but of the taxes raised in the province, and assumed a prolonged period in which the armies of the successor kingdoms were still financed by the tax system.[15] These discussions brought some differentiations, and clarified that neither Gaupp's nor Goffart's model are fully tenable. It is hardly plausible that the *hospitalitas* system could have served as a platform for actual divisions of land. Andrea Trisciuoglio, in his chapter in the present volume, emphasises its inbuilt limits, under which the hosts only provided accommodation but no food or other commodities (*salgamum*), which had to come from the system of military provisions (*annona*) or from local authorities.

As Pierfranco Porena argues in his chapter, after 414, in the West 'neither the Theodosian Code nor the western *Novellae*, nor the substantial collection of official documents of the Ostrogoth kings in the *Variae* by Cassiodorus (507–537 AD), nor the Acts of the councils, nor other sources such as historians, written correspondence and homilies make any reference to the institution of *hospitalitas / metatum* described in the Late Roman regulations as temporary accommodation during travel in provincial territories.'[16] As Porena argues, after the Roman defeat at Adrianople in 378 and the instalment of a strong Gothic army under its own leader in a Roman province (initially, in Thrace) the imperial authorities faced the dilemma of pacifying these barbarian troops and keeping them from plundering the province, but avoiding endowing them with land of their own which would have provided them with permanent control over resources. It took 40 restless years until the Visigoths were settled in the *Gallia Narbonensis* and obtained the means to build their kingdom. The

14 Ernst Theodor Gaupp, *Die germanischen Ansiedlungen und Landtheilungen in den Provinzen des Römischen Westreiches in ihrer völkerrechtlichen Eigenthümlichkeit und mit Rücksicht auf verwandte Erscheinungen der alten Welt und des späteren Mittelalters* (Breslau 1844).

15 Walter Goffart, *Barbarians and Romans A.D. 418–584: the Techniques of Accommodation* (Princeton 1980). For an overview of the debate, see the chapter by Pierfranco Porena, in this volume.

16 Porena at p. §§.

terms *hospes* and *hospitalitas* could now be used to describe the new barbarian defenders of the province and the modalities by which they had received a share of the land.

The Ostrogothic settlement in Italy, negotiated with some of the most influential senators in the country, relied, as Porena argues, to at least some degree on expropriation. As I have proposed, such transfers of property were not necessarily actual partitions of the land. The Ravenna papyri provide ample evidence for 5th- and 6th-century property transfers between absentee landlords which were calculated in shares of twelfths (*unciae*) of the income from the estates. What was sold and bought were shares of the rent owed by the farmers who worked the land. It would have been easy enough to transfer one third, four ounces, of these property shares to a Gothic or Longobard soldier, who might also remain garrisoned in the city.[17] Of course, this solution was only viable in productive areas with a functioning system of production and distribution, controlled by *actores* who could serve as intermediaries between city-based landlords and the producers, and could be guaranteed by documentation in the *Gesta municipalia*. As Porena concludes, procedures of settlement differed between the regna, and actual conditions also within them.

In fact, the Roman army and its structures did not simply disappear with the deposition of the last Western emperor. It was a regular Roman army under the command of an officer of barbarian origin, Odoacer, that overthrew the child emperor Romulus Augustulus and his father Orestes in 476. Odoacer and Orestes knew each other from Attila's court, where Orestes had been Attila's secretary and Odoacer's father Edecon an influential Hunnic aristocrat. Odoacer maintained what was left from the Italian army; one of his trusted commanders was the Roman general Pierius, who led the last Roman army ever to march to the Norican Danube in 487/88, defeating the Rugians and evacuating the provincial population of the forts along the river. In 489–93, the Ostrogoths under Theodoric conquered Italy in a long war – unlike the Romans in 452 (when Attila invaded Italy) and 568 (when the Longobards came), Odoacer's army was capable of offering stubborn resistance. The Gothic army, *exercitus Gothorum*, swiftly replaced Odoacer's Roman army within an administrative system that still relied much on imperial traditions, as Cassiodorus was eager to demonstrate in his *Variae*. When the Eastern Roman army attacked in 535, Roman propaganda styled this as a war of Romans against barbarians, but it was more a conflict between similar late-Roman regimes. The Roman army was mostly composed of barbarians, and loyalties even among the Roman senators

17 Walter Pohl, '*Per hospites divisi*. Wirtschaftliche Grundlagen der langobardischen Ansiedlung in Italien', in *Römische Historische Mitteilungen* 43 (2001), pp. 179–226.

were divided. The arrival of Eastern Roman tax collectors in 540, when the war seemed won, prompted a revolt, and the war dragged on for 12 more years until the Gothic regime collapsed.

Paradoxically, it was the invasion of an imperial army and the war that followed that put an end to many elements of the late-antique system in the former heartland of the Empire. Attempts to restore it after the war only succeeded to a degree in some coastal areas. In many other regions, the Longobards had to seek integration on a more modest level. Their invasion in 568 met no coordinated resistance, and found no partners to negotiate an overall settlement, as the Ostrogoths had done. Unlike most other founders of post-Roman kingdoms, the Longobards had relatively little experience in Roman military service and in dealing with substantial Roman population and infrastructure. The conquering army soon fell apart, and its component parts, commanded by one or several dukes, variously plundered in Frankish Gaul, founded duchies in Central and Southern Italy, were bought over by the Romans or entrenched themselves in northern cities. It took many years before the Longobard kingdom could reassert itself on the basis of its integration in the late Roman society.

Stefano Gasparri's chapter addresses the legal and social status of the Longobard warriors in their kingdom.[18] In sources of the period, both the Latin term *exercitales* and the Longobard term *arimanni* were used. For a long time, scholars believed that the *arimanni* had been settled on fiscal land in military colonies, for which the term *arimanniae* was adopted from late sources, a model that Gasparri rejects on good grounds. He also discusses a recent article by Paolo Delogu which posits the existence of a coherent class of Longobard *exercitales* until the end of the kingdom.[19] Although he agrees with Delogu's contention that the Longobard kingdom was constructed on a military basis, he doubts that these *exercitales* were a distinctive group legally separated from the ordinary *liberi homines*, or that they constituted some sort of standing army.

The Longobard example shows *ex negativo* that long experience within and with the Roman army and the structures that fed into it, but also in negotiations with Roman functionaries, aristocrats and bishops was an asset for establishing a kingdom on Roman territory. Unlike Goths or Franks, the Longobards

18 See also Stefano Gasparri, 'I barbari, l'impero, l'esercito e il caso dei Longobardi', in *Civitas, iura, arma*, ed. Botta and Loschiavo, pp. 91–102.

19 Paolo Delogu, 'Ritorno ai Longobardi', in *Desiderio. Il progetto politico dell'ultimo re longobardo* (Atti del I Congresso internazionale di studio), ed. G. Archetti (Spoleto 2015), pp. 19–50.

had been much less involved in the process of adaptation to Roman ways that had taken place in the military sphere, where generations of barbarians had been trained and had learnt to interact with their Roman environment. However, in the troubled years between 568 and the 590s they learnt to live under the new circumstances.

The new elites of the successor kingdoms, then, may have caused trouble, especially in the transition period, but they could also offer benefits to the provincial population. They gradually abandoned the tax system, they interfered less in Church affairs than the emperors in Constantinople, and they generally maintained the existent social order from which Roman landowners continued to profit. Overall, they reduced the complexity of society that had increased in Late Antiquity. And, as many chapters in this volume make clear, they respected and sometimes adapted the Roman legal system. On the downside, they reduced most Romans to a lower legal status, who were protected by lower compensation if they were victims of a violent act, and whose options to acquire full freedom were limited. Thus, being Roman gradually lost its attraction after the dissolution of the Western Empire.[20] Neither Gregory of Tours in the late sixth nor Isidore of Seville in the early seventh centuries showcased their Roman identity. Gregory emphasised his senatorial descent and was proud of the many bishops in his family, and Isidore described the Roman provincials in Spain as *Spani*, while the Byzantine enemies were regarded as Romans.[21] On the other hand, the Roman Empire, both the existent one and the one of glorious memory, and many of its features retained their prestige. Among them were the Latin legal language and Roman law.

3 Personality of the Law or Legal Pluralism

Historians of law have long treated Roman and Germanic law as separate entities, the former building on sophisticated written laws and the latter on popular norms originally embedded in oral traditions.[22] Yet on the one hand, as Stefan Esders has recently emphasised, Roman law developed in early medieval Europe in manifold ways, was transmitted in different collections reflecting

20 *Transformations of Romanness in the Early Middle Ages: Regions and Identities*, ed. Clemens Gantner, Cinzia Grifoni, Walter Pohl and Marianne Pollheimer (Berlin / New York 2018).

21 Helmut Reimitz, *History, Frankish Identity and the Framing of Western Ethnicity* (Cambridge 2015).

22 See, most recently, Luca Loschiavo, 'Il più antico 'processo' longobardo: per una rilettura', in *Reti Medievali Rivista*, 22.2 (2021), pp. 141–172.

particular interests, and was modified according to the needs of the Church or even supplemented by extracts from the lawbooks of the successor kingdoms.[23] On the other hand, the law codes of the regna did not constitute alien, 'Germanic' legal systems. Of course, we should not deny that Germanic peoples had their ways of settling disputes and of sanctioning transgressions, but that had never been disconnected from what happened in the Roman world. They surely drew on the experience of the late Roman army where generations of barbarian soldiers had been socialised. This is the central issue addressed in the chapters of this volume from different angles.

One of the cornerstones of the traditional history of Germanic law was the conviction that the personality of the law was a Germanic legal principle which clearly distinguished it from the territoriality of Roman law. Following the chapter by Luca Loschiavo in this volume, there are several reasons why this juxtaposition is flawed. First, both classical Roman law and the laws of the successor kingdoms allowed for legal pluralism, which is a more adequate category than 'personality' to describe the legal situation in the West after the dissolution of the Empire.[24] In all kingdoms, the law-codes of the respective peoples and the law of the Romans, mostly building on the *Codex Theodosianus*, were both in use.[25] In the Visigothic and the Burgundian kingdoms, the Theodosian Code was excerpted for the use of the Romans of the kingdom. Both compilations also circulated beyond the respective kingdoms.[26] As the Lombard laws contain no references to Romans (except one to a slave), we may assume that the Roman population also continued to use Roman law.[27]

---

23 Stefan Esders, 'Roman Law in the regnum Italiae under the Emperor Lothar I (817–855): Epitomes, Manuscripts, and Carolingian Legislation', in *Between Ostrogothic and Carolingian Italy. Survivals, Revivals, Ruptures*, ed. F. Oppedisano (Firenze 2022), pp. 19–39.

24 Stefan Esders and Helmut Reimitz, 'Diversity and Convergence: The Accommodation of Ethnic and Legal Pluralism in the Carolingian Empire', in *Empires and Communities in the Post-Roman and Islamic World, c. 400–1000 CE*, eds. W. Pohl and R. Kramer (Oxford and New York 2021) pp. 227–252.

25 Stefan Esders, 'Roman law as an identity marker in post-Roman Gaul (5th–9th centuries)', in *Transformations of Romanness. Early Medieval Regions and Identities,* eds. W. Pohl, C. Gantner, C. Grifoni and M. Pollheimer-Mohaupt (Berlin - New York 2018), pp. 325–344; *The Theodosian Code: Studies in the Imperial Law of Late Antiquity*, eds. J.D. Harries and I.N. Wood (Ithaca 1994); Charles M. Radding and Antonio Ciaralli, *The Corpus Iuris Civilis in the Middle Ages. Manuscripts and Transmission from the Sixth Century to the Juristic Revival* (Leiden / Boston, 2007), pp. 35–76.

26 *Le Bréviaire d'Alaric*, eds. M. Rouche and B. Dumézil (Paris 2008), pp. 105–278.

27 Walter Pohl, 'Legal Pluralism in Lombard Italy', in *Legal Pluralism and Social Change in Late Antiquity and the Middle Ages. A Conference in Honor of John Haldon*, eds. W. Brandes, H. Reimitz, J. Tannous (Frankfurt/Main 2024), pp.401-21. See also (with some different

Second, however, the option to choose between different Germanic laws according to the origin of the person is only really attested in the Carolingian period. The *professio iuris*, the legal statement of an individual under which law he should be judged, is first attested in Carolingian Italy in the 820s.[28] To argue that it was an authentic Germanic legal heritage, legal historians used a passage in the 'History of the Longobards' by Paul the Deacon:[29] After the conquest of parts of Italy in 568/72, he writes, the Saxons, who had taken part in the expedition, went home again because the Longobards would not allow them '*in proprio iure subsistere*', to remain under their own *ius*. This has always been understood as meaning that they were not allowed to use their own law, and was supposed to prove that a 'Germanic' idea of the personality of the law had predated the arrival on Roman territory. However, Paul the Deacon – and the Lombard laws – never use *ius* as meaning 'law', but only for rights of rule or property. Very probably, what the Saxons wanted was to build their own kingdom in Italy, as the Suebi had established theirs alongside the Vandal and Visigothic realms in Spain.[30] In fact, in the many charters transmitted from Longobard Italy before 774, the only laws that are mentioned are those of the Longobards, and in some cases of the Romans; professions of different laws in charters only begin in the Carolingian period.[31]

Third, we should also go beyond notions of duality between 'Roman' and 'Germanic' law, as Luca Loschiavo makes clear in his chapter in this volume. He argues that even in the Ostrogothic and Burgundian lawbooks that seem to confirm the duality between 'Germanic' and Roman law, both the codes are to some extent aimed at both groups (and cover conflicts between them), and both display influences of Roman law, if to a different extent. Ian Wood's observations, for instance about the role of legally-trained Roman advisors in the

views) Ennio Cortese, *Le grandi linee della storia giuridica medievale* (Rome 2000), pp. 44–51.

28 Esders, 'Roman Law in the regnum Italiae', p. 31.

29 Paulus Diaconus, *Historia Langobardorum*, ed. L. Bethmann – G. Waitz, MGH, *Script. rerum Lang.* (Hannoverae, 1878), pp. 12–187.

30 Walter Pohl, 'Probleme einer Sinngeschichte ethnischer Gemeinschaften – Identität und Tradition', in *Leges – Gentes – Regna. Zur Rolle von germanischen Rechtsgewohnheiten und lateinischer Schrifttradition bei der Ausbildung der frühmittelalterlichen Rechtskultur*, eds. G. Dilcher and E.-M. Distler (Berlin 2006), at pp. 56–57, with a reply by Gerhard Dilcher, 'Zur Entstehungs- und Wirkungsgeschichte der mittelalterlichen Rechtskultur', in *Leges – Gentes – Regna*, at pp. 622–628. See my overview of the controversy in Walter Pohl, 'Legal Pluralism'.

31 Brigitte Pohl-Resl, 'Ethnische Bezeichnungen und Rechtsbekenntnisse in langobardischen Urkunden', in *Ethnogenese und Überlieferung*, eds. K. Brunner and B. Merta (Vienna 1994), at p. 167.

compilation of the Burgundian and Visigothic laws, confirm this picture.[32] In issues such as treason, desertion, espionage or disobedience, Roman military law was particularly relevant for maintaining discipline and settling conflicts within the barbarian armies that had grabbed power in Roman provinces.

Fourth, the distinction between the military and the civil sphere, important as it is to understand the transformation from the empire to the kingdoms, was not clear-cut, and did not fully overlap with the barbarian-Roman divide. Fabio Botta and Luca Loschiavo, in their introduction to the valuable volume 'Civitas, Iura, Arma', have already maintained that the 'civil dimension of the Roman army' and its influence on the legal and social level survived the disappearance of the formal imperial structures.[33] This impact grew in the regna in the context of the decline of the civilian-administrative elite in the empire and its provinces, which had already begun under imperial rule. Military commanders on one side and bishops on the other side took on administrative and juridical responsibilities that had before been the task of the senatorial aristocracy and civil officials. This process facilitated the transfer of juridical norms from the military to the civil sphere, whose boundaries were blurred further in the regna.

4 From *Ius Speciale* to the Laws of the Kingdoms

If we consider the impact of Roman military law on the successor kingdoms, we should not conceptualise it as a unilateral development. Forms of military self-representation had already shifted away from classical models to a more hybrid military culture. Roman war-lords of the fifth century such as Boniface or Aetius afforded a barbarian (often Hunnic) bodyguard, and inspired a sense of loyalty in their troops that was less to the Roman state than to themselves.[34] The Roman Army was an institution in which barbarian soldiers could be taught Roman ways; but in Late Antiquity, this was not simply a one-way process of Romanisation, rather, as Andrea Verardi argues in his chapter, a space for reciprocal acculturation. Grave finds in army camps along the Rhine do not differ much from those north of the Danube or the Black Sea; archaeologists

32 Ian Wood, 'Burgundian law-making, 451–534', in *Italian Review of Legal History*, 3 (2017), n. 15, pp. 1–27.

33 Fabio Botta and Luca Loschiavo, 'Presentazione', in *Civitas, Iura, Arma*, pp. 5–14. See also the Introduction to this volume.

34 Dick Whittaker, 'Landlords and Warlords in the later Roman Empire', in *War and Society in the Roman World*, eds. J. Rich and G. Shipley (London / New York 1993), pp. 277–302.

attribute them to barbarian *foederati*, but chances are that this was the way in which soldiers preferred to style themselves.[35] The 'Strategicon of Maurice', an Eastern Roman military treatise compiled c. 600, lists numerous elements of equipment that the Roman army had adopted from barbarians, such as Gothic tunics and shoes, Bulgarian coats, Slavic javelins, Avar cavalry lances, breastplates and tents:[36] 'For how splendid a soldier is in his armour, with so much commitment will he attack, and so much fear will he instil in his enemies,' as it states, and this was an incentive for adopting barbarian splendour.[37] As Philipp von Rummel has shown, the 'habitus barbarus' despised by late antique authors and occasionally banned form some social fields by legal measures, largely corresponded to military attire.[38]

The chapter by Wolfgang Haubrichs provides many examples for early borrowings in the military sphere from Germanic languages into Late Spoken Latin/ Early Romance languages, for instance, *heri-berge* 'lodging, quarters (of the army)' which is at the root of Italian *albergo* or French *auberge* and *héberger*, or Italian *sperone*, spur, from Germanic **spurnan*. Not surprisingly, the terms *anno* 'soldier's pay', from Latin *annona*, and *karkara* 'prison', from Latin *carcer*, entered the Gothic language, as Chiara Falluomini shows in her chapter. The overlapping relationship between military men and civilians on the one hand, and between barbarians and Romans on the other hand, allowed for integration of new groups in late-Roman society, if to different degrees; but it also provided room for exchanges. We should be prepared to acknowledge the 'barbarian' contribution to the legal systems of the successor kingdoms as well.

An impressive example for the logic of the adoption and adaptation of legal measures for the Roman military sphere in the regna is the *centena*, presented in the chapter by Stefan Esders. Building on previous work by Alexander Callender Murray, Esders breaks with a centenary tradition of regarding the +Frankish *centena* as a Germanic institution.[39] The title *centurio* current in the republic and the principate had partly been replaced by the late antique

---

35 Philipp von Rummel, *Habitus barbarus. Kleidung und Repräsentation spätantiker Eliten im 4. und 5. Jahrhundert* (RGA Erg. Bd. 55 – Berlin / New York 2007).

36 *Das Strategikon des Maurikios*, 12.1; 12.5; ed. G.T. Dennis and E. Gamillscheg, p. 419; p. 423; pp. 79–83 – Engl. transl. G.T. *Dennis, Maurice's Strategikon. Handbook of Byzantine Military Strategy* (Philadelphia 1984).

37 *Das Strategikon des Maurikios*, 1.2, ed. Dennis and Gamillscheg, p. 79.

38 Von Rummel, *Habitus barbarus.*

39 Alexander Callander Murray, 'From Roman to Frankish Gaul: *Centenarii* and *centenae* in the administration of the Merovingian kingdom', in *Traditio* 44 (1988), pp. 59–100.

centenarius, derived from a unit of (notionally) one hundred men.[40] It was also known in the Byzantine army as *hekatontarches* or *kentarchos* – curiously, but perhaps realistically the ninth-century Arab geographer Ibn Khurdadbīh defines the latter as a commander of 40 men.[41] This Roman decimal organisation was complemented by the *decanus/dekarches* (for units of ten) and (only attested in the Gothic and Vandal realms) the *millenarius/chiliarchos* for a thousand.[42] In most successor kingdoms, the titles in the military hierarchy remained late Roman, although their functions differed: *dux, comes, tribunus, centenarius, decanus*.[43] However, as Carla Falluomini mentions in her chapter in this volume, the Gothic translations for the millenarius, ***þusundifaþs***, and the centenarius/centurio, *hundafaþs*, are well attested in the Gothic translation of the Bible.

The way in which the *centena* changed its meaning in the Frankish kingdoms is typical for the creative use of Roman models in the successor kingdoms. As Esders shows, the initial military unit was superimposed on existing territorial and administrative units, whose manifold functions only emerge more clearly in the Carolingian period. Primarily, these units were endowed with military functions, as two legal provisions from the mid-sixth and early-seventh centuries demonstrate. Both address the problem of itinerant robbers and deserters; to track them down and judicate them, a regional militia was instituted which was also called *centena* and was entrusted with guarding the *quies communis*, the common peace. It also received extraordinary competences to inflict capital punishments on robbers, which were modelled on the late antique *ius militiae* and other legal/institutional models.

40 Arnold Hugh Martin Jones, *The Later Roman Empire 284–602: A Social, Economic, and Administrative Survey* (Oxford 1964), vol. 1, p. 634.

41 John Haldon, *Warfare, State and Society in the Byzantine World, 565–1204* (London 1999), pp. 109–110.

42 Jones, *The Later Roman Empire*, p. 258; see also the chapter by Esperanza Osaba, in this volume.

43 Germanic titles for military commanders are rather rare, such as the Visigothic *thiufadus* – Herwig Wolfram, *History of the Goths* (Berkeley/Los Angeles/London 1987) at pp. 218–219 – a title which Clara Falluomini, in her chapter in this book, very plausibly derives from *þiudafaþs 'head of a people' (= multitude). For the role of the *thiufadus* (listed among the inferiors in Erwig's laws) and of other military commanders in the Visigothic kingdom, see also the chapter by Esperanza Osaba. For the Visigothic *gardingus*, a retainer or guardsman, see Wolfram, *History of the Goths*, p. 242. The evidence that the Longobard *gastaldus* originally was a military commander, as assumed by Pier Maria Conti, *Il Ducato di Spoleto e la storia istituzionale dei Longobardi* (Spoleto 1982), is inconclusive, see Stefano Gasparri, *Italia Longobarda. Il regno, i Franchi, il papato* (Bari 2012), pp. 36–37, and his chapter in the present volume.

The example of the *centena* is only one among several connections between late Roman military and legal practices that Stefan Esders has diligently traced from the late Empire to Frankish uses. An important line of appropriation was the Frankish oath of loyalty to the king to be taken by his envoys (*missi*) in each *centena*, which was modelled on the Roman military oath.[44] This could entail the acceptance of legal norms, for instance of Carolingian capitularies, by the population, but also the acknowledgement of local customs by the king. Another, less obvious strand is the continued significance of the Roman system of changing horses for messengers along main roads, the *paraveredus*, which was visible enough to become the root of the modern German word *Pferd*, horse.[45] The Bavarian law code also contains significant traces of Roman military law, for instance in the legislation protecting the role of the dux against lèse-majesté (*laesa maiestas*), *seditio*, inciting discord (*scandalum*) within the army or within the duke's court, or pillage and arson during military campaigns in the province (*praedatio*).[46] Several chapters of the present volume adduce further evidence for the ways in which Roman military law and practices shaped legal conditions and institutional set-up in the successor kingdoms.

Francesco Castagnino, in his chapter, outlines how the conditions for barbarians in military service changed in the course of the imperial period. In the principate, soldiers generally did not have the right to marry. Barbarians also remained *peregrini*, foreigners, unless they were specifically granted Roman citizenship. I agree with Castagnino (and with Kent Rigsby and also Valerio Marotta in this volume) that the well-known late-antique inscription from Aquincum, '*Francus ego cives* (*sic*) *Romanus miles in armis*', should be understood as 'I, a Frank, a Roman citizen and a soldier in arms' – he was proud to have acquired Roman citizenship, not to be a Frankish citizen (whatever that may have meant). As the need for soldiers grew in Late Antiquity, their conditions were improved. Yet a ban on marriage between Romans and barbarians

44 Stefan Esders, 'Rechtliche Grundlagen frühmittelalterlicher Staatlichkeit: der allgemeine Treueid', in *Der frühmittelalterliche Staat - europäische Perspektiven*, eds. W. Pohl and V. Wieser (Vienna 2009), pp. 423–432; Id., "Faithful believers': Oaths of Allegiance in Post-Roman Societies as Evidence for Eastern and Western 'Visions of Community'', in *Visions of Community in the Post-Roman World*, eds. W. Pohl, C. Gantner and R. Payne ((Farnham / Burlington 2012), pp. 357–374.

45 Stefan Esders, 'The 'Staffelsee inventory'. Carolingian manorial economy, mobility of peasants, and 'pockets of functional continuity' in the transition from Antiquity to the Middle Ages', in *The Journal of European Economic History* 49 (2020), pp. 206–250.

46 Stefan Esders, 'Spätrömisches Militärrecht in der Lex Baiuvariorum', in *Civitas, Iura, Arma*, pp. 43–78 (English version: https://doi.org/10.35562/cliothemis.1168).

remained; only citizenship, often conferred on retirement, allowed for *connubium* with Roman women. As Valerio Marotta argues in his chapter, the distinction between Romans and *peregrini*, foreigners, was in fact maintained after the 3rd-century *Constitutio Antoniniana*, which did not ease access to Roman citizenship for later generations as much as often presumed. Only Justinian's *Novellae* in the 530s bridged the legal distinction between Romans and foreigners. The marriage ban, fixed in the *Codex Theodosianus*, was variously taken up by the legislation of the successor kingdoms, and that may have been intended to limit access to the military caste now in power, or also to curb unions between Catholics and Homoians. This interpretation, however, hinges on the definition of *barbari* (or *gentiles*) in the respective lawcode; it may well be that in some cases the ruling *gens* was excepted.[47]

Since Julius Ceasar and the emperor Titus, the *libera testamenti factio* had been a privilege of Roman soldiers, the topic of Iolanda Ruggiero's chapter. As a *ius speciale*, the *testamentum militum* offered considerable freedom in drawing up a will. In the army, often stationed far from their home countries, the soldiers had mostly lost touch with their families, and as long as they were active did not have the opportunity to found a family themselves, so they could decide to pass their inheritance to a brother in arms. In the successor kingdoms, further interests had potential bearing on inheritance law. In a society where the maintenance of the army depended on landownership, sons could not be bypassed too easily in inheriting their father's land. That had already been the principle guiding late-antique legislation for the *limitanei*, frontier troops equipped with land and slaves to maintain themselves under the condition that their sons would join the army and inherit the land.[48] On the other hand, the Church favoured the option to endow ecclesiastic institutions with estates in a testament.[49] The *Leges* navigate through these divergent interests in different ways. Yet, as Iolanda Ruggiero maintains, provisions such as *Pactus*

47 Hagith Sivan, 'The appropriation of Roman law in barbarian hands', in *Strategies of Distinction. The Construction of Ethnic Communities, 300–800*, eds. W. Pohl and H. Reimitz (The Transformation of the Roman World 2, Leiden / New York / Cologne 1998), pp. 189–204; Walter Pohl, 'Appropriating the discourse about 'barbarians' in the early medieval West', in *Indigeneity and Empire: Comparing the Roman and American Experiences*, ed. M. Maas, forthcoming.

48 Benjamin Isaac, 'The Meaning of the Terms Limes and Limitanei', in *Journal of Roman Studies* 78 (1988), pp.125–147.

49 Peter Brown, *Through the Eye of a Needle. Wealth, the Fall of Rome, and the Making of Christianity in the West, 350–550 AD* (Princeton 2012).

legis Salicae 46 rely to a considerable extent on the options offered by Roman military law, clad into new ritual forms.[50]

A judicial ritual is also at the core of Soazick Kerneis's chapter: the ordeal. As its occurrence in the *Leges* has no precedent in Roman Law, it has usually been attributed to Germanic or Christian influence. Kerneis uses the evidence of a curse tablet from Trier which alludes to the nails, a known element of late-antique torture, and the cauldron, which is probably a reference to the hot-water ordeal known from late evidence from the successor kingdoms. As part of the text is in Celtic, it may well denote a Celtic/Irish origin of the procedure (an Irish contingent was stationed nearby). It was, however, employed by Roman military judges, who were faced with the problem that barbarians swearing by foreign deities were considered less trustworthy, but who were not allowed to submit soldiers in Roman service to torture. The ordeal as a test of truthfulness could provide a way out of the dilemma.

The detailed study by Esperanza Osaba deals with a key problem of military law, desertion. It had become more virulent in the regna because their armies relied on the willingness of the soldiers to follow the call to arms, whereas a standing army could be set in motion more easily. In Roman legislation against desertion, the main focus was on mutiny or on inciting discord in the army, also addressed, for instance, in the Bavarian Code.[51] In Visigothic law, the thrust of legislation was different, and several seventh-century kings legislated against those who did not go to war or returned home. These laws distinguished between different grades of commanders or simple soldiers who ignored the call or let others off duty without sufficient reason. They addressed bribery that may have been involved, and specified exemptions. The sanctions also differed: death penalties, also current in Roman military law, appear in earlier laws, later replaced by severe flogging, mostly used against persons of lower status in the Roman Empire, but also in the army, and shaming (decalvation).[52] High compensation payments to the king, up to one pound of gold, or loss of freedom were the most frequent penalties. By comparison, the penalty of 20 solidi for refusing to join the army in Rothari's Edict seems rather light.[53] Some of the rhetoric used about the defence of the *patria* is taken from the

50 *Pactus Legis Salicae* 46, ed. Eckhardt, pp. 176–177. For the genesis of the *Lex Salica*, the best current study is Karl Ubl, *Sinnstiftungen eines Rechtsbuchs. Die* Lex Salica *im Frankenreich* (Ostfildern, 2016). But see also Jean-Pierre Poly, 'Sous les chênes de Salaheim. La loi salique, l'armée romaine et le bilan de la barbarie', in *Civitas, Iura, Arma*, pp. 193–220.

51 Sara Elise Phang, *Roman Military Service. Ideologies of Discipline in the Late Republic and Early Principate* (Cambridge 2008); Esders, 'Spätrömisches Militärrecht'.

52 Jill Harries, *Law and Empire in Late Antiquity* (Cambridge 1999), pp. 141–142.

53 *Leges Langobardorum, Edictus Rothari* c. 21, eds. C. Azzara and S. Gasparri, p. 20.

Theodosian Code. Both in Rome and in the Visigothic kingdom, desertion was regarded as a breach of the military oath.

A fairly direct continuity between Roman law and the *Leges* can be detected in the complex of *contubernium*, the involvement of entire groups in crimes or misdeeds, explored in Fabio Botta's chapter. In Roman Law – there is no specific treatment of the issue in military law in the strict sense – each man involved risked undergoing the full penalty for the committed act; if soldiers united for committing a crime, this could also be regarded as factionalism or conspiracy. Several provisions in the *Lex Salica* address group liability in different cases. Clause 19 in the Longobard Edict of Rothari[54] even uses the term *exercitus*, army, for groups of up to four perpetrators committing acts of violence to take revenge, suggesting that this was not least considered a problem of military discipline. Some gradation in the sums to be paid by the person who had committed the murder or another criminal act and his accomplices occurs in many cases. The structure of these groups mostly remains obscure, and Francesco Borri's critique of the concept of *Gefolgschaft* in his chapter in this volume may warn against basing too many assumptions on the relatively few occurrences of deeds committed by armed bands contained in the *Leges*.

A material element of the symbolical language of military status and identity is the Roman military belt, the *cingulum militiae*, a conspicuous part of a well-structured form of Roman self-representation; its use was extended to civil officials in the fourth century and to high dignitaries. This is the subject of Andrea Verardi's chapter. The *cingulum* (or *balteus*) is repeatedly mentioned in texts in the Ostrogothic and Frankish kingdoms, mostly in connection with its concession to high officials by the king. Interestingly, most of the cases Verardi discusses are men with Roman names. Belt fittings are also prominent in the early medieval archaeological record. A splendid specimen was found in the grave of king Childeric († 482) at Tournai, who had also been appointed as a high commander by the Empire. Still, belts as a symbol of dignity were an intercultural phenomenon at the time. Multipartite belts with metal plaques or single fittings of them are found in graves in many parts of Europe, often, but not always together with weapons. They are frequent in Longobard warriors' graves, but, as Andrea Verardi states, find hardly any mention in the written sources. In the eighth-century Avar realm, cast bronze belt-sets in very homogeneous forms are particularly wide-spread, which corresponds to their role in Eastern Central Asian steppe empires as attested in Chinese texts.[55] A fresco in

54 *Leges Langobardorum, Edictus Rothari* c. 19, eds. C. Azzara and S. Gasparri, p. 20.

55 Walter Pohl, *The Avars. A Steppe Empire in Central Europe, 567–822* (Ithaca 2018), pp. 236–239.

Santa Maria Antiqua in Rome shows the young nephew of an eighth-century pope with a multipartite belt, which resembles one found in the grave of a Slavic nobleman in Carantania.[56]

5 Conclusion

Roman military law had many elements of *ius speciale* or *ius singulare*, aimed at a circumscribed group of persons within the Roman legal system. It established a privileged, in some respects also more regulated status of soldiers within Roman society. This could be valuable in the post-imperial societies where the new military elites increased their privileges as a ruling class and, if to a different extent, limited access to this privileged status. Roman law had also defined the role of barbarians in Roman society, which was legally restricted in some respects, for instance through the ban on marriages between Romans and barbarians. This ban was continued in several regna, possibly to control access to elite status in the other direction. All these social boundaries could be crossed in both ways thanks to personal privileges. Particular marriages could be allowed, barbarians could become Roman citizens, and Romans in the Frankish kingdom could achieve superior status by becoming the king's table companions, or become Franks though manumission by 'penny-throw' in the presence of the king.[57] Roman military law and practices also provided useful models for many questions of organisation, for the settlement of disputes and the control of violent behaviour, or for symbolic communication and ritualised procedures, for instance the oath of loyalty, the ordeal and the military belt, the organisation of police forces or the rules for inheritance. The chapters of this volume deal with these issues in a balanced way. They expose the potential, but also the limits of explaining many features of the successor kingdoms by the model of Roman military law, and by the experiences of the new military elites in the Roman army.

56 Falko Daim, ',Byzantinische' Gürtelgarnituren des 8. Jahrhunderts', in *Die Awaren am Rand der byzantinischen Welt. Studien zu Diplomatie, Handel und Technologietransfer im Frühmittelalter*, ed. F. Daim (Innsbruck 2000), pp. 77–204. On S. Maria Antiqua and its paintings: *Santa Maria Antiqua. The Sistine Chapel of the Early Middle Ages*, eds. E. Rubery, G. Bordi and J. Osborne (Turnhout 2021).

57 *Pactus legis Salicae*, XLI, 1; 8; 9, ed. K.A. Eckhardt, in *MGH Leges nat. Germ.*, 4.1, pp. 154–57; *Lex Ribuaria* 64, ed. F. Beyerle – R. Buchner, in *MGH Leges nat. Germ.* 3.2, p. 117; Stefan Esders, 'Roman law as an identity marker'.

Bibliography

Primary Sources

Cassiodorus Senator, *Variae*, ed. Th. Mommsen [MGH, *Scriptores – Auct. ant.* XII] (Berlin, 1894) – Engl. transl. by M.S. Bjornlie, *The Variae. The Complete Translation.* (Oakland, CA, 2019)

Codex Theodosianus: *Theodosiani libri XVI cum constitutionibus Sirmondianis et leges Novellae ad Theodosianum pertinentes*, ed. Th. Mommsen – P.M. Meyer, 2 vols. (Berlin, 1905) – Engl. transl. by C. Pharr *The Theodosian Code and Novels and the Sirmondian Constitutions. A Translation with commentary, glossary, and bibliography*, [Princeton, 1952], 10th Print (Union, N.J., 2001)

Edictum Rothari, ed. C. Azzara and S. Gasparri, *Le leggi dei Longobardi. Storia, memorie e diritto di un popolo germanico* (Rome 2005)

Lex Ribvaria, ed. F. Beyerle – R. Buchner [MGH, *Leges nat. Germ.* III.2] (Hannoverae, 1954)

Mauricius, *Strategikon*, ed. G.T.Dennis and E. Gamillscheg, *Das Strategikon des Maurikios* (Corpus Fontium Historiae Byzantinae XVII – Vienna 1981) – Engl. transl. G.T. *Dennis, Maurice's Strategikon. Handbook of Byzantine Military Strategy (Philadelphia*, 1984)

Pactus legis Salicae, ed. K.A. Eckhardt [MGH, *Leg. Nat. Germ.*, 4.1] (Hannoverae, 1962)

Paulus Diaconus, *Historia Langobardorum, Historia Langobardorum*, ed. L. Bethmann – G. Waitz, MGH, *Script. rerum Lang.* (Hannoverae, 1878), pp. 12–187

Secondary Literature

Althoff, Gerd, *Die Ottonen. Königsherrschaft ohne Staat* (Stuttgart, Berlin and Cologne 2000)

Ascheri, Mario, 'Un ordine giuridico senza Stato? Tra miti e disconoscimenti', in *Rechtshistorisches Journal* 15 (1996) pp. 360–372

Botta, Fabio and Luca Loschiavo, 'Presentazione', in *Civitas, Iura, Arma*, pp. 5–14

Brown, Peter, *Through the Eye of a Needle. Wealth, the Fall of Rome, and the Making of Christianity in the West, 350–550 AD* (Princeton 2012)

Civitas, Iura, Arma. Organizzazioni militari, istituzioni giuridiche e strutture sociali alle origini dell'Europa (*secc. III–VIII*), eds. F. Botta and L. Loschiavo (Lecce 2015)

Conti, Pier Maria, *Il Ducato di Spoleto e la storia istituzionale dei Longobardi* (Spoleto 1982)

Cortese, Ennio, 'Nostalgie di romanità: leggi e legislatori nell'alto medioevo barbarico', in *Ideologie e pratiche del reimpiego nell'alto Medioevo* (Settimane CISAM, 46 – Sopleto 1999), pp. 485–510

Cortese, Ennio, *Le grandi line della storia giuridica medievale* (Rome 2000)

Daim, Falco, ',Byzantinische' Gürtelgarnituren des 8. Jahrhunderts', in *Die Awaren am Rand der byzantinischen Welt. Studien zu Diplomatie, Handel und Technologietransfer im Frühmittelalter*, ed. F. Daim (Innsbruck 2000), pp. 77–204.

Delogu, Paolo, 'Ritorno ai Longobardi', in *Desiderio. Il progetto politico dell'ultimo re longobardo* (Atti del I Congresso internazionale di studio), ed. G. Archetti (Spoleto 2015), pp. 19–50

Demandt, Alexander, *Die Spätantike. Römische Geschichte von Diocletian bis Justinian 284–565 n. Chr.* (Handbuch der Altertumswissenschaft 3/6 – Munich 2007)

Der frühmittelalterliche Staat – Europäische Perspektiven, eds. W. Pohl and V. Wieser (Forschungen zur Geschichte des Mittelalters 16 – Vienna 2009)

Dilcher Gerhard, 'Zur Entstehungs- und Wirkungsgeschichte der mittelalterlichen Rechtskultur', in *Leges – Gentes – Regna. Zur Rolle von germanischen Rechtsgewohnheiten und lateinischer Schrifttradition bei der Ausbildung der frühmittelalterlichen Rechtskultur*, eds. G. Dilcher and E.-M. Distler (Berlin 2006), pp. 603–637

Early medieval militarisation, eds. E. Bennett, G. M. Berndt, S. Esders, L. Sarti (Manchester 2021)

Esders, Stefan, 'Rechtliche Grundlagen frühmittelalterlicher Staatlichkeit: der allgemeine Treueid', in *Der frühmittelalterliche Staat – europäische Perspektiven*, eds. W. Pohl and V. Wieser (Vienna 2009), pp. 423–432

Esders, Stefan, "Faithful believers': Oaths of Allegiance in Post-Roman Societies as Evidence for Eastern and Western 'Visions of Community'", in *Visions of Community in the Post-Roman World*, eds. W. Pohl, C. Gantner and R. Payne (Farnham / Burlington 2012), pp. 357–374

Esders, Stefan, 'Spätrömisches Militärrecht in der Lex Baiuvariorum', in *Civitas, Iura, Arma*, pp. 43–78 (English version: *Late Roman Military Law in the Bavarian Code*, Clio@Themis 10 (2016), https://doi.org/10.35562/cliothemis.1168)

Esders, Stefan, 'Roman law as an identity marker in post-Roman Gaul (5th–9th centuries)', in *Transformations of Romanness*, pp. 325–344

Esders, Stefan, 'The 'Staffelsee inventory'. Carolingian manorial economy, mobility of peasants, and 'pockets of functional continuity' in the transition from Antiquity to the Middle Ages, in *The Journal of European Economic History* 49 (2020), pp. 206–250

Esders, Stefan and Helmut Reimitz, 'Diversity and Convergence: The Accommodation of Ethnic and Legal Pluralism in the Carolingian Empire', in *Empires and Communities in the Post-Roman and Islamic world, c. 400–1000 CE*, eds. W. Pohl and R. Kramer (Oxford and New York 2021) pp. 227–252

Esders, Stefan, 'Roman Law in the regnum Italiae under the Emperor Lothar I (817–855): Epitomes, Manuscripts, and Carolingian Legislation', © Author(s), CC BY 4.0, DOI 10.36253/978-88-5518-664-3.05, in *Between Ostrogothic and Carolingian Italy. Survivals, Revivals, Ruptures*, ed. F. Oppedisano (Firenze 2022), pp. 19–39.

Gasparri, Stefano, *Italia Longobarda. Il regno, i Franchi, il papato* (Bari 2012)

Gasparri, Stefano, 'I barbari, l'impero, l'esercito e il caso dei Longobardi', in *Civitas, iura, arma*, pp. 91–102

Gaupp, Ernst Theodor, *Die germanischen Ansiedlungen und Landtheilungen in den Provinzen des Römischen Westreiches in ihrer völkerrechtlichen Eigenthümlichkeit und mit Rücksicht auf verwandte Erscheinungen der alten Welt und des späteren Mittelalters* (Breslau 1844)

Giuffrè, Vincenzo, *"Iura" e "arma". Intorno al VII libro del Codice Teodosiano* (Napoli 1979)

Goffart, Walter, *Barbarians and Romans A.D. 418–584: the Techniques of Accommodation* (Princeton 1980)

Goffart, Walter A., *Barbarian Tides: The Migration Age and the Later Roman Empire* (Philadelphia 2006)

Grossi, Paolo, *L'ordine giuridico medievale* (Roma-Bari 1995)

Grossi, Paolo, 'Un diritto senza Stato (La nozione di autonomia come fondamento della costituzione giuridica medievale)', in *Quaderni fiorentini per la storia del pensiero giuridico moderno* 25 (1996), pp. 267–284

Haldon, John, *Warfare, State and Society in the Byzantine World, 565–1204* (London 1999)

Halsall, Guy, *Barbarian migrations and the Roman West, 376–568* (Cambridge 2007)

Harries, Jill, *Law and Empire in Late Antiquity* (Cambridge 1999)

Heather, Peter J., *The Fall of the Roman Empire: A New History* (London / Oxford 2005)

Isaac, Benjamin, 'The Meaning of the Terms Limes and Limitanei', in *Journal of Roman Studies* 78 (1988), pp.125–147

Jones, Arnold Hugh Martin, *The Later Roman Empire*, 3 vols. (Oxford 1964)

Kingdoms of the Empire. The Integration of Barbarians in Late Antiquity, ed. Walter Pohl (*The Transformation of the Roman World* 1 – Leiden/New York/Köln 1997)

Le Bréviaire d'Alaric, eds. M. Rouche and B. Dumézil (Paris 2008)

Loschiavo, Luca, *L'età del passaggio. All'alba del diritto comune europeo (secoli III–VII)* (Turin, 2nd ed. 2019)

Loschiavo, Luca, 'Il più antico 'processo' longobardo: per una rilettura', in *Reti Medievali Rivista*, 22.2 (2021), pp. 141–172

Meier, Mischa, *Geschichte der Völkerwanderung. Europa, Asien und Afrika vom 3. bis zum 8. Jahrhundert n. Chr.* (Munich 2019)

Murray, Alexander Callander, 'From Roman to Frankish Gaul: *Centenarii* and *centenae* in the administration of the Merovingian kingdom', in *Traditio* 44 (1988), pp. 59–100

Palazzi, Massimo G., *Foedus cum natione. I trattati alle radici dell'Europa. Aspetti delle relazioni fra Impero romano e popolazioni barbariche fra II e V secolo* (Gallarate 2014)

Phang, Sara E., *Roman Military Service. Ideologies of Discipline in the Late Republic and Early Principate* (Cambridge 2008)

Pohl, Walter, '*Per hospites divisi*. Wirtschaftliche Grundlagen der langobardischen Ansiedlung in Italien', in Römische Historische Mitteilungen 43 (2001), pp. 179–226

Pohl, Walter, 'Die Anfänge des Mittelalters – alte Probleme, neue Perspektiven', in *Mediävistik im 21. Jahrhundert. Stand und Perspektiven der internationalen und interdisziplinären Mittelalterforschung*, eds. H.-W. Goetz and J. Jarnut (Munich 2003) pp. 361–378

Pohl, Walter, 'Probleme einer Sinngeschichte ethnischer Gemeinschaften – Identität und Tradition', in *Leges – Gentes – Regna. Zur Rolle von germanischen Rechtsgewohnheiten und lateinischer Schrifttradition bei der Ausbildung der frühmittelalterlichen Rechtskultur*, eds. G. Dilcher and E.-M. Distler (Berlin 2006), pp. 51–68

Pohl, Walter, 'Rome and the Barbarians in the Fifth Century', in *Antiquité Tardive* 16 (2008) pp. 93–101

Pohl, Walter, 'Christian and barbarian identities in the early medieval West: introduction', in: *Post-Roman Transitions: Christian and Barbarian Identities in the Early Medieval West*, eds. W. Pohl and G. Heydemann (Turnhout 2013), pp. 1–46.

Pohl, Walter, *The Avars. A Steppe Empire in Central Europe, 567–822* (Ithaca 2018)

Pohl, Walter, 'The Emergence of New Polities in the Break-up of the Western Roman Empire', in *Empires and Communities in the Post-Roman and Islamic World, c. 400–1000*, eds. W. Pohl and R. Kramer (New York / Oxford 2021), pp. 28–63

Pohl, Walter, 'Legal Pluralism in Lombard Italy', in *Legal Pluralism and Social Change in Late Antiquity and the Middle Ages. A Conference in Honor of John Haldon*, eds. W. Brandes, Helmut Reimitz, Jack Tannous (Frankfurt/Main 2024), pp. 401-21

Pohl, Walter, 'Appropriating the discourse about 'barbarians' in the early medieval West', in *Indigeneity and Empire: Comparing the Roman and American Experiences*, ed. M. Maas (forthcoming)

Pohl-Resl, Brigitte, 'Ethnische Bezeichnungen und Rechtsbekenntnisse in langobardischen Urkunden', in *Ethnogenese und Überlieferung*, eds. K. Brunner and B. Merta (Vienna 1994), pp. 163–171

Poly, Jean-Pierre, 'Sous les chênes de Salaheim. La loi salique, l'armée romaine et le bilan de la barbarie', in *Civitas, Iura, Arma*, pp. 193–220

Porena, Pierfrancesco, 'La posizione dell'elemento militare nell'impero romano e i 'regni romano-barbarici'', in *Civitas, Iura, Arma*, pp. 221–257

Radding Charles M. and Antonio Ciaralli, *The Corpus Iuris Civilis in the Middle Ages. Manuscripts and Transmission from the Sixth Century to the Juristic Revival* (Leiden / Boston, 2007)

Reimitz, Helmut, *History, Frankish Identity and the Framing of Western Ethnicity* (Cambridge 2015)

Santa Maria Antiqua. The Sistine Chapel of the Early Middle Ages, eds. Eileene Rubery, Giulia Bordi and John Osborne (Turnhout 2021)

Laury Sarti, *Perceiving War and the Military in Early Christian Gaul* (*ca. 400–700* A.D.) (Leiden / Boston 2013)

Sivan, Hagith, 'The appropriation of Roman law in barbarian hands', in *Strategies of Distinction. The Construction of Ethnic Communities, 300–800*, eds. W. Pohl and H. Reimitz (*The Transformation of the Roman World* 2 – Leiden / New York / Cologne 1998), pp. 189–204

The Theodosian Code: Studies in the Imperial Law of Late Antiquity, eds. J.D. Harries and I.N. Wood (Ithaca 1994)

Transformations of Romanness in the Early Middle Ages: Regions and Identities, eds. C. Gantner, C. Grifoni, W. Pohl and M. Pollheimer (Berlin / New York 2018)

Ubl, Karl, *Sinnstiftungen eines Rechtsbuchs. Die* Lex Salica *im Frankenreich* (Ostfildern, 2016)

von Rummel, Philipp, *Habitus barbarus. Kleidung und Repräsentation spätantiker Eliten im 4. und 5. Jahrhundert* (RGA Erg. Bd. 55 – Berlin / New York 2007)

Whittaker, Dick, 'Landlords and Warlords in the later Roman Empire', in *War and Society in the Roman World*, eds. J. Rich and G. Shipley (London / New York 1993), pp. 277–302

Wickham, Chris, *Framing the Early Middle Ages. Europe and the Mediterranean, 400–800* (Oxford 2006)

Wiemer, Ulrich, 'Die Goten in Italien. Wandlungen und Zerfall einer Gewaltgemeinschaft', in *Historische Zeitschrift* 296 (2013), pp. 593–628

Wolfram, Herwig, *History of the Goths* (Berkeley / Los Angeles / London 1987)

Wood, Ian, *The Transformation of the Roman West* (Leeds 2018)

Wood, Ian, 'Burgundian law-making, 451–534', in *Italian Review of Legal History*, 3 (2017), pp. 1–27

Index of Names and Subjects

This index includes names of ancient and medieval authors and historical figures as well titles of laws and legal commentaries, institutional offices, concepts.